T0345914

VIEWS OF THE CORDILLERAS AND
MONUMENTS OF THE INDIGENOUS PEOPLES
OF THE AMERICAS

ALEXANDER
VON HUMBOLDT
IN ENGLISH

A series edited by Vera M. Kutzinski and Ottmar Ette

VIEWS OF THE CORDILLERAS AND MONUMENTS OF THE INDIGENOUS PEOPLES OF THE AMERICAS

A Critical Edition

ALEXANDER VON HUMBOLDT

*Edited with an Introduction
by Vera M. Kutzinski and Ottmar Ette*

Translated by J. Ryan Poynter

*With Annotations by Giorleny D. Altamirano Rayo
and Tobias Kraft*

THE UNIVERSITY OF CHICAGO PRESS   CHICAGO AND LONDON

Vera M. Kutzinski is the Martha Rivers Ingram Professor of English
and professor of comparative literature at Vanderbilt University.
Ottmar Ette is professor of Romance literatures at University of Potsdam, Germany.
J. Ryan Poynter is assistant vice provost for undergraduate academic affairs at New York University.
Giorleny D. Altamirano Rayo is HiE's assistant editor. Tobias Kraft is an instructor in
Romance languages at the University of Potsdam, Germany.

The University of Chicago Press, Chicago 60637
The University of Chicago Press, Ltd., London
© 2012 by The University of Chicago
All rights reserved. Published 2012.
Printed in the United States of America

33 32 31 30 29 28 27 26 25 24   2 3 4 5

ISBN-13: 978-0-226-86506-5
ISBN-10: 0-226-86506-1

Library of Congress Cataloging-in-Publication Data

Humboldt, Alexander von, 1769–1859, author.
 [Vues des Cordillères, et monumens des peuples indigènes de l'Amérique. English]
 Views of the Cordilleras and monuments of the indigenous peoples of the Americas : a critical
edition / Alexander von Humboldt ; edited with an introduction by Vera M. Kutzinski and
Ottmar Ette ; translated by J. Ryan Poynter ; with annotations by Giorleny D. Altamirano Rayo
and Tobias Kraft.
  pages cm — (Alexander von Humboldt in English)
 Translated from French.
 Includes bibliographical references and index.
 ISBN-13: 978-0-226-86506-5 (cloth : alkaline paper)
 ISBN-10: 0-226-86506-1 (cloth : alkaline paper) 1. Indians of Mexico—Antiquities.
2. Mexico—Antiquities. 3. Peru—Antiquities. 4. Andes Region—Antiquities. 5. Mexico—
Description and travel. 6. Peru—Description and travel. 7. Indians of South America—Andes
Region—Antiquities. I. Kutzinski, Vera M., 1956– editor. II. Ette, Ottmar, editor. III. Poynter,
J. Ryan, translator. IV. Altamirano Rayo, Giorleny D. V. Kraft, Tobias, 1978– VI. Title.
VII. Series: Humboldt, Alexander von, 1769–1859. Works. English.
 F1219.H913 2012
 972'.01—dc23

2011047458

CONTENTS

The Art of Science:
Alexander von Humboldt's Views
of the Cultures of the World

AN INTRODUCTION BY
VERA M. KUTZINSKI AND OTTMAR ETTE

Since the turn of the century, a happy revolution has taken place in our conception of the civilizations of different peoples, and of the factors that either obstruct or encourage progress. We have come to know certain peoples whose customs, institutions, and arts differ as much from those of the Greeks and Romans as the original forms of extinct animals differ from those species that are the focus of descriptive natural history. The Asiatick Society of Calcutta has cast a vivid light on the history of the peoples of Asia. The monuments of Egypt, which are nowadays described with admirable exactitude, have been compared to monuments in the most distant lands, and my study of the indigenous peoples of the Americas appears at a time when we no longer consider as unworthy of our attention anything that diverges from the style that the Greeks bequeathed to us through their inimitable models. (2)

With these remarks in the 1813 introduction to his *Views of the Cordilleras and Monuments of the Indigenous Peoples of the Americas*, Alexander von Humboldt took a clear and contentious stand within the centuries-long polemic known as the "Dispute on the New World." This debate began with the so-called discovery and conquest of America and became more pointed during the course of the eighteenth century in the writings of the Comte de Buffon, Cornelius de Pauw, the Abbot Raynal, and many others. Much to Humboldt's chagrin, G. W. F. Hegel's work extended it into the philosophical discussions of the nineteenth century. The New World as portrayed in this dispute was largely a continent without history. It was "new" even in a geological sense, a landmass that had only recently risen from the waters of enormous lowland streams, inhabited by uncultured hordes without a notable past or a real future. Invoking Francisco Javier Clavijero's *History of Mexico* (1780–81), Humboldt advanced in his own *Views of the Cordilleras*, which appeared in seven installments (or *livres*) in Paris between 1810 and 1813, a critical perspective on the presumption of inferiority that characterized the discourse on the Americas in the work of authors who advanced seemingly self-evident judgments about the New World without ever having bothered to travel there. The "happy revolution," as Humboldt called it, changed all that through a pronounced reappraisal of eighteenth-century Enlightenment philosophy in Europe, especially in France. Ironically perhaps, the one who contributed most to this felicitous change in the perception of cultural differences was none other than Alexander von Humboldt himself. In the wake of his and Aimé Bonpland's five-year travels in the Americas (1799–1804), Humboldt presented to the public his monumental *Voyage to the Equinoctial Regions of the New Continent*, which appeared between 1805 and 1838 in thirty lavish folio editions and numerous smaller and abridged editions.

Views of the Cordilleras occupies a special place within this immense travel work, which set in motion a vast international machinery of reception and translation (see Editorial Note). The author of many books during the course of his long life (1769–1859), Humboldt chose distinctive forms of writing and presentation for each of his book projects. In *Views of the Cordilleras*, as in his earlier *Views of Nature* (1808), he decided to weave together textual fragments in a way that is best described as a rhizome, a network of roots that we will discuss in some detail later on. But in contrast to *Views of Nature*, which he published in German, Humboldt included in *Views of the Cordilleras*, which he wrote in French, a large number of costly and artistically ambitious *planches* (Plates). *Views of the Cordilleras and Monuments of the Indigenous Peoples of the Americas* became an exceedingly complex ensemble of sixty-nine Plates and sixty-two text fragments.

Before examining the multiple layers of this book, which has never before been printed in English in its entirety, we will show how innovative, indeed revolutionary, Humboldt's discourse on the Americas really was in projecting both a hemispheric interrelatedness and a new order of the world's cultures.

A New Discourse on the New World

The title *Views of the Cordilleras and Monuments of the Indigenous Peoples of the Americas* already signals Humboldt's intent to work against (mis)understanding the New World as a historyless "realm of nature" populated by uncivilized roving hordes. His purpose was to develop an alternative perspective on the interrelations of nature and culture in the Americas, a view, as it were, whose worldwide entanglements would not permit Europeans to define "America" as their Other. *Views of the Cordilleras* opens with an introduction whose arguments take on the eighteenth-century theory of the New World's belatedness and inferiority. Adopting a planetary perspective, Humboldt insisted that both the "old" and the "new" world are of the same geological age.

> In examining closely the geological makeup of the Americas, in reflecting upon the balance of fluids that are spread across the surface of the earth, one would be hard-pressed to claim that the new continent emerged from the waters at some later point than the old one did. One observes there the same succession of rocky layers as in our hemisphere, and it is likely that the granite, the micaceous schist, or the different gypsum and sandstone formations in the mountains of Peru date from the same periods as their counterparts in the Swiss Alps. The entire globe appears to have experienced the same catastrophes. (5)

To argue that "[t]here is no proof that the existence of humankind is a much more recent phenomenon in the Americas than in other continents," Humboldt mustered anthropological evidence alongside natural scientific data. On the basis of comparable, albeit different, developments that also included plant and animal life, Humboldt began to foreground his perspective on "the American race" by exploring the first populations, their migrations, and, above all, their cultural traits such as language:

> The American race, the least numerous of all races, nevertheless occupies the largest territory on earth. That territory stretches across the two hemispheres, from 68 degrees northern latitude to 55 degrees southern latitude. It is the only one of the races that dwells both on the steaming plains bordering the ocean and on the mountainside, where it reaches heights that exceed that of the Peak of Tenerife [Mount Teide] by 200 toises.
>
> The number of distinct languages among the small indigenous tribes in the new continent appears to be much larger than in Africa, where . . . more than one hundred and forty languages are spoken. . . .
>
> When we consider that several hundred languages exist in a continent with a population smaller than that of France, we must acknowledge the differences between languages that have the same relationships between them as do, if not German and Dutch or Italian and Spanish, then at least Danish and German, Chaldean and Arabic, or Greek and Latin. The further one penetrates into the labyrinth of American languages, the more one senses that although several of them can be grouped into families, a large number remain isolated, like Basque among the European languages and Japanese among the Asiatic ones. . . .
>
> The majority of American languages, including even those belonging to groups as distinctive as the Germanic, Celtic, and Slavic languages, exhibit a certain degree of conformity in their overall structure. . . . The uniform tendency of these languages suggests either a common origin or, at the very

least, an extreme uniformity in the intellectual aptitudes of the American peoples from Greenland to the lands of Magellan. (6–7)

These passages show that Humboldt applied the concept of hemisphere both longitudinally and latitudinally, differentiating the Southern from the Northern Hemisphere and the Old World from the New. In this way, he had the benefit of a dual web of analogies: for one, there is a web of intra-American—what we have come to know as inter-American or "hemispheric"—relations that encompass the entire continent; for another, there is also a web of external relations that enabled Humboldt to connect American phenomena with events and experiences from very different regions of the world without running the risk of homogenizing the wealth of distinct cultural developments. A description of the cultures of the Americas as part of the web of world cultures had to feature fragments and resort to discontinuity, lest it were to end up with nothing more than a universal vision of a great "family of man." Whenever Humboldt examined local particularities, he always saw them as parts of a network of dual and doubled reference points. In this way, he could address the specific relations within the American hemisphere *and* embed those relations within *trans*regional global contexts. In his discussion of possible reasons for the construction of earthen pyramids and hills in the two Americas, for example, he drew on new research by George Macartney in China, since "[a]n established custom in eastern Asia might shed some light on this important question" (388). His comparative analytical design allowed Humboldt to explain the cultural, and especially linguistic, diversity of the Americas as the result of the limiting impact that distinct environmental factors had had on contact among the different cultures. This disconnectedness was still in evidence at the time of Humboldt's visit. For various geographical and political reasons, the different Spanish viceroyalties were barely in touch with each other, being linked mainly through the Spanish metropole. As a result, few actual exchanges occurred among the major cities of the Americas.

While Humboldt's comparative cultural studies cross regions, his understanding of the cultures themselves still originated in a very specific cultural perspective: the traditions of so-called Western civilization. As our initially cited passage makes clear, Humboldt did cast a net across the cultures of the world, a mesh of maps whose cultural meridian passed through Greek antiquity and its "inimitable models" (2). This occidental orientation—informed by the ideas about Greek antiquity that prevailed in Germany and France at the time—was based on a concept of freedom and of the individual that characterized the universalist ideals of the French Revolution no less than it did the views of antiquity held during German classicism. But Humboldt saw something different in the great cultures of the early Americas:

> By studying on site those Peruvians who have retained their national physiognomy throughout the centuries, one learns to understand the true value of Manco Capac's code of laws and the effects that it produced on customs and public happiness. There was at once general welfare and little private happiness; more resignation to the decrees of the sovereign than love of the fatherland; passive and spineless obedience without courage for daring feats; a sense of order that meticulously regulated even the pettiest actions in one's life; and no reach into the realm of ideas, no elevation in that of character. The most complex political institutions in the history of human society smothered the germ of individual liberty; and the founder of the empire of Cusco, who congratulated himself for his success in forcing people to be happy, reduced the latter to the state of simple machines. Peruvian theocracy was certainly less oppressive than the rule of the Mexica kings. But both contributed to giving to the monuments, worship, and mythology of two mountain peoples the gloomy and somber aspect that so contrasts with the arts and sweet fictions of the peoples of Greece. (11–12)

Ending his introduction on this note put into relief the extent to which Humboldt's new discourse on the New World, its fundamental innovations notwithstanding, remained indebted to the idea of Europe as a cultural and political center. At the same time, however, he quite explicitly thematized this Eurocentrism, always turning it into an occasion for critical self-reflection. As we shall see, doing so was a vital aspect of Humboldt's new way of ordering the cultures of the world.

Provisional Knowledge

The incompleteness of all knowledge and the resulting openness of all processes of knowing characterize Humboldt's concept of science above all else. No other of Humboldt's works demonstrates as memorably and resolutely as *Views of the Cordilleras* does the scientific openness beyond all system thinking and a deep-seated awareness of the provisionality of all knowledge. His European biases notwithstanding, Humboldt understood himself not as a single vanishing point but as representative of a transitional phase within an ever-accelerating historical process that had lost its messianic meanings. Humboldt's new way of ordering scientific discourse was fundamentally based on the observer's ability to perceive patterns of movement among all sorts of different phenomena, be they natural or cultural. This ability, which inevitably extends to his readers, is a function of curiosity, skepticism toward received ideas, and an inexhaustible willingness to be surprised. When writing about the "Aztec Hieroglyphic Manuscript Preserved in the Vatican Library" (Plate XIII), for instance, Humboldt registers his own astonishment in the form of a series of (rhetorical) questions:

> It would be absurd to assume the existence of Egyptian colonies everywhere that we find pyramidal monuments and symbolic paintings; but how can one not be struck by the points of resemblance between the vast tableau of customs, arts, languages, and traditions found today among the most far removed peoples? How can one not draw attention to structural parallels among languages, in style among monuments, and in fiction among cosmogonies, wherever such parallels reveal themselves, even when one is unable to identify the unknown causes of these resemblances, and when no historical phenomenon dates back to the period of contact among the inhabitants of these different climes? (79)

Sentiments of being surprised, intrigued, astounded, and struck in different ways by unexpected parallels and combinations pervade Humboldt's *Views*. As this passage demonstrates, the parallels he draws do not derive from rash, reductive conclusions based on superficial similarities between cultures, for it is crucial for him to distinguish with care "what is certain from what is merely plausible" (215). Nor was Humboldt so caught up in charting resemblances between the cultures of the world that he lost sight of irreducible differences. His insight into Americans' participation in planetary history did not tempt him to neglect the different characteristics of the hemisphere and its past inhabitants. "For a European," Humboldt remarked, "there is nothing more striking about the Aztec, Mexica, or Nahuatl language than the excessive length of the words," offering as an example "[t]he most astonishing actual compound that I know . . . the word *amatlacuilolitquitcatlaxtlahuilli*" (389). To Humboldt, this marvelous compound word exemplifies "[t]he genius of the American languages," which is "to convey a large number of ideas in a single word" (244).

Even though the globalization of human history in *Views of the Cordilleras* is inscribed, however ambivalently, with the universalizing gaze of the European traveler, this did not preclude Humboldt from historicizing the position of the observer (his own and that of others before him) by referring to his own temporality ("today," "nowadays," etc.) and the limits of his own knowledge. The scientific information he amassed in his wide-ranging interpretation of the Aztec calendar, for example, is not presented as authoritative or finite but, rather, as "everything we know up to now about the division of time among the Mexica" (215). Time and again, Humboldt poses unanswered questions about "what we so vaguely call the state of civilization" (296), emphasizing that "[w]e cannot presume to be able to resolve these questions with the current state of our knowledge" (209). But he was also clear that an inability as yet to "identify . . . unknown causes" must not translate into an unwillingness to ask and explore questions that are "not unimportant for the philosophical history of humanity" (236).

A Poetics of the Fragment

In aesthetic terms, *Views of the Cordilleras and Monuments of the Indigenous Peoples of the Americas* is Alexander von Humboldt's most experimental book. Its beauty lies in the very absence of a single ordering principle. The overall arrangement of the plates and the commentaries of unequal length is neither historical nor geographical, neither thematic nor cultural. One might say that the Humboldtian Cosmos as we glimpse it in this book is not a fixed, tidy constellation suffused with an innate order and beauty but a universe that includes vital elements of restlessness and motion—and thus the dimension of chaos—in its own living structure. But chaos does not mean disorder; it refers us to a *different* way of ordering things and mapping the world.

Humboldt variously remarked on the forms of writing that he developed in his books, essays, and letters. His statements show a dual indebtedness to literary and scientific models, both French and German, and an emphasis on writing that is also informed by personal acquaintance with its objects and subject through travel. For Humboldt, "views" were opinions and theories shored up by firsthand experience. Defending the explorer Gemelli Careri against those who did not believe that he had ever left Italy, Humboldt insisted that "Gemelli's descriptions have that local color that constitutes the main charm of travel narratives, even when they are written by the least enlightened of men, and that only those who have had the advantage of seeing with their own eyes can provide" (257). Significantly, there are also countless references to what one might call the poetics of the Humboldtian fragment. Humboldt's writing, which aimed at portraying total, or full, impressions, uses fragment in the sense of *modèles réduits* that can be read separately, for each contains all the key components of the work as a whole.

Views of the Cordilleras had initially been intended as a *"Picturesque Atlas"* for the actual travelogue of Humboldt and Aimé Bonpland's voyage to the Americas, the *Relation historique*, or *Personal Narrative*, which had not yet been published at the time Humboldt crafted *Views*. Unlike the *Relation historique*, which follows the travelers' itinerary despite countless digressions and detours, *Views* does not have a discernible linear structure. This is apparent even from a cursory glance at the first and last Plates. The book opens with two images of Chalchihutlicue (Nahuatl: she of the jade skirt), the Mexica goddess of flowing waters, and closes with a drawing of the Dragon Tree of Orotava in the Canary Islands, which Humboldt and Bonpland had visited in 1799, at the beginning of their Atlantic crossing. The volume is arranged in such a way that the reader is subjected to a ceaseless jumping back and forth between the individual phases, places, and phenomena of Humboldt and Bonpland's travels, which, in contrast to a traditional travel report, produces an unsteady, erratic, almost nomadic movement. The nonchronological order creates a simultaneity of things observed in different places at different times, and this simultaneity facilitates the links with other observations worldwide. A visual metaphor serves to illustrate this. When describing Cotopaxi (Plate X), "the highest of the volcanoes in the Andes to have erupted in recent memory," Humboldt imaginatively measures its absolute elevation by stacking other volcanoes on top of each other, so that Cotopaxi is "eight hundred meters higher than Vesuvius would be if it were stacked on top of the Peak of Tenerife" (62). Such spatiotemporal overlap and layering, which keeps the discontinuous structure of *Views of the Cordilleras* in the forefront of the reader's mind, does not, however, mean dehistoricization. Perhaps paradoxically, this foregrounded structure leads to a heightened historical (and historiographical) sensibility. By largely freeing himself from the actual time of the voyage, which dominated and still dominates travel literature, Humboldt could make the twin temporalities of human history and planetary history the principal frame for his writing.

Views of the Cordilleras may be read primarily as a geographic engagement with the Mesoamerican objects that Humboldt encountered during his eleven-month stay in New Spain in 1803–4. Humboldt himself concedes at the outset of *Views* that a geographical order would have been "useful"; yet he also claims that "the difficulty of both gathering and finishing a large number of Plates in Italy, Germany, and France made it impossible for me to adhere to his principle," suggesting that "the benefit of variety" (2) would compensate for this seeming lack of order, at least to some extent. Indeed, the incredible material, thematic, geographical, historical, and especially cultural heterogeneity of the images, documents, references, and reflections that Humboldt brought into play in *Views* compels readers to link together in their own minds what at first glance appear to be rather incompatible objects and subjects—incompatible not only because of cultural and generic differences but also because they occupy different geographical and historical spaces. Humboldt himself was well aware of the difficulties this posed to his readers, who were (and still are) very much used to ordering systems that guide the process of interpretation. As a cue—and a red herring—he offers his readers an inventory of the represented objects, but one from which no single ordering principle can be gleaned, not to mention the fact that seven of the sixty-nine plates are missing from this initial list. That Humboldt himself did not follow this list in assembling his book may turn it into a joke at the expense of readers who expect a reliable roadmap. At the same time, this introductory inventory may also represent a ground zero from which we move to a different order of things.

Art, nature, and science effectively permeate each other in what is no doubt Alexander von Humboldt's formally most daring book project. Instead of settling for either a purely illustrative work—the "Atlas" he had initially planned—or a more conventional travelogue, Humboldt returned in *Views* to the aesthetic principles that had informed his earlier *Geography of Plants* (1807), where he had already begun to explore an *inter*medial approach to science. This method allowed him to mix different media while also interrelating text and image on what might be called a *trans*medial plane. In *Geography of Plants*, Humboldt had used these techniques to make natural phenomena more exciting to the eye. In *Views of the Cordilleras*, he wanted to apply the same method to cultural phenomena. Doing so required an aestheticization of the mountainscapes of the Cordilleras and the cultural artifacts he offered up for contemplation. When describing the "majestic spectacle" of the great Tequendama waterfalls (Plate VI), for example, Humboldt tellingly intertwines sublime awe with scientific precision: "The impression that waterfalls make upon the observer's soul depends upon the combination of several circumstances: the volume of rushing water must be proportional to the height of the falls, and the surrounding landscape must have a romantic and untamed character" (39). The combination of seemingly unrelated categories of experience is typical of Humboldt's "tableaux," a hybrid form of representation, almost like a mixed-genre painting, that seeks to combine scientific information (tables) with visual art (drawings and paintings).

Volcanoes occupy a singular place of interest among the natural spectacles that fascinated Humboldt. His vivid accounts of climbing volcanoes, including some that were (and are) still active, rhythmically punctuate *Views of the Cordilleras*. While this thematic preoccupation gestured toward the passion for mountaineering that swept Europe in the early to mid-nineteenth century, Humboldt also used his mountain-climbing episodes as a formal literary device. Their purpose was to merge the various lines and trajectories of Humboldtian thought, from plant geography (with its different elevations and layers of plant growth) to landscape painting (with its excessively steep cliffs), and from climatology (with its temperature gradients) to the poeticization of what is now known as Volcano Alley. From where Humboldt and Bonpland were staying in Quito in 1802, "one can observe all at once, and in frightening proximity, the colossal volcano of Cotopaxi, the slender peaks of Iliniza, and the Nevado de Quilindaña. It is one of the most majestic and impressive sites that I have seen in either hemisphere" (66). The presence of Tenerife's Teide, which Humboldt climbed during his stopover in the Canary Islands, in a work that was ostensibly devoted to the Americas makes good sense. It was, after all, in the Canaries that the young scholar first grasped the complex intersections of nature and culture as he experienced volcanological phenomena in tandem with the aesthetic effects of sublime wonder. The brief description of the Ecuadorian stratovolcano Chimborazo, which occupied a special place in Humboldt's heart, perhaps because of his failure to reach its summit, encapsulates his aesthetic investment of nature in a suggestive simile: the majestic mountain "appears to the observer like a cloud on the horizon; it stands out from the neighboring peaks; it looms above the entire Andes range just as that majestic dome, Michelangelo's work of genius, looms above the ancient monuments that surround Capitoline Hill" (127). If Humboldt's fieldwork inspired scientists such as Charles Darwin and Richard and Robert Schomburgk, his "humble sketches" of the Andean landscape would equally stir the curiosity of "travelers with a passion for art to visit the regions that [he] traversed" (16), notably nineteenth-century painters such as Johann Moritz Rugendas and Frederick Edwin Church.

For Humboldt, volcanoes were at once the epitome of aestheticized nature and key features of a distinct cultural landscape. As the book's title suggests, natural and cultural phenomena share the same space here. Humboldt's frequent meditations on the fact that the indigenous peoples had lived and produced different cultural forms at different elevations—on the coast, in the lowlands, and especially in the mountains—weave together aspects of natural and cultural space with natural and cultural scientific knowledge. He notes "a small rock mass on the southwest side [of Cotopaxi], half-hidden under the snow and spiked with points, and which the natives call the *Head of the Inca*," and comments on its ties to a local legend about an omen of the fall of the Inca Tupac Yupanqui (65). Similarly, he tells us that among the Mexica "a volcano is called *a speaking mountain*," connecting this metaphor to hieroglyphic images of a bird that, according to a local legend of the great flood, "distributes thirty-three different tongues" to the mute ancestors of humanity. Mexica paintings, Humboldt explains, "depict the latter as a cone with several tongues floating above it because of the underground noise that one sometimes hears in the vicinity of a volcano" (75). The "thin plume of smoke above the crater of Cotopaxi" (66), which Humboldt took the liberty of including in his sketch even though he had not actually seen any smoke that day, resonates with both the metaphor and the hieroglyphic image.

The Myth of Western Civilization

Logos and mythos go hand in hand in *Views of the Cordilleras*. Not only was Humboldt aware of the lasting impressions that visual images left; he was also conscious of the enlivening power of narrative. He tested this literary power in examples of indigenous myths as much as in the narrative fragments of a travelogue that did not yet exist at the time; it was slated to appear between November 1814 and April 1831. From his contacts with and research about the American peoples, Humboldt found out about a great number of myths that he liked to recount in his writings. Fiction and imagination thus come to play an essential role in his narratives, ranging from the founding myth of Tenochtitlan, today's Mexico City, to the myth of Bochica in what later became Colombia. Scattered throughout *Views of the Cordilleras*, these narrative fragments supplement philosophical discussions and reports about the individual legs of Humboldt's American journey. They are literary fragments that keep rhythmic time, as it were, in Humboldt's scientific discourse.

As in the myths of the so-called Indians, kinship metaphors play a weighty role in Humboldt's narrative. In his comments on the *Codex Mendoza*, for instance, Humboldt enthusiastically agrees with Nils Gustaf Palin, whose work he had treated more critically elsewhere, "that it is a beautiful and fruitful idea to consider all the peoples of the earth as belonging to the same family, and to identify in the Chinese, Egyptian, Persian, and American symbols examples of a language of signs that is common, as it were, to the entire species, and that is the natural product of the intellectual faculties of humankind" (341). Roland Barthes has shown in his famous *Mythologies* (1957) that such talk about the great "family of man" is but an old Western myth that functions to lay claim to the unity of humankind, even to the very "essence" of what it is to be human, whenever palpable ethnic, social, or cultural differences come into view. Humboldt escapes Barthes's reproach that the whole point of the myth is, in the end, to naturalize complex histories, at least to some extent because Humboldt is concerned precisely with the historicity of cultural differences and with fact-bound analysis. Yet the genealogies in Humboldt's (American) discourse are undeniable traces of Western thought: regardless of his emancipatory goals with respect to the peoples of the Americas, he still placed individual members of this large human family at different points in a developmental spectrum that ranges from barbarism to civilization. In fact, Humboldt used the adjective *barbarous* with quite some frequency in his descriptions of the peoples of the Americas and of certain cultural practices, notably human sacrifice. Yet he was rather ambivalent about these inherited categories, even as he applied them. From the Dispute on the New World he had certainly learned what terrible effects the West's exclusionary mechanisms could have.

> This people, who based their festivals upon the movements of the stars and who engraved their celebrations upon a public monument, had likely reached a higher level of civilization than that accorded to them by [de] Pauw, Raynal, and even Robertson, the most judicious of all the historians of the Americas. These authors regard as barbarous any state of humanity that diverges from the notion of culture that they have established, based on their own systematic ideas. We simply cannot accept such sharp distinctions between barbarous and civilized peoples. (216)

Although Humboldt repeatedly mentions the violence that the Spanish visited upon indigenous populations in the Americas, he did not get caught up in the tradition of the *leyenda negra*, the "black legend" that was replete with sterile condemnations of the atrocities committed during the Spanish conquest. In one of his commentaries on the *Codex Vaticanus anonymous A*, for instance, Humboldt levels characteristically caustic criticisms in a seemingly offhand manner: "[T]his manuscript contains copies of hieroglyphic paintings made after the conquest: here one sees natives hung from trees, holding crosses in their hands; a number of Cortés's soldiers on horseback setting fire to a village; friars who are baptizing unfortunate Indians at the very moment when the latter are put to death by being cast into the water" (237). Humboldt always called attention to the barbarity within civilization itself, notably to the barbaric aspects of so-called Western civilization, in full awareness that presumably gentle and peaceful Christianity got along rather well with colonialism and slavery in his day. He distanced himself from these implications of the civilizational process, adopting a critical attitude toward the myth of Western progress.

An Alternative Library

Thanks to the extensiveness of his American travels, Humboldt can be regarded as one of the few who, at the turn of the nineteenth century, could compare and relate from personal familiarity detailed information about the late-colonial societies in the Americas. The spectrum of his social experience in the Spanish colonial viceroyalties of New Spain and New Granada ranged from contacts with native guides and carriers (*cargueros*) in the Andes and many indigenous groups in the mountains and lowlands of today's Venezuela, Colombia, Ecuador, Peru, and Mexico to the highest echelons of clergy, scientists, and government administrators. The encounters with colonial leaders, the importance of which he emphasized from the outset and which were facilitated by support from the Spanish crown, gave Humboldt ready access to archives, libraries, and document collections. The roles of educated elites, especially the Indian and Mestizo clerics, intrigued him. The research of the colonies' scientific leaders, who would become major supporters of the independence movements in the Spanish colonies only a few years later, enabled Humboldt to give historical depth to the societal rifts in Spain's former possessions. Always careful not to cover over differences among the very distinct societies and cultures of the Americas, Humboldt did not limit his attention to the genealogies and histories of the rulers but also, significantly, focused on social organization and stratification as reflected in myths and legends. We learn, for instance, that "[t]he Totonac, who had adopted the entire Toltec and Aztec mythology, made a racial distinction between, on the one hand, the deities who demanded bloody worship and, on the other, the goddess of the fields" (117). In this way, a highly nuanced picture of both Mesoamerican and colonial societies emerges in the interstices between narrative and pictorial fragments.

Humboldt's work benefited notably from the advancements in knowledge generated in the colonial capitals and the European metropoles. His tireless travels even after his return from the Americas, which included visits to various archives, museums, and collections across the Old World, gave him an even more extensive knowledge of early American art and culture. "My own travels in the various parts of the Americas and of Europe have given me the advantage of being able to examine a greater number of Mexica manuscripts than could Zoëga, Clavijero, Gama, the Abbé Hervás, Count Rinaldo Carli . . . , and other scholars who, following Boturini, wrote about these monuments of the ancient civilization of the Americas" (88). Throughout his stay in New Spain and New Granada, Humboldt avidly explored vestiges of American cultures in the archives of the viceroyalty, and he published parts of them. As a result, *Views of the Cordilleras* came to include a broad array of sources and documents, among them fragments of different codices. This made Humboldt's book something of an annotated anthology of hieroglyphic paintings, landscape drawings, and other documents, along with often copious citations and references or allusions to the work of others. A stranger to nationalistic urges, Humboldt conceived his writing in dialogue with many other scholars and scientists. He saw himself as part of the international republic of letters of his day and participated in the impassioned debates of this (then largely European) community.

With *Views of the Cordilleras*, Humboldt created a complex space that may also be understood as a library. (We have attempted to represent this aspect of his work by including in this edition a detailed bibliography of his sources.) This library placed the protagonists of the Dispute on the New World side by side with references to the latest scientific research about non-European cultures and languages, including work by Johann Christoph Adelung, Jean Joseph-Marie Amiot, Johann Friedrich Blumenbach, Dominique Denon, Adam Johann von Krusenstern, Charles Marie de La Condamine, Joseph Lafitau, Nils Gustaf Palin, Friedrich Schlegel, Antoine Silvestre de Sacy, Melchisédec Thévenot, Johann Severin Vater, Ennio Quirino Visconti, and William Warburton, among many others. These are joined by copious references to authors from antiquity, such as Apollonius, Aristophanes, Aristotle, Cicero, Eratosthenes, Herodotus, Hesiod, Homer, Origenes, Plato, Pliny, Plutarch, Polybius, Ptolemy, Seneca, Strabo, Suetonius, Virgil, and Vitruvius, not to mention Dante Alighieri, whose writings appear in conjunction with mentions of scientific studies in areas such as anatomy, arithmetic, astronomy, botany, geology, geomorphology, history, linguistics, mineralogy, mathematics, philosophy, and zoology. For the history of the Americas and its conquest by the Spanish, Humboldt resorted to the *relaciones* by conquistadores such as Hernán Cortés, Bernal Díaz del Castillo, and Gonzalo Jiménez de Quesada, as well as to the writings of official Spanish chroniclers, missionaries, and travelers, including Father José de Acosta, José Antonio Alzate y Ramírez, Toribio de Benavente, Francisco Clavijero, José Domingo Duquesne, Antonio de León y Gama, Francisco López de Gómara, Andrés de Olmos, Lucas Fernández de Piedrahita, Bernardino de Sahagún, Carlos de Sigüenza y Góngora, and Juan de Torquemada. Of particular importance to Humboldt's own research were the publications and remnants of collections by Lorenzo Boturini Benaducci, an Italian traveler whose gathering of early American materials had been scattered and partly destroyed by the Spanish colonial authorities decades before Humboldt arrived in New Spain.

While Humboldt's new discourse on the New World did not exclude those authors who had traditionally legitimized European views on non-European subjects, *Views of the Cordilleras* specifically, and uniquely, called attention to others whose work either was unknown in Europe at the time or else had been discredited and thus had not entered the debates on the New World. Notable among these are the Mestizos Garcilaso de la Vega el Inca and Cristóbal del Castillo, the Nahua historians Alvarado Tezozomoc and Chimalpahin Quauhtlehuanitzin, and Alva Ixtlilxochitl and Netzahualcoyotl, a poet-ruler "as memorable for the culture of his mind as he is for the wisdom of his laws" (393). Humboldt was among the very first to bring to Europeans' attention a library in which the former historyless objects of Europe's impe-rial gaze became subjects whose testimonies and testimonials provided other perspectives on the *conquista* and continue to do so today. Contributing to making the views of the conquered public knowledge was a fundamental part of Humboldt's intellectual agenda.

A Museum of World Cultures

We have seen that *Views of the Cordilleras* crosses many traditional genres of literary and scientific writing, ranging from travelogue and essay to scientific treatise, (art) historiography, and anthology. Although Humboldtian writing is distinctive, it is, as the recurring vestiges of the travel narrative show, not without precedent. Humboldt's desired connection of sensual experience with intellectual clarity, of scientific groundwork with aesthetic pleasure, de-manded a high degree of artistic shaping.

In addition to being a library, Humboldt's image-text may be understood as a *collection* composed of many individual pieces at a remove of their original cultural contexts. The Prussian knew about the collection of early American cultural objects that Bernardino de Sahagún had created not long after the conquest in the sixteenth century. He also knew of the work of New Spain's great collector Carlos de Sigüenza y Góngora in the seventeenth century and admired the selfless research and collecting of the already mentioned Bot-urini during the eighteenth century. He tried to trace these collections in the libraries and archives he visited, and when Antonio de León y Gama's collec-tion was dissolved, Humboldt was lucky enough to acquire a few pieces that had originally been part of Boturini's *Museo histórico indiano*, probably the most extensive collection of indigenous art and the foundation for its scien-tific study. With *Views of the Cordilleras* Humboldt wrote himself into the "American" genealogy of collectors that begins with Sahagún and Sigüenza y Góngora and extends from Boturini to León y Gama and Clavijero. But Hum-boldt modified this traditional line of descent in that he, unlike Clavijero, who had been expelled from New Spain for being a Jesuit, was concerned not with historical narrative but with the idea of a museum in which each exhibit could be viewed at once from many different angles and independently of the others. Rather than placing the cultures of the world in separate spaces, Hum-boldt brings them together in the space of a museum, combining European landscape painting of non-European nature scenes with Aztec pictograms.

The exhibits in this imaginary museum are based on the multifaceted interplay of image-text and text-image in which the actual plates are never merely illustrations that accompany the text. Image and writing come together in a scientific-artistic hybrid. If one tries to separate them, as happened in the first German translation of this book (see Editorial Note), much is lost. *Views of the Cordilleras* demands active readers who are willing to explore all sorts of the corridors and directions. The visitors to Humboldt's virtual museum have many different choices: they may follow the examples of European landscape painting (based largely on Humboldt's own sketches) or sample the Mexica hieroglyphs; alternatively, they may prefer either the research on the different calendars and monuments—what Visconti called "an entirely new branch of archaeology" (372)—or the descriptions of amazing natural phenomena, such as volcanoes. As in the physical space of a museum, there are many possible relations among the individual exhibits, depending on the order in which they are viewed. Even though the number of exhibits is finite, the number of possible relations among them, and with the visitors, is nearly infinite.

The space of Humboldt's multilayered image-text, however, is best imagined not as that of an actual museum, which is static, but as a dynamic hypertext that continuously repositions the user. Humboldt's open arrangement of knowledge in *Views of the Cordilleras* uncannily anticipates today's interactive hypermedia. Nothing is really closed off in such a structure. As in the space we now call the Internet, visitors to Humboldt's virtual space can choose their interpretive pathways as they stroll (or scroll) freely. This virtual museum-in-a-book has the same advantage that we hope the Internet does: it cannot be dismantled and scattered by any colonial power.

Globalization and World-Openness

Terms such as *world trade*, *world communication*, and *world history* were key for Humboldt. In *Cosmos* he would add the neologism *Weltbewußtsein*, world consciousness. Even though Humboldt's cosmopolitanism has a deeply European imprint, it invites and depends on dialogues with non-European cultures. His objective was not what we now know as regional or area studies research but an understanding of global cultural phenomena that would take in the interrelations, homologies, and parallels among far-flung cultures. He regarded the development of the cultures that had first flourished in the Americas as being as much part of the order of the world's cultures as were the advances of the Egyptians, Etruscans, Indians, Chinese, Greeks, and Romans.

An important facet of Humboldt's global vision was also to bring the cultures marginalized as non-Western into conversation with each other. In his comprehensive comparisons of calendrical systems, for instance, we find out about the relativity of ideas of time. We also come to understand that in this book it is impossible to avoid leaps and breaks, or, for that matter, to filter out experiences of discontinuity and heterogeneity. In its performative presentation of a potentially discomfiting global relationality, Humboldt's universalist and, at the same time, historicist cultural theory reproduces the dynamic of his own creative curiosity as a reading (and viewing) experience:

> I am pleased to have called attention to a remarkable monument of Mexica sculpture and to have provided new details about a calendar that neither Robertson nor the illustrious author of *Histoire de l'astronomie* seems to have approached with as much curiosity and care as it merits. People's curiosity will be stoked by the information we shall provide below about the Mexica tradition of the *four ages*, or four suns, which bears a striking similarity to the *yugas* and the *kalpas* of the Hindus, and to the ingenious methods that the Muisca Indians, a mountain people of New Granada, employed to correct their lunar years. (154).

In his in-depth analysis of the Aztec Calendar Stone that had been found in the main plaza of the capital of Mexico in 1790, Humboldt not only highlights (once again) just how provisional all knowledge necessarily is but also stresses the need to reassemble scattered source documents. "It is impossible," he writes, "to determine the accuracy of this theory until a larger number of Mexican paintings in Europe and the Americas has been consulted; for—I cannot repeat this often enough—everything that we have learned up to now about the former state of the peoples of the new continent will pale by comparison with the insights on this subject that will one day be reported once all the materials that are scattered throughout the two worlds, and that have survived centuries of ignorance and barbarity, are successfully gathered together in one place" (211). The process of scientific inquiry that Humboldt adumbrates here is analogous to the movements that bring together the artworks of Europe and of the Americas in the space of his virtual museum of the world's cultures. For Humboldt, knowing presupposes movement, including movement that is physical and geographic. No matter how hard he tried to collect all relevant data, he was always aware that the worldwide network of knowledge production he created and maintained also included significant gaps, the size and significance of which could not even be estimated. As the above passage implies, Humboldt was not interested only in continuous epistemological advancement but also in processes of loss and forgetting, even the destruction of extensive inventories of knowledge. Throughout *Views of the Cordilleras* he tells of many "curious objects," such as the mysteriously lost volume of the Ucayali that "bore a perfect resemblance to our *quarto* books" (93); documents such as the Velletri manuscript that fell into the hands of household servants, who, "unaware of the price that a collection of hideous figures might have fetched, handed it over to their children" (110); and Mexica images burned by fanatical monks and insouciant conquistadores. What the scientific observer knows, then, and at which time, is always understood as contingent and transitional. It is part and parcel of a larger process that Humboldt saw as moving toward freedom. It is for good reason, then, that he distanced his views of culture from all system, or doctrinal, thinking, mocking "the assertion by Pauw and other, equally system-bound writers who claim that none of the indigenous peoples of the new continent is able to count beyond three in its own language" (295).

It is quite likely that Humboldt self-consciously availed himself of the semantic flexibility of the titular word *monument*, which can mean testimony, document, memorial, and work of art. Some passages, especially from the comments on the initial plates, may create the impression that Humboldt was mainly, or even exclusively, interested in the historical-documentary character of the "monuments" of the indigenous peoples of the Americas—that is, in "monuments produced by peoples who did not attain a high level of intellectual culture, or who, for either religious or political reasons or because of the nature of their societal organization, seemed less appreciative of the beauty of forms, [which] can only be considered historical monuments" (13). Yet there are many comments throughout *Views of the Cordilleras* to suggest that he was not in favor of exiling indigenous *monumens* from the realm of the arts. Already in his introduction he also stresses the "connections . . . between *artworks* from Mexico and Peru and from the Old World" (14, emphasis added). He reiterates this point in his later discussion of the possibilities of Inca architecture (Plate LXIII): "[T]he primary goal of this work is to give an exact idea of the state of the arts among the civilized peoples of the Americas" (356). While these inconsistencies register a certain hesitation in according American "monuments" the same status as the anointed masterpieces of Western art, it is also worth recalling that at the time, Mexica, Inca, or Maya painting, sculpture, and architecture were not even recognized as cultural products, let alone as artifacts or artworks. Humboldt was the first to call attention to both the historical *and* the artistic significance of Aztec and Inca cultures—much as John Lloyd Stevens did for the Maya in the early 1840s—laying the groundwork for the research of generations of archaeologists and art historians.

At the same time, the seeming contradictions in Humboldt's arguments alert us to the fact that relativizing of apparently unequivocal statements is an integral feature of Humboldtian science. Self-critical passages and lengthy quotations of opinions critical of his own serve the same purpose. Visconti's letter at the end of *Views of the Cordilleras* is a case in point, as is Humboldt's endnote correcting an error he had committed earlier: "According to information from Mexico, which I have received since the publication of the first part of this work, this remarkable sculpture was not found in Oaxaca, which I claimed in error (67–71), but farther to the south, near Guatemala, in the old *Quauhtemallan*" (393). Humboldt advocated and created an open research climate in which new scientific results and insights are included as quickly as possible, and in which knowledge is not the static possession of a single individual but the dynamic process of a collectivity. Offering a variety of snapshots of scientific processes as they unfold within a global community, Humboldt grants his readers insight into how knowledge is actually produced. In this way he historicizes both the objects of knowledge and the processes of knowing themselves. Humboldt's interpretations of the Aztec pictographs from the Royal Library in Dresden (Plate XLV) characteristically begin with the aim of charting the most comprehensive interrelations among certain cultural artifacts within their own cultural system and in relation to other cultures worldwide.

> Following this very principle, namely, that an explanation for one monument can be found in another, and that in order to enter into the history of a people in greater depth, one must have in front of one all of the works infused with the character of that people, I decided to commission engravings (on Plates XLV–XLVIII) of fragments taken from the Mexica manuscripts of Dresden and Vienna. The first of these manuscripts was entirely unknown to me when the printing of these pages began. It is not easy to provide a complete explanation for the hieroglyphic paintings that escaped the destruction with which they were threatened during the discovery of the Americas by the monks' fanaticism and the foolish insouciance of the first conquerors. Mr. Böttiger, an antiquarian who has conducted scholarly studies on the arts, mythology, and domestic life of the Greeks and the Romans, first brought to my attention the *Codex mexicanus* of the Royal Library at Dresden. He mentioned it quite recently in a work that offers the most advanced theories on the painting of the barbarous peoples, as well as on that of the Hindus, the Persians, the Chinese, the Egyptians, and the Greeks. (311)

Humboldt built networks within disciplines and beyond familiar disciplinary boundaries. The speed with which he integrated the information he obtained in this way into his own publications demonstrates the rapid growth of knowledge and of the forms of communication within the Humboldtian circle of correspondents and colleagues. In no way does he understate the fast-paced changes in his own (published) knowledge, but he always highlights the character of his books as *works in progress* that seek to reflect the most up-to-date thinking. The repeated staging of the provisionality and incompleteness of all research results is indisputably a sign of Humboldt's intellectual integrity.

Much like Humboldt's other writings, *Views of the Cordilleras* is, as we saw above, a veritable thicket of references, suggesting ever-changing directions of reading and a capacity to link together highly diverse fragments of knowledge. It is not accidental that the very last of Humboldt's endnotes refers back to the first two plates, the "Bust of an Aztec Priestess," with which he opens his book. Rather than providing closure, this gesture serves as an invitation for rereading what is now a familiar text to discover new associations. That many of these connections are subterranean, as it were, might be expected from this former mining inspector, who was eminently familiar with galleries below the surface. Several of the Aztec monuments that Humboldt examines in his book were, in fact, buried below the surface of the Great Plaza in Mexico City, below the stones of the *zócalo* in front of the great cathedral depicted in Plate III. And below the textual and pictorial surface of the book itself the testimonies of Mexica culture and art join up to create a rhizomatic structure of internal references. This root system links the Aztec calendar stone with an entire pantheon of Mexica deities, including the statue of the goddesses Teoyamiqui and Coatlicue as well as the "monstrous idol" in Plate XXIX. Likewise, the tomb of the sacred wolf unearthed in the same location points to comments on the different zodiacs by way of the Mexica sign *itzcuintli*, meaning (feral) dog.

In addition to the structure of internal, or intratextual, relations integrating the individual pieces into a collection that is both a library and a virtual museum, there are frequent examples of *inter*texual references that identify *Views of the Cordilleras* as part of the vast travel work that Humboldt was then in the process of compiling. These include references to volumes that had already been published or were just about to appear in print, such as his *Political Essay about the Kingdom of New Spain* (1811), and allusions to future publications that he would not complete during his lifetime. This specific intertextual dimension shows the extent to which Humboldt's very distinctive book projects were conceived as parts of a work in progress, a large, dynamic network. This idea of an open, fundamentally chaotic system forms the basis for an innovative discourse on the Americas that was grounded in his fieldwork and served as a polemical weapon in his struggle against the then-dominant discourse on the Americas. This openness also enabled him to glean from a plethora of observational details developments for which he, like his contemporaries, did not yet have a suitable (cultural studies) terminology. For example, he writes in "Dress of the Indians of Michoacan" (Plates LII und LIII) that "one is struck by the peculiar combination of the ancient Indian attire with the garb introduced by the Spanish colonists" (328). Humboldt's order of the cultures of the world is to be understood as a dynamic process that connects all sorts of disparate elements and dimensions.

Humboldt's own astonishment drew his attention to the asynchronicity of social versus artistic developments in Asia and Europe. There is the "contrast between social perfection and artistic infancy" in several peoples who played a role on the world stage (particularly the peoples of Central and Eastern Asia); similarly, we find an "unrefined state [of the arts], among a people [the Mexica] whose political existence had for centuries suggested a certain degree of civilization" (243). Recognizing this pattern prevented Humboldt from casually applying to the Mexica the familiar civilizational sequence that begins with human and animal sacrifice to end with fire and harvest rituals. The logical consequence of Humboltian thinking is, of course, that the envisioned order of the world's cultures is also provisional, always changing.

Views of the Cordilleras and Monuments of the Indigenous Peoples of the Americas leaves ample room for new insights into and fresh perspectives on European civilization and its own forms of barbarism. By bringing into view the similarity of developments that did not happen at the same precise (historical) time, Humboldt made it very clear that Western civilization, should it desire to derive any lessons from the vestiges of barbarity within itself, must remain acutely conscious of the provisionality and incompleteness of its civilizational processes and achievements.

Nashville-Potsdam, March 2010

Note on the Text

The translation in this edition is based on the full text of the original French folio edition of Alexander von Humboldt's *Vues des Cordillères et Monumens des Peuples Indigènes de l'Amérique* (1810–13). It includes all references, tables, and plates. The page numbers in the outer margins of each page refer back to that edition. Any additions and corrections are placed in [square brackets], reserving the use of parentheses entirely for Humboldt himself.

Throughout, we also keep intact Humboldt's seemingly capricious use of capitalization and italics and all words, phrases, and quotations originally written in Spanish and other languages (such as Latin and Greek). Only where meanings are not evident from context do we add our own translations of foreign words and phrases in [square brackets]. Humboldt himself rarely converted currencies and other units of measure, illustrating their variety across different parts of the world. We follow his practice, offering explanations at the beginning of our annotations of currencies and units of measure no longer in use. To avoid encumbering this already densely textured work with endnote numbers, we use ˅ to signal annotations related to a name, concept, or historical event. The annotations, which follow the text of the translation, are ordered in accordance with the translation's pagination. To help readers trace at least part of the vast and intricate scientific network that Humboldt created, we provide a bibliography of the sources he acknowledges and implies in this work.

The Editorial Note offers more information about the thinking behind this translation and about the historical and scholarly contexts for *Views of the Cordilleras and Monuments of the Indigenous Peoples of the Americas*. Select materials from these contexts can be found at www.press.uchicago.edu/humboldt.

The HiE Team

Picturesque Views of the Cordilleras
and Monuments of the Indigenous Peoples of the Americas

À Monsieur

Ennius-Quirinus-Visconti

Membre de l'Institut de France

A. de Humboldt   Aimé-Bonpland

Introduction

BY ALEXANDER VON HUMBOLDT

I have brought together in this work everything that relates to the origin and early technical and artistic advances of the indigenous peoples of the Americas. Two-thirds of the Plates included here depict architectural and sculptural relics, historical scenes, and hieroglyphs relating to the division of time and to the calendar system. Images of monuments relevant to the philosophical study of man are combined here with picturesque views of the most remarkable sites of the new continent. This arrangement was motivated by factors that are stated within the general considerations at the head of this Essay.

Inasmuch as the nature of the subject permits, the description of each Plate constitutes a separate account, and I have given more elaboration to those that might one day cast light on the similarities between the inhabitants of the two hemispheres. It was surprising to discover, toward the end of the fifteenth century, and in a world that we call new, the very kinds of ancient institutions, religious ideas, and shapes of buildings that in Asia seem to date back to the dawn of civilization. It would seem that the characteristics of peoples, not unlike the internal structure of plants, are disseminated across the surface of the earth. The imprint of an original type appears everywhere, despite the differences produced by the nature of the climate, the makeup of the soil, and the combination of several contingent factors.

At the beginning of the conquest of the Americas, the attention of Europe was singularly focused on the gigantic constructions of Cusco [Quechua: Qusqu], the great roads traced through the heart of the Cordilleras, the terraced pyramids, the religion, and the symbolic script of the Mexica. Nowadays, the area surrounding Port Jackson in New Holland and the island of Tahiti are not written about more often than many parts of Mexico and Peru were at that time. One needs to have visited these places to appreciate the naiveté and the authentic local color that characterizes the accounts of the first Spanish travelers. When studying their writings, however, one regrets that the latter were not accompanied by images that might give an exact idea of the many monuments destroyed by fanaticism or fallen into ruin as a result of shameful neglect.

The fervor with which people devoted themselves to studying the Americas began to fade at the start of the seventeenth century. The Spanish colonies, within whose bounds lie the only regions formerly inhabited by civilized peoples, remained closed to foreign nations, and when the ˟Abbot Clavijero recently published his *Storia antica del Messico* in Italy, a number of events to which a crowd of eyewitnesses (including several mutual enemies) attested were viewed with great suspicion. Some famous writers, who were struck more by contrasts than by the harmony of nature, had indulged in portraying the Americas as a swampy land unfavorable for raising animals and newly inhabited by hordes as uncivilized as the inhabitants of the South Seas. In historical studies of the Americans, healthy criticism gave way to absolute skepticism. The declamatory descriptions of ˟Solís and a few other writers who had not left Europe were accorded the same value as the simple and true accounts of the first travelers; it seemed to be the duty of a philosopher to repudiate everything that had been observed by missionaries.

Since the turn of the century, a happy revolution has taken place in our conception of the civilizations of different peoples, and of the factors that either obstruct or encourage progress. We have come to know certain peoples whose customs, institutions, and arts differ as much from those of the Greeks and Romans as the original forms of extinct animals differ from those species that are the focus of descriptive natural history. The ˟Asiatick Society of Calcutta has cast a vivid light on the history of the peoples of Asia. The monuments of Egypt, which are nowadays described with admirable exactitude, have been compared to monuments in the most distant lands, and my study of the indigenous peoples of the Americas appears at a time when we no longer consider as unworthy of our attention anything that diverges from the style that the Greeks bequeathed to us through their inimitable models.

It might have been useful to arrange the materials in this work in geographical order. But the difficulty of both gathering and finishing a large number of Plates in Italy, Germany, and France made it impossible for me to adhere to this principle. To a certain extent, this lack of order is compensated for by the benefit of variety and is, moreover, less objectionable in descriptions contained within a picturesque Atlas than in a formal discourse. I will attempt to remedy this lack through an overview in which the Plates are organized according to the nature of the objects they depict.

I. Monuments

A. *Mexica*

    Bust of an Aztec Priestess, Plates I and II, pages 18–19

    Pyramid of Cholula, Plate VII, page 44

v   I have attempted to render the objects in these engravings with the greatest possible precision. Those engaged in the practical aspects of the arts know how difficult it is to oversee the large number of Plates that make up a picturesque Atlas. If some of these are less perfect than experts may desire, this imperfection should not be attributed to the artists entrusted with carrying out my work under my supervision but, rather, to the sketches that I made on site in often trying circumstances. Many of the landscapes have been colored, because in this type of engraving snow stands out much better against a sky background, and because the reproduction of Mexica paintings made the use of both colored and black-ink Plates indispensable. We have experienced great difficulty in giving the former the vigor that we so admire in ˅Mr. Daniell's *Oriental Scenerys*.

In my description of the monuments of the Americas, I have proposed to hold to a happy medium between the two paths followed by scholars who have conducted investigations into the monuments, the languages, and the traditions of these peoples. Some scholars have devoted their time to generating theories that, although brilliant, are founded on shaky ground, and they have therefore drawn general conclusions from a small number of isolated facts. They have seen both Chinese and Egyptian colonies in the Americas; they have found both Celtic dialects and the Phoenician alphabet. Although it is still unknown whether the Osci, the Goths, or the Celts were actually transplanted peoples from Asia, some have nonetheless pronounced on the origin of all of the hordes of the new continent. Other scholars have accumulated materials without rising to any general idea, a method as sterile in the history of peoples as it is in the physical sciences. May I have been fortunate enough to avoid the pitfalls that I have just described! A few peoples, very distant from one another—the Etruscans, the Egyptians, the Tibetans, and the Aztecs—exhibit striking parallels in their buildings, their religious institutions, their division of time, their cycles of regeneration, and their mystical ideas. It is the duty of an historian to draw attention to these similarities, which are as difficult to explain as the connections between Sanskrit, Persian, Greek, and the Germanic languages. But in trying to generalize, one must know to stop at the very point where precise data are missing. With these principles in mind, I will bring to light the conclusions to which the knowledge I have hitherto acquired about the indigenous people of the new world appears to lead.

In examining closely the geological makeup of the Americas, in reflecting upon the balance of fluids that are spread across the surface of the earth, one would be hard-pressed to claim that the new continent emerged from the waters at some later point than the old one did. One observes there the same succession of rocky layers as in our hemisphere, and it is likely that the granite, the micaceous schist, or the different gypsum and sandstone formations in the mountains of Peru date from the same periods as their counterparts in the Swiss Alps. The entire globe appears to have undergone the same catastrophes. Fossilized pelagic shells are suspended on the crests of the Andes, at a height exceeding that of Mont Blanc. Fossils of elephant remains are scattered throughout the equinoctial regions; what is particularly remarkable is that they are not found in the steaming plains of the Orinoco but, rather, on the highest and coldest plateaus of the Cordilleras. In the new world, as in the old, generations of now-extinct species preceded those that today populate the earth, the sea, and the air.

There is no proof that the existence of humankind is a much more recent phenomenon in the Americas than in other continents. In the tropics, the vigorous plant life, the size of the rivers, and the partial floods have raised powerful barriers to the migration of peoples. Vast swaths of northern Asia are as thinly populated as the savannahs of New Mexico and Paraguay, and one need not suppose that the lands that have been inhabited the longest are necessarily those with the largest number of inhabitants.

The problem of the original population of the Americas resides no more within the province of history than questions about the origin of plants and animals, and about the distribution of organic germs, lie within the domain of the natural sciences. Venturing back to the earliest times, historical research reveals to us that nearly every part of the earth was once occupied by men who believed themselves to be aboriginal because they were unaware of their filiations. Amidst a multitude of peoples who have succeeded one another and have intermixed, it is impossible to determine what exactly was the initial base of the population, that original layer beyond which begins the domain of cosmogonic tradition.

The peoples of the Americas—with the exception of those bordering the Arctic Circle—form a single race characterized by skull shape, skin color, extremely scarce facial hair, and limp, smooth hair. The American race has unmistakable connections with that of the Mongol people, which includes the descendants of the Xiongnu (formerly known as the Huns), the Khalkha, the Kalmyk, and the Buryats. Recent observations have proved, furthermore, that not only the inhabitants of Unalaska but also many small tribes of South America offer evidence, in the form of certain osteological characteristics of the head, of a transition from the American race to the Mongol race. Once we have more thoroughly studied the brown men of Africa and that swarm of peoples who inhabit the interior and northeast of Asia and to whom traveling categorizers have referred, using the vague terms Tartars and Uralians, then the Caucasian, Mongol, American, Malay, and Negro races will appear less isolated, and we will recognize in this great family of humankind one single organic type, modified by circumstances, which will perhaps remain forever unknown to us.

Although the indigenous peoples of the new continent are united by close ties, they exhibit, in their facial expressions, in their more or less swarthy skin tone, and in their tall stature, differences as striking as those between Arabs, Persians, and Slavs, all of whom belong to the Caucasian race. Nevertheless, the hordes that cross the steaming plains of the equinoctial regions do not have a darker skin tone than the mountain peoples or the inhabitants of the temperate zone, either because, for humankind and for the majority of animals, there is a certain period of organic life after which the influence of climate and diet is more or less nil, or because deviation from the original type becomes noticeable only after a long series of centuries. Moreover, all the evidence suggests that Americans, like the Mongol peoples, have a less mutable physical makeup than other peoples of Asia and of Europe do.

The American race, the least numerous of all, nevertheless occupies the largest territory on earth. That territory stretches across the two hemispheres, from 68 degrees northern latitude to 55 degrees southern latitude. It is the only one of the races that dwells both on the steaming plains bordering the ocean and on the mountainside, where it reaches heights that exceed that of the ˙Peak of Tenerife [Mount Teide] by 200 toises.

The number of distinct languages among the small indigenous tribes in the new continent appears to be much larger than in Africa, where, according to the recent study by ˙Mr. Seetzen and Mr. Vater, more than one hundred and forty languages are spoken. In this respect, the Americas as a whole

VIII

resemble the Caucasus, Italy prior to the Roman conquest, or Asia Minor, when that small stretch of terrain was home to the Cilicians of the Semitic race, the Phrygians of Thracian origin, the Lydians, and the Celts. The lay of the land, the vigorous plant life, and the fear that mountain peoples in the tropics have of exposing themselves to the heat of the plains obstruct communication and thereby contribute to the astonishing variety of American languages. One also observes that this variety is less evident in the savannahs and in the northern forests that hunters can cross freely, on the banks of the great rivers, along the Ocean coasts, and wherever the Inca had spread their theocracy by force of arms.

When we consider that several hundred languages exist in a continent with a population smaller than that of France, we must acknowledge the differences between languages that have the same relationships between them as do, if not German and Dutch or Italian and Spanish, then at least Danish and German, Chaldean and Arabic, or Greek and Latin. The further one penetrates into the labyrinth of American languages, the more one senses that although several of them can be grouped into families, a large number remain isolated, like Basque among the European languages and Japanese among the Asiatic ones. This isolation is perhaps only superficial, and one has reason to suspect that languages that seem resistant to ethnographic classification are related either to other languages that have been extinct for quite some time or to the languages of peoples whom travelers have not yet encountered.

The majority of American languages, including even those belonging to groups as distinctive as the Germanic, Celtic, and Slavic languages, exhibit a certain degree of conformity in their overall structure, for example in the complexity of their grammatical forms, in the modifications that the verbs undergo according to the nature of the object, and in the multiplicity of additive particles (*affixa* and *suffixa*). The uniform tendency of these languages suggests either a common origin or, at the very least, an extreme uniformity in the intellectual aptitudes of the American peoples from Greenland to the lands of Magellan.

A number of studies conducted with extreme care and following a method that was not formerly employed in the field of etymology have proved that there are a small number of words that are common to all the languages of the two continents. In the eighty-three American languages that ˈMr. Barton and Mr. Vater analyzed, approximately one hundred and seventy words have been identified that appear to have the same roots; and one is easily persuaded that this similarity is not coincidental, and that its basis lies neither in mimetic harmony nor in the uniform organ shape that makes the first sounds uttered by infants more or less identical. Of the one hundred and seventy words that betray some degree of similarity, three-fifths recall Manchu, Tungus, Mongol, Samoyed, and two-fifths recall the Celtic and Tschud languages, as well as Basque, Coptic, and Congolese. These words were found by comparing all the American languages with all the languages of the old world, for we do not yet know of any language of the Americas that, more than any other, appears to be linked to one of the numerous groups of Asiatic, African, or European languages. What some scholars, relying upon abstract theories, have suggested in regard to the supposed poverty of American languages and the extreme imperfection of their numerical system is as baseless as the claims concerning the weakness and stupidity of humankind in the new continent, the shrinking of the natural realm, and the degeneration of the animals brought from one hemisphere to the other.

Many languages that are the exclusive heritage of barbarian peoples today seem to be the vestiges of rich, supple languages indicative of an advanced culture. We shall not discuss whether the original state of humankind was one of brutish mindlessness or whether the savage hordes descend from peoples whose intellectual faculties were at the same developmental stage as the languages that reflect such faculties. We will merely remark that the little we know of the history of the Americans tends to suggest that the tribes whose migrations led them from north to south already exhibited, in the northernmost countries, the same variety of languages that we find in the Torrid Zone today. From this we can conclude by analogy that the ramification—or, to use an expression not tethered to any system of thought, the multiplicity—of languages is a very old phenomenon. It may be that what we refer to as American languages belong no more to the Americas than Magyar (or Hungarian) and Tschud (or Finnish) belong to Europe.

One cannot deny that the process of comparing the languages of the two continents has not yet led to general conclusions. But we must not lose hope that this investigation might become more fruitful once scholars are able to apply their wisdom to a larger collection of materials. How many languages exist in the Americas and in Central and East Asia whose central mechanism is still as unknown to us as that of Tyrrhenian, Oscan, and Sabine! Among the peoples who disappeared in the old world, there may well be several from which a number of small tribes have been preserved in the vast solitudes of the Americas.

While languages offer only scant proof of prior contact between the two worlds, this contact is undoubtedly evident in the cosmogonies, monuments, hieroglyphs, and institutions of the peoples of both the Americas and Asia. I may flatter myself, perhaps, to think that the following pages will justify this assertion by adding several new pieces of evidence to those that have long been known. An attempt has been made to distinguish carefully between, on the one hand, indices of a common origin and, on the other, the effects experienced by all peoples who find themselves in the situation of beginning to improve their social conditions. XII

It has hitherto been impossible to determine the period of contact between the inhabitants of the two worlds. It would be rash to suggest a particular group of peoples from the old continent to whom the Toltec, Aztecs, Muisca, or the Peruvians appear most closely connected, for such connections manifest themselves in traditions, monuments, and customs that may actually predate the current division of the Asians into Mongols, Hindus, Tungus, and Chinese.

At the time of the discovery of the new world—or, rather, at the time of the first invasion by the Spanish—the most culturally advanced of the American peoples were the mountain peoples. People born on the temperate plains had followed the ridge of the Cordilleras, which increase in elevation as they approach the Equator. In those lofty regions, they found both a climate and plants resembling those of their native lands.

The faculties develop more easily wherever humans, settled on less fertile land and forced to fight against the obstacles that nature places in their path, do not succumb to this protracted struggle. In the Caucasus and Central Asia, the arid mountains offer a refuge to free barbarian peoples. In the equinoctial regions of the Americas, where verdant savannahs are suspended high above the clouds, we have found only civilized peoples at the heart of the Cordilleras. Their early technical and artistic advances were as old as the bizarre forms of government that did not favor individual liberty.

Like Asia and Africa, the new continent has centers of an original civilization, the mutual connections of which are as unknown to us as those of Meroë, Tibet, and China. Mexico inherited its culture from a country situated to the north; in South America, the great buildings of Tiahuanaco XIII served as models for the monuments that the Inca erected in Cusco. Amid the vast plains of Upper Canada, in Florida, and in the desert bounded by the Orinoco, the Casiquiare, and the Guainía, embankments of a considerable length, bronze weaponry, and sculpted stone suggest that industrious peoples once inhabited the very lands that hordes of savage hunters cross today.

The unequal distribution of animals across the globe has had a profound impact upon the lot of peoples and their more or less rapid march toward civilization. In the old continent, pastoral life marked the transition from hunting to farming. Ruminants, which are very easy to acclimate to any environment, followed the African Negro as well as the Mongol, the Malay, and the peoples of the Caucasian race. Although several quadrupeds and a larger number of

plants are common to the northernmost regions of both worlds, the only bo-
vine creatures native to the Americas are the bison and the musk ox, two animals difficult to domesticate, and whose females produce little milk, despite
the richness of the pastures. The tending of herds and the habits of pastoral
life did not prepare the American hunter for agriculture. The Andes dweller
was never tempted to milk the llama, the alpaca, or the guanaco. Dairy products were formerly unknown among Americans, just as they were among
some peoples of East Asia.

Nowhere has the free savage roaming the forests of the temperate zone
been seen to abandon voluntarily the life of the hunter to embrace that of the
farmer. Only the force of circumstance can bring about this transition, both
the most difficult and the most important in the history of human societies.
When, during their long migrations, the hordes of hunters, driven by other
warlike hordes, reached the plains of the equinoctial region, the dense forests
and abundant plant life brought about a change in both their habits and their
character. There are lands between the Orinoco, the Ucayali, and the Amazon where the only open space, so to speak, that can be found is in the form
of rivers and lakes. Established on the riverbanks, the most savage tribes surrounded their huts with banana plants, jatropha [*Jatropha curcas*], and other
food-bearing plants.

Neither historical events nor any legends link the peoples of South America with those who dwell north of the Isthmus of Panama. The annals of the
Mexica Empire appear to date back to the sixth century of our time. One finds
there the epochs of the migrations, the causes that brought these about, as well
as the names of the chiefs issuing from the illustrious ʼCitin family, which led
the northern peoples from the unknown regions of Aztlan and Teocolhuacan
to the plains of Anahuac. The founding of Tenochtitlan, like that of Rome,
falls in heroic times; and it is only since the twelfth century that the Aztec annals, like those of the Chinese and the Tibetans, report without interrup-
tion the secular feasts, the genealogy of kings, the tributes imposed upon the
conquered, the founding of cities, celestial phenomena, and even the most
minute events that had an influence on the state of these nascent societies.

Although their legends do not suggest any direct connection between the
peoples of the two Americas, their history nonetheless offers a number of
striking links in their respective political and religious upheavals, from which
date the civilizations of the Aztecs, Muisca, and Peruvians. Men with beards
and lighter skin than that of the natives of Anahuac, Cundinamarca, and the
plateau of Cusco appeared, without anyone being able to determine their
place of birth. High priests, legislators, lovers of peace and the arts fostered
by the latter, these men rapidly changed the condition of the peoples who

welcomed them with veneration. Quetzalcoatl, Bochica, and ʿManco Capac are the sacred names of these mysterious beings. Quetzalcoatl, clothed in black, in priestly habits, came from Panuco, from the shores of the Gulf of Mexico. Bochica, the Buddha of the Muisca, revealed himself in the high plains of Bogotá, where he arrived from the savannahs east of the Cordilleras. The story of these legislators, which I have attempted to develop in this work, is laced with marvels, religious fictions, and features suggestive of allegory. A few scholars have suggested that these strangers are, in fact, shipwrecked Europeans or descendants of those Scandinavians who have visited Greenland, Newfoundland, and perhaps even Nova Scotia since the eleventh century. But one has only to consider the period of the first Toltec migrations, as well as monastic institutions, religious symbols, the calendar, and the form of the monuments of Cholula, Sogamoso, and Cusco, to realize that Quetzalcoatl, Bochica, and Manco Capac did not draw their code of laws from the north of Europe. Everything seems to point toward East Asia, toward those peoples who were in contact with the Tibetans, the Shamanist Tartars, and the bearded Ainu [Aynu or Aino] of the islands of Yezo [Hokkaido] and Sakhalin.

In using the words *monuments of the new world*, *advances in the art of drawing*, and *intellectual culture* throughout the course of this study, I have not intended to suggest a state of affairs that would indicate what is vaguely referred to as a highly advanced civilization. Nothing is more difficult than comparing nations that have taken different paths in their social development. The Mexica and the Peruvians cannot be judged according to principles drawn from the history of peoples whom our own education ceaselessly brings to mind. They are as different from the Greeks and the Romans as they are similar to the Etruscans and the Tibetans. Although the theocratic government of the Peruvians favored the progress of industry, public works, and everything indicative of mass civilization, so to speak, it nevertheless hindered the development of individual faculties. On the contrary, the development of the Greeks, which was so free and so rapid before the time of Pericles, did not correspond to the slow progress of mass civilization. The Inca Empire resembled a great monastic establishment in which some means of contributing to the common good was prescribed to each member of the congregation. By studying on site those Peruvians who have retained their national physiognomy throughout the centuries, one learns to understand the true value of Manco Capac's code of laws and the effects that it produced on customs and public happiness. There was at once general welfare and little private happiness; more resignation to the decrees of the sovereign than love of the fatherland; passive and spineless obedience without courage for daring feats; a sense of order that meticulously regulated even the pettiest actions in one's life; and no reach into the realm of ideas, no elevation in that of character. The most complex political institutions in the history of human society smothered the germ of individual liberty; and the founder of the empire of Cusco, who congratulated himself for his success in forcing people to be happy, reduced the latter to the state of simple machines. Peruvian theocracy was certainly less oppressive than the rule of the Mexica kings. But both contributed to giving to the monuments, worship, and mythology of two mountain peoples the gloomy and somber aspect that so contrasts with the arts and sweet fictions of the peoples of Greece.

Paris, April 1813

Picturesque Views of the Cordilleras
and Monuments of the Indigenous Peoples
of the Americas

There are two very different ways in which the monuments of peoples from whom we are separated by centuries can command our attention. If the works of art that have reached us belong to peoples of a highly advanced civilization, then what elicits our admiration is the harmony and beauty of their form and the genius with which they were conceived. Even if no inscription identified ʼthe conqueror of Arbela, the bust of Alexander from the gardens of Pison would still be recognized as a precious relic of antiquity. An engraved stone or a coin from the golden age of Greece interests the art lover because of the austerity of its style and finish, even in the absence of any legend or monogram linking such objects to a specific historical period. Such is the privilege of everything produced under the skies of Asia Minor and in the regions of southern Europe. 1

On the other hand, monuments produced by peoples who did not attain a high level of intellectual culture, or who, for either religious or political reasons or because of the nature of their societal organization, seemed less appreciative of the beauty of forms, can only be considered historical monuments. To this category belong the sculptural relics scattered throughout the vast lands stretching from the banks of the Euphrates to the eastern coasts of Asia. The idols of Tibet and Hindustan and those found on the central plateau of Mongolia attract our attention because they shed light upon the ancient contacts between peoples and the common origin of their mythological traditions.

The crudest works, the most bizarre forms, the masses of sculpted rocks impressive only for their majesty and for the antiquity we attribute to them, and the colossal pyramids that bring to mind teeming throngs of workers— these are all connected to the philosophical study of history. 2

For this very reason, the scant vestiges of the arts, or, rather, the industry, of the peoples of the new continent merit our attention. Convinced of this truth, I have gathered throughout my travels all that my lively curiosity has led me to discover in countries where, throughout centuries of barbarism, intolerance has led to the destruction of everything related to the customs and religion of the ancient inhabitants; where buildings were dismantled in order to extract stones or seek treasures hidden in them. The connections that I intend to draw between artworks from Mexico and Peru and from the Old World will generate some interest in my own research and in the picturesque Atlas that contains my findings. While I am far removed from any ˅systematic approach, I will nevertheless call attention to naturally occurring parallels, distinguishing the ones that appear to suggest similarities in terms of race from those that most likely result only from strictly internal causes and from the resemblance all peoples exhibit as their intellectual faculties develop. I must limit myself here to a succinct description of the objects depicted in these engravings. The conclusions to which the study of these monuments as a group appears to lead can only be discussed in the narrative of my journey. Since the peoples to whom we attribute these buildings and sculptures still exist, both their physiognomy and an understanding of their customs will serve to illuminate the history of their migrations.

Studies of the monuments erected by peoples still only halfway emerged from barbarism are of interest for yet another reason that we might call psychological: they place before our eyes the spectacle of the uniform and progressive advancement of the human mind. The works of the first inhabitants of Mexico occupy a middle ground between those of the Scythian peoples and the ancient monuments of Hindustan. What an impressive sight the genius of humanity affords us as it covers the space extending from the tombs of Tinian and the statues of Easter Island to the monuments of the Mexica temple of Mitla, and from the crude idols inside this temple to the masterpieces of the chisel of ˅Praxiteles and of Lysippus!

3   Let us not be surprised by the crudeness of style and the inaccurate contours within the works of the peoples of the Americas. Isolated perhaps at an early stage from the rest of humanity and roaming across a land where humans had to struggle, for a long time, against a wild and forever restless natural realm, these peoples, left to their own devices, were able to develop only slowly. Eastern Asia and western and northern Europe exhibit the same phenomena. In acknowledging these, I shall not undertake to pronounce on the mysterious causes by which the seed of the fine arts has sprouted in only a very small part of the globe. How many peoples of the old continent have lived in a climate similar to that of Greece, surrounded by all that might stir the imagination, without rising to a sense of the beauty of forms, a sense that developed only in those areas where the arts were inspired by the genius of

the Greeks? These reflections suffice to establish the objective that I have set for myself in publishing these fragments of American monuments. Their study may become useful like that of the most imperfect languages, the interest of which lies not only in their similarity to known languages but also in the intimate connection that exists between their structure and the degree of ability of humans more or less removed from civilization.

By presenting in a single work the crude monuments of the indigenous peoples of the Americas and the picturesque views of the mountainous lands that these peoples inhabited, I believe to have brought together objects whose connections have not escaped the sagacity of those who are devoted to the philosophical study of the human mind. Although the customs of peoples, the development of their intellectual faculties, and the specific character inscribed in their works all depend at once upon a large number of causes that are not exclusively local, we cannot deny that climate, the lay of the land, the physiognomy of plants, and the prospect of either a cultivated or a wild natural environment have influenced their technical and artistic advances, as well as the style that distinguishes their works. This influence is more appreciable the further removed humans are from civilization. What a contrast between the architecture of a people that dwells inside vast and gloomy caverns and that of the nomadic hordes whose bold monuments recall the slender trunks of desert palms in the shafts of their columns! In order to understand properly the origin of the arts, one must study the nature of the place that witnessed their birth. The only American peoples in whose midst we find remarkable monuments are the mountain peoples. Isolated in the cloud regions on the world's highest plateaus, surrounded by volcanoes whose craters are ringed with eternal ice, they seem to admire, in the solitude of these deserts, only whatever strikes the imagination through the sheer grandeur of its dimensions. The works that they have produced bear the imprint of the wild nature of the Cordilleras.

Part One of this Atlas is intended to be an introduction to the grand scenes that this natural realm presents. I was less interested in depicting those scenes and their picturesque effect than in representing the exact contours of the mountains, the valleys that furrow their sides, and the impressive waterfalls formed by plunging torrents. The Andes are to the High Alps what the Alps are to the Pyrenees. Everything romantic or grand that I have seen on the banks of the Saverne in northern Germany, in the Euganean Hills, in the central mountain range of Europe, or on the steep slopes of the volcano of Tenerife—this is all combined in the Cordilleras of the new world. Several centuries would not be enough time to observe their beauties and to discover the wonders that nature has lavished upon an expanse of two thousand five hundred leagues, from the granite mountains of the Strait of Magellan to the coasts that neighbor eastern Asia. I will consider my goal fulfilled if the humble sketches in this book inspire travelers with a passion for art to visit the regions that I traversed in order to depict faithfully these majestic sites, which cannot be compared to those of the old continent.

PLATES I AND II
Bust of an Aztec Priestess

I have placed at the head of my picturesque Atlas the precious remnants of an Aztec sculpture. It is a bust in basalt preserved in Mexico City at the home of an enlightened amateur, ˅Mr. Dupé, a captain in the service of His Catholic Majesty. This educated officer, who developed a love for the arts while in Italy as a young man, has made several trips to the interior of New Spain to study Mexica monuments. He has drawn with special care the reliefs of the pyramid of Papantla, on which subject he may one day publish a very interesting study.

The bust, depicted in its original size and from two sides (Plates I and II), is especially striking for a kind of headdress that bears some resemblance to the veil or ˅*calantica* on the Isis heads, the Sphinx, the Antinous, and a large number of other Egyptian statues. One must nevertheless observe that in the Egyptian veil, the two ends that extend below the ears are most often very slight and folded crosswise. In a ˅statue of Apis at the Capitoline Museum, the ends are convex in the front with vertical pleats, while the back section, which touches the collar, is flat, not rounded as it is in the Mexica headdress. The latter exhibits the greatest similarity to the pleated drapery covering the heads embedded in the capitals of the columns of ˅*Tentyris*, as one sees in the faithful renderings that Mr. Denon included in his *Voyage en Égypte*.[1]

Perhaps the fluted bourrelets that extend toward the shoulders in the Mexica bust are, in fact, a hair arrangement similar to the plaits on Isis in a Greek statue in the library of the Villa Ludovisi in Rome. This unusual arrangement is especially striking on the back of the bust in the second Plate, which shows a very large bag attached in the middle by a knot. The famous ˅Zoëga, of whom death has recently robbed the sciences, assured me that he had seen a perfectly similar bag on a small bronze statue of Osiris at the museum of ˅Cardinal Borgia in Velletri.

5

1. Denon, *Voyage [en Égypte]*, Plates 39, 40, 60 (numbers 7 and 8).

Plate I

Plate II

The Aztec priestess's forehead is adorned with a line of pearls edging a very narrow headband. These pearls have not been observed on any Egyptian statue. They confirm that contact occurred between the city of Tenochtitlan, the precursor to Mexico City, and the coasts of California, where pearls were then harvested in large quantities. The collar is draped in a triangular neck-erchief, from which hang twenty-two small bells [or tassels] in almost perfect symmetry. Like the headdress, these small bells are found in a large number of Mexica statues, bas-reliefs, and hieroglyphic paintings. They recall the tiny apples and pomegranates that were attached to the robe of the high priest of the Hebrews. On the front of the bust and at a height of one-half decime-ter above the base, we notice a set of toes on either side, though the lack of hands is a sign of the infancy of the art. From the back, it seems that the figure is seated or even squatting. There is reason to be surprised that the eyes are without pupils, even though we find them indicated in the bas-relief recently discovered in Oaxaca (Plate XI).

The basalt of this sculpture is very hard and of a lovely black color; it is true basalt, with a few grains of peridot [green gemstone], not Lydian stone or the grünstein-based porphyry that antiquarians generally call Egyptian ba-salt. The folds of the headdress, and especially the pearls, have a very smooth finish, although the artist, lacking steel chisels and likely working with the same kinds of copper and tin tools that I brought back from Peru, must have had a difficult time perfecting them. This bust has been very accurately ren-dered by a student at the academy of painting in Mexico City, under the su-pervision of Mr. Dupé. It is 0.38 m high and 0.19 m wide. I have retained the title that the locals use: *Bust of a Priestess*. It may well be, however, that it represents some Mexica deity and that it was originally classed among the household Gods. The headdress and the pearls, which are also found in an idol discovered in the ruins of Tetzcoco [Nahuatl: Tetzcohco] and deposited in the cabinets of the King of Prussia in Berlin, justify this conclusion: the or-namentation on the collar and the non-monstrous form of the head make it more likely that the bust simply represents an Aztec woman. If this assump-tion is correct, the fluted bourrelets that extend toward the chest could not be plaits, for all the virgins who devoted themselves to the service of the temple were shorn by the high priest or *Tepanteohuatzin*.

A certain resemblance between the *calantica* of the Isis heads and the Mexica headdresses, the many-terraced pyramids, similar to those in *Fai-yum* and *Sakkarah*, the frequent use of hieroglyphic painting, and the five supplementary days added to the end of the Mexica year (which recall the ʾepagomenal days of the Memphian year) constitute remarkable points of similarity between the peoples of the new and the old continent. We are nevertheless quite far from espousing theories that would be as vague and as unsubstantiated as those by which the Chinese have been turned into a colony of Egypt and the Basque language into a dialect of Hebrew. Most of these similarities fade once we examine the facts in isolation. For example, despite its epagomenal days, the Mexica year differs completely from that of the Egyptians. ⸱A great geometrician, who was gracious enough to examine the fragments that I brought back with me, has observed, using the Mexica intercalation, that the length of the Aztec tropical year is almost identical to the length determined by the astronomers of Al-Ma'mun.[1]

If we venture back to the earliest times, history shows us many centers of civilization, the mutual connections of which are unknown to us, such as Meroë, Egypt, the banks of the Euphrates, Hindustan, and China. Other sources of enlightenment, even more remote, were perhaps situated on the central Asian plateau; and it is to their glint that we are tempted to attribute the beginnings of American civilization.

7

1. Laplace, *Exposition du système du monde*, [third] edition, p. 554.

PLATE III

View of the Main Square of Mexico City

The city of Tenochtitlan, capital of Anahuac, founded in 1325 on a small group of islets situated in the western part of the salt lake of Texcoco, was completely destroyed during the siege that the Spanish laid to it in 1521, which lasted seventy-five days. The new city, which counts nearly one hundred and forty thousand inhabitants, was built by ˅Cortés upon the ruins of the former, following the same street pattern; but the canals that crossed these streets were filled in little by little, and Mexico City, radically refurbished by the viceroy, the ˅Count of Revillagigedo, is today comparable to the most beautiful cities of Europe. The main square, depicted on the third Plate, is the site formerly occupied by the ˅great temple of Mexitli, which, like all the *teocalli*, or dwellings of the Mexica gods, was a pyramidal building, similar to the Babylonian monument dedicated to Jupiter Belus. On the right-hand side, we see the palace of the viceroy of New Spain, a building of simple design originally belonging to the Cortés family, which is that of the *Marquis of the Valle de Oaxaca, the Duke of Monteleone*. In the center of the engraving is the cathedral, one part of which (*el sagrario*) is in the ancient Indian or Moorish style commonly called Gothic. Behind the cupola of the *sagrario*, where *Indio triste* runs into Tacuba street, lies King Axayacatl's palace, where ˅Montezuma lodged the Spanish when they arrived in Tenochtitlan. Montezuma's own palace was to the right of the cathedral, across from the present viceroy's palace. I felt it necessary to indicate these locations since they are not without interest to those who study the history of the conquest of Mexico.

Since 1803 the *Plaza mayor*, which must not be confused with the main market of Tlatelolco (which Cortés described in his letters to ˅Emperor Charles V), has been adorned with the equestrian statue of King Charles IV commissioned by the ˅viceroy Marquis of Branciforte. This bronze statue is of great stylistic purity and beautifully made: it was designed, modeled, cast, and put in place by the same artist, ˅Don Manuel Tolsa, a native of Valencia, Spain, and director of the sculpture class at the fine arts academy of Mexico City. We do not know what we should admire most: this artist's talent

Plate III

or the courage and perseverance that he displayed in a country where everything was yet to be created, and where he had many obstacles to overcome. This beautiful work was a success from the first casting. The statue weighs nearly twenty-three thousand kilograms; it is two decimeters taller than the equestrian statue of Louis XIV in the Place Vendôme in Paris. They had the good taste not to gild the horse and were content to coat it with a brownish olive-colored varnish. As the buildings that border the square are generally low, we see the statue projected against the sky, which, on the ridge of the Cordilleras where the atmosphere is a very deep blue, produces a most picturesque effect. I assisted in the transfer of this enormous piece from the site of its casting to the *Plaza mayor*. It crossed a distance of around sixteen hundred meters in five days. The mechanics that Mr. Tolsa employed to raise it up on a pedestal of exquisite Mexican marble are quite ingenious and merit detailed description.

Today, the main square of Mexico City is of an irregular shape since, contrary to Cortés's plan, they built the square that housed the Parian [set of shops at the southwest corner]. To make the square look less asymmetrical, it was deemed necessary to place the equestrian statue, which the Indians know only by the name of the *great horse*, in a special enclosure. This enclosure is paved in porphyry tile and is more than fifteen decimeters higher than the adjacent streets. The oval, whose major axis is one hundred meters long, is surrounded by four fountains and, to the great dismay of the natives, is closed off by four gates with bronze-decorated grating.

9   The engraving that I include here is the faithful copy of a drawing made, on a larger scale, by ˙Mr. Ximeno, a distinguished artist who directs the painting class at the academy of Mexico City. In the figures placed outside the enclosure, this drawing shows the dress of the Guachinangos, the Mexican lower classes.[1]

1. See my *Essai politique sur le royaume de la Nouvelle-Espagne*, pp. 119, 168, 186.

PLATE IV

Natural Bridges of Icononzo

Among the majestic and varied scenes of the Cordilleras, what most inspires the imagination of the European traveler is the valleys. The enormous height of the mountains can be fully grasped only from a considerable distance and from a vantage point on the plains that extend from the coasts to the foot of the central range. The plateaus that surround the glacier-covered summits are, for the most part, at an elevation of two thousand five hundred to three thousand meters above sea level. To some extent, this circumstance detracts from the impression of grandeur that the colossal massifs of Chimborazo, Cotopaxi, and Antisana produce when viewed from the plateaus of Riobamba and Quito. But what goes for the mountains does not go for the valleys. Deeper and narrower than those of the Alps and the Pyrenees, the valleys of the Cordilleras contain some of the wildest sites, which fill the soul with both awe and dread. The bottoms and the edges of these crevices are adorned with vigorous plant life, and their depth is often so great that Vesuvius and Puy-de-Dôme could be placed in them without their summits rising above the curtain of the neighboring mountains. The interesting travels of ˙Mr. Ramond have given a clearer idea of the Ordesa valley, which descends from Monte Perdido and has a mean depth of nearly nine hundred meters (four hundred fifty-nine toises). Traveling on the ridge of the Andes from Pasto to the *Villa de Ibarra* and descending from Loja toward the banks of the Amazon, ˙Mr. Bonpland and I crossed the famous clefts of Chota and Cutaco, which have a perpendicular depth of more than fifteen hundred and thirteen hundred meters, respectively. To give a more complete idea of the immensity of these geological formations: the bottom of these crevices has an elevation above sea level that is only one-quarter less than that of the St. Gotthard and Mont Cenis passes.

10

Plate IV

The valley of Icononzo (or Pandi), part of which is shown on the fourth Plate, is less remarkable for its dimensions than for the extraordinary shape of its boulders, which look as though they were carved by human hands. Their dry, barren peaks contrast delightfully with the clusters of trees and shrubs that cover the edges of the cleft. The little torrent that has cut its path through the Icononzo valley bears the name *Río de la Sumapaz*. It descends from the western range of the Andes, which, in the kingdom of New Granada, separates the Magdalena river basin from the vast plains of the Meta, the Guaviare, and the Orinoco. This torrent, confined within a nearly inaccessible bed, could not be crossed without great difficulty had nature herself not created two rock bridges there, which are rightfully seen in the country as one of the sights most deserving of travelers' attention. In September of 1801 we crossed these natural bridges of Icononzo en route from Santa Fé de Bogotá to Popayán and Quito.

Icononzo is the name of an ancient Muisca Indian village set on the southern edge of the valley; only a few scattered huts remain of it. Today, the inhabited place closest to this remarkable site is the small village of *Pandi* or *Mercadillo*, at a distance of one-quarter league toward the northeast. The road from Santa Fé to Fusagasuga (lat. 4° 20' 21" north, long. 5° 7' 14"), and from there to Pandi, is one of the most difficult and most poorly cleared that one finds in the Cordilleras. One must passionately love the beauties of nature not to prefer the regular road that leads from the Bogotá plateau toward the natural bridge of Icononzo, through the Mesa de Juan Díaz, to the banks of the Magdalena, to the perilous descent from the *Páramo* de San Fortunato and the mountains of Fusagasuga.

The deep gorge through which the torrent of the Sumapaz gushes is at the center of the Pandi valley. Near the bridge, the gorge maintains its east-west direction for more than four thousand meters. There are two spectacular waterfalls, one at the point where the river enters the chasm to the west of Doa and another where it emerges in its descent toward Melgar. It is quite likely that this crevice resulted from an earthquake: it resembles an enormous vein from which miners have extracted the gangue. The surrounding mountains are of sandstone with clay cement: this formation, which rests on the primitive schist (*thonschiefer*) of Villeta, stretches from the rock salt mountain of Zipaquirá to the Magdalena river basin. It also contains the coal strata of Canoas or Chipa, which are mined near the great Tequendama waterfall (Plate VI).

The sandstone in the Icononzo valley is composed of two distinct kinds of rock. A very dense quartziferous sandstone with scarce cement and almost no stratification fissures rests on finely grained schistose sandstone (*Sandstein-schiefer*) divided into infinite strata, which are small, very thin, and nearly horizontal. We may believe that at the time when the crevice was formed, the compact and quartziferous bed resisted the force that tore the mountains apart, and that it is the uninterrupted continuation of this bed that serves as a bridge for crossing from one part of the valley to the other. This natural arc is fourteen and a half meters long and 12,7 meters wide; it is 2,4 meters thick at the center. A number of carefully conducted gravity experiments with a ʼBerthoud chronometer gave us 97,7 meters for the height of the upper bridge above the level of the torrent waters. Prior to our arrival, ʼDon Jorge Lozano, an enlightened person who owns a pleasant estate in the beautiful valley of Fusagasuga, had measured this very height by means of a sounding line; he found it to be one hundred twelve *varas* (93,4 meters). The torrent's mean depth appears to be six meters. For the safety of travelers, who are quite rare in this deserted land, the Indians of Pandi have built a small parapet of reeds that extends toward the path leading to the upper bridge.

Ten toises below this first natural bridge lies another one, which we reached by following a narrow track leading down to the edge of the gorge. Three enormous rocks have fallen into a mutually supportive position: the middle one forms the keystone, an accident that might have led the natives to conceive of masonry arcs, which were unknown to the peoples of the new world, as they had been to the inhabitants of ancient Egypt. I shall not decide whether these pieces of rock were flung from afar or whether they are merely fragments of an arc that was destroyed in place but had originally been similar to the upper natural bridge. A similar phenomenon in the Coliseum of Rome makes the latter scenario probable; there, in a half-collapsed wall, we see several stones arrested in the midst of their descent because of the arc that they accidentally formed when they fell. In the middle of the second bridge of Icononzo there is a hole of over eight square meters, through which one can see the bottom of the abyss: this is where we conducted gravity experiments. The torrent appears to flow into a dark cave. The doleful noise one hears is due to a seemingly infinite number of nocturnal birds that live in the crevice and that one is initially tempted to take for the gigantic bats so common in the equinoctial regions. One can make out thousands of them gliding above the water.

The Indians assured us that these birds are the size of chickens, with owl eyes and hooked beaks. They call them *cacas*, and the solid color of their plumage, which is of a brownish gray, leads me to believe that they do not belong to the genus *Caprimulgus*, of which there is quite a variety in the Cordilleras. Because of the depth of the valley, it is impossible to catch one of them. We were able to examine them only by throwing flares into the crevices in order to light up the walls.

The natural bridge of Icononzo has an elevation of eight hundred ninety-three meters (four hundred fifty-eight toises) above sea level. In the mountains of Virginia, in the county of *Rockbridge*, there exists a phenomenon similar to the upper bridge that I have just described. It has been examined by ˙Mr. Jefferson with the care that distinguishes that excellent naturalist's observations.[1] The natural bridge of *Cedar Creek* in Virginia is a limestone arch with an aperture of twenty-seven meters; its elevation above the river waters is seventy meters. The earthen bridge (*Rumichaca*) that we found on the slope of the porphyritic mountains of Chumban in the province of *Los Pastos*; the bridge of *Madre de Dios*, called *Danto*, near Totonilco in Mexico; and the pierced rock near Grândola in the province of Alentejo in Portugal are geological phenomena that all bear some resemblance to the bridge of Icononzo. But I doubt that anything as extraordinary as the three rocks that support one another by forming a natural arc has been found in any part of the globe to this day. 13

I have sketched the bridges of Icononzo from the northern part of the valley and from a point where the arch is viewed in profile. The galley proofs of this Plate erroneously list the engraver as ˙Mr. Gmelin of Rome, instead of Mr. Bouquet of Paris.

---

1. *Notes on the State of Virginia*, p. 56.

PLATE V

Quindiu Pass in the Cordillera of the Andes

In the kingdom of New Granada, from 2° 30' to 5° 15' of northern latitude, the Cordillera of the Andes is divided into three parallel ranges, of which only the two lateral ranges are covered, at extreme heights, with sandstone and other secondary formations. The *eastern range* separates the Magdalena River valley from the plains of the Río Meta. The natural bridges of Icononzo lie on its western slope. Its highest peaks are the Páramo de la *Sumapaz* and that of *Chingaza*. Neither of them reaches the regions of the eternal snows. The middle range divides the waters between the Magdalena River basin and that of the Río Cauca. It often reaches the perpetual snow line, and it surpasses the latter by far in the colossal peaks of *Guanacas*, *Baragan*, and *Quindiu*. At sunrise and sunset, this middle range presents a magnificent spectacle to the inhabitants of Santa Fé; it recalls the view of the Swiss Alps, but on a more impressive scale.

The *western range* of the Andes separates the valley of Cauca from the province of Chocó, as well as from the coasts of the South Sea. Its elevation is barely fifteen hundred meters, and it falls to such a low height between the headwaters of the Río Atrato and those of the Río San Juan that one has difficulty tracing it to the Isthmus of Panama.

These three mountain ranges merge toward the north, at 6° and 7° northern latitude. They form a single group to the south of Popayán in the province of Pastos. We must not, however, mistake them for the division of the Cordilleras that ˙Bouguer and La Condamine observed in the kingdom of Quito, from the equator to 2° of southern latitude.

The city of Santa Fé de Bogotá is located to the west of the Páramo de *Chingasa*, on a plateau at an absolute altitude of two thousand six hundred fifty meters that extends to the ridge of the *eastern Cordillera*. The result of the Andes' peculiar structure in these parts is that to reach Popayán and the banks of the Cauca from Santa Fé, one must either descend the *eastern range* by the *Mesa* and the *Tocaima* or cross the Magdalena River valley by way of

14

the natural bridges of *Icononzo* and then scale the middle range. The most heavily traveled pass is that of the *Páramo de Guanacas*, which Bouguer described upon his return from Quito to Cartagena de Indias. Following this path, the traveler crosses the ridge of the middle Cordillera in a single day through an inhabited area. We preferred the pass of the *mountain of Quindiu* (or *Quindío*), between the cities of Ibagué and Cartago, to that of Guanacas. The entry to the first pass is depicted in Plate V. It seemed to me indispensable to provide these geographical details in order to give a better sense of the position of a place that one would seek in vain on the best maps of South America, for example, that of ʼLa Cruz.

Quindiu Mountain (lat. 4° 36', long. 5° 12') is reported to be the most difficult passage in the entire Cordillera of the Andes. It is a dense, completely uninhabited forest that cannot be crossed in less than ten to twelve days, even in good weather. There are no huts and no means of livelihood: at all times of the year, travelers stock up with a month's worth of supplies, for it often happens that they find themselves cut off by thawing snow and the sudden swelling of the torrents, unable to descend from either the Cartago or the Ibagué side. The highest point on this path, the Garito del Páramo, is three thousand five hundred meters above sea level. As the foot of the mountain is only nine hundred sixty meters high near the banks of the Cauca, the climate there is appreciably mild and temperate. The trail by which one ascends the Cordillera is so narrow that its average width is only four to five decimeters: for the most part, it resembles a hollowed-out, open-air gallery. In this and nearly every part of the Andes the rock is coated in a thick layer of clay. Water trickling down from the mountain has hollowed out ravines six to seven meters deep. One walks in these mud-filled crevices, whose darkness is exacerbated by thick vegetation that stretches across their openings. The bodies of oxen, the beasts of burden commonly used here, pass with difficulty through these galleries, which can be up to two thousand meters long. If one has the misfortune of encountering one of these animals on the trail, there is no other way of avoiding it than to turn back or to climb onto the earthen wall that runs alongside the gorge and then hold oneself up by hanging on to the roots that reach there from the surface of the soil.

Crossing Quindiu Mountain in October of 1801, on foot and followed by twelve oxen that carried our instruments and collections, we suffered greatly from the continual downpours to which we were exposed throughout the three or four final days of our descent of the western slope of the Cordillera. The path passes through swampy terrain covered with bamboo. The prickles that arm the roots of these gigantic grasses had torn our shoes, forcing us to go barefoot, like all those travelers who do not want to be carried on *man-*

15

Plate V

back. This peculiar circumstance, plus the constant humidity; the length of the crossing; the muscle strength that one must employ to walk in thick, muddy clay; and the need for wading through deep torrents of icy water, makes this an extremely tiring voyage. But however difficult it may be, one encounters none of the dangers that frighten gullible travelers. Though the path is quite narrow, it skirts precipices only very infrequently. As the oxen have the habit of always stepping in the same tracks, a series of small ditches form in and across the path, separated from one another by very narrow protuberances of earth. During heavy rains these thresholds remain hidden underwater, and the traveler's step is doubly uncertain, for he does not know whether he steps on the dam or in the ditch.

Since few wealthy people in these climes are in the habit of walking on foot and on such difficult paths for fifteen to twenty days in a row, they have themselves carried by men who have chairs tied to their backs; given the current state of the Quindiu pass, it would be impossible to go by mule. In this country one hears the phrase *to go on man-back* (*andar en carguero*), just as one says *to go on horseback*. There is no stigma attached to the job of the *cargueros*. The men who devote their lives to it are not Indians but, rather, Mestizos and sometimes even Whites. One is often surprised to hear naked men, dedicated to what in our eyes must be a withering profession, arguing in the middle of a forest because one of them refused the other, who claims to have whiter skin, the pompous title of either *Don* or *Su Merced* [Sir or Your Excellency]. The *cargueros* generally carry six to seven *arrobas* (seventy-five to eighty-eight kilograms); the hardiest of them carry up to nine *arrobas*. When one reflects on the enormous fatigue to which these miserable men are exposed as they walk for eight to nine hours a day in mountainous terrain; when one knows that their backs are sometimes bruised like the backs of beasts of burden and that travelers are often so cruel as to abandon them in the forest when they fall ill; when one considers that for a trip from Ibagué to Cartago they earn no more than 12 to 14 piasters (60 to 70 francs) for fifteen, sometimes even twenty-five to thirty, days, one struggles to understand how this profession of *cargueros*, one of the most arduous to which a man might devote himself, is so readily embraced by all the hardy young people who live at the foot of these mountains. The taste for a rootless and vagabond life, as well as the idea of a certain independence in the midst of the forests, causes them to prefer this arduous occupation to sedentary and monotonous employment in the towns and cities.

The Quindiu Mountain pass is not the only part of South America in which one may travel on *man-back*. An entire province, Antioquia, is

16

hemmed in by mountains that are so difficult to cross that those who do not wish to entrust themselves to the nimbleness of a *carguero*, and who are not sufficiently robust to brave the path from Santa Fé de Antioquia to the Boca de Nares or to Río Samana on foot, must simply stay home. I met an inhabitant of this province who was positively enormous: he had only ever encountered two Mestizos capable of carrying him, and it would have been impossible for him to return home had these two *cargueros* died while he was on the banks of the Magdalena, at Mompox or Honda. There are so many young people who perform the job of beasts of burden in Chocó, Ibagué, and Medellín that one sometimes encounters lines fifty or sixty men deep. A few years ago, when the project was conceived of building a mountain path passable for mules from the village of Nares to Antioquia, the *cargueros* formally protested against the improvement of the roads, and the government was weak enough to give in to their demands. It is useful to recall here that in the mines of Mexico there are men whose sole occupation is carrying others on their backs. The sloth of whites in these climes is so great that every mine director has in his employ one or two Indians called his *horses* (*caballitos*), because they are saddled every morning and, by leaning on a small cane and thrusting their body forward, they carry their master from one part of the mine to another. The *caballitos* and *cargueros* who have sure footing and a gentle, steady step are singled out and recommended to travelers. It is sad to hear a man's qualities characterized in terms of the gait of horses and mules. Those who have themselves carried in a *carguero*'s chair must sit completely still and lean back for several hours. The least movement would be enough to cause the man carrying them to fall, and such falls are all the more dangerous since the *carguero*, overconfident in his own dexterity, often chooses the steepest slopes or crosses a torrent on a narrow and slippery tree trunk. Accidents are nevertheless very rare, and those that do happen should be attributed to the carelessness of frightened travelers who have jumped from on top of their chairs.

The fifth Plate depicts a quaint site at the entrance to Quindiu Mountain, near Ibagué, at a post called the foot of the Cuesta. The truncated cone of Tolima, covered in perpetual snows, its shape recalling that of Cotopaxi and Cayambe, appears above a mass of granitic rocks. The small river of Combeima, which blends its waters with those of the Río Cuello, winds through a narrow valley and cuts its path across a palm grove. In the background one can make out a part of the city of Ibagué, the great Magdalena River valley, and the eastern range of the Andes. In the foreground a group of *cargueros* is heading to the mountain. One can see here the distinctive way in which the bamboo chair is tied to the shoulders and balanced by a headpiece similar to that worn by horses and oxen. The rolled-up bundle in the third *cargue-*

ro's hand is the roof or, rather, the mobile house that the traveler uses as he crosses the forests of Quindiu.

Upon arrival in Ibagué, and once preparations for the trip are under way, several hundred *vijao* leaves are cut from the neighboring mountains. This plant, which belongs to the same family as the banana tree, forms a new genus related to Thalia and must not be confused with Heliconia bihai. These leaves, membranous and glossy like those of the Musa, have an oval shape with a length of fifty-four centimeters (twenty inches) and a width of thirty-seven centimeters (fourteen inches). They have a silvery white underside covered in a floury substance that flakes off in scales. It is this peculiar *varnish* that makes it resistant to rain for long periods of time. In gathering them, people make an incision at the central vein, which is the extension of the petiole: this incision serves as a hook from which they are hung when it is time to set up the mobile roof; the leaves are then spread out and rolled carefully into a cylindrical bundle. It takes fifty kilograms of leaves to cover a six- to eight-person hut. When one arrives at a place in the middle of the forest where the soil is dry and where one expects to spend the night, the *cargueros* cut a few tree branches that they assemble into a tent shape. In the space of a few minutes this light framework is divided into panes, with lianas or agave threads stretched in parallel lines three or four decimeters apart. During this time the bundle of *vijao* leaves is unrolled, and several people then busy themselves with arranging them on the trellis so that they overlap one another like house tiles. These hastily constructed huts are very cool and comfortable. If, during the night, the traveler feels the rain trickling in, he simply points out the spot with the leak; a single leaf suffices to fix this annoyance. We spent several days in the Boquia valley in one of these leaf tents without getting wet, although the rain was very heavy and almost interminable. Quindiu Mountain is among the richest places in terms of useful and interesting plants. It was there that we found a palm tree (*Ceroxylon andicola*) with a trunk covered in vegetal wax, passion flowers in the trees, and the superb Mutisia grandiflora, whose scarlet flowers are sixteen centimeters (six inches) long.

19

PLATE VI
Tequendama Falls

The plateau on which the city of Santa Fé de Bogotá lies resembles the high-land of the Mexican lakes in many respects. Both are at a higher elevation than the convent of St. Bernard: the first is two thousand six hundred sixty meters high; the second is two thousand two hundred seventy-seven meters above sea level. Surrounded by an enclosing wall of porphyritic mountains, the Valley of Mexico is covered in water at its center; until the Europeans built the canal of Huehuetoca, none of the numerous torrents that plunge into the valley found an opening through which to flow out. The Bogotá pla-teau is also surrounded by high mountains: everything there—from its per-fectly level terrain to its geological constitution and the shape of the boulders of Suba and Facatativá, which rise like islets in the midst of the savannahs—seems to point to the existence of a former lake. The Funza River, generally called the Río de Bogotá, into which all the waters of the valley flow, has forged a path across the mountains to the southwest of the city of Santa Fé. It emerges from the valley near the farm of Tequendama, plunging through a narrow opening into a crevice that descends towards the Magdalena River basin. If one attempted to seal this gap, the sole opening in the Bogotá valley, the fertile plains would be converted, little by little, into a lake similar to the Mexica lakes.

It is easy to perceive the influence that these geological phenomena had upon the traditions of these lands' ancient inhabitants. We shall not decide whether the appearance of such places led peoples who were not far removed from civilization to hypothesize about the first revolutions of the globe, or whether the great floods of the Bogotá valley were recent enough for their memory to have been preserved among humans. Historical traditions blend everywhere with religious notions, and it is interesting to recall here those which the conquistador of this country, ˅Gonzalo Jiménez de Quesada,

20

Plate VI

found to be widespread among the Muisca, Pancha, and Natagaima Indians when he first penetrated into the mountains of Cundinamarca.[1]

According to Muisca (or Mozca) Indian myth, in the earliest times, before the moon joined the earth, the inhabitants of the Bogotá plateau lived like barbarians: naked, without agriculture, laws, or religion. A man suddenly appeared in their midst who came from the plains that lie east of the Cordillera of Chingaza: he seemed to be of a different race from the natives, for he had a long, bushy beard. He was known by three different names: *Bochica*, *Nemquehteba*, and *Zuhe*. Like Manco Capac, this old man taught people how to clothe themselves, build huts, plow the earth, and gather together in communities. He brought with him a woman, to whom tradition still gives three names: *Chia*, *Yubecayguaya*, and *Huythaca*. Graced with a rare beauty but also afflicted with terrible malice, this woman thwarted her husband in everything that he undertook for the sake of people's happiness. Using her magic art, she caused the Funza River to swell, and its waters flooded the entire Bogotá valley. The majority of the inhabitants perished in this flood, and only a few of them escaped to the nearby mountain peaks. Furious, the old man chased the lovely Huythaca far from the earth; she became the moon, which, from that time on, began to cast light upon the earth at night. Feeling pity for the people scattered on the mountains, Bochica then rent with his powerful hand the boulders that sealed the valley on the side of Canoas and Tequendama. He made the waters of the Funza Lake flow through this opening, gathered the peoples together again in the Bogotá valley, built the cities, introduced the worship of the sun, named two leaders, between whom he divided ecclesiastical and secular powers, and withdrew, under the name *Idacanzas*, to the holy valley of Iraca, near Tunja, where he lived, strictly observing the most austere penance, for two thousand years.

This Indian fable, which attributes the Tequendama waterfall to the founder of the *Zaque* empire, displays a large number of traits that we find throughout the religious legends of several peoples of the ancient continent. We seem to recognize the notions of good and evil personified in the old man, Bochica, and his wife, Huythaca. The distant time, prior to the moon's existence, recalls the Arcadians' claim concerning the ancientness of their origin. The moon is portrayed as a wicked being that raises the earth's humidity, while Bochica, child of the sun, dries the soil, protects agriculture, and becomes the benefactor of the Muisca, as the first Inca was for the Peruvians.

Travelers who have seen firsthand the impressive site of the great Tequendama waterfall will not be surprised that such simple peoples attributed a

1. See ▼Lucas Fernández Piedrahita, Obispo de Panamá, *Historia general del Nuevo Reyno de Granada*, p. 17, a work composed from Quesada's manuscripts.

miraculous origin to its boulders, which seem carved by human hands; to its narrow chasm, into which plunges a river that collects all the waters of the Bogotá valley; to its rainbows, which shine with the most beautiful colors and change shape at every moment; and to its vapor column, which rises like a thick cloud that one can see from five leagues away while strolling in the city of Santa Fé. The sixth Plate can give but a modest idea of this majestic spectacle. If it is difficult to describe the beauty of these waterfalls, it is even more so to convey it in a drawing. The impression that waterfalls make upon the observer's soul depends on the combination of several circumstances: the volume of rushing water must be proportional to the height of the falls, and the surrounding landscape must have a romantic and untamed character. The Pissevache and the Staubbach Falls in Switzerland are very tall but have a relatively insignificant water volume. The Niagara and the Rhine Falls, on the other hand, have an enormous water volume, but their height does not surpass fifty meters. A waterfall surrounded by low hills produces a lesser effect than what we see in the deep, narrow valleys of the Alps, the Pyrenees, and especially the Cordillera of the Andes. In addition to the height and the volume of the water column, not to mention the lay of the land and the appearance of the rocks, what gives a unique character to these grand natural scenes is the vigor and form of the trees and herbaceous plants and their distribution into scattered groups or clusters, as well as the contrast between the rocky masses and the freshness of the vegetation. The Niagara Falls would be even more beautiful if, instead of the pines and oaks typical of its northerly region, its surroundings were adorned with Heliconia, palms, and tree ferns.

Assembled within the falls (*salto*) of Tequendama are all the elements that make for an eminently picturesque site. By no means is it the highest waterfall on the globe, as the locals believe[1] and as physicists have repeated in Europe: the river does not plunge, as Bouguer says, into a chasm of five to six hundred meters of perpendicular depth; but there is not a single other waterfall at such a considerable height that collects such a large volume of water. After watering the swamps that lie between the villages of Facatativá and Fontibón, near Canoas (just above the *salto*), the Río Bogotá still maintains a width of forty-four meters, which is half that of the Seine in Paris between the Louvre and the Palais des Arts. The river narrows considerably near the waterfall itself, where the gorge, which appears to have been formed by an earthquake, has an opening of only ten to twelve meters. During great droughts the volume of water, which plunges to a depth of one hundred seventy-five meters in two bounds, still has a profile of ninety square meters. We have added the

22

---

1. Piedrahita, [*Historia general del Nuevo Reyno de Granada,*] p. 19; ▾Julian, *La Perla de la América, provincia de Santa Marta*, 1787, p. 9.

figures of two men to the drawing in order to provide a sense of the scale of the *salto*'s total height. The upper bank, where the men are placed, is at an elevation of two thousand four hundred sixty-seven meters above sea level. From this point to the Magdalena River, the little Bogotá River still has a descent of more than two thousand one hundred meters, which is more than one hundred forty meters per *common league*.

The path that leads from the city of Santa Fé to the *salto* of Tequendama passes through the village of Soacha and the large farm of Canoas, renowned for its impressive wheat harvests. It is believed that the enormous vapor mass, which rises daily from the waterfall and is precipitated by contact with the cold air, contributes a great deal to the fertility of this part of the Bogotá plain. From the hill of Chipa, a short distance from Canoas, the traveler enjoys a magnificent view, which is stunning for the contrasts it presents. One has just left behind fields sown with wheat and barley; now, in addition to the ivy, the alstonia theæformis, the begonia, and the yellow cinchona (*Cinchona cordifolia*, Mutis), one is surrounded by oaks, alders, and other plants with shapes recalling the vegetation of Europe. Suddenly one discovers, as though from a terrace—at one's feet, so to speak—a country where palms, banana trees, and sugarcane flourish. Since the crevice into which the Río Bogotá plunges is linked to the plains of the hot region (*tierra caliente*), a few palms have pushed all the way to the base of the waterfall. Because of this peculiar circumstance, the inhabitants of Santa Fé say that the Tequendama Falls are so high that the water plunges in a single bound from the cold country (*tierra fría*) into the hot country. One senses that a height difference of one hundred seventy five meters is not significant enough to have an appreciable influence on the air temperature. The contrast between the vegetation of the Canoas plateau and that of the ravine is not at all due to the ground height, for if the rock of Tequendama (which is a clay-based sandstone) were not cut so steeply, and if the Canoas plateau were as sheltered as the crevice, the palms that grow at the base of the waterfall would probably have pursued the course of their migration all the way to the river's upper level. The inhabitants of the Bogotá valley are all the more intrigued by the appearance of this vegetation since they live in a climate where the thermometer frequently falls to the freezing point.

I managed to carry instruments right into the crevice, all the way to the base of the waterfall. This descent takes three hours along a narrow track (*camino de la Culebra*) that leads to the Povasa ravine. Although the river, as it falls, loses a large portion of its water, which is reduced to vapor, the speed of the lower current obliges the observer to remain at a distance of almost one hundred forty meters from the basin hollowed out by the impact of the water. The bottom of this crevice is only dimly lit by sunlight. The solitude of the place, the richness of its vegetation, and the terrifying noise one hears there make the base of the Tequendama waterfall one of the wildest sites of the Cordilleras.

PLATE VII
Pyramid of Cholula

Despite their political divisions, five of the peoples who appeared successively on Mexican soil from the seventh to the twelfth century of our time—namely the ˅Toltecs, the Chichimecs, the Acolhua, the Tlaxcalteca, and the Aztecs—spoke the same language, worshiped the same gods, and built pyramidal structures, which they regarded as *teocalli*, that is, as the dwelling places of their gods. Although these structures were of very different dimensions, they all had same shape: terraced pyramids with sides that exactly followed the meridian and the parallel of the site. The teocalli [god-dwelling] was raised in the middle of a vast square enclosure surrounded by a wall. This enclosure, comparable to the Greeks' περίβολος [*peribolos*, enclosing wall] contained gardens, fountains, the priests' residences, sometimes even armories; like the ˅ancient temple of Baal-berith burned by Abimelech, the dwelling place of every Mexica god was also a fortress. A tall staircase led to the top of the truncated pyramid; on top of this platform were one or two chapels built like towers, which housed colossal idols of the particular deity to which the teocalli was dedicated. This part of the structure must be regarded as the most important; it is the ναός [*naos*, temple room] or, rather, the σηκός [*sekos*, sacred area] of Greek temples. This is also the place where the priests kept the sacred flame. The distinctive layout of the structure, which we have just described, made it possible for the ˅sacrificer to be seen by a large crowd of people at the same time. One could see the procession of the *teopixqui* from a distance as they climbed or descended the steps of the pyramid. The structure's interior served as the sepulcher for the kings and the other Mexica dignitaries. It is impossible to read the descriptions of the ˅temple of Jupiter Belus, which Herodotus and Diodorus Siculus bequeathed to us, without being struck by the points of similarity between this Babylonian monument and the teocalli of Anahuac.

When the Mexica or Aztecs, one of the seven tribes of the *Anahuatlaca* (*riverine* people), arrived in the equinoctial region of New Spain in the year 1190, they encountered the pyramidal monuments of *Teotihuacan, Cholula* (or *Cholollan*), and *Papantla*. They attributed these great constructions to the ˅Toltec, a powerful and civilized people who had lived in Mexico five hundred years earlier, had used hieroglyphic script, and had a calendar year and a chronology that were more exact than those of most of the peoples of the old continent. The Aztecs did not know for certain whether other tribes had inhabited the country of Anahuac before the Toltec. Considering the god-dwellings of Teotihuacan and Cholollan to be the work of the Toltec, they attributed to the latter the most ancient origin they could imagine. It is possible, however, that they were built before the Toltec invasion, that is, before the year 648 of the ˅common era. Let us not be surprised that the history of an American people might begin before the seventh century, and that the history of the Toltec might be as uncertain as that of the ˅Pelasgians and the Ausonians. An insightful scholar, ˅Mr. Schlözer, has provided ample evidence that the history of northern Europe does not go back any further than the tenth century, a period when the Mexican plateau supported a civilization that was much more advanced than that of Denmark, Sweden, or Russia.

The *teocalli* of Mexico City was dedicated to ˅Tezcatlipoca, the most important Aztec deity after Teotl, the supreme and invisible Being, and to Huitzilopochtli, the god of war. It was built by the Aztecs, following the model of the pyramids of Teotihuacan, only six years before the discovery of the Americas by Christopher Columbus. This truncated pyramid, which Cortés called the main Temple, was ninety-seven meters wide at its base and fifty-four meters high. It is not surprising that a structure of such dimensions could be destroyed a few years after the siege of Mexico City. In Egypt, only a few vestiges remain of the enormous pyramids that rose from the waters of ˅Lake Moeris and that, Herodotus claimed, were adorned with colossal statues. In Etruria, ˅the pyramids of Porsena—the description of which seems a bit fanciful, and four of which, according to Varro, were more than eighty meters high—also disappeared.[1]

But whereas the European conquerors toppled the Aztecs' *teocalli*, they did not also succeed in destroying the most ancient monuments, those we attribute to the Toltec people. We shall give a succinct description of these monuments, which are remarkable for both their form and their size.

The pyramid group of *Teotihuacan* is located in the Valley of Mexico at a distance of eight leagues to the northeast of the capital and in a plain that bears the name *Mixcoatl*, or *Pathway of the Dead*. There one can still see two great pyramids[2] dedicated to the sun (*Tonatiuh*) and the moon (*Meztli*) and

26

1. ˅Pliny, [*Historia naturalis*,] XXXVI, 19.
2. Clarification by ˅Mr. Langlès to Norden, *Voyage [d'Égypte et de Nubie]*, Vol. III, p. 327, number 2.

Plate VII

surrounded by several hundred small pyramids that form roads between them, the direction of which is exactly north-south and east-west. One of the two great *teocalli* has a perpendicular elevation of fifty-five meters, while the other has forty-four meters. Given that the base of the former, the Tonatiuh Yztaqual, is two hundred eight meters long, if we take into consideration the measurements ˙Mr. Oteyza made in 1803, it would appear that this structure is taller than the Mycerinus, the third of the great pyramids of Giza, and that the length of its base is more or less equal to that of ˙Chephren. The small pyramids surrounding the great dwellings of the moon and the sun are barely nine to ten meters high: according to the natives' legends, they served as the tribal chiefs' sepulcher. Similarly, around Cheops and Mycerinus in Egypt, one can make out eight small pyramids arranged in a very symmetrical fashion and parallel to the sides of the large ones. The two *teocalli* of Teotihuacan had four main terraces; each of these was subdivided into small steps, the edges of which can still be made out. Their core is of clay mixed with small rocks and covered with a thick wall of *tezontli*, porous amygdaloids. This construction recalls one of the Egyptian pyramids of Sakkarah, which has six terraces and which, according to ˙Pococke's narrative,[1] is a mass of stones and yellow mortar covered with rough stones on the outside. At the top of the great Mexica *teocalli* were two colossal statues of the sun and the moon. They were made of stone and coated in strips of gold; these strips were removed by Cortés's soldiers. When ˙Bishop Zumárraga, a Franciscan monk, undertook to destroy everything related to the religion, the history, and the antiquities of the indigenous peoples of the Americas, he also had the idols of the Mixcoatl plain burned. One can still see the remains of a staircase built of large cut stones that formerly led to the platform of the *teocalli*.

To the east of the Teotihuacan pyramid group, as one descends from the Cordillera toward the Gulf of Mexico, the Papantla pyramid rises in the midst of a dense forest called *Tajín*. It was only by accident that Spanish hunters discovered it less than thirty years ago, for the Indians prefer to hide from the whites everything that they have long venerated. The shape of this *teocalli*, which had six, perhaps even seven, levels, is slenderer than that of any other monuments of this kind: although it is about eighteen meters high, it is only twenty-five meters long at its base. It is, therefore, almost half as high as ˙the pyramid of Gaius Cestius in Rome, which is thirty-three meters. This small structure is built entirely of extraordinarily large dressed stones with an exquisite, very regular cut. Three staircases lead to its top; the surface of its terraces is adorned with hieroglyphic sculptures and small recesses

---

1. Pococke, *Voyage* (Neuchâtel edition, 1772), Vol. 1, p. 147.

arranged very symmetrically. The number of these recesses seems to allude to the three hundred eighteen simple and compound signs of the days of the *Cempohualilhuitl*, the Toltec civil calendar.

The greatest, most ancient, and most famous of all the pyramidal monuments of Anahuac is the *teocalli* of Cholula. Today it is called *the handmade mountain* (*monte hecho a mano*). Seeing it from afar, one would indeed be tempted to take it for a natural hill covered in vegetation. It is in its current state of degradation that this pyramid is depicted on the seventh Plate.

A vast plain, that of Puebla, is separated from the Valley of Mexico by the volcanic range that extends from Popocatepetl toward Río Frío and the peak of Telapón.[1] This fertile but treeless plain is rich in relics of Mexica history: it contains the capitals of the three republics of Tlaxcala, Huejotzingo, and Cholula, which, despite their continual strife, nonetheless resisted the despotism and the usurpatory spirit of the Aztec kings.

'The small city of Cholula, which Cortés compared to the most densely populated cities of Spain in his letters to Emperor Charles V, has barely sixteen thousand inhabitants today. The pyramid is located to the east of the city on the road that leads from Cholula to Puebla. It is very well preserved on its western flank, and it is this side that is depicted in the engraving that we are including here. The Cholula plain exhibits the barren character particular to plateaus with an elevation of two thousand two hundred meters above sea level. In the foreground one can make out a few agave stalks and dragon trees, while in the background one sees the snow-capped top of the Orizaba volcano, a colossal mountain with an absolute elevation of five thousand two hundred ninety-five meters. I have published a sketch of this mountain in my Mexican Atlas, Plate XVII.

'The *teocalli* of Cholula has five terraces of equal height. It seems to have been oriented in exact alignment with the four cardinal points; but since the edges of the terraces are not very distinct, it is difficult to discern their original direction. This pyramidal monument has a broader base than any similar structure in the old continent. I have measured it carefully and have confirmed that its perpendicular height is no more than fifty-five meters but that each side of its base is four hundred thirty-nine meters long. 'Torquemada judged its height to be seventy-seven meters; Betancourt, sixty-five; and Clavijero, sixty-one. 'Bernal Díaz del Castillo, an ordinary soldier in Cortés's expedition, found a source of distraction in counting the steps in the staircases leading to the platforms of the *teocalli*; he found one hundred fourteen at the great temple of Tenochtitlan, one hundred seventeen at Tetzcoco, and one hun-

1. See my *Mexican Atlas*, Plates III and IX.

dred twenty at Cholula. The base of the Cholula pyramid is twice as large as that of Cheops, but it is only slightly higher than the Mycerinus pyramid. A comparison between the dimensions of the dwelling place of the sun at Teotihuacan and those of the Cholula pyramid suggests that the people who constructed these remarkable monuments intended to build them to the same height, but with bases at a length ratio of one to two. The proportion between the base and the height is also different among the various monuments. The three great pyramids of Giza have a height-to-base ratio of one to one and seven-tenths; for the hieroglyph-covered pyramid of Papantla, this ratio is one to one and four-tenths; for the great pyramid of Teotihuacan, it is one to three and seven-tenths; and for Cholula, it is one to seven and eight-tenths. The last of these monuments is constructed of alternating layers of unfired bricks (*xamilli*) and clay. A number of Cholula Indians assured me that the interior of the pyramid is hollow and that during Cortés's stay in their city, their ancestors had hidden a large number of warriors there in order to swoop down unexpectedly upon the Spanish. Both the material from which this *teocalli* is constructed and the silence of the contemporary historians on this matter[1] seem to cast doubt upon the validity of this claim.

What is indisputable is that in the interior of this pyramid, as in other teocalli, there were considerable cavities that served as the natives' sepulcher; a peculiar circumstance occasioned their discovery. Seven or eight years ago, the road from Puebla to Mexico City, which formerly passed by the north side of the pyramid, was redirected; to align this road, they bore straight through the lowest terrace, which left one-eighth of the terrace isolated like a pile of bricks. While boring this hole, they found in the pyramid's interior a square house built of stone and supported by beams of bald cypress (*cupressus disticha*); it contained two corpses, idols in basalt, and a large number of varnished, artistically painted vases. Although they did not bother to preserve these objects, they claim to have verified carefully that this house, covered in bricks and layers of clay, had no exit whatsoever. If one supposes that the pyramid was built not by the Toltec, the first inhabitants of Cholula, but by the prisoners whom the Cholulans took from among the neighboring peoples, it seems plausible that the corpses were those of a few unfortunate slaves deliberately left to die inside the teocalli. We have examined the ruins of this subterranean house and have identified a peculiar arrangement of the bricks, which tends to reduce the amount of pressure exerted on the roof. Since the natives did not know how to construct vaulted roofs, ˇthey placed especially wide bricks in a horizontal position so that those on top extended past the

29

1. Hernán Cortés, *Cartas* (Mexico, 1770), p. 69.

lower ones. The result was a tiered assembly that functioned somewhat like a Gothic arch; similar vestiges have also been found in several Egyptian structures. ˅It would be interesting to hollow out a gallery through the teocalli of Cholula in order to examine its internal construction; remarkably, the desire for finding hidden treasures has not yet prompted such an undertaking. During my trip to Peru, when I visited the vast ruins of the city of Chimu, near Manische, I entered the interior of the famous *Huaca de Toledo*, the tomb of a Peruvian prince, in which ˅García Gutiérrez de Toledo discovered more than five million francs' worth of solid gold while making a gallery in 1576, as is confirmed by the ledgers preserved in the town hall of Trujillo.

At the top of the great teocalli of Cholula, also called the mountain of unfired bricks (*Tlalchihualtepec*), was an altar dedicated to ˅Quetzalcoatl, the god of the air. This Quetzalcoatl (whose name means snake dressed in green feathers, from *coatl*, snake, and *quetzalli*, green feather) is probably the most mysterious figure in all of Mexica mythology. He was a white, bearded man, like the Muisca's Bochica, whom we mentioned above in the description of the Tequendama waterfall. He was the high priest at Tula (*Tollan*), a legislator, the head of a religious sect who, like the ˅Sannyasin and the Buddhists of Hindustan, imposed a very cruel penance upon himself: he introduced the custom of piercing the lips and the ears and mortifying the rest of the body with spines from agave leaves and cacti, while inserting reeds into the wounds so that blood streamed out in visible abundance. In a Mexica drawing preserved in the Vatican library,[1] I have seen a figure that represents Quetzalcoatl appeasing the gods' wrath through his penance when, thirteen thousand sixty years after the creation of the world (I am following the extremely vague chronology reported by ˅Father Ríos), there was a great famine in the province of Culan. The saint had retired near Tlaxapuchicalco, on the volcano of Catcitepetl (*the talking mountain*), where he walked barefoot on agave leaves armed with spines. One has the impression of gazing at one of the Rishi [seers or divine scribes], the hermits of the Ganges, whose pious austerity is celebrated in the Puranas.[2]

Quetzalcoatl's reign was the golden age of the peoples of Anahuac. At that time, all animals, even humans, lived in peace; the earth yielded the richest harvests without any need for farming; and the sky was filled with a multitude of birds admired for their song and for the beauty of their plumage. But this reign, like that of Saturn, did not last long, nor did the happiness of the world: the Great Spirit Tezcatlipoca, the Brahma of the peoples of Anahuac,

---

1. ˅*Codex [Vaticanus] anonymous*, number 3738, folio 8.
2. ˅Schlegel, *Über Sprache und Weisheit der Indier*, p. 132.

offered Quetzalcoatl a drink that, while rendering him immortal, made him crave travel, giving him in particular an irresistible desire to visit a faraway land that legend calls Tlalpallan.[1] The similarity between this name and that of Huehuetlapallan, the Toltec's homeland, does not appear to be accidental; but how is it plausible that this white man, the priest of Tula, headed, as we will see shortly, for the *southeast*, toward the plains of Cholula, and from there to the eastern coasts of Mexico, only to arrive at this *northerly* land that his ancestors had left in the year 596 of our era?

As Quetzalcoatl crossed the territory of Cholula, he gave in to the pleas of the inhabitants, who offered him the reins of government. He stayed with them for twenty years, taught them how to smelt metal, ordained the great forty-day fasts, and devised the intercalations of the Toltec year. He exhorted men to live in peace and forbade all offerings to the deity other than the first harvests. From Cholula, Quetzalcoatl journeyed to the mouth of the Goatzacoalcos River, where he disappeared after sending word to the Cholulans (*Chololtecatles*) that he would return after a period of time to govern them again and renew their happiness.

It was for the descendants of this saint that the unfortunate Montezuma mistook Cortés's companions. "We know from our books," he said in his first interview with the Spanish general, "that all those who inhabit this country, including myself, are not natives but, rather, foreigners who have come from far away. We also know that the leader who had brought our ancestors returned to his native land for a period of time, and that he came back in search of those who had settled here. He found them married to the women of this land, with numerous offspring, and living in the cities they had built. Our people did not wish to obey their former master, and he went back alone. We have always believed that his descendants would one day come to take possession of this land. Considering that you come from the place where the sun is born, and that, as you claim, you have known us for a long time, I cannot doubt that the king who sends you is our natural master."[2]

Among the Indians of Cholula a different, quite remarkable legend persists to this day, according to which the great pyramid had not originally been designed for use in the worship of Quetzalcoatl. After my return to Europe I examined the Mexica manuscripts in the Vatican library in Rome and found a record of this very legend in a manuscript by Pedro de los Ríos, a Dominican monk who, in 1566, had copied on site all of the hieroglyphic paintings he could procure. "Before the great flood (*apachihuiliztli*) that happened four

1. Clavijero, *Storia [antica] del Messico*, Vol. II, p. 12.
2. First letter by Cortés, § XXI and XXIX.

thousand eight years after the creation of the world, the land of Anahuac was inhabited by giants (*Tzocuillixeque*). All those who did not perish were turned into fish, with the exception of seven, who took refuge in caves. When the waters subsided, one of these giants, Xelhua, known as the architect, went to Cholollan where, in memory of Tlaloc Mountain, which had served as a shelter for him and six of his brothers, he built an artificial hill in the shape of a pyramid. He had bricks made in the province of Tlalmanalco at the foot of the Sierra de Cocotl, and in order to transport them to Cholula, he placed a line of men who passed them along by hand. The gods were incensed by the sight of this structure, whose top was meant to reach the clouds; enraged by Xelhua's audacity, they cast fire on the pyramid. Many workers perished, the work was discontinued, and it was afterward consecrated to the god of air, Quetzalcoatl."

This story recalls the ancient legends of the East, which the Hebrews recorded in their holy books. In Cortés's time, the Cholulans preserved a stone, shrouded in a globe of fire, that had fallen from the heavens onto the top of the pyramid; this aerolite had the shape of a toad. To prove the ancient origin of the Xelhua fable, Father Ríos noted that it was contained within a canticle that the Cholulans sang at their feasts as they danced around the teocalli, and that this canticle began with the words *Tulanian hululaez*, which are not of any current Mexica language. In all corners of the globe, on the ridge of the Cordilleras as on the island of Samothrace in the Aegean Sea, fragments of original languages have been preserved in religious rites.

The platform of the Cholula pyramid, from which I have conducted a great number of astronomical observations, has a surface area of four thousand two hundred square meters. From there one has a magnificent view of Popocatepetl, Iztaccihuatl, Pico de Orizaba, and the Sierra de Tlaxcala, famous for the storms that brew around its summit. It is possible to see in one glance three mountains that are all higher than Mont Blanc, two of them active volcanoes. A small chapel surrounded by cypresses and dedicated to Our Lady de los Remedios has replaced the temple of the god of air, the Mexica Indra; a cleric of the Indian race celebrates mass daily on the top of this ancient monument.

33   In Cortés's time, Cholula was regarded as a holy city: nowhere could one find a larger number of teocalli, more priests and religious orders (*tlamaca-zque*), more magnificence in worship, and more austerity in fasting and penance. Since the introduction of Christianity among the Indians, the symbols of a new religion have not entirely supplanted the memory of the former; crowds of people come from afar to the top of the pyramid to celebrate the feast of the Virgin. A secret fear and a religious respect seize the natives at the

sight of this immense pile of bricks covered in shrubs and perennially green grass.

I have pointed out above the great similarity in construction that can be observed between the Mexica teocalli and the temple of Bel (or Belus) in Babylon. This similarity had already struck Mr. Zoëga, although he was only able to obtain very incomplete descriptions of the pyramid group of Teotihuacan.[1] According to Herodotus, who visited Babylon and saw the temple of Belus, this pyramidal monument had eight terraces; it was one ʼstadium high, and the width of the base was equal to its height. The wall that formed the exterior enclosure, the περίβολος [*peribolos*, enclosing wall], was two square stadia (a common Olympic stadium equaled one hundred eighty-three meters: the Egyptian stadium equaled only ninety-eight).[2] The pyramid was built of bricks and asphalt; it had a temple (ναός [*naos*]) at its top and another near its base. According to ʼHerodotus, the former was without statues; there was only a golden table and a bed on which lay a woman chosen by the god Belus.[3] Diodorus Siculus, on the other hand, insisted that this upper temple contained an altar and three statues, to which he gave the names Jupiter, Juno, and Rhea, following concepts taken from Greek religion.[4] But neither these statues nor the monument in its entirety were still in existence at the time of Diodorus and Strabo. In the Mexica teocalli, as in the temple of Bel, the lower *naos* was distinguished from the one located on the platform of the pyramid. This very distinction is clearly noted in Cortés's Letters and in the History of the conquest, written by Bernal Díaz, who resided for several months in the palace of King Axayacatl, and thus opposite the teocalli of Huitzilopochtli.

None of the ancient authors—neither Herodotus nor Strabo,[5] nor Diodorus, nor ʼPausanias,[6] nor Arrian,[7] nor Quintus Curtius[8]—mentions that the temple of Belus was oriented in accordance with the four cardinal points, as the Egyptian and Mexica pyramids are. Pliny merely observes that Belus was regarded as the inventor of astronomy: *Inventor hic fuit sideralis scientiae* [He was the inventor of the science of the stars].[9] Diodorus reports that the Babylonian temple served as an observatory for the Chaldeans: "It is ac-

34

1. Zoëga, *De origine [et usu] obeliscorum*, p. 380.

2. ʼVincent, *Voyage de Néarque*, p. 56.

3. Herodotus, [*Histoire,*] Book 1, ch. CLXXXI–CLXXXIII, pp. 181–3.

4. Diodorus Siculus, [*Bibliothecae historicae*] (ed. Wesseling), Vol. 1, Book II, p. 123.

5. ʼStrabo, [*Géographie,*] Book XVI, p. 211.

6. Pausanias, [*Voyage de la Grèce,*] Book VIII, ed. Xylandri, p. 509, note 31.

7. Arrian, [*Expéditions d'Alexandre*], Book VII, p. 17.

8. Quintus Curtius, [*The Historie of Quintus Curtius,*] Book V, pp. 1 and 37.

9. Pliny, [*Historia naturalis,*] Book VI, p. 30.

knowledged," he says, "that this construction was extraordinarily tall, and that the Chaldeans used it to conduct observations of the stars, whose rising and setting could be discerned with great exactitude because of the elevation of the building." The Mexica priests (*teopixqui*) also observed the position of the stars from the top of the teocalli and announced the hours of the night to the people by the sound of the horn.[1] These teocalli were built in the interval between Muhammad's time and the reign of Ferdinand and Isabella, and it is surprising to learn that American structures, whose form is nearly identical to that of one of the most ancient monuments on the banks of the Euphrates, belong to a relatively recent period.

If one compares the pyramidal monuments of Egypt, Asia, and the new continent, one sees that despite their similarity in form, their purpose was very different. The pyramid groups at Giza and Sakkarah in Egypt; the triangular pyramid of Zarina, queen of the Scythians, which was one stadium high and three stadia wide and adorned with a colossal figure[2]; the fourteen Etruscan pyramids that were said to be contained within King Porsena's labyrinth at Clusium—all of these had been built to serve as sepulchers for illustrious personages. Nothing is more natural to humans than to mark the final resting place of those whose memory they cherish. These were initially simple piles of earth, and afterward *tumuli* of surprising height. Those of the Chinese and the Tibetans are only a few meters high,[3] but the dimensions increase as one moves west: in Lydia, the ʼ*tumulus* of Alyattes, the father of Croesus, was six stadia in diameter, while that of Ninus was more than ten stadia.[4] In the north of Europe the sepulchers of the Scandinavian King Gormus and Queen Daneboda are covered in piles of earth three hundred meters wide and thirty meters high. These *tumuli* are found in both hemispheres, in Virginia, in Canada, as well as in Peru, where numerous stone galleries with interlinking shafts fill the insides of the ʼ*huacas*, or artificial hills. The Asian sense for extravagance is reflected in the ornamentation, which nevertheless maintains the original form of these rustic monuments; the tombs of Pergamum are earthen cones raised upon an enclosing wall that appears to have formerly been covered in marble.[5]

The teocalli or Mexica pyramids were both temples and tombs. We have

35

---

1. ʼGama, *Descripción cronológica de la piedra calendaría*, Mexico, 1792, p. 15.

2. Diodorus Siculus, [*Bibliothecae historicae*,] Book II, ch. XXXIV, p. 34.

3. ʼDu Halde, *Description de la Chine*, Vol. II, p. 126. [ʼJohn Shore, "Account of the Kingdom of Nepal,"] *Asiatick Researches*, [1801,] Vol. II, p. 314.

4. Herodotus, [*Histoire*,] Book I, C. XCIII. Ctesias [of Cnidus] in *Diod[ori] sicul[i]*, Book II, ch. VII.

5. ʼChoiseul Gouffier, *Voyage pittoresque de la Grèce*, Vol. II, pp. 27–31.

noted above that the name of the plain where the dwelling places of the sun and the moon of Tenochtitlan rise is the *Pathway of the dead*; but the essential, main section of the teocalli was the chapel, the *naos*, at the top of the structure. At the dawn of civilization, peoples chose high places for making sacrifices to the gods. The first altars and temples were erected on mountains. If the mountains were isolated, they preferred to give them regular shapes by cutting terraces into them and making steps in order to climb to the summit more easily. Both continents offer numerous examples of hills that have been terraced and dressed in brick or stone walls. The teocalli appear to me nothing more than artificial hills raised in the middle of a plain and designed to serve as a base for altars. Indeed, there is nothing more impressive than a sacrifice that can be seen by an entire people at the same time! The pagodas of Hindustan have nothing in common with the Mexica temples; that of Tanjore [Thanjavur], of which Mr. Daniell made such superb sketches,[1] is a tower with several terraces; but the altar is not found at the top of the monument.

The pyramid of Bel was at once the temple and the tomb of this god. 36 Strabo does not even refer to this monument as a temple; he calls it simply *tomb of Belus*. In Arcadia, the top of the tumulus ($\chi\tilde{\omega}\mu\alpha$[*chōma*, funeral mound]) containing the ashes of ʽCallisto bore a temple to Diana; Pausanias[2] described it as a man-made cone covered with ancient vegetation. Here is an example of a remarkable monument in which the temple is more than just an accidental ornament; it serves as a transition, so to speak, between the pyramids of Sakkarah and the Mexica teocalli.[3]

1. *Oriental Scenerys*, Plate XVII.

2. Pausanias, [*Voyage de la Grèce,*] Book VIII, ch. XXXV (p. 35).

3. See my *Essai politique sur le royaume de la Nouvelle-Espagne* [1811], pp. 169, 187, 239, and 274.

PLATE VIII

Detached Section from the Cholula Pyramid

The Cholula pyramid is covered in vegetation so thick that it is very difficult to examine the great terraces. The Spanish historians of the sixteenth century, many of whom visited Mexico either during Montezuma's lifetime or just a few years after his death, reported that the whole structure had been built of bricks. As I noted above, while perusing Father Pedro de los Ríos's manuscript[1] in the Vatican library in Rome, I learned that following an ancient legend, the residents of Cholula believed that the bricks used for the teocalli had been made in the province of Tlalmanalco, at the foot of Cocotl mountain, and that prisoners had been lined up to pass the bricks by hand over a distance of several leagues, from Cocotl to Cholula. This legend, which recalls the most fanciful elements of Arabic tales, is also found among the Peruvians. Those of the Cusco plateau, who regard themselves as the inhabitants of a holy place, claim that when the ʼInca Tupac Yupanqui seized the kingdom of Quito (*Quitu*), he ordered that enormous cut stones be brought there from the quarries around Cusco in order to construct temples to the sun in the newly conquered lands.

I was able to discern the internal structure of the Cholula pyramid from two different locations: near the summit, on the flank facing the volcano Popocatepetl, and from the north side, where the lowest terrace is traversed by the new road that leads from Puebla to Mexico City. It was during the digging of this road that the outer edge of the terrace became detached from rest of the structure. The eighth Plate shows this detached section: one can identify the alternating layers of bricks and clay. The bricks are generally eight centimeters high and forty long. It seemed to me that they were not fired but merely dried in the sun; it is possible, however, that they underwent a light firing and that the humidity of the air made them crumbly. In the interior of the pyramid, the layers of clay separating the bricks are perhaps not

37

---

1. *Codex Vaticanus anonymous*, number 3738, folio 10.

found in the sections that support the enormous weight of the external mass. Mr. Zoëga[1] had assumed, incorrectly, that the teocalli of Cholula was a true ($\chi\tilde{\omega}\mu\alpha$ [*chōma*]), a pile of earth coated on the outside with a layer of bricks; already ˙Gemelli, whom Robertson and other first-rate historians have—perhaps somewhat undeservedly—accused of inaccuracy, refers to this structure as an earthen pyramid.[2]

As we have observed above, the construction of the teocalli recalls the most ancient monuments of human civilization. The temple of Jupiter Belus, who is referred to by the name Bali[3] in Hindu mythology, the pyramids of Meidum and Dahshur, and several of the Sakkarah group in Egypt were also only immense piles of bricks whose ruins have been preserved to this day— that is, over a period of thirty centuries.

1. *De [origine e uso] obeliscurom*, p. 380.
2. *Giro del Mondo*, Vol. VI, p. 135.
3. ˙Fra. Paolini da S. Bartolomeo, *Viaggio alle Indie Orientali*, p. 241.

Plate VIII

PLATE IX

ʻXochicalco Monument

This remarkable monument, of which only a sculpture-laden fragment is shown on this Plate, is regarded by the locals as a *military monument*. To the southeast of the city of Cuernavaca (the former Quauhnahuac), on the western slope of the Cordillera of Anahuac, in the fortunate region called the *tierra templada* (temperate region) by the locals because of its perpetual spring weather, there rises an isolated hill, which, according to ʻMr. Alzate's barometric measurements, is one hundred seventeen meters tall from base to top. This hill is located to the west of the road leading from Cuernavaca to the village of Miacatlan. The Indians call it, in the Mexica (or Aztec) language, *Xochicalco, House of flowers*. As we will see below, the etymology of this name is as uncertain as the date of the monument's construction, which is attributed to the Toltec. This people represents to Mexican antiquarians what the Pelasgian colonists long represented to their counterparts in Italy. Everything that is lost in the mists of time is seen as the work of a people to whom we trace the first seeds of civilization.

The hill of Xochicalco is a mass of rocks that human hands have shaped into a fairly regular cone; it is divided into five terraces, each of which is covered in masonry work. The terraces have a perpendicular elevation of around twenty meters. They narrow toward the top, as do the teocalli, or Aztec pyramids, whose summit was adorned with an altar. All of the terraces slope toward the southwest, perhaps in order to facilitate the flow of rainwater, which is very abundant in this region. The hill is surrounded by a rather deep and very wide moat, making the whole entrenchment nearly four thousand meters in circumference. The size of these dimensions should not surprise us: on the ridge of the Cordilleras of Peru, at heights almost equaling that of the ʻPeak of Tenerife [Teide], Mr. Bonpland and I saw even more impressive monuments. The plains of Canada exhibit lines of defense and entrenchments of extraordinary length. All of these American works bear a resemblance to the ones encountered daily in the eastern part of Asia, where peoples of the Mongol race, especially the ones most advanced in civilization, have built great walls separating entire provinces.

Plate IX

The summit of the hill of Xochicalco has an oblong platform that is seventy-two meters long from east to west and eighty-six meters from north to south. This platform is surrounded by a wall made of cut stones more than two meters high and used by combatants as a means of defense. In the center of this spacious parade ground are the remnants of a pyramidal monument that once had five terraces and a shape resembling that of the teocalli that we have just described above. Only its lowest terrace, a drawing of which is found on the ninth plate, has been preserved. The proprietors of a neighboring sugar refinery were so barbarous as to destroy the pyramid by extracting stones to build their ovens. The Indians of Tetlama insist that the five terraces were still intact in 1750; following the dimensions of the lowest step, we may suppose that the structure was twenty meters high. Its flanks are oriented in exact alignment with the four cardinal points. The base of the structure is 20,7 meters long and 17,4 meters wide. Especially striking is the fact that one does not come across any vestiges of a staircase leading to the top of the pyramid, where they claim to have once found a stone seat (*ximotlalli*) adorned with hieroglyphs.

Travelers who have examined firsthand this work by the indigenous peoples of the Americas are captivated by the polish and the cut of the stones, all of which have the shape of parallelepipeds [a three-dimensional figure formed by six parallelograms]; the care with which they were joined together, without any cement to fill the joints; and the execution of the reliefs that adorn the terraces. Each figure stretches across several stones, and since the contours are not interrupted by the joints of the stones, we may suppose that the reliefs were sculpted after the construction of the monument had been completed. Among the hieroglyphic decorations on the Xochicalco pyramid one can make out crocodile heads that spout water and seated human figures with legs crossed in the manner of the Asian peoples. Considering that the structure is located on a plateau over thirteen hundred meters above sea level and that crocodiles live only in coastal rivers, it is surprising to see that instead of reproducing the plants and the animals known to the mountain peoples, the architect portrayed the gigantic creatures of the Torrid Zone in these reliefs—and with a particular meticulousness.

The moat surrounding the hill; the surface of the terraces; the large number of underground chambers hollowed out in the rock on the north side; and the wall protecting the platform from approach—together these details lend the monument of Xochicalco the character of a military monument. To this day the natives still refer to the ruins of the pyramid that rises in the middle of the platform by a name that translates as "fortified castle" or "citadel." The great similarity in shape between this supposed citadel and the *dwelling places of the Aztec gods* (teocalli) leads me to believe that the hill of Xochicalco was nothing more than a *fortified temple*. The pyramid of Mexitli, the great temple of Tenochtitlan, also contained an arsenal within its walls and served as a fortification during the siege, at some points for the Mexica and at others for the Spanish. The Hebrews' holy books tell us that in early antiquity the temples of Asia (for example, that of Baal-berith at Shechem in Canaan) were at once structures consecrated for worship and entrenchments where city dwellers took cover from the enemy's attacks. Indeed, there is nothing more natural to humans than to fortify the places in which they keep the country's tutelary gods; when the res publica is in danger, there is nothing more reassuring than to take refuge at the foot of altars and to fight under their immediate protection! For peoples whose temples had preserved one of the most ancient forms, that of the pyramid of Belus, the monument's construction could meet the dual use of worship and defense. In Greek temples the lone wall that formed the περίβολος [*peribolos*, ring wall] offered refuge to the besieged.

The natives of the neighboring village of Tetlama have a geographical map that was created before the arrival of the Spanish and to which a few names have been added since the conquest: on this map, at the place marking the location of the monument of Xochicalco, there is an image of two warriors fighting with clubs, one of whom is named Xochicatli and the other Xicatetli. We shall not follow the Mexican antiquarians' etymological discussions to determine whether one of these warriors gave his name to the hill of Xochicalco; whether the image of the two combatants simply refers to a battle between two neighboring peoples; or, finally, whether the name *House of flowers* was given to the pyramidal monuments because the Toltec, like the Peruvians, offered only fruits, flowers, and incense to the deity. Thirty years ago they also found, near Xochicalco, an isolated stone bearing an image in relief of an eagle tearing apart a captive, a likely allusion to a victory the Aztecs won over some neighboring people.

The drawing of the relief on the lowest terrace was copied from an engraving published in Mexico City in 1791. I did not have the opportunity to visit this remarkable monument myself. When, en route to New Spain from the South Sea, I traveled from Acapulco to Cuernavaca, I was unaware of the existence of the hill of Xochicalco, and I regret not having been able to verify with my own eyes the description[1] by Mr. Alzate, a corresponding member of the Academy of Sciences in Paris. Since an indication of scale was mistakenly omitted from Plate IX, I must note that the height of the seated figures with crossed legs is 1,03 meters.

---

1. [José Antonio Alzate y Ramírez,] *Descripción de las antigüedades de Xochicalco* (Mexico City, 1791). ▼Pietro Márquez, *Due antichi monumenti di architettura messicana illustrati* (Rome, 1804).

PLATE X

Cotopaxi Volcano

In my descriptions above of the Iconozo valley, I noted that the huge elevation of the plateaus surrounding the high peaks of the Cordilleras detracts
somewhat from the impression that these great rock masses make upon the
soul of a traveler accustomed to the majestic scenes of the Alps and the Pyrenees. In all climes what gives a mountainous landscape its particular character is not so much the absolute height of the mountains as their appearance,
their shape, and their grouping.

It is the physiognomy of these mountains that I have attempted to depict in
a series of drawings, a few of which have already appeared in the *Geographical and Physical Atlas* that accompanies my [*Political*] *Essay on the Kingdom
of New Spain*. It seemed to be of great interest to geologists to be able to compare the shapes of mountains in the most remote parts of the globe, just as
one compares vegetation from different climates. To date, very few materials
have been gathered for this important work. Without the help of geodesic
instruments for measuring very small angles, it is practically impossible to
determine contours with any great precision. At the same time as I was conducting these measurements in the southern hemisphere on the ridge of the
Cordillera of the Andes, ꞌMr. Osterwald (with the assistance of Mr. Tralles,
a distinguished geometrician) was following a similar method in drawing the
Swiss Alps as seen from the banks of Lake Neufchâtel. This view, which has
just been published, is so exact that if the distance to each mountaintop were
known, one would be able to find a mountain's relative height merely by fac-
toring into one's calculation the simple measurement of the contours in the
drawing. Mr. Tralles used a ꞌrepeating circle. The angles by which I determined the dimensions of the different parts of a mountain were taken with
one of ꞌRamsden's sextants, the limb of which reliably indicated six to eight
seconds. If this work were repeated century after century, we would eventually come to understand the random changes that the surface of the globe un-
dergoes. In a country susceptible to earthquakes and thrown into upheaval
by volcanoes, it is extremely difficult to resolve the question of whether ejec-

tions of ashes and scoria lead to imperceptible decreases or increases in the size of the mountains. This question will be resolved more effectively by taking simple height angles in select stations than by conducting complete trigonometric measurements, the results of which are affected by the errors that one might make in measuring not only the base but also the oblique angles.

A comparison of the mountains of both continents reveals a similarity in shape that would be unimaginable if one reflected upon the combination of forces that, in the primal world, had a tumultuous effect on the soft surface of our planet. The volcanoes' fire spews cones of ash and pumice stone out through its craters; bubbles that resemble extraordinarily tall domes appear to have been produced by the expansive force of elastic fumes alone; earthquakes either raised or straightened seashell-filled layers; ocean currents furrowed the bottoms of basins that today form circular valleys or plateaus surrounded by mountains. Each region of the globe has its specific physiognomy; but even when one is surrounded by characteristic features that give the natural realm such a rich and varied appearance, one is struck by a resemblance in shape that is based on both similar causes and local circumstances. As one navigates among the Canary Islands and observes the basalt cones of Lanzarote, Alegranza, and Graciosa, one has the impression of seeing the Euganean Hills [in Italy] or the trappean hills in Bohemia. The granite, micaceous schist, ancient sandstone, and limestone formations to which mineralogists have given the formation names *Jura*, *High Alps*, or *transition limestone* lend a peculiar character to the contour of these tall masses and to the rifts in the ridge of the Andes, the Pyrenees, and the Urals. The nature of the rocks has shaped the outward appearance of mountains everywhere.

Cotopaxi, whose top is shown on the tenth Plate, is the highest of the volcanoes in the Andes to have erupted in recent memory. Its absolute elevation is five thousand seven hundred fifty-four meters (two thousand nine hundred fifty-two toises), or double the height of Mount Canigou [in the Pyrenees]. It is therefore eight hundred meters higher than Vesuvius would be if it were stacked on top of the Peak of Tenerife [Teide]. Cotopaxi is also the most feared of all the volcanoes in the kingdom of Quito, the one with the most frequent and most devastating explosions. Considering the amount of scoria and rock pieces ejected from this volcano and strewn across the surrounding valleys over an area of several square leagues, one imagines that their combined mass would form a colossal mountain. In 1738 the flames of Cotopaxi rose to a height of nine hundred meters above the edge of the crater. In 1744 the roaring of the volcano could be heard as far away as Honda, a city located two hundred common leagues away on the banks of the Magdalena River. On April 4, 1768, the quantity of ash spewed from the mouth of Cotopaxi was so

Plate X

great that in the cities of Ambato and Latacunga it was still nighttime at three o'clock in the afternoon, forcing the inhabitants to carry lanterns with them in the streets. The explosion that took place in January of 1803 was preceded by a terrifying phenomenon: the sudden melting of the snows that cap the mountain. For more than twenty years, no smoke, no visible vapor, had been emitted from the crater, but in a single night, the subterranean fire became so intense that by sunrise the outer walls of the cone had been raised to such a high temperature that they appeared bare and of the blackish color that is peculiar to vitrified scoria. In the port of Guayaquil, fifty-two leagues, as the crow flies, from the edge of the crater, we heard the volcano roaring night and day, like the repeated discharge of a battery; we could make out the dreadful noise even in the South Sea to the southwest of Puná Island.

44

Cotopaxi is located twelve leagues to the south-southeast of the city of Quito, between Rumiñahui Mountain, whose crest, spiked with small, isolated boulders, stretches forth like an enormously high wall, and the Quilindaña volcano [in Ecuador], which soars to the region of eternal snows. In this part of the Andes, a valley running lengthwise divides the Cordilleras into two parallel ranges. Since the bottom of this valley is still three thousand meters above ocean level, when viewed from from the Licán and Mulaló plateaus, Chimborazo and Cotopaxi appear only as high as Col du Géant [in the Mont Blanc Massif] and Cramont [in the Swiss Alps], as measured by ˅Saussure. As there is reason to suppose that the proximity of the Ocean helps to stoke volcanic fire, the geologist is surprised to note that the most active volcanoes in the kingdom of Quito—Cotopaxi, Tungurahua, and Sangay—belong to the eastern range of the Andes, thus to the range farther from the coast. With the exception of Rucu-Pichincha, the peaks that crown the western Cordilleras all appear to be volcanoes that have been extinguished for several centuries. The mountain depicted in our drawing, which is 2° 2' away from the nearest coasts—those of La Esmeralda and San Mateo Bay—nevertheless shoots out sprays of fire periodically, desolating the surrounding plains.

The shape of Cotopaxi is the most beautiful and the most regular of all the colossal peaks in the upper Andes. It is a perfect cone that, cloaked in an enormous layer of snow, shines dazzlingly at sunset and stands out delightfully against the azure sky. This snow cover hides the smallest irregularities in the ground from the observer: neither boulder tips nor rocky masses pierce through these eternal ice fields to disturb the regularity of the conical shape. The summit of Cotopaxi resembles the sugarloaf (*pan de azúcar*) at the top of the Teide, but its cone is six times higher than that of the great volcano of the island of Tenerife.

45

It is only near the edge of the crater that one notices the rock banks that are never covered in snow and that from a distance seem like deep black lines. What causes this phenomenon is probably the steep slope of this part of the cone and the crevices through which currents of hot air escape. This crater, similar to that of the Peak of Tenerife, is surrounded by a short enclosing wall that, when examined with good binoculars, looks like a parapet; one can make it out particularly well from the southern slope, when one is positioned either on the *Mountain of Lions* (Puma Urcu) or on the banks of the tiny lake of Yuracoche. To acquaint my readers with the volcano's peculiar structure, I have added to the bottom of the Plate the view of the crater's southern edge, exactly as I sketched it near the perpetual snow line (at an absolute elevation of four thousand four hundred eleven meters) at Suniguaicu, on the porphyric mountain ridge that joins Cotopaxi to the Nevado de Quilindaña.

The conical part of the Peak of Tenerife is very accessible; it rises in the midst of a pumice-covered plain, where a few clusters of Spartium supranubium grow. When scaling the volcano of Cotopaxi, however, it is extremely difficult to reach the lower edge of the perpetual snow line. We experienced this difficulty ourselves during an excursion we undertook in May of 1802. The cone is surrounded by deep crevices that during eruptions carry scoria, pumice, water, and blocks of ice toward the Río Napo and the Río de los Alaques. When one has examined the summit of Cotopaxi firsthand, one can say with some degree of certainty that it would be impossible to reach the edge of the crater.

Given the regularity of the cone's shape, it is all the more striking to find a small rock mass on the southwest side, half-hidden under the snow and spiked with points, which the natives call the *Head of the Inca*. The origin of this bizarre moniker is uncertain. According to a folk belief that exists in the country, this isolated boulder was once part of Cotopaxi's summit. The Indians claim that when the volcano first erupted, it hurled forth a rocky mass that had covered the enormous cavity containing the underground fire like the calotte of a dome. Some insist that this extraordinary catastrophe took place a short time after the Inca Tupac Yupanqui had invaded the kingdom of Quito and that the rock piece that one can make out on the tenth Plate to the left of the volcano is called the Head of the Inca because its fall was the sinister omen of the conqueror's death. Others, even more gullible, insist that this mass of *pechstein*-based porphyry was dislodged in an explosion that happened at the very moment that the Inca Atahualpa was strangled by the Spanish at Cajamarca. It seems, in fact, quite certain that there was an eruption at Cotopaxi when ʼPedro Alvarado's army corps left Puerto Viejo for the Quito plateau, although ʼPiedro de Cieza [de León][1] and Garcilaso de la

46

1. *Crónica del Perú*, 1554, Ch. XLI, folio 109.

Vega[1] refer only vaguely to a mountain that spewed out ashes, whose sudden fall startled the Spanish. But to believe that the boulder called *Cabeza del Inca* took its current position at that time, one must assume that Cotopaxi had not undergone any prior eruptions. This assumption is all the more flawed in that the walls of the Inca's palace at Callo, built by ˇHuayna Capac, contain rocks of volcanic origin hurled from the mouth of Cotopaxi. We shall discuss elsewhere the important question of whether this volcano had likely already reached its current height at the moment when the underground fire broke through its top, or whether several geological facts do not jointly prove that its cone, like the Vesuvius's *Somma*, is composed of a large number of lava layers stacked on top of one another.

I sketched Cotopaxi and the *Head of the Inca*, to the west of the volcano, from the tenanted farm of *La Ciénaga*, on the terrace of a lovely country house belonging to our friend the ˇyoung marquis of Maenza, who has just inherited his grandeeship and the title of Puñonrostro. In these views of the Andes summits, to distinguish the mountains that are still active volcanoes from those that no longer erupt, I took the liberty of including a thin plume of smoke above the crater of Cotopaxi, although I did not see any smoke emitted when I made this sketch. Built by a close acquaintance of Mr. La Condamine, the house at La Ciénaga is located in the vast plain that stretches between the two branches of the Cordilleras, from the hills of Chisinche and Tiopullo all the way to Ambato. There one can observe all at once, and in frightening proximity, the colossal volcano of Cotopaxi, the slender peaks of Iliniza, and the Nevado de Quilindaña. It is one of the most majestic and impressive sites that I have seen in either hemisphere.[2]

47

1. *Comentarios reales*, Book II, ch. II, Vol. II, p. 59.

2. *Géographie des Plantes*, p. 147; *Nivellement barométrique*, p. 29; *Tableau de la Nature*, Vol. II, p. 24; *Essai politique sur [le royaume de] la Nouvelle Espagne*, pp. LXXVI–LXXX.

PLATE XI

'Mexica Relief Found in Oaxaca

This relief, one of the most curious relics of Mexica sculpture, was found a few years ago near the city of Oaxaca. The drawing was passed on to me by a distinguished naturalist, 'Mr. Cervantes, a professor of botany in Mexico City, to whom we owe the knowledge of the new genera Cheirosteomon and Guardiola and of many other plants that will be published in the Flora of New Spain [*Flora Mexicana*] by 'Mr. Sessé and Mr. Monziño. The individuals who sent this drawing to Mr. Cervantes assured him that it had been copied with the greatest care and that the relief, sculpted in an extremely hard blackish rock, was over a meter high.

Those who have closely studied Toltec and Aztec monuments must be struck by both the similarity and the contrasts between the Oaxaca relief and the images that we find repeated in hieroglyphic manuscripts, in idols, and on the surfaces of several teocalli. Instead of stocky human figures barely five heads high and reminiscent of the most ancient Etruscan style, one can make out, in the relief shown on the eleventh Plate, a group of slender figures whose realistic design heralds an emergence from artistic immaturity. The fear, of course, is that the Spanish painter who copied this structure in Oaxaca corrected the contours here and there, perhaps unintentionally, especially in the design of the hands and toes; but is one allowed to assume that he changed the proportion of entire figures? Does this assumption not lose all probability if one examines the painstaking care with which the shape of the heads, the eyes, and, in particular, the decorative details on the helmets were rendered? These details, among which one can identify feathers, ribbons, and flowers, as well as extraordinarily large noses, can also be found in the Mexica paintings preserved in Rome, Velletri, and Berlin. It is only by comparing everything that was produced in the same period and by peoples of a common origin that one can develop an exact idea of the style that characterizes the different monuments—that is, if one can use the term "style" to describe the perceptible connections among a multitude of fantastic and bizarre shapes.

One might still ask whether the Oaxaca relief does not date from a period subsequent to the first landing of the Spanish, when the Indian sculptors already had knowledge of a few European artworks. To discuss this question, one must recall that three or four years before Cortés made himself master of the land of Anahuac, and before the missionaries forbade the natives to sculpt anything other than figures of saints, ⸱Hernández de Córdoba, Antonio Alaminos, and Grijalva had visited the Mexican coasts, from Cozumel Island to Cabo Catoche on the Yucatán Peninsula all the way to the mouth of the Pánuco River. Everywhere these conquistadores went, they interacted with the inhabitants, whom they found to be well dressed, gathered in densely populated cities, and with an infinitely more advanced civilization than all the other peoples of the new continent. It is likely that these military expeditions left crosses, rosaries, and some images revered by Christians in the hands of the inhabitants. It is also possible that such images then passed from hand to hand, from the coasts all the way into the interior, to the lands in the mountains of Oaxaca; but can one assume that the sight of a few correctly drawn images led them to abandon forms consecrated by centuries-old customs? A Mexica sculptor would probably have copied faithfully the image of an apostle; but in a country where, as in Hindustan and China, the natives cling to the customs, habits, and arts of their ancestors with the greatest tenacity, would they have dared to represent an Aztec hero or deity in strange new forms? Moreover, the historical paintings that Mexica painters created subsequent to the arrival of the Spanish (several of which can be found in the remnants of the ⸱Boturini collection in Mexico City) demonstrate, of course, that the influence of the European arts on the tastes of the peoples of the Americas and on the accuracy of their drawings took effect only very slowly.

49 It seemed imperative to voice the doubts that might be raised about the origin of the Oaxaca relief. I had it engraved in Rome, based on the drawing that was passed on to me; but I am quite far from being able to pronounce upon such an extraordinary monument, one that I did not have the opportunity to examine myself. The architecture of the palace of Mitla and the elegance of the *Greek frets* and the labyrinths adorning its walls prove that the civilization of the Zapoteca peoples was superior to that of the inhabitants of the Valley of Mexico. With this consideration in mind, we should be less surprised that the relief that captures our attention was found in Oaxaca, the former *Huaxyacac*, which was the capital of the land of the Zapoteca. If I were to express my personal opinion, I would say that it seems easier to attribute this monument to the Americans who had not yet had contact with whites than to assume that some Spanish sculptor who had followed Cortés's army had, for his own enjoyment, made this relief in the Mexica style in honor of

Plate XI

the vanquished people. The natives of the northwest coast of the Americas have never been counted among the highly civilized peoples, yet they have succeeded in making drawings whose accurate proportions elicited the admiration of English travelers.[1]

Be that as it may, it seems certain that the Oaxaca relief represents a warrior leaving a battle scene arrayed in the spoils of his enemies. Two slaves are placed at the victor's feet. Most striking in this composition are the enormous noses repeated in each of the six heads viewed in profile. These noses are a characteristic feature of Mexica sculpture. In the hieroglyphic paintings preserved in Vienna, Rome, Velletri, and in the viceroy's palace in Mexico City, all the deities, the heroes, and even the priests are represented with long, aquiline noses, often pierced near the tip and adorned with the amphisbaena, the mysterious two-headed serpent. It is quite possible that this extraordinary physiognomy points to some race of humans very different from those who inhabit these regions today and who are characterized by wide, flat noses of medium length. It may also be that the Aztecs, like the prince of philosophers,[2] believed that there is something majestic and royal (βασιλικόν [basilikon, royal]) about a large nose, and that in their reliefs and paintings they would have considered it to be the symbol of power and moral greatness.

The pointed shape of the heads is no less striking in Mexica drawings than is the size of the noses. As I have already observed elsewhere, an osteological examination of the skulls of the natives of the Americas reveals that there is no race in the world whose frontal bone is more depressed in the back or who has a smaller forehead.[3] This extraordinary flatness is found among peoples of the copper-colored race, who have never had the custom of producing artificial deformities, as proven by the skulls of Mexican, Peruvian, and Atures Indians that Mr. Bonpland and I brought back with us, many of which we deposited at the Museum of Natural History in Paris. The Blacks give preference to the thickest, most protruding lips, while the Kalmyk prefer turned-up noses. An illustrious scholar, ˅M. Cuvier,[4] observes that in their statues of heroes, Greek artists artificially raised the *facial line* from eighty-five to one hundred degrees. I am inclined to believe that the barbaric custom of squeezing children's heads between two boards, introduced among a few savage hordes of the Americas, originated from the idea that beauty consists in this

50

1. ˅Dixon, *Voyage [autour du monde]*, p. 272.
2. Plato, *De republica*, Book V.
3. ˅Blumenbach, *Decas quinta craniorum*, 1808, p. 14, tableau 46.
4. *Leçons d'anatomie comparée*, Vol. II, p. 6.

extraordinary flatness of the frontal bone, by which nature has characterized the American race. It is probably based on this very principle of beauty that even the Aztecs, who never disfigured the heads of children, depicted their heroes and their main deities with a head much flatter than that of any of the 'Caribs I encountered in the Lower Orinoco.

The warrior represented in the Oaxaca relief displays an extraordinary mixture of clothing styles. The decorative details on both his headgear, which is in the shape of a helmet, and the standard (*signum*) that he is holding in his left hand, on which a bird is visible (as on the standard of Ocotelolco), are found in all Aztec paintings. His doublet, with its long, narrow sleeves, recalls the piece of clothing that the Mexica called *ichcahuepilli* [quilted cotton vest]; but the netting covering his shoulders is a detail that is no longer found among the Indians. Under his belt appears the striped hide of a jaguar with an undocked tail. The Spanish historians report that to appear more terrifying in battle, the Mexica warriors wore enormous wooden helmets in the shape of tigers' heads, with mouths armed with the teeth of this animal. Two skulls, probably those of vanquished enemies, are attached to the victor's belt. His feet are sheathed in a kind of boot that recalls the σκελέαί [*skeleai*, leggings] or caligae [boots] of the Greeks and the Romans. 51

The slaves depicted sitting cross-legged at the victor's feet are remarkable for both their bearing and their nudity. The one on the left resembles the image of the saints that are frequently found in Hindu paintings and that the navigator 'Roblet discovered on the northwestern coast of the Americas among the hieroglyphic paintings of the natives of the 'Cox Canal.[1] In this relief, one might easily recognize the Phrygian cap and the apron (περίζωμα [*perisoma*]) of Egyptian statues, if one were inclined to follow the lead of one scholar[2] who, carried away by his fervent imagination, believed himself to have found Carthaginian inscriptions and Phoenician monuments on the new continent.[3]

1. Marchand, *Voyage [autour du monde]*, Vol. I, p. 312.

2. 'Court de Gébelin.

3. See *Archaeologia, or Miscellaneous Tracts relating to Antiquity;* published by the Society of Antiquarians of London, Vol. VIII, p. 290.

PLATE XII

Genealogy of the Princes of Azcapotzalco

On this Plate we have included two fragments of hieroglyphic pictures, both of which were painted after the arrival of the Spanish on the coasts of Anahuac. ᵛThe originals from which these drawings were made are part of the Aztec manuscripts that I brought back from New Spain and deposited in the royal library of Berlin. This engraving, which was printed using several exchangeable spare copper plates, perfectly renders not only the drawing itself but also the color of Mexica paper. It recalls the famous mummy wrapping that was preserved for some time at the home of a private individual in Strasbourg and that has recently been added to the large, valuable collections of the Egyptian Institute [in Cairo].

The paper used in the hieroglyphic paintings of the Aztec peoples bears a great deal of similarity to the ancient Egyptian paper made with reed fibers (Cyberus papyrus). The plant used in Mexico for making paper is what we in our gardens generally refer to as aloe. It is the century plant (Agave americana), which the Aztecs called *metl* or *maguey*. The techniques employed in the making of this paper were more or less the same as those employed in the South Sea islands to make paper from the bark of the paper mulberry (*Broussonetia papyrifera*). I have seen pieces of it three meters long and two meters wide. Today, they grow agave not to make paper but, rather, to make from its juice, at the point when it develops a mast and flowers, the intoxicating drink known as *octli* or *pulque*, for the century plant (*metl*) performs the roles of the Asian hemp, the Egyptian paper reed, and the European grapevine.

The painting whose duplicate appears at the bottom of the twelfth Plate is five decimeters long and three decimeters wide. It seems that this fragment of hieroglyphic script, which I purchased in Mexico City at the sale of Mr. [León y] Gama's collections, was once part of the museum of the Chevalier Boturini Benaducci. This Milanese traveler had crossed the seas with no other objective than to study on site the history of the indigenous peoples of the Americas. While traversing the country to examine monuments and to conduct research on its antiquities, he had the misfortune of arousing the suspicion of the Spanish government. After stripping him of the fruits of his labor, they sent him to Madrid as prisoner of state in 1736. The king of Spain declared him innocent, but this declaration did not result in his property being returned to him. Although Boturini included a catalog of these collections as an addendum to his *Essay on the Ancient History of New Spain* [*Idea de una nueva historia general de la América Septentrional*], printed in Madrid, they remained buried in the archives of the viceroyalty in Mexico City. These precious remains of Aztec culture were so carelessly maintained, however, that barely one-eighth of the hieroglyphic manuscripts seized from the Italian traveler still exist today.

Those who, before Boturini, were in possession of the genealogical tableau that I am including here have added some explanatory notes, sometimes in Mexica [Nahuatl], at other times in Spanish. These notes reveal that the family whose genealogy is represented in the drawing is that of the lords (*tlatoani*) of Azcapotzalco. These princes' small territory, to which the Tepanecs gave the rather pompous name "kingdom," was located in the Valley of Mexico near the western bank of Lake Texcoco to the north of the Escapuzalco River. Torquemada said that these princes, fiercely proud of their ancient nobility, traced their origin all the way back to the first century of our era. They were not of the Mexica (or Aztec) race; they considered themselves to be the descendants of King Alcolhua, who had ruled Anahuac before the arrival of the Aztecs. The latter had made the princes of Azcapotzalco tributaries in the eleventh calli of the Mexica era, which corresponds to the year 1425 of the Christian era. 53

The genealogical tableau appears to contain twenty-four generations, indicated by a corresponding number of heads placed one below the other. It should not be surprising that only one male child is ever visible, since among even the poorest tribute-paying Indians, all inheritance is based on primogeniture.[1] The genealogy begins with a prince named Tixlpitzin, who should not be confused with Tecpaltzin, the leader of the Aztecs during their first migration from Aztlan, nor with Topiltzin, the last Toltec king. It is perhaps surprising to find the name Tixlpitzin, rather than that of Acolhuatzin, the first king of Azcapotzalco and a member of the *Citin* family, who, according to the natives' legend, reigned over a distant land in northern Mexico. Next to the fourteenth head is written the name Vitznahuatl. Were this prince identical to a king of Huexotla, whom Mexican historians also call Vitznahuatl and who was alive in 1430, then the genealogy of the family of Azcapotzalco would go as far back as the year 1010 of the Common Era, if we allow thirty years for each generation. If this is true, then how can we explain the ten subsequent

1. Gómara, *Historia de la Conquista de Mexico*, 1553, folio CXXI.

Plate XII

generations, given that the drawing appears to have been made toward the end of the sixteenth century? I shall not decide here why the year 1565 is shown between the names of the two princes Anahuacatzin and Quauhtemoztin. We know that the latter name is that of the unfortunate Aztec king whom ˅Gómara erroneously calls Quahutimoc, and who, on Cortés's orders, was hung by his feet in 1521, as is shown in an invaluable hieroglyphic account preserved in the San Felipe Neri convent in Mexico City.[1] But why does this king, a nephew of Montezuma, appear among the family of the lords or *tlatoanis* of Azcapotzalco? 54

What is certain is that when the last of these princes commissioned the genealogical tableau of his ancestors, his father and grandfather were still alive. This circumstance is clearly indicated by the small *tongues* placed at some distance from the mouth. A deceased man, say the natives, is reduced to eternal silence; in their view, to live is to speak, and as we will shortly see, to speak at length is a sign of power and nobility. These tongue images can also be found in the Mexica painting of the great flood, which Gemelli published from ˅Sigüenza's manuscript. Here we see men, born mute, who spread far and wide to repopulate the earth, and a bird who distributes thirty-three different tongues to them. Similarly, the Mexica depict the latter as a cone with several tongues floating above it because of the underground noise that one sometimes hears in the vicinity of a volcano; a volcano is called *a speaking mountain*.

It is certainly worth noting that the Mexica painter only gave the diadem (*copilli*), a sign of sovereignty, to the three persons who were his contemporaries. We find this same head covering, but without the knot extending toward the back, in the images of the kings of the Aztec dynasty published by the Abbot Clavijero. The last offspring of the lords of Azcapotzalco is shown seated on an Indian chair, with his feet uncovered. The deceased kings, on the other hand, are depicted not only without tongues but also with their feet wrapped in the royal mantle (*xiuhtilmatli*), which makes these images closely resemble the Egyptian mummies. There is almost no reason to recall here the general observation that in all Mexica paintings, the objects linked to a head by means of a thread indicate, to those who speak the natives' language, the names of the persons whom the artist intended to designate. The natives pronounce these names as soon as they see the hieroglyph. Chimalpopoca means steaming shield; Acamapitzin, a hand holding reeds. To indicate the names of these two kings, predecessors of Montezuma, the Mexica painted a shield and a closed hand, linked by a thread to two heads adorned with the royal headband. In paintings made after the conquest, I noted that the valorous 55

1. See my *Essai politique sur [le royaume de] la Nouvelle-Espagne*, p. 185.

Pedro Alvarado was depicted with two keys placed behind the nape of his neck, probably an allusion to the keys of St. Peter, whose image people saw everywhere in the Christians' churches. I do not know what is meant by the footprints that appear behind the heads in the genealogical tableau. In other Aztec paintings this hieroglyph refers to paths, migrations, and sometimes the direction of a movement.

A Trial Document in Hieroglyphic Script

A large number of the paintings that the first conquerors found were intended to serve as documentary evidence in court cases. The fragment attached to the genealogy of the lords of Azcapotzalco is an example of this category. It is a document from a trial over the possession of an Indian estate.

The lawyer's profession was unknown in Mexico under the dynasty of the Aztec kings. The opposing parties would appear in person to plead their case either before the local judge, called *Teuctli*, or before the High Courts of Justice referred to as either *Tlacatecatl* or *Cihuacoatl*. Given that the sentence was not pronounced immediately after both parties had been heard, it was in their best interest to leave a hieroglyphic painting in the possession of the judges to remind the latter of the main object under dispute. Whenever the king presided over the judges' assembly, which took place every twenty or, in some cases, every eighty days, these trial documents were shown to the monarch. In criminal cases the painting represented the accused not only at the moment that the crime was committed but also in the various situations in his life that had occurred prior to this action. When he pronounced a death sentence, with the tip of his spear the king would trace a line that pierced through the head of the accused person depicted in the painting.

The use of these paintings as trial documents was maintained in the Spanish courts for quite some time after the conquest. Since the natives were able to speak to the judge only through an interpreter, they saw the use of hieroglyphs as doubly necessary. The latter were presented to the different courts of justice in New Spain—the *Real Audiencia*, the *Sala del Crimen*, and the *Juzgado de Indios*—until the beginning of the seventeenth century. When Emperor Charles V, having resolved to nurture the development of the sciences and the arts in these distant regions, founded the ʼUniversity of Mexico in 1553, three chairs were established: one for the teaching of the Aztec language, another for that of the Otomi language, and a third for the explanation of hieroglyphic paintings. For a long time, it was considered indispensable to have lawyers, prosecutors, and judges who were capable of reading trial documents, genealogical paintings, the former code of laws, and the list

56

of taxes (*tributos*) that each territory was expected to pay to its suzerain. To this day there are still two professors of Indian languages in Mexico City, but the chair dedicated to the study of Aztec antiquity has been eliminated. The use of paintings has been completely lost, not because the Spanish language has made great inroads among the natives but, rather, because the latter know how much more useful it is, within the current organization of the courts, to appeal to lawyers to defend their cases before the judges.

The painting represented on the twelfth Plate appears to show a trial pitting natives against Spaniards. The object in litigation is a tenanted farm; one can see the blueprint represented in the painting. One can recognize the main road marked by footprints; houses drawn in profile; an Indian whose name is represented by a bow; and some Spanish judges sitting in chairs, with the legal documents before their eyes. The Spaniard placed immediately below the Indian is probably named *Aquaverde*, since the hieroglyph for water, painted in green, appears behind his head. The *tongues* are distributed quite unequally in this painting. Everything seems indicative of a conquered land: the native barely dares to defend his own case, while foreigners with long beards speak at length and loudly, like the descendants of a conquering people.

PLATE XIII

Aztec Hieroglyphic Manuscript Preserved in the Vatican Library

Mexica paintings (a very small number of which have been passed down to us) are interesting for two reasons: for the light they shed on the mythology and the history of the first inhabitants of the Americas as well as for the connections that people claim to have discovered with the hieroglyphic script of certain peoples of the old continent. To collect in this work all that might enlighten us as to the contact that appears to have occurred in the earliest times between groups of peoples separated by steppes, mountains, or seas, we shall record here the results of our research on the hieroglyphic paintings of the Americas.

In Ethiopia one finds characters that bear a stunning resemblance to those of ancient Sanskrit, especially to the inscriptions in the Canarah caves, the construction of which dates back prior to the known periods of Indian history.[1] The arts seem to have flourished in Meroë and in Axum, one of the most ancient cities in Ethiopia, long before Egypt had emerged from barbarism. 'Sir William Jones,[2] a famous writer with a vast knowledge of Indian history, was convinced that the Ethiopians of Meroë, the first Egyptians, and the Hindus all formed a single people. On the other hand, it is almost certain that the Abyssinians, who should not be confused with the *autochthonous* Ethiopians, were an Arab tribe, and Mr. Langlès suggests that the same 'Himyarite characters that can be found in eastern Africa still adorned the gates of the city of Samarkand in the fourteenth century of the Common Era. Relationships undoubtedly existed between Habesh, the ancient Ethiopia [now part of Eritrea], and the plateau of central Asia.

A prolonged struggle between two religious sects, the Brahmans and the Buddhists, resulted in the 'Shramanas' emigration to Tibet, Mongolia,

---

1. Notes by Mr. Langlès to Norden, *Voyage [d'Égypte et de Nubie]*, Vol. III, pp. 299–349.

2. ["The Eighth Anniversary Discourse, Delivered 24 February 1791,"] *Asiatick Recherches*, Vol. III, p. 5.

China, and Japan. If, as etymological studies[1] seem to show, tribes of the Tartar race journeyed to the northwest coast of the Americas and from there southward and eastward toward the banks of the Gila and the Missouri, it should not be surprising to find among the semibarbarous peoples of the new continent idols and monuments of architecture, hieroglyphic script, a precise knowledge of the length of the year, and legends relating to the original state of the world—all of which recall the knowledge, arts, and religious beliefs of the Asiatic peoples.

As it is with the study of the history of humankind, so it is with the study of the immense number of languages spread across the surface of the globe. To attribute a common origin to such a large number of races and different languages would be tantamount to getting lost within a maze of conjectures. Neither the roots of Sanskrit found in the Persian language nor the great number of Persian, and even Pahlavi, roots that can be found in languages of Germanic origin[2] give us the right to regard Sanskrit, Pahlavi (the ancient language of the Medes), Persian, and German as deriving from a *single* common source. It would be absurd to assume the existence of Egyptian colonies everywhere that one finds pyramidal monuments and symbolic paintings; but how can one not be struck by the points of resemblance between the vast tableau of customs, arts, languages, and traditions found today among the most far removed peoples? How can one not draw attention to structural parallels among languages, in style among monuments, and in fiction among cosmogonies, wherever such parallels reveal themselves, even when one is unable to identify the unknown causes of these resemblances, and when no historical phenomenon dates back to the period of contact among the inhabitants of these different climes?

Looking closely at the graphic means that peoples have used to express their ideas, we find true hieroglyphs, either curiologic or figurative, like those whose use spread from Ethiopia into Egypt; symbolic figures composed of several keys, intended to speak to the eyes rather than to the ear, and expressing entire words, as Chinese characters do; syllabaries, like those of the Manchu Tartars, in which the vowels are merged with the consonants, but which can be broken down into simple letters; finally, true alphabets, which exhibit the highest degree of perfection in the analysis of sounds, and some of which, such as Korean—as Mr. Langlès has astutely observed[3]—still seem to evidence the transition from hieroglyphs to alphabetical script.

58

---

1. Vater, *Über Amerikas Bevölkerung*, pp. 155–69.
2. *Adelung, *Mithridates*, Vol. 1, p. 277. Schlegel, *Über Sprache und Weisheit der Indier*, p. 7.
3. Norden, *Voyage [d'Égypte et de Nubie]* (ed. Langlès), Vol. III, p. 296.

Within its immense expanse, the new continent reveals peoples who have attained a certain degree of civilization. One can identify forms of government and institutions there that could have resulted only from a prolonged struggle between the prince and the peoples, between the priesthood and the magistracy. One can find languages there, some of which—such as Greenlandic, Cora, Tamanac, Totonac, and Quechua[1]—exhibit a wealth of grammatical forms that cannot be found anywhere in the old continent, except in the Congo and among the Basques, who are the remnants of the ancient Cantabria. But even considering these traces of culture and this perfecting of languages, it is remarkable that not a single indigenous people of the Americas rose to the level of sound analysis, which leads to the most admirable, one might say the most marvelous, of all inventions: an alphabet.

We see that the use of hieroglyphic paintings was common to the Toltecs, the Tlaxcalteca, the Aztecs, and many other tribes that from the seventh century of our era appeared successively on the Anahuac plateau; nowhere do we find alphabetical characters. One might suppose that the development of symbolic signs, in addition to the ease with which subjects were painted, had stymied the introduction of letters. One might cite in support of this opinion the example of the Chinese, who for thousands of years have contented themselves with eighty thousand figures composed of two hundred fourteen keys, or radical hieroglyphs. But do we not see with the Egyptians the simultaneous use of an alphabet and of hieroglyphic script, as is incontrovertibly proved by the precious papyrus scrolls found in the wrappings of several mummies and depicted in Mr. Denon's picturesque Atlas?[2]

ˈKalm reports in his *Voyage to America* that in 1746, Mr. de Verandier had discovered, on the steppes of Canada, nine hundred leagues west of Montréal, a stone tablet embedded in a sculpted pillar and containing lines that were taken for a Tartar inscription. Several Jesuits in Québec City assured the Swedish traveler that they had once been in possession of this tablet, which the ˈChevalier de Beauharnais, then governor of Canada, had had sent to Mr. de Maurepas in France.[3] It is deeply regrettable that we have no idea of what has since happened to this monument, which is of such great interest to the history of humankind. But were there people in Québec City capable of assessing an alphabet's character? And if this alleged inscription had actually been acknowledged in France as a Tartar inscription, why would an enlightened and arts-loving minister not have made this public knowledge?

59

60

---

1. [Vater, "Proben Amerikanischer Sprachen,"] *Archiv für Ethnographie*, Vol. I, p. 345. Vater, [*Über Amerikas Bevölkerung,*] p. 206.

2. Denon, *Voyage en Égypte*, plates 136 and 137.

3. Kalm, *Reise*, Vol. III, p. 416.

Anglo-American antiquarians have brought to light an inscription, assumed to be Phoenician, that is engraved on the rocks at Dighton on Narragansett Bay near the banks of the Taunton River, twelve leagues south of Boston. From the end of the seventeenth century to the present day, ˅Danforth, Mather, Greenwood, and Sewall have, one after the other, produced drawings of it, which one can barely identify as copies of the same original. The natives who inhabited these lands at the time of the first European settlements preserved an ancient legend, according to which a number of foreigners, navigating in wooden houses, had sailed up Taunton River, formerly called Assonet. After conquering the red men, these foreigners had engraved lines into the rock, which is today covered by the river waters. Like the scholar ˅Dr. Stiles, Court de Gébelin does not hesitate to regard these lines as a Carthaginian inscription. He states, with characteristic enthusiasm that is so detrimental to discussions of this kind, that "this inscription has just emerged from the new world, with the express purpose of confirming his own ideas on the origin of peoples, and that one *evidently* sees a Phoenician monument, a tableau whose foreground represents an alliance between the American peoples and the foreign people arriving by the *northern winds* from a rich and industrious country."

I have carefully examined the four drawings of the famous Taunton River stone, which ˅Mr. Lort[1] published in London in the *Journal of the Society of Antiquaries*. Far from identifying a symmetrical arrangement of simple letters or syllabic characters, I can see only a barely sketched-out drawing similar to those found on the rocks of Norway[2] and in almost all of the countries inhabited by Scandinavian peoples. Through the shapes of the heads, it is possible to make out five human figures surrounding an animal with horns, whose fore section is much higher than its hind parts.

In the journey that Mr. Bonpland and I undertook to ascertain the linkages between the Orinoco and the Amazon River, we also learned of an inscription claimed to have been found in the range of granite mountains that, at seven degrees latitude, stretches from the Indian village of Uruana, or Urbana, to the western banks of the Caura. ˅Ramón Bueno, a missionary and Franciscan monk who happened to have taken shelter in a cave formed by the separation of two rock banks, saw in the middle of this cave a large block of granite, on which he claims to have identified characters assembled into several groups and arranged on a single line. The difficult circumstances in which we found ourselves on the return trip from the Río Negro to Santo Tomé in Guiana did

61

1. [Michael] Lort, "Account of an Ancient Inscription [in North America]," *Archaeologia*, Vol. III, p. 290.

2. ˅Suhm, *Samliger til den Danske Historie*, Vol. II, p. 215.

not, unfortunately, permit us to verify this observation ourselves. The missionary gave me a duplicate of one part of these characters, the engraving of which I am including here:

Z 9/ᑭﻠᏋﾟ:⫯

One might point out the resemblance between these characters and the Phoenician alphabet, but I very much doubt that the good monk, who appeared to have had little interest in this alleged inscription, copied it very carefully. It is quite remarkable that none of the seven characters is repeated here more than once; I had them engraved only in order to direct the attention of those scholars who might one day visit the forests of Guiana to such a noteworthy object.

It is, moreover, quite remarkable that this same wild and deserted land, where Father Bueno thought he saw letters engraved on granite, has a large number of boulders that, at extraordinary heights, are covered with animal figures; representations of the sun, the moon, and stars; and other signs that may well be hieroglyphic. The natives say that at the time of the great waters, their ancestors reached the top of these mountains by canoe, and that the stones were then in such a softened state that men could draw lines on them with their fingers. This legend suggests that the culture of this group is quite different from that of the preceding people; it reveals a complete ignorance of the use of the chisel and other metal tools.

62   Together these facts suggest that there is no certain proof that the Americans had knowledge of an alphabet. In studies of this kind one cannot be mindful enough to avoid confusing what is due to chance and idle whims with actual letters or syllabic characters. ˅Mr. Truter[1] reports that at the southern tip of Africa, where the Bechuana dwell, he saw children using a sharp instrument to draw characters on a rock, which bore the most perfect resemblance to the P and the M of the Roman alphabet; yet these crude peoples are quite far from having a knowledge of writing.

This lack of letters, observed in the new continent at the moment of its second discovery by ˅Christopher Columbus, leads one to assume that the tribes of the Tartar or Mongol race, who presumably came to the Americas from eastern Asia, did not have an alphabetical script either, or, less likely,

1. ˅Bertuch, ["John Barrows Auszug," *Allgemeine*] *Geographische Ephemeriden*, Vol. XII, p. 267.

that having regressed into barbarism under the influence of a climate unfavorable to the development of the mind, they had lost that marvelous art which had been known only to very few individuals. We shall not debate here the question of whether the ˚Devanāgarī alphabet has been in use on the banks of the Indus and the Ganges since early antiquity or whether, as Strabo claims,[1] following ˚Megasthenēs, the Hindus had no knowledge of writing prior to being conquered by Alexander. Farther to the east and to the north, in the region of monosyllabic languages, as well as in that of the Tartar, Samoyed, Ostyak, and Kamchadal languages, the use of letters, wherever it is found today, was introduced very late. It even appears quite likely that it was Nestorian Christianity[2] that gave the ˚Estrangelo alphabet to the Uyghur and the Manchu Tartars, an alphabet that is even more recent in the northern regions of Asia than are runic characters in the north of Europe. One need not, therefore, assume that the contact between eastern Asia and the Americas dates back to the earliest periods of antiquity in order to understand how this part of the world was unable to receive an art that for many centuries was known[3] only in Egypt, in the Phoenician and Greek colonies, and in the small land area bound by the Mediterranean, the Oxus River [Amu Darya], and the Persian Gulf.

63

A glance through the history of unlettered peoples reveals that nearly everywhere, in both hemispheres, humans have attempted to paint the subjects that have struck their imagination, to represent things by showing a part in place of the whole, to compose tableaus by assembling figures or the parts that evoke them, and thereby to perpetuate the memory of a few remarkable events. As they walk through the woods, Delaware Indians carve lines in tree bark to announce the number of enemy men and women whom they have killed; the conventional sign indicating the scalp ripped from a woman's head differs by only a simple line from the one that represents a man's hair. As Mr. Zoëga rightly observes, if one chooses to label all pictorial representations of ideas in the form of objects as hieroglyphic, there is no corner of the globe where one would not find hieroglyphs; but this very scholar, who produced an in-depth study of Mexica paintings, also observes that hieroglyphic script should not be confused with the representation of an event, nor with paintings in which subjects are in a relationship of action with one another.

1. Strabon, [*Géographie*,] Book XV, pp. 1035–44.
2. Langlès, *Dictionnaire tartare-mantchou*, p. 18. [William Jones, "V.ᵉ Discours anniversaire,"] *Recherches Asiatiques*, Vol. II, p. 62, note d.
3. Zoëga, *De origine [et usu] obeliscorum*, p. 551.

The first monks to visit the Americas, ˙Valadés and Acosta,[1] already deemed Aztec painting "a script similar to that of the Egyptians." If, afterward, ˙Kircher, Warburton, and other scholars disputed the accuracy of this claim, it is because they did not distinguish between *mixed-genre paintings*, in which true hieroglyphs, either curiologic or figurative, are added to the natural representation of an action, and *simple hieroglyphic script*, as we find it not on the *pyramidion* but on the large sides of the obelisks. The famous inscription of Thebes, cited by ˙Plutarch and by Clement of Alexandria,[2] the only one whose explanation has reached us, articulated the following sentence through the hieroglyphs of a child, an old man, a vulture, a fish, and a hippopotamus: "You who are born and who must die know that the Eternal loathes impudence." To render the same idea, a Mexica would have painted the great spirit Teotl punishing a criminal; certain characters placed above the two heads would have sufficed to indicate the ages of the child and the old man. He would have *individualized* the action, but the style of his hieroglyphic paintings would not have furnished him with the means for expressing the feelings of hate and vengeance more generally.

64

According to the ideas the ancients have passed down to us about the ˙hieroglyphic inscriptions of the Egyptians, it is highly likely that they could be read in the same way that Chinese books are read. The collections that we quite incorrectly call Mexica *manuscripts* contain a large number of paintings that might be interpreted or explained like the reliefs on ˙Trajan's column; but one sees only very few legible characters. The Aztec peoples had true simple hieroglyphs for water, land, air, wind, day, night, middle of the night, speech, and movement; they had them for the numbers and for the days and months of the solar year. When added to the painting of an event, these signs showed, rather ingeniously, whether the action occurred during the day or at night, what was the age of the persons whom one intended to represent, whether the latter had spoken, and which one of them had spoken the most. Among the Mexica we even find vestiges of the genre of hieroglyphs that are called *phonetic* and that signal relationships not with objects but, rather, with spoken language. Among semibarbarous peoples the names of individuals, as well as those of cities and mountains, generally allude to things that strike the senses, such as the shapes of plants and animals, fire, air, or land. This situation furnished the Aztec peoples with the means of *writing* the names

1. Diego Valadés, *Rhetorica Christiana*, Rome, 1579, Part II, ch. 27, p. 93. Acosta, [*Histoire naturelle et morale des Indes*,] Book VI, ch. 7.

2. Plutarch, *De Iside [et Osiride]*, ed. Parisina, 1624, Vol. II, p. 363, Clement [of] Alexandria, *Stromata*, Book V, ch. 7 (ed. Potter, Oxon, 1715), Vol. II, line 30.

of cities and of their sovereigns. The verbal translation of *Axayacatl* is *face of water*, that of *Ilhuicamina, arrow that pierces the sky*; and it happens that to represent the kings ʹMotecuhzoma Ilhuicamina and Axayacatl, the painter connected the hieroglyphs for water and sky to the image of a head and an arrow. The names of the cities of Macuilxochitl, Quauhtinchan, and Tehuilojoccan mean *five flowers, house of the eagle,* and *place of mirrors*, respectively: to indicate these three cities, they painted a flower on top of five dots, a house with an eagle's head sticking out, and an obsidian mirror. The connecting of several hieroglyphs thus indicated compounds of signs that spoke to the eyes and to the ear; often, the characters that represented cities and provinces were also taken either from local vegetation or from the local industry.

    As a result of these fine details, the Mexica paintings that have been preserved to this day bear a great resemblance not to the hieroglyphic script of the Egyptians but, rather, to the papyrus scrolls found in mummy wrappings, which must also be considered as *mixed-genre paintings* because isolated symbolic characters are added there to the representation of an action. In these papyri one can identify initiations, sacrifices, allusions to the state of the soul after death, tributes paid to victors, the beneficial effects of the Nile floods, and agricultural work. Among a large number of figures represented in action or in contact with one another, one can see true hieroglyphs, those isolated characters that belonged to writing. One finds traces of the mixed genre that merges painting and writing, not only on papyri and on mummy wrappings but even on obelisks. Both the lower part and the tip of Egyptian obelisks typically show a group of two figures in contact with one another, which should not be confused[3] with the isolated characters of symbolic script.

    In comparing Mexica paintings with the hieroglyphs that adorned the temples, the obelisks, and perhaps even the pyramids of Egypt, and in reflecting on the progressive advance that the human mind appears to have made in the invention of graphic means capable of expressing ideas, we see that the peoples of the Americas were quite far from the perfection that the Egyptians had attained. Indeed, the Aztecs still knew only very few simple hieroglyphs; although they had hieroglyphs for the elements and for relations of time and space, it is only when such characters occur in large number and lend themselves to being used *in isolation* that the *painting* of ideas becomes an easy feat and approximates *writing*. We find among the Aztecs the seeds of phonetic characters: they knew how to *write* names by connecting a few signs that recalled sounds. This artifice might have led them to the beautiful discovery of a syllabary; it might have incited them to *alphabetize* their simple hiero-

65

3. Zoëga, [*De origine et usu obeliscorum,*] p. 438.

66 glyphs. But how many centuries would have to pass before these mountain peoples, who clung to their own customs with a tenacity reminiscent of the Chinese, the Japanese, and the Hindus, rose to the level of breaking down words, analyzing sounds, and inventing an alphabet!

Despite the extreme shortcomings of Mexica hieroglyphic script, the use of paintings compensated rather well for the lack of books, manuscripts, and alphabetical characters. In Montezuma's time thousands of people were occupied with painting, either composing new works or copying paintings that already existed. The ease with which paper was made, using the leaves of the maguey, or century plant (agave), probably contributed to making the use of painting so ubiquitous. In the old continent, paper reed (*Cyperus papyrus*) grows only in humid, temperate areas. The century plant, on the other hand, grows equally well in the plains and on the highest mountains; it grows both in the hottest regions of the earth and on plateaus where the thermometer drops to the point of freezing. Some of the Mexica manuscripts (*codices Mexicai*) that have been preserved are painted on deerskins, while others are painted on canvases of cotton or maguey paper. It is highly likely that among the Americans (as with the Greeks and other peoples of the old continent) the use of tanned, prepared skins preceded that of paper; the Toltec, at least, seem to have used hieroglyphic painting in the remote period when they dwelled in northerly provinces where the climate is adverse to the cultivation of agave.

Among the peoples of Mexico, images and symbolic characters were not drawn on separate leaves. Regardless of the material used for the manuscripts, the drawings were very rarely intended to form scrolls; they were almost always folded zigzag, in a peculiar manner, more or less like the paper or the fabric in our fans. Two tablets made of a light wood were glued to the ends, one on top, the other on the bottom, so that before the painting was unfolded, the whole thing bore a perfect resemblance to our bound books. The result of this arrangement was that in opening a Mexica manuscript like we open our own books, one sees no more than half the characters, namely those painted on the same side of the skin or the maguey paper. To examine all of

67 the pages—if we can call pages the various tucks within a band often twelve to fifteen meters long—one must spread out the whole manuscript first from left to right, and then from right to left. In this respect the Mexica paintings correspond most closely to the Siamese manuscripts preserved at the imperial library in Paris, which are also folded zigzag.

The volumes that the first missionaries in New Spain called, quite incorrectly, Mexica books contained writings on a wide range of subjects. These

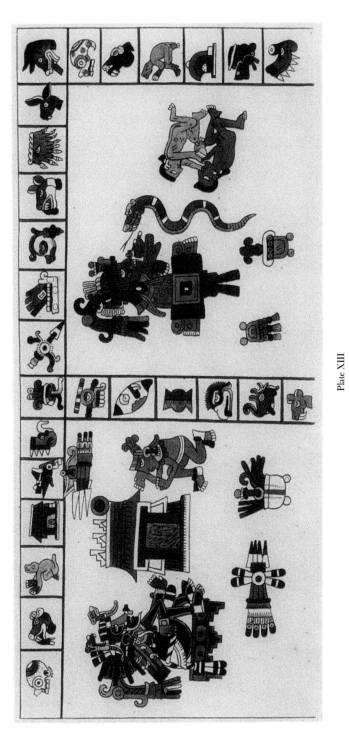

Plate XIII

included historical annals of the Mexica empire; rituals indicating the month and the day when sacrifices were to be made to one deity or another; cosmogonic and astrological representations; trial documents; documents relating to the land register or to the division of property within a community; lists of the tributes that were to be paid at various points of the year; genealogical tables according to which inheritances and the order of succession in families were determined; calendars showing the intercalations of the calendar year and the holy year; and, finally, paintings that evoked the penalties by which judges were to punish offenses. My own travels in the various parts of the Americas and of Europe have given me the advantage of being able to examine a greater number of Mexica manuscripts than could Zoëga, Clavijero, Gama, the ʼAbbé Hervás, Count Rinaldo Carli (the insightful author of the *Lettere americane*), and other scholars who, following Boturini, wrote about these monuments of the ancient civilization of the Americas. In the precious collection preserved in the viceroy's palace in Mexico City, I have seen fragments of paintings relating to each of the subjects that I have just enumerated.

One must be struck by the great resemblance between the Mexica manuscripts preserved in Velletri, Rome, Bologna, Vienna, and Mexico City; at first glance, one might think that they were copied from one another. They all exhibit highly inaccurate contours, details executed with scrupulous care, and extremely vivid colors arranged so as to produce the most arresting contrasts. The bodies of the human figures are generally stocky, like those in Etruscan reliefs; in terms of the faithfulness of their design, they are not up to even the most imperfect examples of Hindu, Tibetan, Chinese, and Japanese painting. In Mexica paintings one can make out heads of enormous size, extremely short bodies, and feet with long toes that make them look like bird claws. Heads were consistently drawn in profile, although the eye was placed as though the figure were viewed from the front. All this is indicative of the infancy of the art, but one must not forget that peoples who express their ideas in paintings, and who are forced by their social order to make frequent use of mixed hieroglyphic script, give as little importance to painting accurately as do the scholars of Europe to using beautiful handwriting in their manuscripts.

One cannot deny that the mountain peoples of Mexico belong to a race of men who, like several Tartar and Mongol hordes, enjoy imitating the shapes of objects. Everywhere in New Spain, as in Quito and in Peru, one sees Indians who know how to paint and sculpt; they manage to copy slavishly everything they see. Although they have learned, since the Europeans' arrival, how to render shapes with greater accuracy, there is no indication that they are filled with a sense for beauty, without which neither painting nor sculpture can rise above the mechanical arts. In this and many other respects, the inhabitants of the new world resemble all the peoples of central Asia.

One can understand, moreover, how the frequent use of mixed hieroglyphic painting might have contributed to ruining a people's taste, accustoming them to the appearance of the most hideous human figures and the most inaccurately proportioned shapes. To show a king who, in whatever year, conquered a neighboring people, the Egyptians, with their perfect writing system, arranged on the same line a small number of isolated hieroglyphs that expressed the entire set of ideas they wanted to articulate, and for the most part, these characters consisted of figures of inanimate objects. To resolve the same problem, the Mexica had to paint a group of two persons: an armed king knocking down a warrior who carries the arms of the conquered city. Yet in order to facilitate the use of these historical paintings, they soon began painting only what was absolutely indispensable to make the subjects identifiable. Why give arms to a human figure represented in a posture in which arms would be of no use? Furthermore, the key forms—those by which they represented a deity, a temple, or a sacrifice—must have been established early on. It would have become extremely difficult to understand paintings had each artist been free to vary at will the representation of the 69 subjects that he was obliged to depict frequently. It follows from this that the civilization of the Mexica could have developed considerably without their being tempted to abandon the inaccurate forms that had been agreed upon for centuries. A mountain people, warlike, robust but extremely ugly according to European ideals of beauty, numbed by despotism, and accustomed to the ceremonies of a bloodthirsty religion, they were already reluctant to rise to the cultivation of the fine arts. The custom of painting instead of writing, the daily appearance of so many hideous and disproportionate figures, and the obligation of preserving the same forms without ever altering them—all of these circumstances must have contributed to perpetuating bad taste among the Mexica.

It is in vain that we seek on the central Asian plateau, or farther to the north and to the east, peoples who have made use of the hieroglyphic painting found in the land of Anahuac since the end of the seventh century. The Kamchadals, the Tungus, and other tribes of Siberia that ˅Strahlenberg described paint images that recall historical events. As we observed above, there are peoples more or less devoted to this genre of painting, but there is a wide chasm between, on the one hand, a plate containing a few characters and, on the other, the Mexica manuscripts, all of which are composed in accordance with a uniform system, and which can be considered the annals of the empire. We do not know whether this system of hieroglyphic painting was invented in the new continent or whether it was due to the out-migration of some Tartar tribe that knew the exact length of the year, and whose civilization was as ancient as that of the Uyghur of the Turfan plateau. If the old continent does not present a single people who used painting as extensively as did the Mexica, this is because neither in Europe nor in Asia do we find an equally advanced civilization without the knowledge of either an alphabet or some characters that compensate for one, like the figures of the Chinese and the Koreans.

Prior to the introduction of hieroglyphic painting, the peoples of Anahuac used the knots and the many-colored threads that the Peruvians call ˈquippu, and that are also found[1] not only among the Canadians but also, in a very early period, among the Chinese. Chevalier Boturini was fortunate enough to obtain true Mexica quippu or *nepohualtzitzin*, found in the land of the Tlaxcalteca. In the great migrations of peoples, those of the Americas spread from north to south, just as the Iberians, Celts, and Pelasgians surged from east to west. Perhaps the ancient inhabitants of Peru had once passed through the Mexica plateau; indeed, ˈUlloa,[2] familiar with the style of Peruvian architecture, had been struck by the strong resemblance that the distribution of doors and recesses in certain ancient buildings in western Louisiana bore to the *tambos* built by the Inca. It seems no less remarkable that according to the legends recorded in Licán, the ancient capital of the kingdom of Quito, quippu were known to the Puruays long before the descendants of Manco Capac had subjugated them.

In Mexico, as in China, the use of writing and hieroglyphs caused the knots, *nepohualtzitzin* [the Mesoamerican abacus], to be forgotten. This change occurred around the year 648 of our era. The Toltec, a northern but very civilized people, appeared in the mountains of Anahuac, to the east of the Gulf of California, saying that they had been driven out of a country to the northwest of the Río Gila, called Huehuetlapallan. They carried with them paintings that charted the events of their migration year by year. They claimed to have left their homeland, the location of which is completely unknown to us, in the year 544, during the same period when the utter collapse of the ˈQin dynasty had occasioned large-scale movements among the peoples of eastern Asia—a remarkable coincidence indeed. What is more, the names that the Toltec bestowed upon the cities they founded were those

70

1. ˈLafitau, *Mœurs des Sauvages*, Vol. I, pp. 233, 503. *Histoire générale des Voyages*, Vol. I, Book X, ch. VIII. Martini, *Histoire de la Chine*, p. 21. Boturini, *Nueva Historia de la América septentrional*, p. 85.

2. Ulloa, *Noticias Americanas*, p. 43.

of the northern country that they had been forced to abandon. If we ever discover in northern America or northern Asia a people who knows the names of Huehetlapallan, Aztlan, Teocolhuacan, Amaquemecan, Tehuajo, and Copalla, we will then know the origin of the Toltec,[1] the Chichimec, the Acolhua, and the Aztecs, the four peoples who all spoke the same language and who arrived in Mexico, one after the other, via the same route.

Up to the 53rd parallel, the temperature of the northwest coast of the Americas is warmer than that of the eastern coasts. It is conceivable that civilization had made very early advances in this clime, and at even higher latitudes. To this day, at 57 degrees, in Cox Canal and in Norfolk Bay, which ˙Marchand called the Gulf of Tchinkintana, the natives still have a penchant for hieroglyphic paintings on wood. I have examined[2] elsewhere whether it is possible that these industrious peoples, of a generally mild and affable nature, are actually Mexica colonists who took refuge in the north following the arrival of the Spanish, or whether they are not, rather, descended from Toltec or Aztec tribes that remained in these northern regions at the moment when peoples burst forth from Aztlan. Through the felicitous combination of several circumstances, humans, in terms of their development as organized beings, rise to a certain level of culture even in the most unfavorable climates; in Iceland, near the Arctic circle, we have seen Scandinavian peoples cultivate the arts and letters since the twelfth century, and with greater success than have the inhabitants of Denmark and Prussia.

A few Toltec tribes appear to have mixed with the peoples who formerly inhabited the country bounded by the eastern banks of the Mississippi and the Atlantic Ocean. The Iroquois and the Huron made hieroglyphic paintings on wood that bear striking connections[3] with those of the Mexica: they indicated the name of the persons they wanted to represent by using the same artifice that I mentioned above in the description of a genealogical tableau. The natives of Virginia had paintings called *sagkokok*, which employed symbolic characters to represent events that had taken place over a sixty-year period; these were large wheels divided into either 60 spokes or as many equal parts. In the Indian village of Pommacomek, ˙Lederer[4] reports having seen one of these hieroglyphic cycles in which the time of the Whites' arrival on the coasts of Virginia was marked by the figure of a swan spewing up fire, in-

1. Clavijero, *Storia [antica del] Messico*, Vol. I, p. 126; Vol. IV, pp. 29 and 46.

2. See my *Essai politique [sur le royaume de la Nouvelle-Espagne,]* pp. 78, 336, 349. Marchand, [*Voyage autour du monde,*] Vol. I, pp. 259, 261, 299, 375.

3. Lapitau, [*Moeurs des Sauvages,*] Vol. II, pp. 43, 225, 416. ˙Lahontan, *Voyage dans l'Amérique septentrionale*, Vol. II, p. 193.

4. *Journal des Savans*, 1681, p. 75.

dicating the color of the Europeans, their arrival by sea, and the harm that their firearms had inflicted upon the red men.

72   In Mexico the use of both paintings and maguey paper extended well beyond the borders of Montezuma's empire, all the way to the banks of Lake Nicaragua, where in their migrations the Toltec had taken their language and their arts. In the kingdom of Guatemala the inhabitants of Teochiapan preserved traditions that dated back to the time of a great flood, after which their ancestors, under the leadership of a man called *Votan*, had come from a country toward the north. In the sixteenth century there were still descendants of the Votan (or Vodan) family (these two names are the same, since neither the Toltec nor the Aztecs had the four consonants *d*, *b*, *r*, and *s* in their language) in the village of Teopixca. Those who have studied the history of the Scandinavian peoples in the heroic age [of the Vikings] must be struck to find in Mexico a name recalling that of *Vodan* or *Odin*, who reigned over the Scythians and whose line, according to Bede's noteworthy claim,[1] "gave kings to a great number of peoples."

If it were true, as many scholars have assumed, that these same Toltec, who were driven out of the Anahuac plateau toward the middle of the eleventh century of our era by the combination of a plague and a severe drought, reappeared in South America as the founders of the Inca empire, then why would the Peruvians not have abandoned their *quippu* in order to adopt the Toltecs' hieroglyphic script? Around the same time, at the beginning of the twelfth century, a Greenlandic bishop had brought a number of Latin books not to the American continent but, rather, to Newfoundland (Vinland), perhaps the very ones that the ᵛZeni[2] brothers found there in 1380.

We do not know whether tribes of the Toltec race penetrated as far into the southern hemisphere, not via the Cordilleras of Quito or Peru but by following the plains that stretch to the east of the Andes toward the banks of the Marañón. During my stay in Lima, I learned of a fascinating event that would appear to confirm this. On the banks of the Ucayali (just north of the mouth of the Sarayacu), among the independent Panoa Indians, ᵛFather Narciso Gilbar, a Franciscan monk highly regarded for both his courage and his inquisitive mind, found sheaves of paintings that in their external shape bore a perfect resemblance to our *quarto* books. Each sheet was three decimeters 73   long and two decimeters wide, while the sheaf covering was made of several palm leaves glued together and a very thick parenchyma. The sheets were

1. ᵛBede, *Historia ecclesiastica [gentis Anglorum]*, Book I, ch. XV. ᵛFrancisco Núñez de la Vega, *Constitutiones synodales*, p. 74.

2. [Placido Zurla,] *Viaggio de' fratelli Zeni*, Venice, 1808, p. 67.

made of pieces of cotton canvas, a very thin material, joined together by century plant threads. When Father Girbal arrived among the Panoans, he found an old man seated at the base of a palm tree, surrounded by several young people to whom he was explaining the content of these books. At first, the savages did not want to allow a white man to approach the old man; using the Manoa Indians (the only ones who understood the Panoans' language) as intermediaries, they informed the missionary "that these paintings contained hidden things that no foreigner should learn." It was only with a great deal of effort that Father Girbal managed to obtain one of the sheaves, which he sent to Lima to show to ⸰Father Cisneros, a scholar and editor of a journal[1] that has been translated in Europe. Several persons of my acquaintance have seen this book from the Ucayali, every page of which was covered with paintings. They were able to make out both human and animal figures, as well as a large number of isolated characters that they believed to be hieroglyphs and that were arranged in lines of admirable order and symmetry. They were particularly captivated by the vivid colors. But since no one in Lima had previously seen a fragment of the Aztec manuscripts, it was impossible to assess the stylistic similarity between paintings found at a distance of eight hundred leagues from one another.

Father Cisneros intended to deposit this book in the convent of the Ocopa missions; but, either because the person to whom he entrusted it lost it while crossing the Cordillera or because it was stolen and furtively sent to Europe, what is certain is that it did not arrive at its original destination. Every search conducted to retrieve this curious object has been in vain, and they regretted too late not having copied these characters. The missionary Narciso Girbal, with whom I struck up a friendship during my stay in Lima, promised me that he would try everything possible to obtain another sheaf of the Panoans' paintings. He knows that several exist in their midst and that the Panoa themselves say that these books were passed down to them *by their forefathers.* The explanation that they give for these paintings appears to be based on an ancient legend perpetuated within a few families. The Manoa Indians, whom Father Girbal entrusted to seek out the meaning of these characters, had reason to suspect that they depicted journeys and ancient wars with neighboring groups.

Today, the Panoans differ very little from the rest of the savages who dwell in these humid and excessively hot forests. Naked and living on bananas and whatever they can fish, they are quite far from having knowledge of painting and from feeling the need to communicate their ideas by graphic signs.

74

1. *El Mercurio peruano.*

Like the majority of the tribes who settled on the banks of the great rivers of South America, they do not appear to be long-established in the places where we find them now. Are they the humble remnants of some civilized people who have regressed to a state of mindlessness, or do they descend from the same Toltec who brought the use of hieroglyphic painting to New Spain and whom we see retreating to the banks of Lake Nicaragua, driven out by other peoples? These are questions of great interest to human history; they are linked to other questions, whose importance has not as yet been sufficiently appreciated.

A number of granitic rocks that rise from the savannahs of Guayanas, between the Casiquiare and the Conorichite, are covered with figures of tigers, crocodiles, and other characters that might be considered symbolic. Drawings of similar designs are found five hundred leagues to the north and to the west on the banks of the Orinoco; near the Encaramada and the Caicara on the banks of the Río Cauca; near Timba, between Cali and Jelima; and, finally, right on the plateau of the Cordilleras, in the Páramo de Guanacas. The indigenous peoples of these regions are not acquainted with the use of metal tools; they all agree that these characters already existed when their ancestors arrived in these parts. Is it to a single industrious people, devoted to sculpting (as were the Toltec, the Aztecs, and the whole group of peoples who emerged from Aztlan), that these traces of an ancient civilization are due? In which region should we locate the source of this culture? Is it to the north of the Río Gila, on the plateau of Mexico, or, rather, in the southern hemisphere, on the elevated Tiahuanaco plains, which when the Inca found them were already covered with ruins of an imposing size and which can be considered the Himalaya and the Tibet of South America? The present state of our knowledge does not enable us to resolve these problems.

We have just examined the connections between Mexica paintings and old-world hieroglyphs; we have tried to shed some light upon the origin and migrations of the peoples who introduced the use of symbolic script and papermaking to New Spain. It remains for me to discuss the manuscripts (*codices mexicani*) that have made their way to Europe since the sixteenth century and have been preserved in both public and private libraries. It will be surprising to learn how rare these precious monuments have become, monuments of a people who in their march toward civilization appear to have struggled against the same obstacles that have impeded the development of the arts among all the peoples of northern and even eastern Asia.

According to my research, it appears that there are ˇonly six collections of Mexica paintings in Europe today: those of the Escorial [in Madrid], Bologna, Velletri, Rome, Vienna, and Berlin. The Jesuit scholar ˇFábrega, who is often cited in Mr. Zoëga's works, and some of whose manuscripts relating to Aztec antiquity were passed on to me by the Chevalier Borgia, nephew of the cardinal by the same name, supposes that the archives of Simancas, in Spain, also contain a few of the hieroglyphic paintings to which Robertson refers by the particularly apt term *picture writings*.

The collection preserved in the Escorial was examined by ˙Mr. Waddilove,[1] the chaplain of the British embassy in Madrid at the time of Lord Grantham's mission. It is in the shape of a *folio* book, which might lead one to suspect that it is merely a duplicate of a Mexica manuscript, for the originals that I have examined all resemble *quarto* volumes. The subjects depicted appear to prove that the Escorial collection and those in Italy and Vienna are either astrology books or true *rituals*, which indicated the religious ceremonies prescribed for one or another day of the month. At the bottom of each page is an explanation in Spanish that was added during the conquest.

The *Bologna* collection was deposited in the library of that city's Institute of Sciences. Its origin is unknown, but one reads on the first page that this painting, which is 326 centimeters (eleven *palmi romani*) long, was sold on December 26, 1665, to the ˙Marquis of Cospi by Count Valerio Zani. The characters, which are drawn on thick, badly prepared skin, seem largely concerned with the forms of the constellations and astrological concepts. There exists one outline copy of this *Codex Mexicanus* of Bologna in Cardinal Borgia's museum in Velletri.

The *Vienna* collection, which has sixty-five pages, has become famous   76 for having attracted the attention of Dr. Robertson, who, in his classic work on the history of the new continent, included a few pages from it, but without color and in simple outlines. One reads on the first page of this Mexica manuscript that "it was sent by ˙King Emmanuel of Portugal to Pope Clement VII," and that "it was afterward in the possession of the cardinals Ippolito de Medici and Capuanus." ˙Lambeck,[2] who had a few figures from the *Codex Vindobonensis* engraved, erroneously observes that since King Emmanuel had died two years before the election of Pope Clement VII, this manuscript could not have been presented as a gift to that pontiff but must instead have been presented to ˙Leo X, to whom the king of Portugal had sent a delegation in 1513. But I would ask: how could they have had Mexica paintings in Europe in 1513, when Hernández de Córdoba discovered the Yucatán coasts only in 1517, and Cortés landed in Veracruz only in 1519? Considering that the

---

1. Robertson, *History of America*, 180[3], Vol. III, p. 403.
2. Lambeck, *Commentarium de [augustissima] Bibliotheca [Caesarea de] Vindobonensi* (ed. 1776), p. 966.

inhabitants of the island of Cuba do not appear to have had any contact with the Mexica, despite the proximity of Cabo Catoche to Cabo San Antonio, is it likely that the Spanish found Mexica paintings on that island? It is true that in the note added to the Vienna collection, the latter is not called *Codex Mexicanus* but, rather, *Codex Indiae Meridionalis*; nevertheless, the perfect similarity between this manuscript and those preserved in Velletri and Rome leaves no doubt about their common origin. King Emmanuel passed away in 1521 and Pope Clement VII in 1534; it seems implausible a Mexica manuscript might have existed in Rome prior to the Spaniards' initial entry into Tenochtitlan (November 8, 1519). Regardless of when it reached Italy, it is certain that after passing from person to person, it was presented to ⸱Emperor Leopold by the Duke of Saxe-Eisenach in 1677.

We have no idea what befell the collection of Mexica paintings that still existed at the end of the seventeenth century in London and was published by ⸱Purchas. This manuscript had been sent to Emperor Charles V by the ⸱first viceroy of Mexico, Antonio de Mendoza, Marquis of Mondéjar. A French vessel took the ship that carried this precious object, and the collection fell into the hands of ⸱André Thévêt, geographer to the king of France, who himself had visited the new continent. After the death of this traveler, ⸱Hakluyt, who was the chaplain at the British embassy in Paris, purchased the manuscript for twenty *crowns*, and from Paris it continued on to London, where ⸱Sir Walter Raleigh intended to have it published. The costs that the engraving of the drawings was to have incurred delayed this publication until 1625, when Purchas, complying with the wishes of the historian ⸱Spelman, included the *Mendoza Collection* in his travel series.[1] These very figures were duplicated by ⸱Thévenot,[2] in his *Relation de divers voyages [curieux]*; but as the Abbot Clavijero astutely observed,[3] this duplicate is chock-full of mistakes; for example, events that occurred during King Ahuizotl's reign are attributed to Montezuma's reign.

Although some authors[4] have declared that the original of the famous ⸱*Mendoza Collection* was preserved at the imperial library in Paris, it seems certain that there have been no Mexica manuscripts there for a century. How would this collection, purchased by Hakluyt and transported to England, have returned to France? We know of no Mexica paintings in Paris today, other than

1. Purchas, [*His*] *Pilgrimes*, Vol. III, p. 1065.

2. Thévenot (1696), [*Relation de divers voyages curieux*,] Vol. II, Plate IV, pp. 1–85.

3. Clavijero, [*Histoire du Mexique*,] Vol. I, p. 23.

4. ⸱Warburton, *Essai sur les hiéroglyphes [des Égyptiens]*, Vol. I, p. 18. Papillon, *Histoire de la gravure en bois*, Vol. I, p. 364.

the copies included in a Spanish manuscript that comes from the ᵛTellier library, of which we will shortly have occasion to speak. This book (very interesting, by the way) is preserved in the superb manuscript collection in the imperial library. It resembles the *Codex anonymous* in the Vatican, number 3738, which is the work of the monk Pedro de los Ríos.[1] Father Kircher commissioned a copy of one part of Purchas's engravings.[2]

The *Mendoza collection* sheds light on the history, the political order, and the private life of the Mexica. It is divided into three sections that, like the *Skandhas* [books] from the *Puranas* of India, are concerned with very different subjects: the first section covers the history of the Aztec dynasty, from the founding of Tenochtitlan in the year 1325 of our era to the death of Montezuma II—or, more correctly, *Motecuhzoma Xocoyotzin*—in 1520; the second section is a list of the tributes that each province and small town paid to the Aztec sovereigns, while the third and final section depicts the domestic life and the customs of the Aztec peoples. The viceroy Mendoza had an explanation in Mexica [Nahuatl] and Spanish added to each page of the collection, so that the whole constitutes a very interesting work for the study of history. Despite their unrealistic shapes, the images show several extremely vivid aspects of their customs: one can see the education of children from birth to the point at which they become members of society as farmers or artisans, warriors or priests. The amount of food suitable for each age and the punishment that must be inflicted upon children of both sexes—for the Mexica, everything was prescribed in the most meticulous detail, not by law but by ancient customs from which it was forbidden to deviate. Enchained by despotism and by barbarous social institutions, lacking any freedom in even the most trivial aspects of their domestic lives, the people as a whole were raised in a sad uniformity of habits and superstitions. The same causes produced the same effects in ancient Egypt, India, China, Mexico, and Peru, everywhere that humanity formed masses animated by the same will, everywhere that laws, religion, and customs thwarted development and individual happiness.

78

Among the paintings in the *Mendoza Collection* one can identify the ceremonies that took place upon the birth of a child. Invoking the god Ometeuctli and the goddess Omecihuatl, who dwell in the abode of the blessed, the midwife would cast water on the newborn's forehead and chest. After having recited various prayers[3] in which water was considered a symbol of the purification of the soul, the midwife would summon children who had been

1. See my description of Plate VII above.
2. Kircher, *Oedipus [Aegyptiacus]*, Vol. III, p. 32.
3. Clavijero, [*Historia antica del Messico,*] Vol. II, p. 86.

invited to name the newborn. In some provinces they lit a fire at this time and pretended to pass the child through the flames, as though to purify it both by water and by fire. This ceremony recalls customs whose Asian origin seems to have been lost in early antiquity.

Other plates from the *Mendoza Collection* depict the often barbarous punishments that parents were expected to inflict on their children, depending on the seriousness of the offense and the age and sex of the child who committed it: a mother exposes her daughter to chili [*ají*] pepper (*Capiscum baccatum*) smoke. A father pricks his eight-year old son with century plant leaves tipped with tough spines; the painting shows in which cases the child may be pricked only upon the hands and in which cases the parents are allowed to extend this painful procedure to the whole body. A priest, *teopixqui*, punishes a novice for spending the night outside the temple walls by throwing burning brands on his head, while another priest is painted seated in a stargazing posture, to indicate the midnight hour. In this Mexica painting one can make out both the hieroglyph for midnight placed above the priest's head and a dotted line that stretches from the eye of the observer toward a star.[1] It is also interesting to see figures that represent women spinning thread with the spindle or weaving in *haute-lice* [high-warp tapestry]; a goldsmith who blows into the coals through a tube; a seventy-year old man permitted by law to become intoxicated, just as a woman is when she becomes a grandmother; a matchmaker, called a *cihuatlanque*, who carries the young virgin on her back to the betrothed's house; and finally, the nuptial blessing, for which the ceremony consisted of the priest's (*teopixqui*) tying together the train of the boy's mantle (*tilmatli*) with the train of the young girl's garment (*huepilli*). The *Mendoza Collection* also contains several images of Mexica temples (*teocalli*) in which one can clearly make out the pyramidal monument, divided into terraces, and the small chapel, νεώς [*neos*, temple], at the top. The most complex and ingenious painting in this *Codex Mexicanus*, however, is the one that depicts a *tlatoani*, or provincial governor, strangled to death because he rebelled against his sovereign. The same painting depicts the governor's offenses, the punishment of his entire family, and the vengeance exacted by his vassals upon the messengers of state bearing orders from the king of Tenochtitlan.[2]

Despite the enormous number of paintings that, at the beginning of the conquest, the bishops and the first missionaries deemed monuments of Mex-

79

---

1. Thévenot, [*Relation de divers voyages curieux,*] Vol. II, Plate IV, figures 49, 51, 55, 61.

2. Thévenot, [*ibid.,*] figures 52, 53, 58, 62.

ica idolatry and condemned to be burned, Chevalier Boturini[1] (whose misfortunes we evoked above) nevertheless succeeded in gathering some five hundred of these hieroglyphic paintings toward the middle of the last century. This collection, the most beautiful and richest of all, was scattered, as was Siguënza's, a few humble remains of which were preserved in the library of San Pedro y San Pablo in Mexico City until ˅the Jesuits' expulsion. A portion of the paintings that Boturini collected was sent to Europe on a Spanish vessel, which was captured by a British corsair. It has never been determined whether these paintings reached England or whether they were cast into the sea as badly painted, coarse-cloth canvases. A very learned traveler assured me that a *Codex Mexicanus* is on display at the Oxford library whose vivid colors make it resemble the one in Vienna. In the most recent edition of his *History of America*, however, Dr. Robertson expressly states that there are no monuments of Mexica industry and civilization in England aside from a golden bowl of Montezuma's owned by ˅Lord Archer. How could such a collection at Oxford have escaped this illustrious Scottish historian?

The majority of Boturini's manuscripts, the part that had been confiscated from him in New Spain, were split up, pillaged, and scattered by persons unaware of their importance. What remains of it today in the viceroy's palace consists of only three bundles, each of which measures seven square decimeters and five decimeters high. They have been stored in one of the humid chambers on the ground floor, from which the viceroy, the count of Revillagigedo, had the government archives moved because the paper was decomposing at an alarming rate. One is seized by a feeling of indignation when one sees the state of extreme neglect in which they store the precious remnants of a collection that cost so much work and care, and that the unfortunate Boturini, graced with that enthusiasm peculiar to all enterprising men, calls, in the preface to his *Historical Essay* [*Idea de una nueva historia*], "the only property that he possesses in the Indies, which he would not trade for all the gold and silver in the new world." I shall not attempt to describe in detail here the paintings preserved in the palace of the viceroyalty; I shall merely observe that there are some, more than six meters long and two meters wide, which depict the migrations of the Aztecs from the Río Gila to the valley of Tenochtitlan, the founding of several cities, and wars with neighboring peoples.

The library of the University of Mexico no longer has any original hieroglyphic paintings. I found only a few copies in outline, colorless and carelessly rendered. The most extensive and most beautiful collection in the capital today is that of ˅Don José Antonio Pichardo, a member of the congre-

---

1. Boturini, [*Idea de una nueva historia*,] Tableau général, pp. 1–96.

gation of San Felipe Neri. The home of this learned and hardworking man was for me what Sigüenza's home was for the traveler Gemelli. Father Pichardo sacrificed his small fortune to collect Aztec paintings and to commission copies of those he himself could not acquire. His friend Gama, the author of several astronomical papers, bequeathed to him all of the most precious hieroglyphic manuscripts in his possession.[1] It is in this manner that in the new continent, as nearly everywhere else, private individuals, usually the less privileged, are able to collect and preserve the objects that are worthy of governments' attention.

I do not know if there are persons in the kingdom of Guatemala or in the interior of Mexico who are spurred on by the same zeal as Father Alzate, Velázquez, and Gama. Today, hieroglyphic paintings are so rare in New Spain that the majority of learned persons who reside there have never seen any, and there is not a single manuscript among the remnants of Boturini's collection as beautiful as the *Codices Mexicani* of Velletri and Rome. I do not doubt, however, that many objects of great importance to the study of history are still in the hands of the Indians who live in the province of Michoacan, in the intendancies of Mexico City, Puebla, and Oaxaca, on the Yucatán Peninsula, and in the kingdom of Guatemala. These are the lands where the peoples who had emerged from Aztlan reached a certain degree of civilization. Even today, three centuries after the conquest and one hundred years after the journey of the Chevalier Boturini, a traveler who was fluent in the Aztec, Tarasco, and Mayan languages and was thus able to win the natives' trust, would succeed in collecting a considerable number of historical Mexica paintings.

The *Codex Mexicanus* in the Borgia museum in *Velletri* is the most beautiful of all the Aztec manuscripts I have examined. We will have occasion to speak of it elsewhere, in the explanation of the fifteenth Plate.

The collection preserved in the Royal Library of *Berlin* includes various Aztec paintings that I acquired during my stay in New Spain. The twelfth Plate shows two fragments from this collection; it contains lists of tributes, genealogies, the history of the Mexica migrations, and a calendar from the beginning of the conquest, in which the simple hieroglyphs for the days are joined to figures of saints painted in Aztec style.

The Vatican library in *Rome* has within its precious manuscript holdings two *Codices Mexicani*, under the numbers 3738 and 3773 in the catalog. Like the Velletri manuscript, these collections were unknown to Dr. Robertson when he cataloged the Mexica paintings preserved in the various libraries of

82

1. See my *Essai politique sur [le royaume de] la Nouvelle-Espagne*, Vol. I, p. 124.

Europe. In his description of the obelisks of Rome, 'Mercati[1] reports that toward the end of the sixteenth century, the Vatican had two collections of original paintings. Most likely one of these collections is entirely lost, unless this is the very one on display at the library of the institute of Bologna; the other one was found by the Jesuit Fábrega in 1785, after fifteen years of searching.

The *Codex Vaticanus* number 3773, which Acosta and Kircher mention,[2] is 7,87 meters, or thirty-one and a half palms, long, with a surface area of 0,19 square meters, or seven square pouces. Its forty-eight folds constitute ninety-six pages, or better, an equal number of divisions drawn on both sides of several deer hides glued together. Although each page is subdivided into two boxes, the entire manuscript includes only one hundred seventy-six of these boxes because the first eight pages contain the simple hieroglyphs for the days, arranged in parallel series positioned close to one another. The thirteenth Plate of the picturesque Atlas presents the exact copy of one of these *folds*, that is, of one page of the *Codex Vaticanus*; since all the pages resemble one another in terms of their general layout, this copy suffices to give an understanding of the book as a whole.

The edge of each fold is divided into twenty-six small boxes that contain the simple day-hieroglyphs; the hieroglyphs that form periodic series are twenty in number. Since the short cycles are thirteen days long, the series of hieroglyphs passes from one cycle to another. The entire *Codex Vaticanus* contains one hundred seventy-six of these short cycles, or two thousand two hundred ninety days. We shall not enter into any detail here regarding these subdivisions of time but propose, rather, to provide an explanation below   83 of the Mexica calendar, one of the most complicated but most ingenious in the history of astronomy. Each page shows two groups of mythological figures, in the two aforementioned subdivisions. One would lose oneself in vain conjectures in attempting to interpret these allegories, since the manuscripts of Rome, Velletri, Bologna, and Vienna are missing the explanatory notes that the viceroy Mendoza had added to the manuscript published by Purchas. It would be desirable for some government to subsidize the publication of these remnants of ancient American civilization: it is only by comparing several monuments that one might succeed in determining the meaning of these allegories, which are partly astronomical and partly mystical. If all that remained to us of Greek and Roman antiquity were a few engraved

---

1. Mercati, *Degli obelischi di Roma*, Ch. II, p. 96.
2. Zoëga, *De origine [et usu] obeliscorum*, p. 531.

stones or isolated coins, the simplest allusions would have escaped the sagacity of scholars. How much light has the study of bas-reliefs shed upon that
of coins!

Zoëga, Fábrega, and other scholars who have studied the Mexica manuscripts in Italy regard both the *Codex Vaticanus* and that of Velletri as *tonalamatl, ritual almanacs*; that is, as books that showed the people, over a mul-
tiyear period, the deities that presided over the short cycles of thirteen days
during which they governed the destinies of men; the religious ceremonies
that were to be performed; and especially the offerings that were to be taken
to the idols.

The left-hand side of the thirteenth Plate of my Atlas, which is a copy of
the ninety-sixth page of the *Codex Vaticanus*, depicts a scene of worship: the
deity is wearing a remarkably detailed helmet and is seated on a low bench,
called *icpalli*, before a temple of which only the top, the small chapel placed
on top of the pyramid, is represented. In Mexico, as in the East, the adora-
tion ceremony consisted of touching the ground with one's right hand and
raising this hand to the mouth. In drawing number I, homage is paid through
genuflection: the pose of the figure prostrated before the temple can also be
found in several Hindu paintings.

Group number II depicts the famous *serpent woman*, *Cihuacohuatl*, also
called Quilaztli or Tonacacihua, *wife [or woman] of our flesh*: she is the companion of Tonacateuctli. The Mexica regarded her as the mother of human-
84 kind, and after the god of *heavenly paradise*, Ometeuctli, she occupied the
first rank among the deities of Anahuac: one always sees her image accompanied by that of a large serpent. Other paintings show a plumed grass snake
cut into pieces by either the Great Spirit Tezcatlipoca or the Sun personified,
the god Tonatiuh. These allegories recall ancient legends of Asia. In the Aztecs' *serpent woman* one has the impression of seeing the Eve of the Semitic
people; in the grass snake cut into pieces, the famous serpent Kaliya, or Ka-
linaga, defeated by Vishnu when he took the form of Krishna. The Mexica's
Tonatiuh also seems to be identical to the Hindus' Krishna, celebrated in the
Bhagavata Purana, as well as to the Persians' Mithras. The most ancient traditions of peoples date back to a state of affairs when the earth, covered in
swamps, was inhabited by grass snakes and other animals of gigantic size.
By drying out the ground, the beneficial star liberated the earth from these
aquatic monsters.

Behind the serpent, which appears to be speaking to the goddess Cihuacohuatl, are two nude figures; they are different in color and appear poised
for battle. It is conceivable that the two vases at the bottom of the painting,
one of which is overturned, hint at the cause of this brawl. In Mexico the *ser-*

pent woman was regarded as the mother of twins; the nude figures are perhaps the children of Cihuacohuatl. They recall Cain and Abel of the Hebraic
traditions. I doubt that the difference in color between the two figures marks
a difference in race, as it does in the Egyptian paintings found in the kings'
tombs at Thebes and in the earthen decorations applied to the caskets of the
mummies at Sakkarah.[1] In studying carefully the historical hieroglyphs of the
Mexica, one realizes that the heads and hands of the figures are painted almost at random, sometimes in yellow, at other times in blue or red.

The Mexica's cosmogony—their beliefs regarding the mother of men,
fallen from her original state of bliss and innocence; the idea of a great flood,
in which a single family escaped on a raft; the story of a pyramidal structure
raised by men's pride and destroyed by the Gods' wrath; the ablution ceremonies performed at the birth of children; the idols made of molded corn
flour and distributed in lots to the people gathered within the temple walls;   85
the penitents' confessions of their sins; the religious associations resembling
our monasteries and convents; the universally held belief that white men with
long beards and saintly morals had changed the peoples' religious and political system—all these circumstances had caused the monks, who accompanied the Spanish army during the conquest, to believe that Christianity had
been preached in the new continent at some very remote time. Some Mexican scholars[2] thought they detected the apostle St. Thomas in the high priest
of Tula, that mysterious character whom the Cholulans know by the name of
Quetzalcoatl. There is no question that ˙Nestorianism, blended with the dogmas of the Buddhists and the Shamans,[3] spread through the Tartary of the
Manchu into northeast Asia. One might thus assume, with some justification,
that Christian ideas were communicated via the same route to the Mexica
peoples, especially to the inhabitants of that northerly region from which the
Toltec emerged, and which we must consider as the *officina virorum* [smithy
of men] of the new world.

Such an assumption would be more acceptable than the theory that the
ancient traditions of the Hebrews and the Christians passed into the Americas via the Scandinavian colonies established on the coasts of Greenland
and Labrador, and perhaps even on the island of Newfoundland, since the
eleventh century. These Europeans certainly visited a part of the continent,
which they called ˙*Drogeo*; they discovered lands to the southwest that were

---

1. Denon, *Voyage en Égypte*, pp. 298–313.

2. Sigüenza, *Opera inédita*; ˙Eguiara, *Biblioteca Mexicana*, p. 78.

3. Langlès, *Rituel des Tartares-Mantchoux*, pp. 9 and 14. Georgi, *Alphabetum tibetanum*, p. 298.

inhabited by cannibalistic peoples living together in densely populated cities. But without examining here whether these cities were those of the Ichiaca and Confachiqui provinces visited by ˇHernando de Soto, the conqueror of Florida, suffice it to note that the religious ceremonies, dogmas, and legends that struck the imagination of the first Spanish missionaries were undoubtedly found in Mexico after the arrival of the Toltec and thus three or four centuries before the Scandinavians' navigations to the eastern coasts of the new continent.

86   The monks who penetrated into Mexico and Peru on the heels of Cortés's and ˇPizarro's armies were naturally inclined to exaggerate the parallels that they thought they could see between Aztec cosmogony and Christian religious dogmas. Imbued with Hebraic traditions and possessing only an imperfect knowledge of the local languages and the meaning of hieroglyphic paintings, they related everything back to the system in which they had been educated, just as the Romans saw among the Germans and the Gauls only their own religion and their own deities. If one employs a healthy critical perspective, one finds among the Americans no reason to assume that Asiatic peoples surged into the new continent subsequent to the establishment of the Christian religion. In no way do I deny here the possibility that later contacts took place, and I am not unaware[1] that the Chukchi cross the Bering Strait annually to wage war on the inhabitants of the Americas' northwest coast. But based on the knowledge we have acquired since the end of the last century about the Hindus' sacred books, I think I can maintain that one need not turn to western Asia, inhabited by peoples of the Semitic race, to explain the similarities in traditions mentioned by the first missionaries, since these very traditions also have a long and venerable history among the followers of the Brahman sect as well as among the Shamans of the eastern plateau of Tartary.

We shall return to this interesting subject, either in describing the Pastos,[2] an American people whose sole source of nourishment was plants and who loathed all meat-eaters, or in explaining the dogma of metempsychosis, widespread among the Tlaxcalteca. We shall examine the Mexica tradition of the four suns, or the four destructions of the world, and the traces of the *trimurti*, the Hindu trinity, discovered in the Peruvians' religion. Despite these striking connections between the peoples of the new continent and the Tartar tribes who adopted the religion of Buddha, I believe I have identified

---

1. See my *Essai politique sur [le royaume de] la Nouvelle-Espagne*, Vol. I, p. 346.
2. Garcilaso, *Comentarios reales*, Vol. I, p. 274.

in the Americans' mythology, the style of their paintings, their languages, and especially their physiognomy the descendants of a race of men who, separated early from the rest of humankind, followed a peculiar path, over many centuries, in the development of their intellectual faculties and in their progress toward civilization.

PLATE XIV
Costumes Drawn by Mexica Painters in Montezuma's Time

These nine images are excerpted from the *Codex anonymous* number 3738, preserved among the Vatican manuscripts, which we have had several occasions to cite. They are copies of paintings made by Mexica painters during Cortés's first stay in Tenochtitlan. In copying these drawings, Father Ríos appears to have been more attentive to the detail in the costumes than to reproducing faithfully the images' contours. Comparing the paintings in Plate XIV to those in the original manuscripts that have been passed down to us, one notes that the images that the Spanish monk copied are overly elongated. Such alterations in shape are found wherever artists have not sufficiently appreciated the importance of maintaining the artistic style characteristic of peoples more or less removed from civilization. What a sharp contrast there is, in terms of the accuracy of contours, between the hieroglyphs published by Norden and those that appear in Zoëga's work on the obelisks or in the description of the monuments of Egypt, with which the institute of Cairo has just enriched the sciences!

Numbers I–V. Four Mexica warriors: The first three are wearing the garment called *ichcahuepilli*, a kind of cotton breastplate over three centimeters thick, which covered the body from the collar to the waist. Cortés's soldiers adopted this armor, calling it *escaupil*, in which barely a trace of the Aztec language can be identified. Although the *ichcahuepilli* was impervious to arrows, it should not be confused with the gold and copper coats of mail worn by the generals, called *lords of the eagles and the tigers* (*Quauhtin* and *Oocelo*) because of their mask-shaped armor. The shields, *chimalli*, numbers I and II have a very different shape from those depicted by Purchas and *Lorenzana.[1] Escutcheon number II has an appendage made of cloth and feathers that

---

1. Purchas, [*His*] *Pilgrimes*, Vol. III, p. 1080, figure L M; p. 1099, figure C; Plate IV, figure F. Lorenzana, *Historia de Nueva España*, p. 177, folios 2, 8, and 9. Military garb.

served to deaden the blow of spears; its shape recalls the shields depicted on   88
several vases from ˙Magna Graecia. The club that warrior number III bears
was hollow and contained stones that were hurled with great force, as though
they were shot from a sling.

Image number IV depicts one of the intrepid soldiers who went off to
fight in a state of near nudity, his body wrapped in wide-mesh netting that he
casts upon the heads of his enemies, like the Roman *retiarii* [net warriors]
in combat against *myrmillo* [saltwater fish] gladiators. Number V is a private
who wears only a mantle of cloth and a very narrow strip of hide, *maxtlatl*,
around his waist.

As the *Codex vaticanus* expressly points out, image number VI depicts the
unfortunate Montezuma II, in courtly dress, as he appeared inside his palace.
His robes, *tlachquauhjo*, are covered in pearls; his hair is gathered at the top
of his head and tied with a red ribbon, the military honor of princes and the
most valiant captains; his collar is adorned with a necklace of gemstones (*coz-
capetlatl*); but he wears no bracelets (*matemecatl*), nor ankle-boots (*cozehu-
atl*), nor earrings (*nacochtli*), nor the emerald-laden ring hung from the lower
lip, which was part of the emperor's formal robes. The author of the *Codex
anonymous* states that "the sovereign is portrayed with flowers in one hand
and holds in the other a rush with a cylinder of fragrant resin attached at the
tip." The vase in the emperor's left hand bears some resemblance to the one
found in the hand of the drunken Indian depicted in the Mendoza collection.[1]
Mexica painters generally depicted both kings and high lords with bare feet
to indicate that they were not supposed to use their legs and that they should
always be carried in a palanquin, on the shoulders of their servants.[2]

Number VII. An inhabitant of Zapoteca, a province that included the
southeastern part of the intendancy of Oaxaca.

Number VIII and IX. Two women of La Huasteca: the second figure's
costume is indisputably Indian, but that of number VIII closely resembles
European dress. Is this a local woman to whom Cortés's soldiers had given   89
a shawl and a rosary? I shall not decide this question but shall instead mention that the triangular neckerchief can also be found in several Mexica paintings made before the arrival of the Spanish, and that the supposed rosary,
which does not have a cross at the end, might well be one of the chains that
have existed since the earliest times in all of eastern Asia, Canada, Mexico,
and Peru.

1. Purchas, [*His Pilgrimes,*] p. 1117, figure F.
2. *Codex [Vaticanus] Anonymous* number 3738, folio 60.

Plate XIV

Although, as we noted above, Father Ríos appears to have elongated these figures somewhat, their extremities and the shape of their eyes, as well as that of their lips—with the upper lip always protruding over the lower lip—prove that he reproduced them faithfully.

PLATE XV

Aztec Hieroglyphs from the Velletri Manuscript

Of all the Mexica manuscripts preserved in Italy, the *Codex Borgianus* of Velletri is both the largest and the most remarkable, because of the brilliance and the tremendous variety of its colors. It is forty-four to forty-five *palmi* (nearly eleven meters) long and has thirty-eight folds, or seventy-six pages. It is a ritual and astrological almanac that perfectly resembles the *Codex Vaticanus* (one page of which is represented in the thirteenth Plate) in the manner in which the simple hieroglyphs for days and the groups of mythological figures are distributed.

The Velletri manuscript appears to have belonged to the ᵛGiustiniani family. We do not know what unfortunate accident resulted in its falling into the hands of the household servants, who, unaware of the price that a collection of hideous figures might have fetched, handed it over to their children. It was snatched from the latter by Cardinal Borgia, an enlightened enthusiast of antiquity, when they attempted to burn a few of the pages, or better, the deer-hide folds on which the paintings are drawn. There is no indication of the age of this manuscript, which is perhaps only an Aztec copy of an even older book. The great freshness of the colors might lead one to suspect that the *Codex Borgianus*, like that of the Vatican, does not date back past the fourteenth or fifteenth century.

90 It is impossible to focus one's attention on these paintings without many interesting questions being brought to mind. Did they have then only copies of the famous *divine book*, called *teoamoxtli*, written in Tula in the year 660 by the ᵛastrologer *Huematzin*, in which appeared the history of the heavens and the earth, the cosmogony, the description of the constellations, the division of time, the migration of peoples, the mythology, and the moral code? Was this Mexica *Purana*, the *teoamoxtli*, the memory of which was preserved over so many centuries within Aztec tradition, one of those books that the monks, in their fanaticism, had ordered to be burned in the Yucatán, and whose loss was bemoaned by Father Acosta, who was better educated and

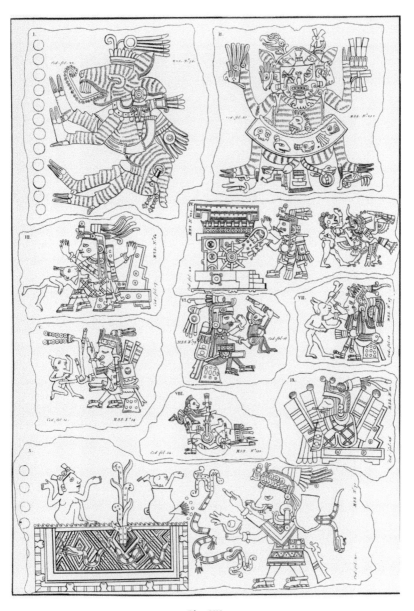

Plate XV

more enlightened than his contemporaries? Is it certain that the Toltec, that hardworking and enterprising people, who are similar in many respects to the Tschud,[1] the ancient inhabitants of Siberia, were the first to introduce painting? Or did the Cuitlateca and the Olmecs, who lived in the Anahuac plateau prior to the irruption of the peoples of Aztlan, and to whom the scholar Sigüenza attributes the construction of the pyramids of Teotihuacan, already record their annals in collections of hieroglyphic paintings? We do not have enough information to answer these important questions, for the darkness that shrouds the origin of the Mongol and Tartar peoples seems to extend across the entire history of the new continent.

The *Codex Borgianus* was the subject of a commentary by the Jesuit Fábrega, who was born in Mexico. During my last stay in Italy in 1805, the Chevalier Borgia, nephew of the cardinal of that name, was gracious enough to have the Mexica manuscript, along with its commentary, brought from Velletri to Rome. I examined both carefully: Father Fábrega's explanations often seemed to me to be arbitrary and quite unfounded. I commissioned engravings of a number of the images that intrigued me the most; to each group represented on the fifteenth Plate, I have added the corresponding quotation, both from the *Codex Borgianus* and from the Italian manuscript intended to serve as a commentary on the latter.

Number I. An unknown animal, adorned with a collar and a kind of harness but pierced by spears: Fábrega calls it the *crowned rabbit*, the *sacred rabbit*. This figure is found in several of the ancient Mexica rituals. According to traditions that have been preserved to this day, it is a symbol of suffering innocence; in this respect, this allegorical representation recalls the lamb of the Hebrews, the mystical idea of an expiatory sacrifice meant to appease the ire of the deity. The incisors, as well as the shape of the head and the tail, seem to suggest that the painter intended to depict an animal of the rodent family. Although its cloven-hoofed feet, each with a dewclaw that does not touch the ground, make it closer to the ruminants, I doubt that it is a *cavia*, a Mexica hare; might it rather be some unknown mammal that lives to the north of the Río Gila, deep inland, toward the northwestern part of the Americas?

It seems to me that this same animal, but with a much longer tail, appears a second time in the *Codex Borgianus*, on the fifty-third sheet: a copy of it can be found on number II of my Plate XV. Mr. Fábrega takes this figure, which is covered with the twenty hieroglyphs for the days, for a stag (*mazatl*). Father Ríos maintains that it is an astrological device for doctors, a painting that teaches that whoever is born on a particular day will experience eye

91

1. *Voyages de Pallas* (Paris translation), Vol. IV, p. 282.

aches, stomachaches, or earaches. Indeed, we see that the twenty simple hieroglyphs for the days are distributed among the different parts of the body.

The sign for the day that began the short period of thirteen days, or the
semilunar month, was regarded as dominant for the entire period, so that a
man born on the day that had an eagle as its hieroglyph had everything to
fear or to expect whenever the eagle presided over the thirteen-day week. Mr.
Zoëga[1] appears to adopt Ríos's explanation; he finds a striking correspondence between this fiction and the ῾iatromathematical ideas of the Egyptians.
One has only to glance at our own almanacs to see that such absurd ideas
have been preserved to this day, because it is often less profitable to educate
the people than to exploit their gullibility. I found this same allegorical figure, which belongs to astrological medicine, in the Codex Borgianus, folio 17
(Mss. number 66), and in the Codex anonymous in the Vatican, folio 54.

Number III, V, VI, VII. A newborn child is depicted four times; the hair
sticking up like two horns indicates that it is a girl. The child is being nursed;
her umbilical cord is cut; she is presented to the deity; her eyes are touched as
a sign of blessing. Fábrega claims that the seated figures, numbers V and VII,
represent two priests; he believes that the helmet worn by figure VII identifies
the latter as the high priest of the god Tonacateuctli.

Number IV. The depiction of a human sacrifice: A priest, who is almost unrecognizable underneath his hideous disguise, rips the heart from
the victim; he bears a club in his right hand. The victim's naked body is
painted; there are a number of spots meant to imitate those of a jaguar's (or
an American tiger's) coat. On the left-hand side is another priest (topiltzin),
who is pouring the blood from the extracted heart onto an image of the sun
placed in the recess of a temple. I would not have had this revolting scene engraved had the sacrificer's disguise not exhibited some remarkable and seemingly nonaccidental connections with the Hindus' Ganesh. The Mexica made
use of helmets with shapes imitating the head of a serpent, a crocodile, or a
jaguar. The sacrificer's mask seems to recall the trunk of an elephant or some
other pachyderm with a similar head shape but with an upper jaw filled with
incisors. The snout of the tapir is certainly a bit longer than that of our pigs,
but there is a pronounced difference between the tapir's snout and the trunk
depicted in the Codex Borgianus. Had the peoples of Aztlan, who originally
came from Asia, preserved some vague notions about elephants, or (what
seems far less likely to me) did their traditions date back to the time when
the Americas were still peopled with these gigantic animals, whose petrified
skeletons are found buried in marly ground on the very ridge of the Mexi-

92

1. Zoëga, [De origine et usu obeliscorum,] pp. 523 and 531.

can Cordilleras? Perhaps there also exists, in areas of the northwestern part of the new continent that neither ˙Hearne, Mackenzie, nor Lewis visited, an unknown pachyderm with a trunk shape that places it squarely between an elephant and a tapir.

The day-hieroglyphs that surround the group depicted on the forty-ninth page of the Velletri *Collection* indicate clearly that this sacrifice was carried out at the end of the year, after the *nemontemi*, complementary days. The temple of the sun recalls the religion of a gentle and humane people, the Peruvians. This religion, in which no other offerings were made to the deity than flowers, incense, and the first harvests, undoubtedly existed in Mexico until the beginning of the fourteenth century. One scholar,[1] who has made a number of felicitous comparisons between the mythologies of different peoples, has advanced the theory that the two religious sects of India, the worshipers of Vishnu and those of Shiva, spread into the Americas and that the Peruvian religion is the worship of Vishnu, when he appears in the form of Krishna or the sun, while the bloodthirsty religion of the Mexica is analogous to the worship of Shiva, when he assumes the character of ˙Stygian Jupiter. In Indian statues and paintings, Shiva's wife, the black goddess Kali or Bhavani,[2] the symbol of death and destruction, wears a necklace of human skulls; the Vedas ordain that human sacrifices be made to her. The ancient worship of Kali, the horrible cruelty of which was tempered by the Buddha's reform, certainly bears strong resemblances to the worship of Mictlancihuatl, the goddess of the underworld, as well as to that of other Mexica deities; but in studying the history of the peoples of Anahuac, one is tempted to regard such resemblances as purely accidental. One cannot assume that such contacts took place whenever one happens to find the worship of the sun or the custom of sacrificing human victims among semibarbarous peoples. Far from having been brought from eastern Asia, this custom may have arisen in the Valley of Mexico itself. Indeed, history teaches us that this bloodthirsty religion, which recalls the worship of Kali, Moloch, and the Gauls' Esus, had existed for only two hundred years when the Spanish arrived at Tenochtitlan.

The peoples who flooded into Mexico from the seventh to the twelfth century—the Toltec, the Chichimecs, the Nahuatlacs, the Alcolhua, the Tlaxcalteca, and the Aztecs—formed a single group united by similar languages and customs, somewhat like the Germans, the Norwegians, the Goths, and the Danes, who merge into a single race, that of the Germanic peoples. As

---

1. ˙Frédéric Léopold, Comte de Stolberg, *Geschichte der Religion Jesu Christi*, Vol. I, p. 426.

2. *Recherches asiatiques*, Vol. I, pp. 203 and 293.

we have noted above, it is likely that other peoples—the Otomi, the Olmecs, the Cuitlateca, the Zacateca, and the Tarascana—appeared in the equinoctial region of New Spain before the Toltec. Whenever peoples have moved in a single direction, the location of the sites in which we find them indicates, to   94
some extent, the chronological order of their migrations. In the case of Europe, is there any doubt that the westernmost peoples, the Iberians and the Cantabrians, arrived before the peoples nearest to Asia did, that is, before the Thracians, the Illyrians, and the Pelasgians?

Whatever the relative antiquity of the various races of humans established in the mountains of Mexico, the Caucasus of the Americas, it seems certain that none of these peoples, from the Olmecs to the Aztecs, observed the barbarous custom of sacrificing human victims for very long. The principal deity of the Toltec was named *Tlalocteuctli*: he was the god of water, the mountains, and storms. In the eyes of this mountain people, thunder was formed mysteriously on the high mountaintops forever shrouded in clouds; it was there that they placed the abode of the Great Spirit Teotl, that invisible being called *Ipalnemohuani* and *Tloque-Nahuaque* because *he exists only by virtue of himself*, and because *he contains everything within him*. Both the gales that destroy their huts and the beneficial rains that invigorate their fields come from this nearly inaccessible region. The Toltec had erected the image of Tlalocteuctli on the top of a high mountain. That crudely sculpted image was made from a white stone regarded as *divine stone* (*teotetl*), for, like the Orientals,[1] this people attached superstitious ideas to the color of certain stones. Tlalocteuctli was represented with a bolt of lightning in his hand, seated on a cube-shaped stone with a vase before him that contained offerings of rubber and seeds. The Aztecs followed this same religion until the year 1317, when war with the inhabitants of the city of Xochimilco provided them with the initial idea of a human sacrifice. Immediately after the capture of Tenochtitlan, those Mexica historians, who wrote in their own language but used the Spanish alphabet, passed down to us the details of this dreadful event.

Since the beginning of the fourteenth century, the Aztecs had lived under the rule of the king of Culhuacan; it was they who contributed the most to that king's victory over the Xochimilca. Once the war was over, they wanted to offer a sacrifice to their principal deity, Huitzilopochtli, or Mexitli, whose wooden image, placed in a reed chair called the *seat of God, Teoicpalli,*   95
and carried on the shoulders of four priests, they had collectively followed throughout their migration. They asked their master, the king of Culhuacan,

1. ▾Mill, *Dissertationes selectae*, p. 309.

to give them a few objects of value to solemnify this sacrifice. The king—if we may use this term for the leader of such a scant horde—sent them a dead bird wrapped in a coarse cloth. To add insult to injury, he proposed to attend the feast himself. The Aztecs pretended to be pleased with this offer, but they resolved at the same time to make a sacrifice that would inspire terror in their masters. After a long dance around the idol, they brought forth four Xochimilca prisoners whom they had kept long hidden. Those unfortunate individuals were immolated, with the ceremonies still observed at the time of the Spanish conquest, on the platform of the great pyramid of Tenochtitlan, which was dedicated to this very god of war, Huitzilopochtli. The Culhua exhibited justifiable horror at this human sacrifice, the first that had ever been made in their country. Fearing the savagery of their slaves and seeing the latter emboldened by their success in the war against the Xochimilca, they granted the Aztecs their freedom and enjoined them to leave the territory of Culhuacan.

The outcome of this first sacrifice was a happy one for the oppressed people; vengeance soon gave rise to the second one. After the founding of Tenochtitlan, an Aztec was wandering around the lakeshore, looking to slay an animal that he might offer to the god Mexitli, when he encountered an inhabitant of Culhuacan, named *Xomimitl*. Furious with his former masters, the Aztec fought the Culhua hand to hand. Beaten, Xomimitl was taken to the new city, where he expired on the fatal stone placed at the foot of the altar.

The circumstances of the third sacrifice are even more tragic. Peace had seemingly been restored between the Aztecs and the inhabitants of Culhuacan; nevertheless, the priests of Mexitli could not contain their hatred for the neighboring people, who had made them grovel in slavery. They plotted a horrific act of revenge: they urged the king of Culhuacan to entrust to them his only daughter, that she might be raised in the temple of Mexitli and, after her death, worshiped there as the mother of that god, the Aztecs' patron deity. They added that it was the idol itself that was declaring its will through their mouths. The gullible king accompanied his daughter and escorted her inside the dark compound of the temple. There the priests separated the girl from her father. An uproar was heard in the sanctuary; the unfortunate king could not make out the groans of his dying daughter. They placed a censer in his hand and, a few moments later, ordered him to light the copal. In the pale light from the rising flame he recognized his child, bound to a post, her chest covered in blood, showing no movement, nor any sign of life. His despair robbed him of his senses for the rest of his days. He was unable to take revenge, and the Culhua did not dare pit their strength against a people who inspired fear by such excesses of barbarism. The immolated girl was placed

96

among the Aztec deities under the name *Teteionan*,[1] *mother of the gods*, or Tocitzin, *our grandmother*, a goddess who must not be confused with Eve, or the *serpent woman*, called *Tonantzin*.

Wherever we find traces of human sacrifice in the old continent, their origin is lost in the mists of time. The history of the Mexica, on the other hand, has preserved for us the story of the events that lent a savage and bloodthirsty character to the religion of a people for whom the original offerings to the deity had been exclusively animals or the first fruits of the season. I felt it necessary to report these legends, which probably have a basis in historical truth; intimately tied to the study of customs and of the moral development of our species, they seem more interesting than the Hindus' puerile stories about the numerous *incarnations* of their deities. I shall not, however, decide whether the sacrifice of the four Xochimilca was indeed the first to be offered to the god Mexitli or whether the Aztecs had not preserved some ancient tradition, following which they imagined that the god of war delighted in the blood of human victims. Mexitli had come into the world with a spear in his right hand, a shield in his left, and his head capped with a helmet adorned with green feathers. When he was born, his first action was to slay his sisters and brothers. Perhaps this terrible god, also called *Tetzahuitl* or the *horror*, had already been venerated by a bloodthirsty religion in other climes; perhaps this worship was interrupted only by a lack of prisoners, and thus of victims, during the people's peaceful procession, under the auspices of Mexitli, from the Tarahumara Mountains to the central plateau of Mexico.

The Aztecs' incessant wars after their settlement on the islets of the salt lake of Texcoco provided them with such a large number of victims that human sacrifices were offered without exception to all their deities, even to Quetzalcoatl,[2] who, like the Hindus' Buddha, had preached against this detestable custom, and to the goddess of the harvests, the Mexica Ceres, called *Centeotl* or *Tonacajohua*, she who *feeds humankind*. The Totonac, who had adopted the entire Toltec and Aztec mythology, made a racial distinction between, on the one hand, the deities who demanded bloody worship and, on the other, the goddess of the fields, who asked only for offerings of flowers and fruits, sheaves of corn, or birds that feed on the kernels of this plant so useful to humanity. An ancient prophecy gave this people hope for a beneficial reform in their religious ceremonies. This prophecy foretold that Centeotl, who is identical to the lovely Shri or Lakshmi of the Hindus, and whom the Aztecs, like the Arcadians, called the *Great Goddess* or the *original God-*

97

---

1. Clavijero, [*Historia antica del Messico*,] Vol. I, pp. 166, 168, 172; Vol. II, p. 22.
2. Gómara, *Crónica general de las Indias* (1553 ed.), Vol. II, folio 134.

dess (*Tzinteotl*), would eventually triumph over the other gods' savagery and that human sacrifices would give way to the innocent offerings of the first harvests. One might see in this Totonac tradition a struggle between two religions: a conflict between the ancient Toltec deity, gentle and humane, like the people who introduced this worship, and the savage gods of that warlike horde, the Aztecs, who brought bloodshed to the fields, temples, and altars.

Reading Cortés's letters to Emperor Charles V, the accounts of Bernal Díaz, ˇMotolinía, and other Spanish authors who observed the Mexica prior to the changes that the latter experienced by virtue of their contacts with Europe, one is surprised that such extreme savagery in their religious ceremonies could be found among a people whose social and political state recalls, in other respects, Chinese and Japanese civilizations. The Aztecs were not content to stain their idols with blood, as is still practiced by Tartar Shamans, who sacrifice only oxen or sheep to their *Nogats*; the Aztecs even devoured a part of the corpse that the priests threw to the base of the teocalli steps, after having ripped out the heart. One cannot consider these subjects without wondering whether these barbaric customs, which are also found in the South Sea islands, among peoples whose gentle ways have been excessively praised, would have ceased of their own accord if the Mexica[1] had continued in their stride toward civilization without having any contact with the Spanish. It is likely that such a beneficial reform in their religion, the triumph of the goddess of the harvests over the gods of carnage, would have taken quite some time to manifest itself.

The most powerful people in South America, the Peruvians, worshiped the sun. The Inca undertook the cruelest wars in order to introduce a gentle, peaceful religion; human sacrifice ceased wherever the descendants of Manco Capac brought their laws, their caste divisions, their languages, and their monastic despotism. In the land of Anahuac, the bloodthirsty worship of Huitzilopochtli became increasingly dominant as the Mexica empire swallowed up all the neighboring states. That empire's greatness was based on a very close coalition between the priestly class and the nobles destined for a military career. The high priest, Teoteuctli (*divine Lord*), was generally a prince of royal blood; no wars could be undertaken without his approval. The priests even went off to battle and were elevated to the highest honors in the army;[2] their influence thereby became as powerful as that of the Roman patricians, who had an exclusive right to consult the oracles and in whom a

98

1. Langlès, *Rituel des Tartares-Mantchoux*, p. 18.
2. *Peintures hiéroglyphiques du recueil de Mendoza.* Thévenot, [*Relations de divers voyages curieux,*] Vol. IV, folio 57.

famous author[1] claims to have detected the traces of a Hindu political institution.

In Mexico, where the number and the power of the priests (*teopixqui*)
and the monks (*tlamacazques*) were almost as great as they are today in Tibet
and Japan, any effect of religious fanaticism could sustain only infinitely slow
changes. History proves that the barbaric custom of human sacrifice was preserved for a long time even among the peoples with the most advanced civilizations. The paintings found in the kings' tombs in Thebes leave no doubt
that such sacrifices were customary among the Egyptians.[2] We have already
observed above that long ago in India, the goddess Kali required human victims, just as Saturn demanded them in Carthage. In Rome, after the ʼbattle
of Cannae, two Gauls, a man and a woman, were buried alive, and Emperor
Claudius was forced to issue a law expressly forbidding human sacrifice in the
Roman Empire.[3] What is more: do we not see in less remote times the barbaric effects of religious intolerance in the midst of one of humanity's great
civilizations and in a period of general moderation of both character and customs? Regardless of the difference in cultural advances among peoples, fanaticism and self-interest retain their fatal power. It will be difficult for posterity
to understand that in a civilized Europe and under the influence of a religion
that, through the nature of its principles, promotes freedom and proclaims
the sacred rights of humanity, there exist laws that sanction the enslavement
of Blacks and permit the colonist to tear a child from its mother's arms in
order to sell it in a faraway land. Such reflections prove—and the implication
is far from comforting—that entire peoples may make rapid strides toward
civilization without their political institutions or their forms of worship completely shedding their former barbarism.

Number VIII shows the ceremony for lighting the new fire during the procession to the summit of a mountain near Iztapalapa, which took place every
fifty-two years.

The intercalation occurred at the end of each cycle of either twelve or thirteen days. In expectation of the fourth destruction of the sun and the earth,
the people extinguished all fires until the beginning of the new cycle, at which
point the priests would light new ones. The painting shows a victim stretched

1. Schlegel, *Weisheit der Indier*, p. 190.

2. Denon, *Voyage en Égypte*, p. 298, Plate CXXIV, number 2. *Décade Égyptienne*,
Vol. III, p. 110.

3. ʼSueton[ius], [*Suetonii Tranquilli Opera*,] Ch. XXV (ed. Wolf), Vol. I, p. 48. Pliny,
Historia naturalis, Book XXXI, Ch. I; Book VIII, Ch. XXII. ʼTertullian, *Apologeticum ad-
versus gentes*, Ch. IX (ed. Palmer, 1684, p. 41). ʼLactantius, *Divinae Institutiones*, Book I,
Ch. XXI.

out on the sacrificial stone with a wooden disc on the chest, which the teopix-qui, using the friction method, is setting on fire. The hieroglyph of the starry sky, which can be discerned on the preceding page of the Borgia Collection, seems to allude to the culmination of the Pleiades. At a later point, in our explanation of the twenty-third plate, we shall return to the relationship that allegedly existed between this culmination and the beginning of the cycle.

The art of making fire by rubbing together two types of wood of different hardness dates back to the earliest times. We find it among peoples of both continents. According to ᵛMr. Visconti, in Homeric times its invention was attributed to Mercury.[1] The disc resting on the victim's body, within which the priest is turning the cylindrical piece of wood, is the Greeks' στορεύς [*storeus*, the undermost of two substances by which a fire is produced].[2] Pliny maintains that of all ligneous matter, ivy is what catches fire best when rubbed against laurel wood.[3] We found these πυρεῖα [*pyreia*, fire tools] among the Indians of the Orinoco. Extremely rapid movements are necessary to raise the temperature to the point of incandescence.

Number IX. The figure of a deceased king, surrounded by four flags, with his eye closed, no hands, and his feet wrapped. The chair is the royal seat, called *tlatocaicpalli*, on which Adam or Tonacateuctli, the *Lord of our flesh*, and Eve or Tonacacihua are shown seated in the *Codex Borgianus* (folio 9). This hieroglyphic character is represented in the ritual almanac, on the page that indicates the thirteen-day cycle during which the sun passes through the zenith of Mexico City.

Number X. An allegory that recalls the purifications in India. A deity, whose enormous nose is adorned with the figure of the two-headed grass snake, the mysterious amphisbaena, carries in its hand a xiquipilli, incense pouch; we see on its back a broken container, from which emerges a serpent. In front of the figure lies another serpent, bleeding and cut into pieces. A third serpent, also cut into pieces, is held within a casket filled with water, from which a plant grows. On the right-hand side we see a man placed in a pot; on the left-hand side, a woman adorned with flowers, probably the voluptuous Tlamezquimilli, who is sometimes depicted blindfolded. On the same page are agaves that bleed when cut. Does this allegory allude to the serpent who poisons the water, the source of all organic life,[4] to Krishna's victory

100

---

1. ᵛHomer, *Hymnus in Mercurium*, line 110.

2. ᵛApollonius, Rhodius, *Argonautica*, Book I, Verse 1184, and *Scholien ad eum*.

3. Pliny, *Historia naturalis*, XVI, 77. ᵛSeneca, *Naturales Quaestiones* II, 22. ᵛTheophrastus, [*Opera omnia*,] Verse 10.

4. Paolini de S. Bartolomeo, *Codices Avenses*, p. 235.

over the dragon Kaliya, to the seduction and purification by fire? It is obvious that the figure of the serpent expresses two very different ideas in Mexica paintings. In the reliefs that show the division of the year and the cycles, this figure signifies only time, *aevum* [eternity]. The serpent depicted either as interacting with the *mother of humanity* (Cihuacohuatl) or as struck down by the Great Spirit Teotl when the latter assumes the form of one of the minor deities, is the evil genius, a genuine κακοδαίμων [*kakodaimon*, evil spirit]. To express this idea, the Egyptians used the hieroglyph of the hippopotamus and not that of the serpent.[1]

Figures without clothing, like those in group number X and the goddess of voluptuousness, called *Ixcuina* or *Tlazolteucihua*,[2] are extremely rare in Mexica paintings. Barbarous peoples generally clothe their statues; it is a mark of artistic refinement to present the naked body in the natural beauty of its form. It is also quite remarkable that Mexica painting exhibits absolutely no traces of the symbol of the generative force, or *lingam* [phallus] worship, which is widespread in India and among all those peoples who have had connections with the Hindus. Mr. Zoëga observed that the phallus emblem is also absent from Egyptian works from the earliest times; he thought it possible to conclude from this that phallus worship is not as ancient as it is assumed to be. This claim is, however, at odds with the concepts that 'Hamilton, Sir William Jones, and Mr. Schlegel have drawn from the Shiva Purana,[3] the Kashi Khanda, and several other works written in the Sanskrit language. There is no question that the worship of the twelve lingams, which came from the summit of the Imaüs (Himavat), dates back to the time of the Hindus' earliest traditions. Given the presence of so many other connections that suggest ancient contacts between eastern Asia and the new continent, it is surprising that in the case of the latter, we do not find even a few traces of phallus worship. Mr. Langlès[4] plainly observes that in India the *Vaishnava*, followers of the Vishnu sect, loathe that emblem of the productive force, which is worshiped in the temples of Shiva and of his wife Bhavani, the goddess of abundance. Could we not assume, then, that there also exists a sect that rejects *lingam* worship among the Buddhists exiled in the northeast of Asia, and that it is this expurgated Buddhism that one finds surviving in a few humble traces among the American peoples?

101

---

1. Zoëga, [*De origine et usu obeliscorum,*] p. 445, note 35.
2. *Codex Borgianus*, Mss. folio 73.
3. *Catalogue des manuscrits sanskrits de la Bibliothèque impériale*, pp. 36 and 50.
4. *Recherches asiatiques*, Vol. I, p. 215.

PLATE XVI
View of Chimborazo and Carihuairazo

At certain points, the Cordillera of the Andes divides into several branches separated from one another by longitudinal valleys, while at other points it forms a single mass spiked with volcanic peaks. In our earlier description of Quindiu pass (Plate V), we attempted to provide a geographical overview of the subdivision of the Cordilleras in the kingdom of New Granada, between 2° 30' and 5° 15' northern latitude. In that section we also observed that the great valleys located between the two lateral branches and the central range form the basin of two sizable rivers; their bottoms are at an even lower elevation above sea level than the bed of the Rhone, whose waters hollowed out the Valley of Sion in the High Alps. Traveling southward from Popayán on the arid plateau of *Los Pastos* province, one sees the three secondary chains of the Andes merge into a single group that extends well past the equator.

This group, in the kingdom of Quito, has a peculiar appearance between the Chota River, which winds through basalt mountains, and the Páramo of Azuay, on which rise a number of memorable remnants of Peruvian architecture. The highest summits are arranged in two lines that form something like a double crest within the Cordillera. These colossal peaks, covered in eternal ice, served as markers in the calculations of French academicians during the measurement of the equatorial degree. Because of their symmetrical distribution into two lines running north to south, Bouguer considered them two mountain chains separated by a longitudinal valley; but what that famous astronomer identifies as the bottom of the valley is, in fact, the very ridge of the Andes, a plateau with an absolute elevation of two thousand seven hundred to two thousand nine hundred meters. One must not mistake a double crest for a genuine subdivision of the Cordilleras.

The pumice-covered plain in the foreground of the drawing we are describing here is part of the plateau that separates the western from the eastern crest of the Andes of Quito. It is in these plains that the population of

Plate XVI

this marvelous country is concentrated; one finds cities there with thirty to fifty thousand inhabitants. When one has lived for a few months on this elevated plateau, where the barometer remains at a height of 0,54 meters (twenty pouces), one experiences an extraordinary, inescapable illusion. Little by little, the observer forgets everything that surrounds him: the villages that testify to the ingenuity of this mountain people, the pastures covered with both llama herds and flocks of sheep from Europe, the orchards lined with hedges bursting with Duranta and Barnadesia, and the carefully plowed fields promising rich grain harvests—everything seems to be suspended in the upper regions of the atmosphere. One barely registers that the ground on which one stands is at a higher elevation relative to the nearby Pacific Ocean coasts than is the summit of Canigou in relation to the Mediterranean basin.

Seeing the ridge of the Cordilleras as a vast plain bounded by curtains of distant mountains accustoms one to the notion that the unevenness within the crest of the Andes results from the presence of innumerable isolated peaks. Although they form a single mass for over half their total elevation, Pichincha, Cayambe, Cotopaxi, and all of the volcanic peaks given individual names appear to the Quito dwellers as distinct mountains that rise from the midst of a plain devoid of forests. This illusion is all the more complete since the jagged outlines of the Cordilleras' double crest extend to the level of the high, settled plains. The Andes, therefore, have the appearance of a single range only when seen from afar—either from the coasts of the Great Ocean or from the savannahs that stretch to the foot of their eastern slope. If one were placed right on the ridge of the Cordilleras, either in the kingdom of Quito or in the province of Los Pastos, or farther north still, in the heart of New Spain, one would see only a cluster of scattered mountaintops, groups of isolated mountains that stand out from the central plateau. The larger the mass of the Cordilleras, the more difficult it is to grasp their structure and form as a whole.

The study of the form or, I would dare say, the physiognomy of mountains is made a great deal easier, however, by the orientation of the high plains that form the ridge of the Andes. Journeying from the city of Quito to the Páramo of Azuay, one sees to the west—one after the other and over a course of thirty-seven leagues—the summits of Casitahua, Pichincha, Atacazo, Corazón, Iliniza, Carihuairazo, Chimborazo, and Cunambay; to the east, the tops of Guamaní, Antisana, Pasochoa, Rumiñahui, Cotopaxi, Quilindaña, Tungurahua, and Capac-Urcu, all of which, except three or four, are higher than Mont Blanc. The result of the arrangement of these mountains is that when seen from the central plateau, rather than overlapping they show themselves in their true form, as though projected against the azure sky. One has the impression of seeing both their summits and their peaks on the same vertical plane; they recall the imposing spectacle of the coasts of New Norfolk and the Cook River. They look like a craggy shore that, rising from the waters, seems all the less distant since there are no objects between the shore and the eye of the observer.

But although the Cordilleras' structure and the form of the central plateau lend themselves particularly well to geological observations, and although they enable travelers to examine closely the contours of the Andes' double crest with ease, the tremendous elevation of this very plateau also diminishes the perceptible size of the peaks that, were they placed on islands scattered throughout the vast seas (like Mauna Loa and the Peak of Tenerife [Teide]), would be all the more imposing for their terrifying height. The plain of Tapia, which can be seen in the foreground of the sixteenth Plate, and on which, near Riobamba Nuevo, I drew the group of Chimborazo and Carihuairazo, has an absolute elevation of two thousand eight hundred ninety-one meters (fourteen hundred eighty-three toises). It is only one-sixth lower than the top of Etna. The summit of Chimborazo thus surpasses the height of this plateau by only three thousand six hundred forty meters, which is eighty-four meters less than the difference in height between the top of Mont Blanc and the priory of Chamonix; for the difference between Chimborazo and Mont Blanc is more or less equal to the difference between the elevation of the Tapia plateau and the bottom of the Chamonix valley. Compared to the city of La Orotava, the top of the Peak of Tenerife is even higher than Chimborazo and Mont Blanc are, in relation to Riobamba and Chamonix.

Mountains that would awe us with their height were they located on the seaside seem mere hills if they rise from the ridge of the Cordilleras. Quito, for example, is adjacent to a small cone called Yavirac, which does not seem higher to the residents of that city than Montmartre or the heights of Meudon seem to the inhabitants of Paris. According to my measurements, however, the cone of Yavirac has three thousand one hundred twenty-one meters (sixteen hundred toises) of absolute elevation; it is almost as high as the summit of Marboré, one of the highest mountaintops in the Pyrenees range.

Despite the effects of the illusion produced by the height of the Quito, Mulaló, and Riobamba plateaus, one would search in vain for a location either near the coasts or on the eastern slope of Chimborazo that might offer as magnificent a view as the one I enjoyed for several weeks in the Tapia valley. When one is on the ridge of the Andes, between the double crests formed by the colossal tops of Chimborazo, Tungurahua, and Cotopaxi, one is still close enough to their summits to view them at very wide height angles. As one descends toward the forests that surround the base of the Cordilleras, how-

105

ever, these angles become very narrow, for due to the mountains' enormous mass, one's distance from the summits increases rapidly the closer one moves to sea level.

I have drawn the contours of Chimborazo and Carihuairazo using the same graphic means that I indicated above when discussing the drawing of Cotopaxi. The line that marks the lower edge of the perpetual snow line is at a height that slightly surpasses that of Mont Blanc, for if the latter mountain were located on the equator, it would only rarely be covered with snow. The result of the constant temperature that reigns in this zone is that the limit of the eternal ice is not as irregular as it is in the Alps and the Pyrenees. The road that leads from Quito to Guayaquil, toward the Pacific Ocean coasts, passes over the northern slope of Chimborazo between that mountain and Carihuairazo. The shape of the snow-covered hillocks that rise on this side recalls that of Dôme du Goûter seen from the Chamonix valley. It was on a narrow ridge jutting out from amidst the snows on the southern slope that Mr. Bonpland, ˈMr. Montúfar, and I attempted, not without danger, to reach the top of Chimborazo. We carried our instruments to a considerable height, although we were surrounded by a thick fog, and the thin air gave us serious trouble. The point where we stopped to observe the incline of the magnetized needle appears to be the highest that any man has reached on this mountain ridge; it is eleven hundred meters higher than the top of Mont Blanc, which that most skilled and fearless of travelers, Mr. de Saussure, had the good fortune to reach, struggling against even greater difficulties than we had to overcome near the top of Chimborazo. These grueling excursions, the results of which generally stir the public's imagination, offer only very few results that are useful to scientific progress, since the traveler finds himself on snow-covered ground, in an air layer with a chemical composition exactly the same as that of the low-lying regions, and in circumstances in which it is impossible to conduct delicate experiments with all the required precision.

By comparing Plates V, X, and XVI from this work with those of the *Geographical and Physical Atlas* that accompanies my *Political Essay on the Kingdom of New Spain*, one can distinguish three main kinds of shapes among the high peaks of the Andes. The volcanoes that are still active, those that have only one crater of an extraordinary width, are conical mountains with more or less truncated summits; this is the shape of Cotopaxi, Popocatepetl, and the Pico de Orizaba. Other volcanoes with summits that have fallen in after a long series of eruptions exhibit crests spiked with points, tilting peaks, and shattered boulders that threaten to collapse. This is the shape of El Altar, or Capac-Urcu, a mountain that was once higher than Chimborazo, and whose destruction marks a memorable period in the physical history of the new

106

continent. It is also the shape of Carihuairazo, a large part of which fell in on the night of July 19, 1698. Torrents of water and muddy ejections then spewed forth from the gaping sides of the mountain and turned the surrounding countryside infertile. This horrible catastrophe was accompanied by an earthquake that swallowed up thousands of residents of the nearby cities of Ambato and Latacunga.

A third shape of the high peaks of the Andes, and the most majestic of all, is that of Chimborazo, with its round summit; it recalls the craterless hillocks raised by the elastic force of fumes in regions where the cavernous crust   107 of the globe is worn by subterranean fires. The appearance of these granite mountains bears only a scant resemblance to that of Chimborazo. The granite summits are oblate hemispheres; trappean porphyries form slender cupolas. On the South Sea coasts, after the long winter rains, when the air has suddenly become more lucid, Chimborazo appears to the observer like a cloud on the horizon; it stands out from the neighboring peaks; it looms above the entire Andes range just as that majestic dome, Michelangelo's work of genius, looms above the ancient monuments that surround Capitoline Hill.

PLATE XVII

The Peruvian Monument of Cañar

The high plains that extend across the ridge of the Cordilleras from the equator to near 3° of southern latitude end at a mass of mountains four thousand eight hundred meters high, which link the eastern and western crests of the Andes of Quito like an enormous embankment. This group of mountains, in which micaceous schist and other rocks of original formation are covered in porphyry, is known as the *Páramo del Azuay*. We were forced to cross it in order to travel from Riobamba to Cuenca and to the beautiful forests of Loja, which are renowned for their abundance in cinchona. The pass of Azuay is especially formidable in the months of June, July, and August, when an immense quantity of snow falls, and when the icy winds from the South blow in these lands. Since the main road, according to the measurements that I took in 1802, nearly reaches the elevation of Mont Blanc, travelers there are exposed to extreme cold, and not a year passes when a few do not perish. Midway through this pass, at an absolute elevation of four thousand meters, one crosses a plain that stretches over more than six square leagues. This plain—and this remarkable fact sheds some light on the formation of high plateaus—is almost at the height of the savannahs surrounding the part of the Antisana volcano that is covered in eternal snows. The plateaus of Azuay and Antisana, the geological compositions of which are so strikingly similar, are nevertheless separated by a distance of over fifty leagues. They contain freshwater lakes of great depth that are lined with dense patches of alpine grass. But there are no fish and hardly any aquatic insects to enliven their solitude.

The soil of the *Llano del Pullal* (the name of the high plains of Azuay) is extremely marshy. We were surprised to find there, at elevations far exceeding the top of the Peak of Tenerife, the magnificent ruins of a trail built by the Inca of Peru. Lined with large cut stones, this roadway can be compared with the most beautiful Roman roads that I have seen in Italy, France, and Spain. It is perfectly straight and maintains the same direction over a distance of six to eight thousand meters. As we saw firsthand, it extends to the vicinity of Cajamarca, one hundred twenty leagues south of Azuay, and people

108

Plate XVII

there believe that it once led all the way to the city of Cusco. Near the Azuay road, at an absolute elevation of four thousand forty-two meters (two thousand seventy-four toises), are the ruins of the Inca Tupac Yupanqui's palace, the shell of which, commonly called *los paredones*, is of relatively low height.

Descending from the Páramo of Azuay toward the south, between the farms of Turche and Burgay, one comes upon another masterpiece of ancient Peruvian architecture, which is known as *Ingapirca*, or the fortress of Cañar. This fortress, if one can use this term for a hill topped with a platform, is much less remarkable for its size than for its perfectly preserved condition. A wall built of thick-cut stones, five to six meters high, forms an extremely regular oval, with a major axis that is thirty-eight meters long. Inside this oval is a talus covered with gorgeous vegetation that adds to the charming impression of the landscape. At the center of the enclosure rises a house that has only two chambers and is nearly seven meters high. Both this house and the enclosure, shown on the sixteenth Plate, belong to a system of walls and fortifications over one hundred fifty meters long; we shall speak of them below. The cut of the stones, the layout of the gates and the recesses, and the perfect similarity between this structure and those of Cusco leave no doubt as to the origin of this *military monument*, which served as lodging for the Inca during those princes' occasional journeys from Peru to the kingdom of Quito. The foundations of a large number of buildings surrounding the enclosure offer evidence that there was once enough space at Cañar to lodge the small army corps that usually accompanied the Inca on their travels. In these foundations I found a very artfully carved stone, which is depicted in the left foreground of the painting; I was unable to ascertain the use of this particular cut.

What is most striking about this small monument, surrounded by a few trunks of *schinus molle* trees, is the shape of its roof, which makes it perfectly resemble European houses. One of the first historians of the Americas, Pedro de Cieza de León, who began writing about his travels in 1541, speaks in detail of several of the Inca's houses in the province of *Los Cañares*. He states plainly[1] that "the structures of Tomabamba have a covering of rushes, which is so well-made that, unless destroyed by fire, it can be preserved for centuries with no deterioration whatsoever." This observation necessarily leads one to suspect that the gable on the house at Cañar was added after the conquest. What appears most to confirm this theory is the existence of window openings in this part of the building, for it is certain that in the structures built by the ancient Peruvians, windows are found as frequently as in the ruins of houses in Pompeii and Herculaneum—that is, never.

109

1. Pedro de Cieza de León, *Crónica del Perú* (Antwerp, 1554), Vol. I, Ch. XLIV, p. 120.

In a very interesting paper on a number of ancient monuments of Peru,[1] *Mr. La Condamine also tends toward the opinion that the gable on the small monument of Cañar is not from the time of the Inca. He says that "it is perhaps of modern construction, and that it is not built of cut stones like the rest of the walls but, rather, of a kind of air-dried brick permeated by straw." Elsewhere, this same scholar adds that the use of these bricks, which the Indians call *tica*, was known to the Peruvians long before the arrival of the Spanish, and that for this reason the top of the gable, although it is made of bricks, might well be of ancient construction.

I deeply regret not having been acquainted with Mr. La Condamine's paper before my journey to the Americas. By no means do I intend to cast doubt upon the observations of that famous traveler, whose work necessitated a lengthy stay in the area around Cañar and who had much more time than I did to examine this monument. I am, however, surprised that while debating on site the question of whether this structure's roof was added at the time of the Spanish, Mr. Bonpland and I were not struck by the difference in construction that allegedly exists between the wall and the top of the gable. I did not see any bricks (*ticas* or *adobes*); I noticed only some cut stones coated with a kind of yellowish stucco that was easily removable and embedded with *ichu*, cut straw. The proprietor of a nearby farm who accompanied us on our excursion to the Cañar ruins boasted that his ancestors had contributed greatly to the destruction of these buildings. He told us that the sloped roof had been covered not with tiles, in the European fashion, but with very thin, well-polished slabs of stone. It was this circumstance in particular that made me, at that time, lean toward the probably baseless theory that with the exception of the four windows, the rest of the structure was the same as it had been at the moment of its construction by the Inca. Be that as it may, one must agree that acute-angle roofs would have been quite useful in a mountainous country with plentiful rain. These sloped roofs are known to the natives in the northwest coast of the Americas; in the earliest times, they were even known in southern Europe, as several Greek and Roman monuments suggest, especially the reliefs of the Trajan column and the landscape paintings found in *Pompeii and formerly preserved in the magnificent collection of Portici. Among the Greeks, the rooftop angle was oblique; it became a right angle among the Romans, who lived under less beautiful skies than those of Greece. The roofs become more sloped the farther one moves to the north.

The drawing whose engraving is shown on the seventeenth Plate is based

110

---

1. *Mémoires de l'académie de Berlin*, 1746, p. 444.

on my own sketch; it was made in Rome by Mr. Gmelin, an artist who is rightly celebrated both for his talent and for the breadth of his knowledge. During my most recent stay in Italy, he honored me by extending his friendship, and I am largely indebted to his care for anything in my work that might not be utterly unworthy of the public's attention.

PLATE XVIII 111

Boulder of Inti-Guaicu

In descending the hill whose summit is crowned by the Cañar fortress, one encounters, in a valley hollowed out by the Gulán River, a number of small tracks carved into the rock. These tracks lead to a crevice that, in the Quechua language, is called *Inti-Guaicu*, the *ravine of the sun*. In this lonely place fringed with lovely, vigorous plant life rises an isolated sandstone mass no more than four to five meters high. One of the faces of this small boulder is remarkably white. It drops off sharply, as though it had been cut by human hands. On this smooth, white background one can make out a number of concentric circles that represent the image of the sun as it was depicted by all the peoples of the earth at the dawn of civilization. The circles are of a blackish-brown color; in their interiors one can identify some half-faded lines that suggest two eyes and a mouth. The base of the boulder is carved into steps that lead to a seat cut into the same stone and positioned in such a way that one can contemplate the image of the sun from the bottom of a hollow.

The natives say that when the Inca Tupac Yupanqui advanced with his army to seize the kingdom of Quito, then ruled by the ʼConchocando of Licán, the priests discovered on the stone the image of the deity whose worship was soon to be introduced among the conquered peoples. The inhabitants of Cusco were convinced that they saw the image of the sun everywhere, just as Christians everywhere saw either crosses or the footprint of the apostle St. Thomas painted on rocks. The prince and the Peruvian soldiers regarded the discovery of the stone of Inti-Guaicu as an excellent omen: it very likely prompted the Inca to build a residence at Cañar, for it is known that the descendants of Manco-Capac saw themselves as the children of the sun, a belief that makes for a remarkable point of comparison between the first legislator of Peru and that of India,[1] who was also called ʼ*Vaivasvata*, son of the sun.

---

1. Menou II or Satyavrata [William Jones, "Sur les dieux de la Grèce, de l'Italie et de l'Inde,"] *Recherches Asiatiques*, Vol. I, p. 170; [Jones, "Sur la chronologie des Hindous," *Recherches Asiatiques*,] Vol. II, p. 172. Paolini, *Systema brachmanique*, p. 141.

Plate XVIII

A close examination of the Inti-Guaicu boulder reveals that the concentric circles are thin filaments of brown iron ore, very common in all sandstone formations. The lines indicating the eyes and the mouth are obviously cut with a metal tool; one must assume that they were added by the Peruvian priests to deceive the people more easily. Upon the arrival of the Spanish, the missionaries found it in their interest to hide from the eyes of the natives everything that the latter had venerated for centuries; one can indeed still make out the traces of the chisel used to deface the image of the sun.

According to Mr. Vater's interesting research, the word *inti*, sun, bears no resemblance to any other language known to the old continent. All in all, of the eighty-three American languages examined by this estimable scholar and by ʼMr. Barton of Philadelphia, only one hundred thirty-seven roots have thus far been identified that are also found in the languages of Asia and Europe—that is, in those of the Manchu Tartars, the Mongols, the Celts, the Basques, and the Estonians. This curious finding appears to confirm what we suggested above in our discussion of Mexica mythology. It is undeniable that the majority of the natives of the Americas belong to a race of men separated from the rest of humanity since the beginning of the world, and who exhibit, in the nature and the diversity of their languages, as in their features and the shape of their skulls, incontrovertible proof of their long and complete isolation.

PLATE XIX

Inga-Chungana near Cañar

To the north of the Cañar ruins rises a hillside that slopes gently down toward the Inca's house, although it drops steeply on the Gulán valley side. According to native legend, this hill was part of the gardens that surrounded the ancient Peruvian fortress. As near the *ravine of the sun*, we identified here a large number of small tracks hollowed out by human hands on the slope of a rock that is only lightly covered with topsoil.

In the gardens of Chapultepec near Mexico City, the European traveler contemplates with interest cypresses[1] whose trunks have a circumference of more than sixteen meters and which, it is believed, were likely planted by the kings of the Aztec dynasty. In the Inca's gardens near Cañar, we searched in vain for a tree at least a half-century old; nothing in those lands suggests that the Inca once resided there, with the possible exception of a small stone monument positioned on the edge of a precipice, the purpose of which is a subject of disagreement among the inhabitants of the country.

This small monument, which they call the *Inca's game*, consists of a small mass of stones. The Peruvians used the same trick to build it that the Egyptians used to sculpt the Sphinx at Giza, of which Pliny plainly states: "*e saxo naturali elaborate* [it is made of natural stone]." The boulder of quartziferous sandstone that serves as its base was eroded, so that after the layers that formed its summit were removed, all that remained was a seat surrounded by a wall, which is depicted on this Plate. It is surprising that a people who had stacked such a prodigious number of cut stones on the beautiful Azuay roadway resorted to such a bizarre method to erect a wall only one meter high. Every Peruvian work exhibits the character of a hardworking people who liked digging into the rock, who sought out challenges in order to show off their skill by overcoming them, and who gave all their buildings, even the most humble ones, a character of permanence, which might lead one to believe that at a different time they might have erected more significant monuments.

1. Cupressus disticha, L[innaeus].

Plate XIX

Seen from afar, *Inga-Chungana* resembles a settee with a back adorned by a chain-shaped arabesque. Entering into the oval-shaped enclosure, one sees that there is room on the seat for only one person, but that this person is very comfortably positioned and enjoys the most delightful view of the Gulán valley. A small river winds through this valley, forming several waterfalls whose froth is visible through clusters of gunnera and melastoma. This rustic seat would not be out of place in the gardens of Ermenonville or Richmond, and the prince who chose this site was not insensitive to the beauties of nature; he belonged to a people whom we do not have the right to call barbarians.

114

In this construction I saw only a backed chair located in a pleasant spot on the edge of a precipice, on the steep slope of a hillside overlooking a valley. Some of the older Indians, who are the local antiquarians, find this explanation too simple; they maintain that the incised chain on the edge of the wall was used to hold small balls that were raced for the prince's amusement. One cannot deny that the edge with the arabesque is somewhat sloped, and that the ball, if cast forcibly, could have rolled up from the point where the wall is visibly lower, just as easily as it had rolled down there. But if this theory were accurate, would one not find some hole that would have held the balls at the end of their race? The lowest part of the surrounding wall, the point opposite the seat, corresponds to an opening in the rock at the edge of the precipice. A narrow track cut into the sandstone leads to this grotto in which, according to native legend, there are treasures hidden by Atahualpa. People maintain that a trickle of water once flowed down this track. Is it there that one should search for the *Inca's game*, and was the wall positioned such that the prince could see comfortably whatever was happening on the steep slope of the boulder? We shall reserve our discussion of this grotto for the narrative of our journey to Peru.

PLATE XX

Interior of the Inca's House at Cañar

This Plate shows the plan and the interior of the small building that occupies the center of the esplanade in the Cañar citadel and that Mr. La Condamine believed to have been built as a guardroom. I took all the more care to render this drawing as precisely as possible, since the relics of Peruvian architecture, scattered across the ridge of the Cordilleras from Cusco to Cayambe or from 13° southern latitude to the equator, all display the same character in the cut of the stones, the shape of the doors, the symmetrical distribution of the re- cesses, and the total absence of external decoration. This uniformity of construction is so complete that all of the inns (*tambos*) located along the major roads, which the locals call the houses or palaces of the Inca, seem to be perfectly identical. Peruvian architecture did not rise above the needs of a mountain people; it had neither pilasters nor columns nor round arches. Born in a land spiked with boulders, on virtually treeless plateaus, it did not imitate the structure of a wooden framework, as did the architecture of the Greeks and the Romans. Simplicity, symmetry, and solidity are the three favorable characteristics that all Peruvian buildings shared.

The Cañar citadel and the square buildings that surround it are not made of the same quartziferous sandstone that covers the clayey schist and the porphyry of Azuay, and that is visible on the surface of the Inca's garden as one descends toward the Gulán valley. Nor are the stones used in the structures at Cañar granite, as Mr. La Condamine believed, but, rather, an extremely hard trappean porphyry within which are embedded both vitreous feldspar and amphibole. Perhaps this porphyry was extracted from the large quarries found at an elevation of four thousand meters near the lake of Culebrilla, more than three leagues from Cañar. It is at least certain that those quarries furnished the gorgeous stone used in the Inca's house in the plain of Pullal at an elevation nearly equal to what Puy-de-Dôme would reach were it placed on top of Canigou.

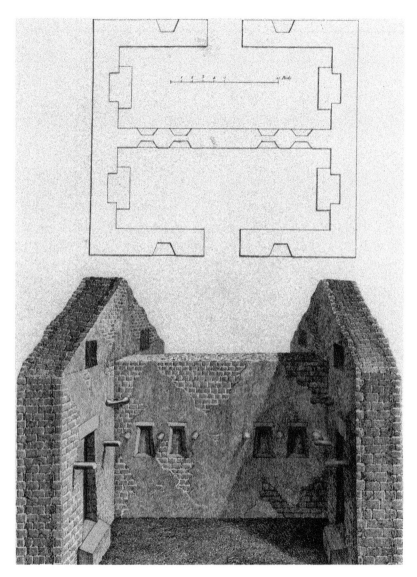

Plate XX

One does not find the enormous stones of the Peruvian structures in Cusco and the adjacent lands in the ruins of Cañar. Acosta measured some at Traquanaco that were twelve meters (thirty-eight feet) long, 5,8 meters (eighteen feet) wide, and 1,9 meters (six feet) thick. Pedro Cieza de León saw the same dimensions in the ruins of Tiahuanaco [Tiwanaku].[1] I did not notice any stones in the Cañar citadel that were over twenty-six decimeters (eight feet) long. They are, in general, much less remarkable for their massive size than for their exquisite cut: the majority of them are joined together with absolutely no appearance of cement, although the latter is visible in some of the buildings that surround the citadel, as well as in the Inca's three houses in Pullal, each one of which is more than fifty-eight meters long. The citadel is made from a mixture of small stones and clayey marl that effervesces upon contact with acid; it is true mortar, and I was able to remove considerable amounts of it using a knife to dig in the cracks between the parallel courses of stones. This fact is worthy of some attention, since all the travelers who had preceded me insisted that the Peruvians were not at all accustomed to using cement; but it was just as wrong to impute an ignorance of cement to them as it was to the ancient Egyptians. The Peruvians did not only use marly mortar: in the great structures of Pacaritambo,[2] they used asphalt (*betún*) cement, a mode of construction that, on the banks of the Euphrates and the Tigris, dates back to the earliest times.

116

The porphyry used in the structures at Cañar is cut into parallelepipeds, with such precision that the joints of the stones would be imperceptible, as Mr. La Condamine justly notes,[3] if their external surface were flat; but the outward-facing side of each stone is slightly convex and beveled on the edges, so that the joints form small flutes that serve as decorations, like the gaps between stones in rustic works. This cut of stone, which Italian architects call ʼ*bugnato*, is also found in the ruins of Callo, near Mulaló [Vallecaucano], where I drew it in detail.[4] It lends the walls of Peruvian structures a strong resemblance to certain Roman constructions, for example, the *muro di Nerva* in Rome.

What is especially characteristic of the monuments of Peruvian architecture is the shape of the doors, which were typically nineteen to twenty decimeters (six to eight feet) high, so that the Inca and other high lords could pass through them even when carried in a litter, on the shoulders of their vas-

1. Cieza, *Crónica del Perú* (Antwerp, 1554), p. 254.
2. Cieza, *Crónica del Perú* (Antwerp, 1554), p. 234.
3. *Mémoires de l'académie de Berlin*, 1746, p. 443.
4. See Plate XXIV.

sals. The jambs of these doors were not parallel but slanted, probably so that stone lintels of a lesser width could be used. The recesses (*hoco*) built into the walls, which served as cabinets, echo the shape of these *porte rastremate*: the slant of their jambs makes the Peruvian structures resemble those of Egypt, in which the lintels are always shorter than the lower door openings. Between the *hocos* are smooth-surfaced, five-decimeter-long cylindrical stones that jut forth from the wall; the natives claimed that these were used to hang weapons or clothes. In addition, strangely shaped struts made of porphyry are visible in the corners of the walls. Mr. La Condamine believes that these were meant to connect the two walls; but I am inclined to think that the ropes of the *hamacs* [hammocks] were looped around these struts, which serve this very purpose at least in all the Indians' huts in the Orinoco, where they are made of wood.

The Peruvians display an incredible ability for cutting the hardest stones. At Cañar there are curved tracks carved into the porphyry to replace the door hinges. In ancient structures built in the time of the Inca, La Condamine and Bouguer saw porphyry decorations that represent animal muzzles, with pierced nostrils bearing mobile rings of the same stone.[1] When, traversing the Cordillera by the Páramo of Azuay, I saw the enormous piles of cut stones extracted from the porphyry quarries of Pullal for the Inca's great roads, I was already skeptical of the claim that the Peruvians had no other tools besides stone axes; I suspected that friction was not the only method they had employed to level stones or give them a regular, uniform convexity. I consequently embraced a theory that was at odds with received wisdom: I hypothesized that the Peruvians had possessed copper tools that obtained great hardness when mixed with tin in a certain proportion. This hypothesis was substantiated by the discovery of an ancient Peruvian chisel found in Vilcabamba [Quechua: Willkapampa, sacred plain] near Cusco, in a silver mine worked during the time of the Inca. This precious instrument, which I owe to the kindness of Father Narciso Girbal and which I was fortunate to bring back with me to Europe, is twelve centimeters long and two centimeters wide. Its material composition was analyzed by ˙Mr. Vauquelin, who detected 0.94 copper and 0.06 tin. The Peruvians' *sharp copper* is almost identical to that of the Gallic axes, which chop wood just as well as steel does.[2] At the dawn of the civilization of peoples, the use of copper mixed with tin (*aes*, χαλκός [*chalkos*, bronze]) prevailed over that of iron everywhere in the old continent, even in areas where iron had long been in use.

---

1. *Mémoires de l'académie de Berlin*, 1746, p. 452, tableau 7, image 4.
2. See my *Essai politique sur [le royaume de] la Nouvelle-Espagne*, Vol. II, p. 485.

PLATE XXI

Aztec Bas-Relief Found in the Main Square of Mexico City

The foundation of the Mexico City cathedral, shown on the third Plate, lies on top of the ruins of the *teocalli*, or *dwelling of the god* Mexitli. This pyramidal monument, built by King Ahuizotl in 1486, was thirty-seven meters tall from its base to its upper platform, from which one enjoyed a magnificent view of the lakes, the neighboring countryside dotted with villages, and the curtain of mountains surrounding the city. This platform, which served as a refuge for combatants, was crowned by two tower-shaped chapels that were each seventeen to eighteen meters tall, so that the teocalli on the whole was fifty-four meters tall. After the siege of Tenochtitlan, the pile of stones that had formed the pyramid of Mexitli was used to build the *Plaza Mayor*. While conducting excavations eight to ten meters deep, they discovered a large number of colossal idols and other remnants of Aztec sculpture. Indeed, three curious monuments that we shall describe in this work—the so-called ʼsacrificial stone, the colossal statue of the goddess Teoyaomiqui, and the *Mexica calendar stone*—were found when the viceroy, the count of Revillagigedo, had the main square of Mexico City leveled, thus lowering the ground. A very trustworthy person, who was responsible for supervising this work, claimed that the cathedral foundations are surrounded by countless idols and reliefs, and that the three masses of porphyry that we mentioned above are the smallest of those they discovered while digging twelve meters down. Near the *Capilla del Sagrario* they discovered a sculpted rock that was seven meters long, six meters wide, and three meters high. Seeing that it was impossible to remove it, the workers wanted to break it up into pieces; but fortunately they were dissuaded by one of the canons of the cathedral, Mr. Gamboa, a learned man and lover of the arts.

Plate XXI

The stone commonly referred to as the sacrificial stone (*piedra de los sac-* 119
rificios) has a cylindrical shape; it is three meters wide and eleven decimeters high and is ringed with a relief in which one can identify twenty pairs
of figures, all of which are depicted in the same posture. One of these figures is always the same: it is a warrior, perhaps a king, whose left hand rests
on the helmet of a man who offers him flowers as a pledge of his obedience.
ˇMr. Dupé, whom I had occasion to cite at the beginning of this work, made
a copy of this relief. While on the premises, I verified the accuracy of his
drawing, one part of which has been engraved on this Plate. I have chosen
the remarkable group that includes a bearded man. One sees that the Mexica
Indians typically have a bit more facial hair than the rest of the natives of the
Americas; it is not rare to see them with mustaches. Was there once a province
whose inhabitants sported long beards? Is the beard in this relief postiche? Is
it one of those fantastic decorations that warriors used to strike fear into the
hearts of their enemies?

Mr. Dupé believes (rightly, in my view) that this sculpture depicts the conquests of an Aztec king. The conqueror is always the same, while the defeated
warrior wears the dress of the people to whom he belongs and whom he represents, as it were; placed behind the defeated man is the hieroglyph that indicates the conquered province. In the *Mendoza Collection* the conquests of
a king are likewise indicated by either a shield or a bundle of arrows placed
between the king, on one side, and, on the other, either the symbolic characters or the arms of the subjugated countries. Since Mexica prisoners were immolated in the temples, it would appear somewhat natural that the triumphs
of a warlike king should be depicted around the fatal stone on which the
topiltzin (sacrificing priest) ripped the heart from the unfortunate victim.
What made this theory especially credible was that the upper surface of the
stone exhibits a groove that is quite deep and appears to have served to drain
off the blood.

Despite this apparent evidence, I am inclined to believe that the so-called
sacrificial stone had never been placed at the top of a *teocalli*; rather, it was
one of those stones called *temalacatl*, on which the *gladiatorial combat*
between the prisoner slated for sacrifice and a Mexica warrior took place. The 120
real sacrificial stone, the one that crowned the platforms of the teocalli, was
green, either jasper or perhaps axstone jade. Its shape was that of a parallelepiped, fifteen to sixteen decimeters long and one meter wide; it had a convex
surface, so that the chest of the victim stretched out on the stone was raised
higher than the rest of the body. There are no reports from historians that this
green stone mass was sculpted: the tremendous hardness of the jasper and
jade stones most likely made it impossible to cut a bas-relief. In comparing

the cylindrical block of porphyry found in the main square of Mexico City with the oblong stones on which the victim was thrown as the *topiltzin* approached, armed with an obsidian knife, one easily sees that these two objects bear no resemblance to one another either in their material or in their shape.

On the other hand, in eyewitness accounts of the *temalacatl*, the stone on which the prisoner slated for sacrifice fought, one can easily recognize the stone whose relief Mr. Dupé sketched. The unknown author of the work published by ᵛRamusio, titled *Relazione d'un gentiluomo di Fernando Cortez*, explicitly states that the *temalacatl* was shaped like a millstone three feet in height, ringed with sculpted figures, and that it was large enough to be used for combat between two persons. The prisoners whose courage or rank had brought them distinction were reserved for the *sacrifice of the gladiators*. Placed on the *temalacatl* and surrounded by an immense crowd of spectators, they had to fight against six Mexica warriors in succession. Were they fortunate enough to win the fights, they were granted freedom and were allowed to return to their homeland. If, on the other hand, the prisoner-turned-gladiator succumbed to the blows of one of his opponents, then a priest, called *Chalchihutepehua*, dragged him dead or alive to the altar to tear out his heart.

It is certainly possible that the stone that was found in the excavations conducted around the cathedral was the very *temalacatl* that Cortés's *gentiluomo* claims to have seen near the wall of the great teocalli of Mexitli. The figures on the relief are nearly sixty decimeters tall. Their footwear is quite remarkable: the winner's right foot is tipped by a sort of nib that appears to be intended for his defense. We may be surprised to find this weapon (of which there is no equivalent among any other peoples) only on the left foot. This same figure, whose stocky body recalls the original Etruscan style, holds the loser by the helmet while squeezing his left hand. In a large number of Mexica paintings that depict battles, we see warriors who are also holding weapons in their left hand: they are shown acting more often with the left than with the right hand.

At first sight one might think that this quirk stems from peculiar customs, but examining a large number of the Mexica's historical hieroglyphs reveals that their paintings placed the weapons sometimes in the right hand and other times in the left, so long as this resulted in a symmetrical arrangement within the groups. I have found striking examples of this in leafing through the *Codex anonymous* at the Vatican, in which we find Spaniards holding their sword in their left hand.[1] It should be noted that this oddity of confusing the right with the left hand is characteristic of the early stages of art. It can be also

1. *Codex Vaticanus anonymous*, folio 86.

seen in a few Egyptian reliefs in which one even finds right hands attached to left arms, as a result of which the thumbs appear to be placed on the outside of the hands. Some learned antiquarians believe they have identified something mysterious in this extraordinary arrangement, which Mr. Zoëga attributes to mere whim or oversight on the part of the artist. I very much doubt that the bas-relief surrounding this *temalacatl* and the numerous other sculptures in basaltic porphyry were executed using only tools made of jade or other extremely hard stones. It is true that I have searched in vain to obtain one of the ancient Mexica metallic chisels similar to the one that I brought back from Peru, but ʼAntonio de Herrera [y Tordesillas], in the tenth book of his General History of the Vast Continent and Islands of America [*Historia general de las Indias*], clearly states that the inhabitants of the coastal province of Zacatollan, situated between Acapulco and Colima, prepared two kinds of copper, one of which was hard and sharp, the other malleable. The hard copper was used to make axes, weapons, and agricultural tools, while the flexible copper was used for vases, boilers, and other household utensils. Considering that the coast of Zacatollan was subject to the kings of Anahuac, it seems improbable that on the outskirts of the capital of the kingdom people continued to sculpt stones using the friction method when they could have obtained metallic chisels. This Mexica sharp copper was probably mixed with tin, like the tool found at Vilcabamba and the Peruvian ax that ʼGodin had sent to Mr. de Maurepas, and that the Count Caylus believed to be *tempered copper*.

PLATE XXII

Basaltic Rocks and the Regla Waterfall

As one changes latitude and climate, one notices a change in the appearance of organic nature, the shape of the animals and the plants, which lends a particular character to each zone; with the exception of a few aquatic plants and cryptogams, the soil of each region is covered in different plants. This is not the case for inorganic nature, that aggregation of earthy substances that covers the surface of our planet. The same decomposed granite on which, in the frigid climate of Lapland, grow Vaccinium, Andromeda, and the lichen that feeds the reindeer is also found in the groves of tree ferns, palm trees, and Heliconia, whose glossy foliage thrives under the influence of the equatorial hot season. When, at the end of a long voyage, after crossing from one hemisphere to the other, the northerner reaches a distant coast, he is surprised to find, amidst a plethora of unknown flora and fauna, the same strata of slate, micaceous schist, and trappean porphyry that form the arid coasts of the old continent, lapped by the glacial Ocean. To the traveler, the earth's rocky crust has the same appearance in all climes; everywhere he ventures in the new world, he recognizes—and not without some degree of emotion—the rocks of his homeland.

The similarities exhibited by inorganic nature extend to the minor phenomena that one would be tempted to attribute to purely local causes. In the Cordilleras, as in the mountains of Europe, granite occasionally exhibits aggregations in the form of flat spheroids divided into concentric layers. In the tropics, as in the temperate zone, one finds within the granite masses abundant in both mica and amphibole that resemble blackish balls set in a mixture of feldspar and milky quartz. Metalloid diallage is found in the serpentine rock of the island of Cuba, just as it is in that of Germany; the amygdaloids and the pearlstone of the plateau of Mexico seem identical to those at the foot of the Carpathian Mountains. The superposition of secondary rocks follows the same laws, even in regions separated by vast distances. In all parts the same monuments attest to the same order in the revolutions that have progressively shaped the surface of the globe.

Plate XXII

If one thinks back to physical causes, one should not find it surprising that travelers have not discovered new rocks in distant regions. Climate influences the shape of animals and plants, because the interplay of affinities that governs the development of the organs is affected both by the temperature of the atmosphere and by the heat generated from the various combinations formed by chemical reactions. Although the unequal distribution of heat, which stems from the oblique angle of the ecliptic, cannot have had any observable influence on the formation of rocks, this very formation must have had a powerful impact on the temperature of the globe and the surrounding air. The phenomenon of large masses moving from a liquid to a solid state cannot take place without being accompanied by an enormous caloric emission. These considerations appear to shed some light on the first migrations of animals and plants. Were I not loath to add to the number of fanciful geological theories, I might be tempted to explain several important problems through reference to this progressive temperature elevation, especially the problem posed by the presence of flora and fauna from the Indies in the soil of the northern lands.

The basaltic rocks of Regla depicted on this Plate constitute undeniable proof of the similarity in shape among rocks from different climes. The traveling mineralogist has only to glance at this drawing to recognize the shape of the basaltic rocks of the Vivarais, the Euganean hills, or the promontory at Antrim, in Ireland. Even the most minor irregularities in the columnar rocks of Europe are also found in the same group of basaltic rocks in Mexico. This high degree of structural similarity suggests that the same causes have affected all climes and at very different times, for the basaltic rocks covered in clayey schist and compact limestone must be of a different age than those resting on coal beds and on pebbles.

124

The small waterfall of Regla is located twenty-five leagues to the northeast of Mexico City, between the famous mines of Real del Monte and the thermal waters of Totonilco. A narrow river, which is used to drive the stamp-mills of the ˇamalgamation factory at Regla (the construction of which cost more than ten million livres tournois), cuts its path through groups of basaltic columns. The sheet of water that plunges here is quite considerable, but this waterfall is only seven to eight meters high. The surrounding boulders (whose grouping is reminiscent of the grotto of Staffa in the Hebrides), the contrasts of the vegetation, the untamed appearance, and the solitude of this site all make this small waterfall quite delightful. From both sides of the ravine rise columnar basalt rocks over thirty meters high, dotted with clumps of cactus and yucca filamentosa. The prisms generally have five to six sides and are sometimes up to twelve decimeters wide; several of them have very regular joints.

Each column has a cylindrical core with a denser mass than the surrounding parts; these cores are seemingly mounted inside the prisms, which exhibit odd convexities in their horizontal fractures. I have indicated this structure (also found in the basaltic rocks of Cape Fair Head) in the left foreground of the drawing.

Most of the columns at Regla are perpendicular, though near the waterfall one also sees a number of them with an incline of 45° toward the east; farther on, some are horizontal. Each group appears to have been affected by a different source of attraction at the moment of its formation. The bulk of these basaltic rocks is very homogeneous; Mr. Bonpland detected cores of either olivine or graniform peridot, surrounded by crystallized mesotype. The prisms—and this fact is worthy of geologists' attention—rest on a bed of slate, under which lies even more basalt. In general, the basalt of Regla is superimposed on the porphyry of Real del Monte, while a compact limestone rock serves as a base for the basalt of Totonilco. This entire basaltic region is at an elevation of two thousand meters above sea level.

PLATE XXIII

Basalt Relief Showing the Mexica Calendar

Among all the monuments that seem to constitute evidence that the peoples of Mexico had attained a certain degree of civilization by the time of the Spaniards' arrival, pride of place can be given to the calendars, the different divisions of time that the Toltec and Aztecs adopted for the use of society at large, for setting the order of the sacrifices or for facilitating astrological calculations. This kind of monument is all the more worthy of our attention in that it attests to the knowledge we have difficulty seeing as the result of observations that mountain peoples made in the uncultivated regions of the new continent. It is tempting to believe that the Aztec calendar is like those lexically and grammatically rich languages found among peoples whose collective ideas are inferior in number to the signs through which such ideas might be conveyed. Those rich, flexible languages and these systems of intercalation, which suggest quite an exact knowledge of the length of the astronomical year, are perhaps only the remnants of a heritage bequeathed to them by peoples who had once been civilized but then regressed into barbarism.

The monks and the other Spanish writers who visited Mexico shortly after the conquest have given only vague and often contradictory information about the different calendars that the peoples of the Toltec and the Aztec races commonly used. Such ideas can be found in the works of Gómara, Valadés, Acosta, and Torquemada. Despite his superstitious gullibility, Torquemada has left us, in his *Monarquía indiana*, with a collection of precious facts that demonstrate a precise knowledge of the localities. Torquemada lived among the Mexica for fifty years and had arrived in the city of Tenochtitlan at a time when the natives still preserved a large number of historical paintings, and when one could still see, in front of the ˙Marquis del Valle's house[1] on the Plaza Mayor, the ruins of the great teocalli[2] dedicated to the

---

1. See p. 22 [this edition], Plate III.
2. The year 1577. Torquemada, [*Monarquía indiana*,] Book VIII, Ch. II, Vol. II, p. 157.

god Huitzilopochtli. Torquemada used the manuscripts of three Franciscan monks—'Bernardino de Sahagún, Andrés de Olmos, and Toribio de Benavente—who had all studied American languages extensively and had gone to New Spain in Cortés's time, before the year 1528. Despite such advantages, this historian of Mexico did not provide as much clarification about the Mexica chronology and calendar as one might have expected, considering his enthusiasm and his learning. He expresses himself with such little regard for accuracy that one reads in his work that the Aztecs' year ended in December and began in February.[1]

For a long time, the convents and public libraries of Mexico City had in their possession materials that were more informative than the accounts of the first Spanish historians. Three Indian authors—'Cristóbal del Castillo, a native of Tetzcoco who died in 1606 at eighty years of age, Fernando de Alvarado Tezozomoc, and Domingo Chimalpahin—left manuscripts composed in the Aztec language on the history and the chronology of their ancestors. These manuscripts, which contain a large number of dates listed in accordance with the Christian era, as well as with the natives' civil and ritual calendar, were fruitfully studied by the scholar Carlos de Sigüenza, professor of mathematics at the University of Mexico; the Milanese traveler Boturini Benaducci; Father Clavijero; and, most recently, Mr. Gama, whose astronomical works I often had occasion to cite with great praise in another work.[2] Finally, in 1790 a stone of enormous dimensions, covered in characters relating to the Mexica calendar, religious festivals, and the days when the sun crossed the zenith over Mexico City, was discovered in the foundations of the ancient teocalli. It was used both to clarify points in question and to call the attention of some learned natives to the Mexica calendar.

I have tried, not only during my stay in the Americas but also since my return to Europe, to conduct a rigorous study of everything that has been published about the Aztecs' division of time and their system of intercalation. I examined on site the famous stone found in the Plaza Mayor, which is shown on the twenty-third Plate. I have distilled some interesting concepts from the hieroglyphic paintings preserved in the convent of San Felipe Neri in Mexico City. In Rome I pored over the handwritten Commentary that Father Fábrega composed on the *Codex Mexicanus* in Velletri. I regret, however, that I do not know Mexica [Nahuatl] well enough to read the works that the natives wrote in their own language, using the Roman alphabet, immediately after the capture of Tenochtitlan. I was, therefore, unable to verify personally

127

---

1. [Torquemada, *Monarquía indiana*,] Book X, Chs. X, XXXIII, XXXIV, and XXVI.
2. *Essai politique sur le royaume de la Nouvelle-Espagne*, p. 124.

all of the claims that Sigüenza, Boturini, Clavijero, and Gama made about the Mexica system of intercalation, by comparing their claims to the manuscripts of Chimalpahin and Tezozomoc, from which those authors supposedly drew their ideas. Whatever doubts may remain about certain points in the minds of those scholars, who are accustomed to subjecting facts to a thorough critique and to accepting only what has been rigorously proved, I am pleased to have called attention to a remarkable monument of Mexica sculpture and to have provided new details about a calendar that neither Robertson nor the illustrious author of *Histoire de l'astronomie* seems to have approached with as much curiosity and care as it merits. People's curiosity will be stoked by the information we shall provide below about the Mexica tradition of the *four ages*, or four suns, which bears a striking similarity to *the *yugas* and the *kalpas* of the Hindus, and to the ingenious methods that the Muisca Indians, a mountain people of New Granada, employed to correct their lunar years through the intercalation of a thirty-seventh moon, called *deaf* [or secret] *cuhupqua*. It is by comparing and contrasting the different systems of American chronology that we shall be able to assess the contacts that appear to have taken place in very remote times between the peoples of India and Tartary and those of the new continent.

The Aztecs' civil year was a solar year of three hundred sixty-five days; it was divided into eighteen months of twenty days each. After these eighteen months, or three hundred sixty days, they added five supplementary days and began a new year. The names *Tonalpohualli* [count of days] and *Cempohualilhuitl*, which distinguish this civil calendar from the ritual calendar, are very much indicative of its primary characteristics. The first of these names means *count of the sun,* in contrast with the ritual calendar, *count of the moon*, *Metzlapohualli*. The second name derives from *cempohualli*, *twenty*, and from *ilhuitl*, *festival*; it alludes either to the twenty days contained within each month or to the twenty solemn festivals celebrated throughout the civil year in the teocalli, *god-dwellings*.

The Aztecs' civil day began at sunrise, like that of the Persians, the Egyptians,[1] the Babylonians, and the rest of the peoples of Asia, with the exception of the Chinese. It was divided into eight intervals, a division that is also found[2] among the Hindus and the Romans. Four of these eight intervals were determined by sunrise, sunset, and the sun's two meridian transits. Sunrise was called *Yquiza Tonatiuh*; midday, *Nepantla Tonatiuh*; sunset, *Onaqui*

128

---

1. *Ideler, *Historische Untersuchungen über die astronomischen Beobachtungen der Alten*, p. 26.

2. Bailly, *Histoire de l'Astronomie ancienne*, p. 296.

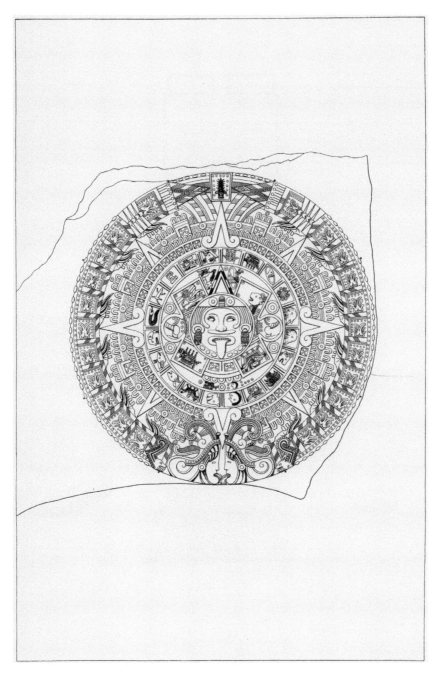

Plate XXIII

Tonatiuh; and midnight, *Yohualnepantla*. The hieroglyph for the day was a circle divided into four parts. Although the length of the day on the parallel of Mexico City varies no more than two hours and twenty-one minutes, it is nevertheless certain that Mexica hours must have originally been unequal, like the Jews' *planetary hours* and all those that the Greek astronomers referred to as καιρικαί [*kairikai*, irregular hours, of unequal duration], in contrast with the ίσημεριναί [*isemerinai*, regular hours, of the same duration], *equinoctial hours.*

The periods of day and night, which correspond more or less to our 3:00, 9:00, 15:00, and 21:00 in astronomical time, had no particular names. To refer to these periods, the Mexica, like our laborers, gestured to the point in the sky where the sun would be positioned as it followed its east-west course. This gesture was accompanied by the remarkable words *iz Teotl, God must be there*, a phrase that recalls the happy time when the peoples who emerged from Aztlan still knew no other deity than the sun and had no bloodthirsty worship whatsoever.[1]

Each Mexica month of twenty days was subdivided into four short five-day periods. It was at the beginning of these short periods that each community held its fair, or *Tianguiztli*. The Muisca, a people of southern America, had three-day weeks. It appears that none of the peoples of the new continent observed the same week, a seven-day cycle, that is found among the Hindus, Chinese, Assyrians, and Egyptians and that, as ˈLe Gentil justly observed,[2] was widespread among most of the peoples of the old world.

One passage in Garcilaso's history of the Inca led ˈMr. Bailly and Mr. Lalande[3] to think that the Peruvians counted in cycles of seven days. According to Garcilaso, "the Peruvians determine the months by the moon; they determine their half-months by the waxing and waning moons; they determine their weeks by the quarter moons, without having any particular names for the days of the week." But Father Acosta, who was more learned than Garcilaso and wrote the first installments of his physical geography of the new continent on site in Peru near the end of the sixteenth century, clearly states that neither the Mexica nor the Peruvians observed the short period of seven days, "since this period," he adds, "depends no more on the course of the moon than on that of the sun. It owes its origin to the number of the planets."[4]

1. See p. 115 [this edition].
2. Le Gentil, *Histoire de l'Académie*, 1772, Vol. II, pp. 207, 209. ˈLaplace, *Exposition du système du Monde*, p. 272.
3. Bailly, *Histoire de l'astronomie*, Book V §17, p. 408. Lalande, *Astronomie*, §1534.
4. Acosta, *Historia natural y moral de las Indias*, Book VI, Ch. III (Barcelona edition, 1591), p. 260.

In reflecting upon the system of the Peruvian calendar, one understands that although the phases of the moon change more or less every seven days, this correspondence is nevertheless insufficiently exact for the seven-day cycles to correspond to the lunar phases over several consecutive lunar months. According to ˅Polo [de Ondegardo] and all the writers of the time, the Peruvians had years (*huata*) of three hundred sixty-five days that, as we shall see below, were based on observations made every month in the city of Cusco. Like nearly all the years observed by the peoples of Asia, the Peruvian year was divided into twelve *moons*, *quilla*, the synodic revolutions of which were completed in three hundred fifty-four days and forty-eight minutes. Adhering to an ancient custom, to correct the lunar year and bring it into alignment with the solar year, they added eleven days, which were distributed across the twelve moons in accordance with the dictates of the Inca. Following this arrangement, it is hardly possible that the four equal periods into which the lunar months would have been divided could have lasted seven days and corresponded to the phases of the moon. The very historian whose account Mr. Bailly cites in support of the theory that the Hindus' week was known to the Americans claims that according to an ancient law set down by the Inca Pachacutec, there had to be three festival and market (*catu*) days in every lunar month and that the people were obliged to work not for seven but for eight days in order to rest on the ninth.[1] This is, without a doubt, a division of a lunar month, or a sidereal revolution of the moon, into three short periods of nine days.

130

We take this opportunity to observe that the Japanese,[2] a people of the Tartar race, did not observe the short seven-day period either, although it is used among the Chinese, who also appear to be originally from Tartary but who have had close contacts with Hindustan[3] and Tibet for some time.

We have seen above that the Mexica year, like that of the Egyptians and the French *new calendar*, had the advantage of being divided into months of equal length. The Mexica referred to the five supplementary days, the *epagomenae* (ἐπαγόμεναι [*epagomenai*, intercalated days]) of the Egyptians, as *nemontemi*, or *empty* days. We shall shortly discover the origin of this term; it suffices to observe here that children born during the five supplementary days were regarded as unlucky and were called *nemoquichtli* or *nencihuatl*, *unfortunate man* or *woman*, in order that these names might remind them throughout their lives what little faith they should place in their star.

1. Garcilaso, [*Comentarios reales*,] Book VI, Ch. XXXV, Vol. I, p. 216.

2. ˅Thunberg, *Voyage au Japon*, p. 317.

3. Sir William Jones, ["Discours anniversaire, Prononcé, le 2 Février 1786, par le Président,"] *Recherches Asiatiques*, Vol. I, p. 420.

Thirteen Mexica years formed a cycle called *tlalpilli*, which was analogous to the Romans' ˅indiction. Four tlalpilli formed a period of fifty-two years, *xiuhmolpilli*, the ˅binding [or bundling] of the years. Finally, two of these fifty-two-year periods formed an *old age, cehuehuetiliztli*. To express myself with greater clarity, I shall use the same terms as several other Spanish authors and will call the *binding* [bundling] a half-century and the *old age* a century. The hieroglyph for a half-century is in keeping with the figurative meaning of the word: it is a bundle of reeds tied by a ribbon. The Mexica regarded a half-century (*xiuhmolpilli*) as a *long year*, and this name most likely prompted Gómara[1] to call the indictions (or the four thirteen-year cycles) *long weeks, las semanas del año*.

The idea of referring to a period by a word that recalls a *binding* of years or moons is also found among the Peruvians. In Quechua, the *lingua del Inga*, a year of three hundred sixty-five days is called a *huata*, a word that clearly derives from *huatani*, to tie, or *huatanan*, a thick rope made of rushes. The Aztecs, it should be noted, did not have any hieroglyphs for an *old age* (that is, a century of one hundred four years), the name of which indicates the life span, as it were, of the elderly.

To summarize what we have just stated concerning the division of time, we find that the Mexica had short periods of five days (half-decades), months of twenty days, civil years of eighteen months, indictions of thirteen years, half-centuries of fifty-two years, and centuries, or *old ages*, of one hundred four years.

According to Mr. Gama's intriguing studies, it seems certain that at the close of a fifty-two-year cycle the civil year of the Toltecs and the Aztecs, like that of the Chinese and the Hindus, ended at the winter solstice, when, as the first missionaries sent to Mexico City wrote simply, "the sun, in its current course, begins its work anew, *cuando desanda lo andado*." This same beginning of the year is also found among the Peruvians, whose calendar alone proves that they do not descend from the Toltec, as several writers have baselessly claimed.[2] According to a legend preserved among the people of Cusco,[3] the first day of the year formerly corresponded to our January 1, until the Inca ˅Titu Manco-Capac, who adopted the epithet *Pachacutec* (*Renewer of Time*),

131

---

1. Gómara, *Conquista de México*, 1553, folio 118.

2. See p. 40 [this edition], and in my "Essai sur la population primitive de l'Amérique," Berlin, [*Neue Berlinische*] *Monatschrift*, March 1806, pp. 177, 208.

3. Acosta, [*Histoire naturelle y morale des Indes*,] p. 260.

ordained that the year begin "when the sun retraces its steps," that is, at the winter solstice.

Among the Spanish authors there is much confusion regarding the names and the order of the eighteen Mexica months. Several of these months had three or four names each, and since some authors have forgotten that the Mexica, when composing a periodical series of signs or hieroglyphs, wrote from *right* to *left*, starting from the bottom edge of the page, they have mistaken the last month for the first. The Aztecs linked together the series of hieroglyphs that indicated the fifty-two year cycle, *xiuhmolpilli*, to form what they called half-century *wheels*. A serpent biting its own tail is wrapped around the wheel and indicates the four *indictions*, tlalpilli, through four knots. This emblem recalls the serpent or dragon that, for the Egyptians as well as for the Persians,[1] represented the century, or one revolution, *aevum* [eternity]. Within this fifty-two-year wheel, the serpent's head designates the beginning of the cycle. This is not at all the case in the *one-year wheel*: there the serpent is not wrapped around the eighteen hieroglyphs for the months and nothing marks the first month of the year.

Since the paper on the Aztec almanac that Mr. Gama published in Mexico City is very difficult to find in Europe, I shall record the series of the months, following that scholar's painstaking research. I shall add here the etymology of the names, all of which are related to the festivals, public works, and climate of Mexico. It goes without saying that *Tititl* is the first month, since 'the Indian Cristóbal del Castillo clearly states in his handwritten history that the *nemontemi*, intercalary days, were added to the end of the month *Atemoztli*. Here are the names of the eighteen months:

1. *Tititl*, perhaps from *titixia*, to glean after the harvest; *Itzcalli*, the month meant for renovating and whitewashing the interiors of houses and temples. From January 9 to January 28, in the first year of the first indiction of the *Xiuhmolpilli* cycle.

2. *Xochilhuitl*. January 29 to February 17.

3. *Xilomanaliztli*; *Atlcahualco*, lacking water or rain; *Quahuitlehua*, month in which the trees begin to sprout; *Cihuailhuitl*, the festival of women. February 18 to March 9.

1. Bailly, [*L'Histoire de l'astronomie ancienne*,] p. 515.

4. *Tlacaxipehualiztli*; the name of this month recalls the horrifying ceremony in which they flayed human victims in order to tan their skins, which were used for priestly garments, as one can see in the hieroglyphic painting shown on Plate XXVII; *Cohuailhuitl*, festival of the grass snake. March 9 to March 29.

5. *Tozoztontli*, month of vigils, because the ministers of the temples were obliged to stay awake during the great festivals celebrated this month. March 30 to April 18.

6. *Hueytozoztli*, the great vigil, the great penance. April 19 to May 8.

7. ˅*Toxcatl*, the month in which *ropes* and garlands of corn were draped around the necks of the idols; *Tepopochuiliztli*, censer [incense holder]. From May 9 to May 28. It was in this month Toxcatl that Cortés's companion in arms, Pedro de Alvarado, the fierce warrior whom the Mexica called the Sun, *Tonatiuh*, because of his blond hair, inflicted horrific carnage upon the Mexica nobility gathered within the walls of the teocalli. This attack was the signal for the civil unrest that led to the death of the unfortunate king Montezuma.

8. *Etzalqualiztli*, a name that may derive from *etzalli*, a particular dish made from corn flour. May 29 to June 17.

9. *Tecuilhuitzintli* [also Tecuilhuitontli], the month or festival of the young warriors. June 18 to July 7.

10. *Hueytecuilhuitl*, the festival of the nobility and of the warriors already advanced in years. July 8 to July 27.

11. *Miccailhuitzintli* [also Miccailhuitontli], the minor festival of the dead; *Tlaxochimaco*, the spreading of flowers. July 28 to August 16.

12. *Hueymiccailhuitl*, the major festival celebrated in memory of the dead; *Xocotlhuetzi*, fall of the fruits, the month in which fruits ripen, corresponding to the end of summer. August 17 to September 5.

13. *Ochpaniztli*, broom, the month dedicated to cleaning the canals and to renovating causeways and roads; *Tenahuitilitzi*. September 6 to September 25.

133

14. *Pachtli*, from the name of a parasitic plant [tree moss] that begins to sprout at this time on the trunks of old oaks; *Ezoztli*; *Teotleco*, the arrival of the gods. September 26 to October 15.

15. *Hueypachtli*, the month in which the *pachtli* plant is already grown; *Tepeilhuitl*, festival of the mountains, or rather, of the rustic deities that governed the mountains. October 16 to November 4.

16. *Quecholli*, the month in which the flamingo (*phoenicopterus*)—a bird that the Mexica called *Teoquechol*, the divine heron, because of the beautiful color of its feathers—arrives on the shores of Lake Texcoco. November 5 to November 24.

134

17. *Panquetzaliztli*, either from the name given to the standard of the god *Huitzilopochtli*, carried in the processions during the famous festival of *Teocualo*, or from the *god eaten by the faithful* in the form of corn flour kneaded with blood. November 25 to December 14.

18. *Atemoztli*, fall of the waters and the snows; the latter begin to cover the mountains surrounding the Valley of Mexico toward the end of December. December 15 to January 3.

In the first year of the cycle, the five supplementary days correspond to January 4, 5, 6, 7, and 8. If a people uses intercalation only once every fifty-two years, the beginning of its year moves back one day every four years, more or less, and therefore twelve to thirteen days by the end of the cycle, *Xiuhmolpilli*. The result of this, as we shall see below, is that the final supplementary day, *nemontemi*, of the final year of the Mexica cycle, corresponds to December 26. Since the five nemontemi were regarded as *vague* and *unlucky* days, people considered the day of the winter solstice, or December 21, the end of the *xiuhmolpilli*. Neither the nemontemi (epagomenae) nor the twelve or thirteen intercalary days belong to the two years between which they fall, and it is for this reason that we referred to the winter solstice above as the end of a fifty-two-year cycle, rather than the beginning.

In the third, fourth, and fifth months, which correspond to our months February, March, and April, solemn festivals were held in honor of *Tlalocteuctli*, the god of water, this being the period of the great droughts that lasts until the end of June and July in the mountainous region. If the priests had neglected the intercalation, the festivals in which they prayed to the gods for a year of plentiful rains would have occurred increasingly close to the time of the harvests. The people would have noticed that the order of the sacrifices had been reversed, and since there were no lunar months, they would not even have been able to accuse the moon, like Aristophanes's gods did,[1] of having sown disorder within the calendar and within their worship. As for the names and the hieroglyphs of the Mexica months, there is nothing to suggest that they were conceived in a more northerly clime. It is true that the word *quahuitlehua* recalls that the trees are covered in young leaves toward the end of February; but this phenomenon, which is not seen in the lowlands of the Torrid Zone, is peculiar to the mountainous region located on 19 and 26 degrees latitude, where the oaks begin to develop new leaves before they have completely shed their old ones.

To this point, we have spoken about the civil calendar called *the count of the sun, Tonalpohualli*. It remains for us to examine the ritual calendar, called *count of the moon, Metztlapohualli*, and *count of the festivals, Cemilhuitl-apohualiztli*, from *tlapohualiztli, count*, and *ilhuitl, festival*. We find traces of this second calendar, the only one used by the priests, in almost all the hieroglyphic paintings preserved to this day; it presents a uniform series of short thirteen-day periods. These short periods can be considered *semilunar months*; they probably derived from the two states of *wakefulness, ixtozoliztli*, and *sleep, cochiliztli*, that the Mexica attributed to the moon, depending on whether this star shone throughout most of the night or whether, appearing on the horizon only in the daytime, it seemed, in the minds of the people, to rest at night. The relationship between the thirteen-day periods and the periods when the moon is visible, before and after opposition, most likely resulted in the name *count of the moon* being given to the ritual calendar; but this name should not prompt us to search for a lunar year in the series of short cycles that uniformly succeed one another and that bear no relation to either the phases or revolutions of the moon.

The multiples of the number 13 contain properties that the Mexica used to maintain the concordance between their ritual and civil almanacs. A civil year of three hundred sixty-five days contains one day more than twenty-eight thirteen-day periods. Since the fifty-two year cycle was divided into four *tlalpilli* of thirteen years, this supernumerary day creates a complete short period at the end of each indiction, and one *tlalpilli* contains three hundred sixty-five of these periods; in other words, it has as many thirteen-day weeks as the year has civil days. One year of the ritual almanac contains twenty *semi-*

---

1. ▾Aristophanes, *Nubes*, line 615.

lunar months, or two hundred sixty days, and this same number of days contains fifty-two half-decades, or short periods of five days. The Mexica thus found their favorite numbers—5, 13, 20, and 52—in the concordance of their two counts: the count of the moon and the count of the sun. A fifty-two-year cycle contained fourteen hundred sixty short periods of thirteen days; and if one adds to this the thirteen intercalary days, one has fourteen hundred sixty-one short periods, a number that happens to coincide with the number of years that constitute the Sothic cycle.

The cycle of nineteen solar years, which corresponds to two hundred thirty-five lunar months (and which the Chinese observed more than sixteen centuries before ʼMeton),[1] has no multiples in either the sixty-year cycle—which is still observed by most of the peoples of eastern Asia and by the Muisca of the Bogotá plateau—or the fifty-two-year cycle adopted by all the peoples of the Toltec, Acolhua, Aztec, and Tlaxcalteca races. It is true that five *old ages* (of one hundred four years each) form a Julian period, give or take one year, and that the double of the Metonic cycle is almost equal to three *indictions* (*tlalpilli*) of the Mexica year; but there is no multiple of thirteen that corresponds exactly to the number of days contained within a period of two hundred thirty-five lunar months. The Metonic cycle contains one hundred thirty-three and a half short thirteen-day cycles, while the Callippic cycle contains two thousand one hundred thirty-four and one-thirteenth. Knowledge of these periods was useful to the peoples of Asia who, like the Peruvians, the Muisca, and other tribes of southern America, had lunar years; but it must have been of no importance whatsoever to the Mexica, since the so-called *count of the moon* (*Metzlapohualli*) was nothing more than an arbitrary division of a long period of thirteen astronomical years into three hundred sixty-five short thirteen-day periods, each of which lasts roughly as long as the *sleep* or *wakefulness* of the moon.

The Mexica maintained annals that dated back eight and a half centuries before the time of Cortés's arrival in the land of Anahuac. We explained above how the subdivisions of these annals showed either a fifty-two-year cycle, a thirteen-year tlalpilli, or a single year of two hundred sixty days divided into twenty short thirteen- day periods, depending on the level of detail within the story. Next to the periodic series of hieroglyphs of either the years or the days are brightly colored paintings—hideous both for their shapes and for their extremely imprecise rendering but often naive and ingenious in their composition—which depict the migrations of peoples, their battles, and the

137

1. Laplace, *Expos[ition du système du monde]*, Vol. II, p. 267.

events that had distinguished each king's reign. It is undeniable that Valadés, Acosta, Torquemada, and more recently Sigüenza, Boturini, and Gama, derived insights from paintings that date back as far as the seventh century. I have held in my own hands paintings in which one could identify the migrations of the Toltecs, but I doubt that the first Spanish conquerors found, as Gómara claims,[1] annals that retraced events *year by year* over a period of eight centuries. The Toltecs disappeared[2] four hundred sixty-eight years before Cortés's arrival; the people whom the Spanish found settled in the Valley of Mexico were of the Aztec race. They could have learned what they knew of the Toltecs only from the paintings the latter had left behind in the land of Anahuac, or else from a few scattered families who, detained by the love of their native soil, had not wished to partake in the fortunes of emigration.

According to Gama, the Aztecs' annals began in a time that corresponds to the year 1091 of our era, a time when, by order of their leader Chalchiuht-latonac, they celebrated the festival of the renewal of the fire in Tlalixco, also known as Acahualtzinco, probably located at 33° or 35° northern latitude. As the Indian historian Chimalpahin unequivocally states, it was only after the year 1091, when they bound together the years for the first time since their departure from Aztlan, that Mexica history shows the greatest degree of order as well as a surprising level of detail in the narration of events.

According to what we have explained up to this point about the *count of the sun* and about the uniform division of the year into eighteen months of equal length, it should have been easy for the Mexica to refer to the period in which specific historical events took place by citing the day of the month and counting the number of years that had passed since the famous sacrifice of Tlalixco. This simple, natural method would probably have been followed had the annals of the empire not been held by the priests, *Teopixqui*. It is true that one occasionally finds a month-hieroglyph to which round dots have been added and arranged into two unequal rows, which demonstrates that the Aztec priests, as we have observed above, ordered the various terms in a series from *right* to *left* and not from *left* to *right*, as do the Hindus and almost all of the peoples who dwell in Europe today. One can still see in Mexico City the copy of a painting that was formerly preserved in the Chevalier Boturini's museum, in which the sign for the month *quecholli*, followed by thirteen dots, is placed next to a Spanish lancer, with the hieroglyph for the city of Tenochtitlan appearing under the feet of the lancer's steed. This painting most certainly depicts the initial entry of the Spanish into Mexico City on day 13 of the

138

1. Gómara, *Conquista de México*, folio CXIX.
2. See p. 43 [this edition].

month quecholli, which, according to Gama, corresponds to November 17, 1519; but it must be admitted that simple month dates, expressed by the number of days elapsed, are only rarely found in the Mexica annals.

As for the years, those which belonged to the same fifty-two-year cycle were never distinguished from one another through the use of numbers; rather, to avoid confusion, they used a peculiar device that we shall describe below and that is all the more curious insofar as it offers points of resemblance between the Mexica's chronological system and that of the peoples of Asia. The *dots*, or numerical signs, are only found added to the bindings of years that indicate fifty-two-year cycles. It was in this manner that the hieroglyph for *Xiuhmolpilli*, followed by four dots placed near the islets on which the temple of Mexitli was built, reminded the Mexica that his ancestors had *bound* the years four times, or, rather, that four times fifty-two years had elapsed since the sacrifice of Tlalixco, before the city of Tenochtitlan was founded in Lake Texcoco. These dots indicated, therefore, that this remarkable event had taken place after the year 1299 and before the year 1351. Let us now examine the ingenious but rather complicated methods that these peoples employed to mark the day and the year of a fifty-two-year cycle.

As we shall explain shortly, this method is identical to the one used by the Hindus, Tibetans, Chinese, Japanese, and other Asian peoples of the Tartar race, who also identify the months and the years through the correspondence between several periodic series with unequal numbers of terms. For the cycle of years, the Mexica used the following four signs, which bear the names of 139

Tochtli, rabbit or hare.
Acatl, reed.
Tecpatl, flint or gun flint.
Calli, dwelling.

One finds these four hieroglyphs in several of the previous plates. For the *rabbit* (tochtli) figure, see Plate XIII, the large-eared animal that appears in the eighth box, counting from the lower right-hand corner; Plate XXIII, the third box at the bottom left; and especially Plate XXVII, number 1, the eighth box. For *reed* (acatl), *flint* (tecpatl), and *dwelling* (calli), see the fifth, tenth, and fifteenth box after the one of the rabbit, from left to right on the circular stone depicted in Plate XXIII. These same shapes are easily recognizable on Plate XXVII, number 1, in boxes thirteen, eighteen, and three, counting in the same row from right to left, and starting from the lower row. The sign *flint* can also be seen on Plate XIII, behind the worshiping figure. On this same plate the *calli* is represented by the complete image of a house, in which one can make out both a door and a high roof.

If one now imagines the cycle, or *half-age*, divided into four tlalpilli of thirteen years each, and the four signs *rabbit, reed, flint,* and *dwelling* added in a periodic series to the fifty-two years contained within a cycle, one will find that two indictions cannot begin with the same sign; rather, the sign placed at the beginning of an indiction must necessarily recur at its conclusion; the same sign cannot recur in conjunction with the same number. Here [on p. 167] is the table of the Mexica cycle, called *binding of years, xiumolpilli.*

140

When placed before the names of the four hieroglyphs for the years, the words *ce, ome,* and *jei* indicate the numbers whose series does not extend past thirteen and which are therefore repeated four times in a *binding of years.* The following table shows the numbers one to thirteen in Mexica (or Aztec [Nahuatl]), the Nootka language, Muisca (or Mosca), Peruvian (or Quechua), Manchu, Uyghur, and Mongol [see table on p. 168].

One might be struck by the extreme dissimilarity between the seven languages in which we have just listed the cardinal numbers. The American languages are just as far removed from one another as they are from the Tartar languages. This lack of similarity should not be cited as evidence against the theory that the American peoples had ancient contacts with eastern Asia. The different groups of Tartar peoples, the Manchu and the Uyghur—the latter of which had emigrated two centuries before our era from the banks of the Selenga to the Turfan plateau located at 43° 30' latitude—speak languages that differ more from one another than German and Latin do. When tribes of the same origin are separated over many centuries by seas and vast deserts, their idioms preserve only a very small number of roots and common forms.

141

Just as the Mexica, when speaking of one year in a cycle, placed one of the cardinal numbers *ce, ome,* and *jei* before the name of one of the four hieroglyphs *rabbit, reed, flint,* and *dwelling* in their paintings, they also connected the sign for this number to the sign for the year. This method was identical to the one used to identify cycles or *bindings of years.* Since there were only thirteen terms in the periodic number series, they had only to add to the hieroglyphs the dots that represented the units.

The symbolic script of the Mexica peoples had simple signs both for twenty and for the second and third power of the same number, which recalls the sum of the fingers and toes. A small standard, or flag, represented twenty units; twenty squared, or four hundred, was represented by a *feather* because specks of gold enclosed within the shaft of a feather served, in some places, as currency or a sign of exchange. The image of a *bag* indicated twenty cubed, or eight thousand, and bore the name *xiquipilli*, also given

First Tlalpilli		Second Tlalpilli		Third Tlalpilli		Fourth Tlalpilli	
Ce Tochtli	1. Rabbit	Ce Acatl	1. Reeds	Ce Tecpatl	1. Flint	Ce Calli	1. Dwelling
Ome Acatl	2. Reeds	Ome Tecpatl	2. Flint	Ome Calli	2. Dwelling	Ome Tochtli	2. Rabbit
Jei Tecpatl	3. Flint	Jei Calli	3. Dwelling	Jei Tochtli	3. Rabbit	Jei Acatl	3. Reeds
Nahui Calli	4. Dwelling	Nahui Tochtli	4. Rabbit	Nahui Acatl	4. Reeds	Nahui Tecpatl	4. Flint
Macuilli Tochtli	5. Rabbit	Macuilli Acatl	5. Reeds	Macuilli Tecpatl	5. Flint	Macuilli Calli	5. Dwelling
Chicuace Acatl	6. Reeds	Chicuace Tecpatl	6. Flint	Chicuace Calli	6. Dwelling	Chicuace Tochtli	6. Rabbit
Chicome Tecpatl	7. Flint	Chicome Calli	7. Dwelling	Chicome Tochtli	7. Rabbit	Chicome Acatl	7. Reeds
Chicuei Calli	8. Dwelling	Chicuei Tochtli	8. Rabbit	Chicuei Acatl	8. Reeds	Chicuei Tecpatl	8. Flint
Chicuhnahui Tochtli	9. Rabbit	Chicuhnahui Acatl	9. Reeds	Chicuhnahui Tecpatl	9. Flint	Chicuhnahui Calli	9. Dwelling
Matlactli Acatl	10. Reeds	Matlactli Tecpatl	10. Flint	Matlactli Calli	10. Dwelling	Matlactli Tochtli	10. Rabbit
Matlactli ozce Tecpatl	11. Flint	Matlactli ozce Calli	11. Dwelling	Matlactli ozce Tochtli	11. Rabbit	Matlactli ozce Acatl	11. Reeds
Matlactli omome Calli	12. Dwelling	Matlactli omome Tochtli	12. Rabbit	Matlactli omome Acatl	12. Reeds	Matlactli omome Tecpatl	12. Flint
Matlactli omey Tochtli	13. Rabbit	Matlactli omey Acatl	13. Reeds	Matlactli omey Tecpatl	13. Flint	Matlactli omey Calli	13. Dwelling

	American Languages				Tartar Languages		
	AZTEC (Mexico)	QUECHUA (Peru)	MUISCA (New Granada)	LANG. OF NOOTKA (Northwest Coast)	MANCHU (Eastern Tartar)	MONGOLIAN (Western Tartar)	UYGHUR (Turfan Plateau)
1. Ce	Huc	Ata	Sahuac	Emou	Neguê	Pir
2. Ome	Iscay	Bosa	Atla	Tchoué	Khour	Iki
3. Jei	Quimza	Mica	Catza	Ilan	Gourbâ	Outche
4. Nahui	Tawa	Muyhica	Nu	Touyin	Durba	Tourou
5. Macuilli	Pichca	Hisca	Sutcha	Soumtcha	Taboî	Pich
6. Chicuace	Zocta	Ta	Nupu	Ningoun	Djourga	Alti
7. Chicome	Canchis	Cuhupqua	Atlipu	Nadan	Dolo	Iti
8. Chicuci	Pussac	Suhuza	Atcual	Tchakoun	Naïma	Sakis
9. Chicuhnahui	Yscon	Aca	Tzahuacuatl	Ouyoun	Youzou	Toukous
10. Matlactli	Chunca	Ubchica	Ayo	Tchouan	Arban	Oun
11. Matlactli ozce	Chunca hucnioc	Quicha ata	Ayo sahuac	Tchouan emou	Arban neguê	Pir oum
12. Matlactli omome	Chunca iscayoc	Quicha bosa	Ayo atla	Tchouan tchoué	Arban khour	Iki oun
13. Matlactli omey	Chunca quimzayoc	Quicha mica	Ayo catza	Tchouan ilan	Arban gourbâ	Outche oum

to a type of purse that held eight thousand cocoa beans. A *standard* divided by two crossed lines and halfway colored-in indicated half of twenty, or ten. If the standard was three-quarters colored in, it referred to fifteen units or three-quarters of twenty. When the Mexica counted, he did not name the multiples of ten that the Arabs call *knots* but, rather, the multiples of twenty. He would say: one twenty, *cem-pohualli*; two twenties, *om-pohualli*; three twenties, *yei-pohualli*; and four twenties, *nahui-pohualli*. This final expression is ᵛidentical to the one used in French. It is almost unnecessary to observe here that the Mexica were unacquainted with the method of giving *position values*[1] to the number signs, an admirable method invented either by the Hindus or by the Tibetans,[2] but also unknown to the Greeks,[3] the Romans, and the civilized peoples of western Asia. The Mexica aligned and tallied their number hieroglyphs more or less in the same way that the Romans repeated the letters of their alphabet, which they used as numerals. It is perhaps unsurprising to note that Mexica arithmetic has only simple hieroglyphs for the hundreds beyond four hundred, if one recalls[4] that until the fifth century of the Hegira, the Arabs were equally lacking in signs for the centenary numbers above four hundred and that to write nine hundred, this people, justifiably famous in the scientific annals, was obliged to place two signs for four hundred next to the sign for one hundred.

From what we have explained about the manner of distinguishing from one another both the *bindings of years* themselves and the years contained within one *binding*, it follows that a period was determined by naming both the number of *bindings*, or cycles, and the two terms that correspond to one another in the two periodic series of thirteen numbers and four signs. The following table shows several noteworthy periods of Mexica history listed in accordance with Aztec chronology. One must recall that these peoples only counted the number of their cycles, *xiuhmolpilli*, from the year 1091, because they had established a new chronological order in their annals, starting from their departure from Aztlan, the beginning of their southward migrations.

142

| Nahui Xiuhmolpilli, ome Calli | 1325. Founding of Tenochtitlan. |
| (4ᵗʰ Cycle, 2nd Dwelling......................) | |

1. Laplace, *Expos[ition du système du monde]*, Vol. II, p. 276.

2. ᵛGeorgi, *Alp[habet] tibet[ain]*, Ch. XXIII, p. 637.

3. ᵛDelambre, "Sur les fonds et les analogues des Grecs," *Œuvres d'Archimède*, by Peyrard, p. 575.

4. ᵛSilvestre de Sacy, *Grammaire arabe*, 1810. Part 1, p. 74.

Macuilli Xiuhmolpilli, ce Calli (5th Cycle, 1st Dwelling)	1389. Crowning of King Huitzilihuitl.
Chicuace Xiuhmolpilli, chicuace Tochtli (6th Cycle, 6th Rabbit)	1446. Great flood of Mexico City.
Chicome Xiuhmolpilli, matlactli omey Tochtli (7th Cycle, 13th Rabbit)	1492. Columbus's arrival in the Antilles.
Chicuei Xiuhmolpilli, ce Acatl (8th Cycle, 1st Reed)	1519. Cortés appears in Tenochtitlan.
Chicuei Xiuhmolpili, ome Tecpatl (8th Cycle, 2nd Flint)	1520. Montezuma's death.
Chicuei Xiuhmolpilli, jei Calli (8th Cycle, 3rd Dwelling)	1521. Taking and destruction of Tenochtitlan.

143   This same device of coordinating two periodic series was used to distinguish the days of a single year. It appears that the Mexica originally gave both a name and a particular sign to each day of the month, as the Persians did. These twenty signs recall the yugas that in the Hindus' astrological almanac are added to the twenty-eight days of the lunar months. In the Metzlapohualli, the Aztecs' *count of the moon*, they were distributed across the short semilunar month cycles, so that a periodic series of thirteen terms, all of which were numerals, was matched with a periodic series of twenty terms, which contained only hieroglyphic signs. Within this day series are also found the four major signs—*rabbit, reed, flint*, and *dwelling*—that were used to designate years belonging to the same cycle, as we have just seen above. Sixteen other lower-order signs were distributed in equal groups of four, which were inserted in between the major signs, separating them from one another.

    If one recalls that each Mexica month was divided into four short five-day periods, one understands that the hieroglyphs *rabbit, reed, flint*, and *dwelling* originally indicated the beginning of these short periods in years where the first day bore one of the four named signs. Indeed, when the first day of the month *tititl* bears the sign *calli*, day six of every subsequent month will be *tochtli*, day eleven will be *acatl*, and day sixteen *tecpatl*. Each month will begin on a Sunday, and these Sundays will fall on the same day of the month throughout the whole year. The Mexica took a particular interest in which-

ever events took place on one of the four days that bore the year-cycle hieroglyphs. We also find traces of this superstition among the Persians, who, in
order to give a sign (*kârkunân*) to each day of the month, added eighteen
lower-order figures to the twelve *heavenly spirits* assigned to the months. The
Mexica viewed the days that bore a particular year's sign as especially auspi-
cious, while the Persians[1] singled out the day over which resided the angel
who ruled over the entire month.

Since the majority of hieroglyphic paintings reproduced on the plates in
this work are related to the sacrifices that were supposed to be made during
each thirteen-day period, one finds the twenty signs for the days repeated 144
several times there. I shall cite only Plates XIII, XXIII, and XXVII here. Here
are the names of these signs:

CALLI, dwelling.

Cuetzpalin, lizard.

Cohuatl, grass snake. This word is also found in Cihuacohuatl,[2] the Ser-
    pent's wife [or woman], the Eve of the Mexica.

Miquiztli, death, skull.

Mazatl, doe or stag.

TOCHTLI, rabbit.

Atl, water.

Itzcuintli, dog.

Ozomatli, monkey.

Malinalli, grass.

ACATL, reed.

Ocelotl, tiger, jaguar.

Quauhtli, eagle.

Cozcaquauhtli, king of the vultures.

Ollin, annual movement of the sun.

TECPATL, flint.

Quiahuitl, rain.

Xochitl, flower.

Cipactli, sea animal: Teocipactli, *fish-god*, is one of the names that the
    Mexica gave to Coxcox, the Noah of the peoples of the Semitic race.

Ehecatl, wind.

Since the numbers thirteen and twenty have no common factors, in the
almanac of the semilunar months terms from each of the two periodic series

1. Langlès, on the Persian Calendar, in ▾Chardin, *Voyage à Ispahan*, Vol. II, p. 265.

2. See p. 102 [in this edition].

can occur jointly only twice after 13 × 20, or two hundred sixty days. In a year in which the first day bears the sign *cipactli*, no semilunar month can begin with the sign *cipactli* within the first thirteen months, but the same combinations of signs and numerals can recur after the month *pachtli*. Faithful to their principle of not revealing the number of short thirteen-day periods, the Mexica turned once again to the periodic series device to avoid this potential source of error. They created a third series of nine signs, called the *lords* or *masters of the night*, which are these:

> *Xiuhteuctli Tletl*, fire, or master of the year.
> *Tecpatl*, flint.
> *Xochitl*, flower.
> *Cinteotl*, goddess of corn.
> *Miquiztli*, death.
> *Atl*, water.
> *Tlazolteotl*, goddess of love.
> *Tepeyollotli*, spirit that dwells inside the mountains.
> *Quiahuitl*, rain.

One might be surprised to find a series of nine terms in a calendar that makes use of only the numbers five, eighteen, twenty, and fifty-two. One might even be tempted to look for some sort of analogy between the nine *lords of the night* of the Mexica and the nine astrological signs of several Asian peoples who add to the seven visible planets two invisible dragons, to which they attribute eclipses. But it was probably only the ease with which the nine *lords of the night* could be distributed forty times across three hundred sixty days that resulted in preference being given to the number nine.

The Mexica referred to the five supplementary days—which the Persians called *furtive days*, *pendjéhi-douzdideh*—as *nemontemi*, *empty days*, because they did not add to them any of the terms from the third series, which the Indian authors saw as the *companions* of the day-signs. It should be noted— and this situation can become awkward within Aztec chronology—that five of these *companions* bear the same name as the day-hieroglyphs, but following the imaginative excesses of the American astrologers, the *spirits* belonging to the series of nine signs rule over the night, while the twenty other signs rule over the day. The Hindus also recognized spirits (*caranas*) who were assigned to the lunar half-day (*ti'thi*).

Since there are twenty day-signs and nine *companions* or *lords of the night*, each companion must recur in conjunction with the same hieroglyph every 9 × 20 or one hundred eighty days; but it is impossible for the same terms from the three different series—that is, the *numbers*, the *day-signs*, and the *companions*, or nocturnal spirits—to coincide more than once in the same three-hundred-sixty-five-day year. In a year that begins with Cipactli:

January 11	corresponds to	3 Calli, xochitl
July 10		1 Calli, xochitl
February 2		12 Cohuatl, tlazolteotl
August 1		10 Cohuatl, tlazolteotl
May 8		3 Xochitl, xochitl
November 4		1 Xochitl, xochitl

The use of the third periodic series, which enabled them to distinguish between two days that had the same number and the same hieroglyph (for example, 1 *Cipactli*, which corresponds to both January 9 and September 26), was unknown to most Spanish historians; it was Mr. Gama who first brought this fact to light, following the Mexica manuscripts of the Indian Cristóbal del Castillo. To identify a day using the Mexica's complicated method, we would say *day four* of a month, which is a *Wednesday* in the Gregorian calendar and a *quintidi* in the ʼRepublican calendar. This expression would indicate the correspondence of specific terms from the three periodic series, namely the thirty or thirty-one days of the month, the seven days of the week, and the ten days of the decade. To resolve any doubts that might remain about the Mexica's chronological system, we are including here a table that displays the divisions of the ritual and civil calendar, as well as their correspondence to the Gregorian calendar. 147

It would be useless to extend this table beyond the first thirty-one days of the Mexica year; but we must recall here that the Indians of Chiapas, who employed the same divisions of time and the same periodic series device, gave the hieroglyphs for the days contained within one month the names of twenty illustrious warriors who had led the first colonists into the *Teochiapan* mountains in the earliest times. Among the day-signs (the Persians' *kârkunân*), the 148 Chiapans, like the Aztecs, set four major and sixteen minor signs apart. The former occurred at the start of the five-day periods; but the Chiapans had replaced in their historical annals the names *dwelling, rabbit, reed*, and *flint* (calli, tochtli, acatl, and tecpatl) with those of *Votan, Lambat, Been*, and *Chinax*, four famous leaders.

We have already brought to the attention of our readers this Votan or Wodan, an American who appears to be of the same family as the Wods or Odins of the Goths and the peoples of Celtic origin. Since, according to Sir William Jones's scholarly studies, Odin and Buddha are probably the same

| | | Metzlapohualli
Ritual and Astrological Calendar | | TONALPOHUALLI CIVIL CALENDAR | MEXICAN MONTHS divided into 5-day units | AGREEMENT With the Gregorian Calendar for the year 1091 |
| | | Periodic Series | | | | |
SHORT PERIODS OF 13 DAYS	SERIES OF 13 NUMBERS	SERIES OF 20 DAY-SIGNS	SERIES OF 9 LORDS OF THE NIGHT			
First Semilunar Month	1	Cipactli	Tletl.	1	Titil	9
	2	Ehecatl	Tecpatl	2		10
	3	Calli	Xochitl	3		11
	4	Cuetzpalin...	Cinteotl	4		12
	5	Cohuatl	Miquiztli	5		13
	6	Miquiztli	Atl	6		14
	7	Mazatl	Tlazolteotl.........	7		15
	8	Tochtli	Tepeyollotli........	8		16
	9	Atl	Quiahuitl..........	9		17
	10	Itzcuintli	Tletl.	10		18
	11	Ozomatli	Tecpatl	11		19
	12	Malinalli	Xochitl	12		20
	13	Acatl	Cinteotl	13		21
Second Semilunar Month	1	Ocelotl......	Miquiztli	14		22
	2	Quauhtli	Atl	15		23
	3	Cozcaquauhtli	Tlazolteotl.........	16		24
	4	Ollin	Tepeyollotli........	17		25
	5	Tecpatl......	Quiahuitl..........	18		26
	6	Quiahuitl....	Tletl.	19		27
	7	Xochitl	Tecpatl	20		28
	8	Cipactli	Xochitl	1	Itzcalli Xochilhuitl	29
	9	Ehecatl	Cinteotl	2		50
	10	Calli	Miquiztli	3		31
	11	Quetzpalin ..	Atl	4		1
	12	Cohuatl	Tlazolteotl.........	5		2
	13	Miquiztli	Tepeyollotli........	6		3
	1	Mazatl	Quiahuitl..........	7		4
	2	Tochtli	Tletl.	8		5
	3	Atl	Tecpatl	9		6
	4	Itzcuintli	Xochitl	10		7
	5	Ozomatli	Cinteotl	11		8

January (first block); February (second block)

person,[1] it is curious to note that in India, Scandinavia, and Mexico, the names *Boud-var*, *Wodans-dag* (Wednes-day), and *Votan* refer to one of the days in a short period. Following the ancient legends collected by Bishop Francisco Núñez de la Vega, "The Chiapans' Wodan was the grandson of the illustrious old man who, during the great flood in which perished the greater part of humankind, was saved in a raft, along with his family." Wodan participated in the building of the great structure that humans had undertaken in order to reach the heavens. The execution of this reckless project was halted; each family thereupon received a different language, and the great spirit *Teotl* commanded Wodan to populate the land of Anahuac. This American legend recalls the Hindus' Manu, the Hebrews' Noah, and the scattering of the Kushites of Shinar. In comparing it either to the Hebraic and Indian legends preserved in Genesis and in two sacred puranas,[2] or to the fable of Xelhua the Cholulan[3] and other events cited throughout the course of this work, it is impossible not to be struck by the similarity between the ancient memories of the peoples of Asia and those of the new continent.

We shall demonstrate here, as we suggested above, that this similarity is especially evident in the division of time, the use of periodic series, and the ingenious (albeit awkward and complicated) method of identifying a day or a year not by numerals but by astrological signs. The Toltec, the Aztecs, the Chiapans, and other peoples of the Mexica race counted by fifty-two-year cycles divided into four periods of thirteen years each, while the Chinese, the Japanese, the Kalmyk, the Moguls, the Manchu, and other Tartar hordes have sixty-year cycles divided into five short twelve-year periods. The peoples of Asia, like those of America, have specific names for the years contained within a cycle; in Lhasa and in Nagasaki they still say, as they once did in Mexico City, that such or such event took place in the year of the *rabbit*, the *tiger*, or the *dog*. None of these peoples has as many names as there are years in the cycle: they must all, therefore, turn to the device of corresponding periodic series. For the Mexica, these series are of thirteen numbers and four hieroglyphic signs. For the peoples of Asia whom we have just mentioned, these series do not contain numerals: they are formed as often by signs that correspond to the twelve constellations of the zodiac as by the names of those elements that contribute ten terms, since each element is con-

149

1. ["Discours anniversaire, Prononcé, le 2 Février 1786, par le Président,"] *Rech[erches] Asiat[iques]*, Vol. I, p. 511; ["Sur l'antiquité du zodiaque indien,"] Vol. II, p. 343.

2. [William Jones, "Discourse the Ninth. On the Origin and Families of Nations," *Asiatick Researches*,] Vol. III, p. 486.

3. See p. 50–51 [this edition].

sidered either male or female. This underlying logic is present in the chronology of the American peoples as well as in that of the Asian peoples. Casting a glance at the table of the years that we have provided above,[1] one sees that the advantage of simplicity is actually on the side of the Mexica. To refer to the period when a Dairi acceded to the throne, the Japanese do not say that it was in the year *ouma* (horse) of the second twelve-year period; they call the nineteenth year of the cycle the year *male water, horse,* placed between the years *female water, ewe,* and *female metal, serpent.* If one is to gain a clearer understanding of the periodic series of the Japanese calendar, one must recall that this people, like the Tibetans, counts five elements: namely, wood (*keno*), fire (*fino*), earth (*tsutsno*), metal or lead (*kanno*), and water (*midsno*). Each element is either male or female, depending on which one of the two syllables *je* and *to* is added, a distinction that was also in use among the Egyptians.[2] To distinguish between the sixty years within a cycle, the Japanese combine the ten elements or earthly principles with the twelve zodiac signs called the heavenly principles. We shall list here only the first two indictions contained within the Japanese cycle.[3]

150

1. *Kino je ne* (rat)	13. *Fino je ne*
2. *Kino to us* (steer)	14. *Fino to us*
3. *Fino je torra* (tiger)	15. *Tsutsno je torra*
4. *Fino to ov* (hare)	16. *Tsutsno to ov*
5. *Tsutsno je tats* (crocodile or dragon)	17. *Kanno je tats*
6. *Tsutsno to mi* (serpent)	18. *Kanno to mi*
7. *Kanno je uma* (horse)	19. *Midsno je uma*
8. *Kanno to tsitsuse* (ewe)	20. *Midsno to tsitsuse*
9. *Midsno je sar* (monkey)	21. *Kino to sar*
10. *Midsno to torri* (chicken)	22. *Kino to torri*
11. *Kino je in* (dog)	23. *Fino je in*
12. *Kino to j* (hog)	24. *Fino to j*

In the Mexica calendar, each of the four thirteen-year indictions begins with a different sign; in the Japanese calendar, however, each twelve-year period is presided over by one of the five male elements. Just as the fourth term of the Mexica number series, *nahui,* can occur jointly with the second term in the sign series, *acatl,* only once in fifty-two years, for the Japanese one of the five male elements can appear only once in a sixty-year cycle alongside

1. See p. 167 [this edition].
2. Seneca, *Quæst[ions] nat[urelles]*, Book 3, ch. 14.
3. ⧫Kaempfer, *Hist[oire] du Japon*, 1729. Vol. I, p. 137, table XV.

one of the twelve zodiac signs. The following table, which contains fourteen Mexica and Japanese years, will make the parallels between the calendars of the peoples of Mexico and of eastern Asia perfectly clear.

NUMBER OF YEARS	JAPANESE CYCLE $\alpha, \alpha', \beta, \beta', \gamma, \gamma'$ . . . are the male and female elements, and a, b, c . . . the celestial signs, hence:		MEXICAN CYCLE $\alpha, \beta, \gamma, \delta$ . . . are the four year-signs and a, b, c . . . the thirteen names of the numbers, hence:	
1	$\alpha,$	a	$a,$	α
2	$\alpha',$	b	$b,$	β
3	$\beta,$	c	$c,$	γ
4	$\beta',$	d	$d,$	δ
5	$\gamma,$	e	$e,$	α
6	$\gamma',$	f	$f,$	β
7	$\delta,$	g	$g,$	γ
8	$\delta',$	h	$h,$	δ
9	$\varepsilon,$	i	$i,$	α
10	$\varepsilon',$	k	$k,$	β
11	$\alpha,$	l	$l,$	γ
12	$\alpha',$	m	$m,$	δ
13	$\beta,$	a	$n,$	α
14	$\beta',$	b	a	β

The custom of periodic series is also found in China, where ten *can* combined with twelve *tchi* is used to identify the days and the years within the sixty-day and sixty-year periods, respectively.[1] Among the Japanese, the Chinese, and the peoples of Mexico, periodic series can be used to identify only fifty-two to sixty years. The Tibetans, on the other hand, complicated the series device to such an extent that they have names for one hundred ninety-two and even up to two hundred fifty-two years. For example, to refer to the memorable period when the great Lama *Kang-ka-gnimbo* assumed both ecclesiastical and secular powers[2] with the consent of the emperor of China, a resident of Lhasa cites the year *male fire, bird* (*me po cia*) of the fourteenth cycle since the flood. He counts fifteen elements: five of masculine gender, five of feminine gender, and five neuters. By combining these fifteen elements with the twelve zodiac signs and referring to the first twelve years of the cycle ex-

1. *Obs[ervations] astr[onomiques]* by ▾[Father] Souciet, published by [Father] Gaubil, Vol. I, p. 26, Vol. II, p. 175.

2. Georgi, *Alph[abet] Tibet[ain]*, p. 516.

clusively by the names of the heavenly signs themselves, he obtains names for 12 × 15 + 12, or one hundred ninety-two years. Finally, by adding sixty years identified by the combination of ten male and female elements with the twelve zodiac signs, he arrives at a long cycle of two hundred fifty-two years. Where a, b, c . . . represent the zodiac signs; α, β, γ . . . the neuter elements; α', β', γ' . . . the male elements; and α'', β'', γ'' . . . the female elements, one has (1) for the first twelve years, a, b, c, d . . . ; (2) for years 13–72, $\alpha\,a$, $\alpha\,b$, $\alpha\,c$ . . . ; $\beta\,a$, $\beta\,b$, $\beta\,c$ . . . ; $\gamma\,a$, $\gamma\,b$, $\gamma\,c$ . . . ; (3) for years 73–132, $\alpha'\,a$, $\alpha'\,b$, $\alpha'\,c$ . . . ; $\beta'\,a$, $\beta'\,b$ . . . ; (4) for years 132–192, $\alpha''\,a$, $\alpha''\,b$, $\alpha''\,c$ . . . ; $\beta''\,a$, $\beta''\,b$, $\beta''\,c$ . . ; (5) for years 193–252, $\alpha'\,a$, $\alpha''\,b$, $\beta'\,c$, $\beta''\,d$, $\gamma'\,e$, $\gamma''\,f$, $\delta'\,g$, $\delta''\,h$, $\varepsilon'\,i$, $\varepsilon''\,k$, $\alpha'\,l$, $\alpha''\,m$, $\beta'\,a$, $\beta''\,a$, $\gamma'\,b$, $\gamma''\,b$ . . . The *Tzihi-chen*, the public reckoners of Lhasa,[1] allege, in favor of the Tibetan chronology, that since years bearing the same name recur only about once every two centuries, the date of a historical event can be determined even when the cycle itself is not indicated. There is greater uncertainty among the Japanese and the Mexica, for whom the same names reappear every sixty or fifty-two years. It is perhaps surprising that the Tibetans, who from the earliest times have used the same numerals and the same numbering system as the Hindus, have not abandoned the complicated method of recurring series. This method originated in the reveries of astrologers: it should have been used only by peoples like the Aztecs and the Toltec who had difficulty expressing very high numbers and whose annals were written in hieroglyphic characters.

We have just seen that the Mexica, the Japanese, the Tibetans, and several other peoples of central Asia have followed the same system in the division of long cycles and in the naming of the years that constitute them. It remains for us to examine a fact that is more directly concerned with the history of the migrations of peoples and that appears to have escaped scholars' research until now. I believe I can prove that a significant portion of the names by which the Mexica referred to the twenty days of their months are those of the signs of a zodiac widely used since the earliest times by the peoples of eastern Asia. In order to demonstrate that this claim is less unjustified than it first appears, I shall bring together in a single table (1) the names of the Mexica hieroglyphs as they were handed down to us by sixteenth-century authors; (2) the names of the twelve signs of the Tartar, Tibetan, and Japanese zodiacs; and (3) the names of the *nakshatras*, or lunar mansions, in the Hindu calendar. I trust that those of my readers who have closely examined this comparative table will become interested in the discussions on the original divisions of the zodiac into which we must now enter.

1. [Ibid.,] p. 469.

| Zodiac Sign | | | | HIEROGLYPHS OF THE DAYS ACCORDING TO THE MEXICAN CALENDAR | NAKSHATRAS OR LUNAR MANSIONS OF THE HINDUS |
HINDUS, GREEK, and Western Peoples	MANCHU-TARTARS	JAPANESE	TIBETANS		
Aquarius	Singueri	Ne	Tchip, rat, *water*	Atl, *water*	
Capricorn	Ouker	Ous	Lang, *steer*	Cipacli, *sea monster*	(*The Mahara is a sea monster*)
Sagittarius	Pars	Torra	Tah, *tiger*	Ocelol, *tiger*	
Scorpion	Taoulaï	Ov	Io, *hare*	Tochtli, *rabbit*	
Libra	Lou	Tats	Brou, *dragon*	Cohuatl, *serpent*	*Serpent*
Virgo	Mogaï	Mi	Proul, *serpent*	(Acatl, *red*)	*Reed*
Leo	Morin	Ouma	Tha, *horse*	(Tecpatl, *flint, knife*)	*Razor*
Cancer	Koin	Tsitsuse	Lon, *goat*	(Ollin, *path of the sun*)	Traces of Vishnu's steps
Gemini	Petchi	Sar	Prchou, *monkey*	Ozomatli, *monkey*	*Monkey*
Taurus	Tukia	Torri	Tcha, *bird*	Quauhtli, *bird*	
Aries	Nokai	In	Ky, *dog*	Itzecuintli, *dog*	*Dog's tail*
Pisces	Gacai	Y	Pah, *hog*	(Calli, *dwelling*)	*Dwelling*

153   From the earliest times the peoples of Asia have observed two divisions of
the ecliptic: one into twenty-seven or twenty-eight lunar mansions or prefec-
tures, the other into twelve parts. It has been mistakenly suggested that the
latter division was found only among the Egyptians. The most ancient masterpieces of [East] Indian literature, the works of 'Kālidāsa and Amarsinh,[1]
mention both the twelve zodiac signs and the twenty-seven *lunar mansions*.
Considering what we know about the contacts that took place between the
peoples of Ethiopia, Upper Egypt, and Hindustan several thousand years
before our era, it is inadmissible to regard everything that the Egyptians transmitted to the peoples of Greece as belonging exclusively to the Egyptians
themselves.

The division of the ecliptic into twenty-seven or twenty-eight lunar mansions is probably[2] more ancient than the division into twelve parts, which
is related to the annual movement of the sun. Phenomena that recur in the
same order every lunar month attract humans' attention much more than do
changes in position, the cycle of which is completed only over the course of
a year. Since the moon is positioned next to the same stars, more or less, in
each lunar month, it seems natural that specific names would have been given
to the twenty-seven or twenty-eight constellations through which it crosses
in one synodic revolution. Little by little, the names of these constellations
passed to the lunar days themselves, and this apparent connection between
sign and day became the primary basis for astrologers' fanciful calculations.

If one examines closely the names that the *nakshatras* or lunar mansions
bear in Hindustan, one recognizes not only almost all of the names of the
Tartar and Tibetan zodiac but also those of several constellations that are
identical with the signs of the Greek zodiac. Each *nakshatra* covers 13° 20',
and 2¼ *nakshatras* correspond to one of our signs. The following table makes
it appear quite probable that the solar zodiac finds its origin in the lunar zodiac, and that the twelve signs of the former were chosen largely from among
the twenty-seven *nakshatras*.

154

LUNAR MANSIONS	ZODIAC SIGNS (*DODECATEMORIA*)
Rat	*Rat*, Aquarius
Gazelle	Steer, Capricorn
Arrow, *bow*	Tiger, *Sagittarius*
Lion's tail	*Leo*
Libra's beam	Dragon, *Libra*

1. *Rech[erches] Asiat[iques]*, Vol. II, p. 346.
2. Le Gentil, [*Voyage dans les Indes,*] Vol. I, p. 261.

Serpent	*Serpent*, Virgo
Horse	*Horse*
Goat	*Ewe*, Cancer
Monkey	*Monkey*, Gemini
Eagle	*Bird*, Taurus
Dog's tail	*Dog*, Aries
Fish	Hog, *Pisces*

In the Arab heavens Orion's Belt is known as the beam of Libra, Mintaka, and it seems all the more remarkable that one of the Hindus' lunar mansions bears the same name, since doubts have been raised about the antiquity of the constellation Libra in the wake of the discovery of the Tentyra zodiac. It is undeniable that the signs that make up the Egyptian, Chaldean, and Greek zodiacs have been known in India since the earliest times, and it is probable that when Julius Caesar added Libra to the Roman zodiac, he did so under the counsel of the ˈastronomer Sosigenes,[1] who, born in Egypt, could not have been ignorant of the ecliptic divisions widely used in the East. There is no need,[2] moreover, to cast doubt upon the antiquity of the sign Libra in order to invalidate the baseless hypothesis that a temple in Upper Egypt was built more than four thousand years before our era.

Intrigued by the similarity between the names of the nakshatras and those of several signs from the Tibetan and the Greek zodiacs, I have examined whether the constellations that bear the same name were located in the same points in the heavens. There was no correspondence between them, either because it is assumed that the first nakshatra, known as the horse, is the horse of the Tibetan zodiac and therefore the lion [Leo] of the Greek zodiac, or because some suppose, as do Mr. Jones and ˈMr. Colebrooke,[3] that the first of the nakshatras is located within the sign of the ram [Aries], which is the dog of the Tibetan zodiac. The latter hypothesis would be plausible only if the lunar mansions happened to be counted in *reverse order* to the *signs*: in that case the six nakshatras named *two faces*, *three footprints of Vishnu*, the *lion's tail*, the *festoon of leaves*, the *arrow*, and the *gazelle's head* would have cor-

155

---

1. ˈButtmann, in Ideler, *Hist[orische] Unt[ersuchungen]*, pp. 372–78.

2. See the scholarly paper by Mr. Visconti ["Notice sommaire des deux zodiaques de Tentyra,"] included in Mr. Larcher's translation of *Herodotus* ([*Histoire d'Hérodote*], second ed.), Vol. II, p. 576; and Visconti, *Miscell[anea] di Musée Pio-Clementin* [*Il Museo Pio-Clementino*], Vol. VI, p. 25, note *c*.

3. ["On the Indian and Arabian Divisions of the Zodiack,"] *Asiat[ick] Resear[ches]*, Vol. IX, p. 300.

responded to our signs Gemini, Cancer, Leo, Virgo, Sagittarius, and Capricorn. But Libra, Leo, and Aries do not appear to be correctly spaced apart in either of the two theories that we have just mentioned. According to scholarly studies by the members of the [Asiatick] Society of Calcutta, the nakshatras *ashvini*, horse; *pushya*, arrow; and *mula*, lion's tail, correspond to α of Aries, δ of Cancer, and γ of Scorpio in the Greek zodiac, or to the dog, the ram, and the hare in the Tartar and Tibetan zodiacs.

At first glance it may seem extraordinary that when creating the twelve signs of the solar zodiac from the twenty-seven or twenty-eight signs of the lunar zodiac, the peoples kept the names of a large number of constellations without taking into account either their absolute position or the order in which they fall. One must not, however, conclude from this that the striking similarity between twelve of the nakshatras and the same number of signs from the Tibetan and Greek zodiacs is purely accidental. Since the names of the lunar mansions gradually passed to the days themselves, it is conceivable that they had become familiar to the people, who were probably unaware of the positions of the stars that constitute the ecliptic divisions. It may well be that certain peoples, having regressed into barbarism, had only a vague recollection of the names of the nakshatras, and that while reforming their calendar, they chose the names of the solar zodiac signs from among these names without following the order that was previously observed. It may also be—and I am inclined to give preference to this latter theory—that the zodiac composed of twelve signs originated in an ancient lunar zodiac in which the order of the nakshatras was more closely aligned to what we see today in the *dodecatemo-ria* of the peoples of Tibet and of Tartary. Indeed, the ecliptic divisions that Sir William Jones, Colebrooke, and ˇSonnerat have brought to light are fundamentally different from one another. The arrow, which one [East] Indian author describes as the eighth nakshatra, is only the twenty-third one according to another. We shall see below, in speaking of a Roman bas-relief described by ˇBianchini, that there once existed solar zodiacs in the East that had the same signs but in a different order. Furthermore, the return of the sun from the tropics toward the equator, as well as the phenomenon of days and nights of equal length, must have prompted humans to make significant changes to the figures of the nakshatras when they used a number of them to create the solar zodiac.

This intimate connection between the lunar mansions and the signs of the zodiac still manifests itself in the names that the Hindus give to the months and the years. According to the intriguing studies by ˇMr. Davis,[1] these names

156

---

1. "On the Indian Cycle of Sixty Years," *Asiat[ick] Res[earches]*, Vol. III, pp. 209–27.

are not those of the *dodecatemoria* of the solar zodiac; they are taken from the nakshatras themselves, with each month bearing the name of the lunar mansion in which the full moon occurred. We have seen above that in Tibet, in China, and among the Tartar peoples, each year within the five indictions of the long cycle bears the name of one of the twelve animals of the solar zodiac. The Hindu years take the name of the nakshatra in which Jupiter is found at its heliacal rising. It is for this reason that *ashvini* (horse) and *magha* (house) are the names of a year, a month, and a *tithi*, or lunar day, just as in Mexico the signs *tochtli* (rabbit) and *calli* (dwelling) can preside over a year, a semi-lunar month, or a day.

Together these considerations suggest that the division of the ecliptic into twelve signs probably originated from its division into twenty-seven or twenty-eight lunar mansions, and that the solar zodiac was originally a lunar zodiac, since each full moon is at an approximate distance of two and a half nakshatras, or 13° 20', from the preceding one. It is in this manner that the earliest astronomy of peoples is tied to the basic movements of the moon. If the twelve zodiac signs bear names that differ radically from those of the nakshatras, one must not conclude from this that the stars themselves were distributed into two separate divisions. For a long time the twelve-sign zodiac was merely an abstract division in eastern Asia,[1] while the zodiac of twenty-seven or twenty-eight nakshatras was the only true starry zodiac. I found it necessary to emphasize the close connection between the two ecliptic divisions in order to demonstrate that one or the other could have given rise to the signs of the Mexica zodiac.

157

Let us first examine the similarity between the names of the Mexica days and those of the Tibetan, Chinese, Tartar, and Mongol zodiac signs. This similarity is striking in the eight hieroglyphs called *atl, cipactli, ocelotl, tochtli, cohuatl, quauhtli, ozomatli,* and *itzcuintli*.

Atl, water, is often represented by a hieroglyph whose parallel, wavy lines are reminiscent of the sign that we use for Aquarius. The first *tse*, or catasterism [placement of the catalog of stars], of the Chinese zodiac, the rat (*chou*), is also frequently represented by the image of water.[2] During the reign of Emperor Tchouen-hiu, there was a great flood, and the heavenly sign hiuen-hiao, whose position corresponds to that of our Aquarius, is the symbol of this reign. As Father Souciet observes in his "Recherches sur les

---

1. Bailly, *[L'Histoire de l']astr[onomie] ind[ienne]*, p. 5; *[L'Histoire de l'] astr[onomie] mod[erne]*, Vol. III, p. 301.

2. *Obs[ervations] mathém[atiques]* by Father Souciet, published by Father Gaubil, Vol. III, p. 33.

cycles et les zodiaques," China and Europe both represent the sign that we call *Amphora* or *Aquarius*, but using different names. For Western peoples, the water that is poured from the vase of *Aquarius* (χύσις ὕδατς [*chysis hydatos*, pouring of water]) formed a separate constellation (ὕδωρ [*hydōr*, water]), to which the beautiful stars *Fomalhaut* and *Beta Ceti* belonged, as several passages by ˙Aratus, Geminus, and the scholiast Germanicus attest.[1]

Cipactli is a sea animal.[2] This hieroglyph exhibits a striking similarity to Capricorn, which the Hindus and other peoples of Asia call *sea monster*. The Mexica sign shows a mythical animal, a cetacean whose forehead is equipped with a horn. Gómara and Torquemada[3] call it *espadarte*, the name by which the Spanish refer to the narwhal, whose long tooth is known as a *unicorn's horn*. Boturini took this horn for a harpoon and incorrectly translated *cipactli* as *serpent armed with harpoons*. Since this sign does not represent a real animal, it is quite natural that its shape should vary more than that of other signs. Sometimes the horn appears to be an extension of the muzzle, as in the famous fish *oxyrinchus* [*Isurus oxyrinchus*, shortfin mako shark] represented in some Indian planispheres[4] under the belly of Capricorn, in the place of the southern fish [*Piscis austrinus* or *australis*]; at other times, the horn is completely missing. Casting a glance at the images on Plates XXIII and XXVII, made from very ancient drawings and reliefs, one sees how mistaken Valadés, Boturini, and Clavijero were in representing the first hieroglyph of the Mexica days as a shark or a lizard. In the manuscript at the Borgia museum, the head of cipactli resembles that of a crocodile, and Sonnerat gives this same name, crocodile, to the tenth sign of the Indian zodiac, which is our Capricorn.

In Mexica mythology, moreover, the idea of the sea animal cipactli is linked to the story of a man who, during the destruction of the fourth sun, saved himself by reaching the top of Culhuacan mountain after having swum in the water for a long time. We have observed above that the Noah of the Aztecs, usually called Coxcox, also bears the name *Teo-Cipactli*, in which the word for *god* or *divine* is added to that of the sign *cipactli*. Looking at the zodiac of the peoples of Asia, we find that the Capricorn of the Hindus is the mythical fish *maharan* or *souro*,[5] famous for his exploits, and has been represented as

158

---

1. Ideler, *Sternnamen*, p. 197.

2. Gama, *Descripc[ión] histór[ica] y cronol[ógica] de dos Piedras* (Mexico City, 1792), pp. 27 and 100.

3. [*Conquista de México,*] folio CXIX. *Mon[arquía] ind[iana]*, Vol. III, p. 223.

4. [˙John Call in] *Philos[ophical] Transact[ions]*, 1772, p. 353.

5. Sonnerat, *Voyage aux Indes*, Vol. I, p. 310. Bailly, *Astr[onomie] ind[ienne]*, p. 210.

a sea monster with a gazelle's head since the earliest times.[1] Given that the inhabitants of India, like the Mexica, often indicate the *nakshatras* (lunar mansions) and the *lagna* (dodecatemoria [the twelve signs of the zodiac]) by only the heads of the animals that compose the lunar and solar zodiacs, one must not be surprised that the Western peoples transformed the *maharan* into Capricorn (αἰγόεκρως [*aigokeros*]) and that Aratus, ʿPtolemy, and the Persian al-Qazwīnī, do not even give it a fish tail. For certain peoples, whose feverish imagination seizes upon even the most tenuous connections, an animal that after having lived for a long time in the sea, takes the form of a gazelle and scales mountains, recalls the ancient legends of Manu, Noah, and the Deucalion, famous among the Scythians and the Thessalians. It is true that according to Germanicus, Deucalion, which might be considered the equivalent of Coxcox or Teo-Cipactli from Mexica mythology, was placed not under the sign of Capricorn but under the sign that immediately follows it, Aquarius (ὑδροχόος [*hydrochoos*]). There is, however, nothing about this particular circumstance that should surprise us; rather, it confirms Mr. Bailly's ingenious theory about the ancient connection between the three signs of Pisces, Aquarius, and Capricorn, or gazelle-fish.[2]

    Ocelotl, tiger, the jaguar (*felis onça* [or *panthera onca*]) of the hot regions of Mexico; *tochtli*, hare; *ozomatli*, female monkey; *itzcuintli*, dog; *cohuatl*, serpent; and *quauhtli*, bird, are catasterisms that appear with the same names in the Tartar and Tibetan zodiacs. In Chinese astronomy, the hare or rabbit is not exclusively used in reference to the fourth *tse*, or zodiac sign: since the distant period of Yao's reign, the moon had been depicted as a disc in which a hare,[3] seated on its hind feet, turns a stick in a vase, as though it were busy making butter, a childish idea that may have arisen in the steppes of Tartary, which are abundant in hares and inhabited by sheepherding peoples. The Mexica monkey, *ozomatli*, corresponds to the Chinese *heou*,[4] the Manchu's *petchi*, and the Tibetans' *prehou*, three names that refer to the same animal. Procyon appears to be the monkey *Hanuman*,[5] so well-known in Hindu mythology, and the position of this star, located on the same line as Gemini and the ecliptic pole, corresponds closely to the place the monkey occupies in the Tartar zodiac between the crab and the bull. Monkeys are also found in the Arabs' heavens: they are stars of the great dog constellation and are called

159

    1. [William Jones, "Sur l'antiquité du zodiaque indien," *Recherches Asiatiques,*], Vol. II, p. 335, number 7.

    2. [*L'Histoire de l'Jastr[onomie]mod[erne]*], Vol. III, p. 297.

    3. ʿGrosier, *Hist[oire] gén[érale] de la Chine*, Vol. I, p. 114.

    4. ʿDe Guignes, *Hist[oire] des Huns*, Vol. I, p. XLVII.

    5. Dupuis, *Origine des Cultes*, Vol. III, p. 363.

Al Kurud[1] in al-Qazwīnī's catalog. I am entering into such detail about the sign *ozomatli* because an animal from the Torrid Zone placed among the constellations of the Mongol, Manchu, Aztec, and Toltec peoples is an important point not only for the history of astronomy but also for that of the migrations of peoples.

The sign *itzcuintli*, dog, corresponds to the penultimate sign of the Tartar zodiac, that is, to the Tibetans' *ky*, the Manchu's *nokaï*, and the Japanese *in*. Father Gaubil teaches us that the dog of the Tartar zodiac is our *dodecatemory* of the *ram* [Aries], and it is quite remarkable that although the Hindus, according to Le Gentil, do not observe the sign series that begins with the rat, they occasionally replace the ram with a *feral dog*. Similarly, for the Mexica, *itzcuintli* refers to a wild dog, for the domesticated dog was called *techichi*. Mexico once abounded in carnivorous quadrupeds[2] that were half-dog and half-wolf, and that Hernández [de Toledo] explained to us only very imperfectly. This race of animals, known variously as *xoloitzcuintli*, *itzcuintepotzotli*, and *tepeitzcuintli*, is not completely extinct; rather, it is likely that these animals have withdrawn into the most deserted, distant forests; in the part of the country that I traversed, I never heard anyone speak of a feral dog. Le Gentil[3] and Bailly were misled when they suggested that the word *mesha*, which refers to our ram, means *feral dog*. This word from the Sanskrit language is the common name for a ram: one finds it employed[4] in quite a poetic way by an Indian author, who describes a fight between two warriors by stating that "in their heads they were two *mesha* (rams); in their arms, two elephants; in their feet, two noble warhorses."

The following table displays the signs of the Tartar zodiac alongside those of the days of the Mexica calendar:

ZODIACS OF THE MANCHU-TARTARS	ZODIACS OF THE MEXICANS
Pars, tiger	*Ocelotl*, tiger
Taoulai, hare	*Tochtli*, hare, rabbit
Mogai, serpent	*Cohuatl*, serpent
Petchi, monkey	*Ozomatli*, monkey
Nokaï, *dog*	*Itzcuintli*, dog
Tukia, bird, chicken	*Quauhtli*, bird, eagle

1. Ideler, *Sternnamen*, pp. 238, 248, 413.
2. See my *Tableaux de la Nature*, Vol. I, p. 117.
3. Le Gentil, *Voyage [dans les Indes]*, Vol. I, p. 247.
4. Observation by Mr. de Chézy.

Even without recalling the hieroglyphs water (*atl*) and sea monster (*cipactli*), which bear a striking similarity to the catasterisms of Aquarius and Capricorn, the six signs of the Tartar zodiac that are also found in the 161 Mexica calendar suffice to make it very likely that the peoples of both continents derived their astrological ideas from a common source. The points of resemblance that we are emphasizing here are not taken from vague or allegorical paintings likely to be interpreted in accordance with the theories one hopes to advance. If one consults the works written at the beginning of the conquest by Spanish or Indian authors unaware of the existence of a Tartar zodiac, one will see that in Mexico, since the seventh century of our era, the days had been called *tiger*, *dog*, *monkey*, *hare* or *rabbit*, just as in all of eastern Asia, the years still bear the same names in Tibetan, Manchu-Tartar, Mogul, Kalmyk, Chinese, Japanese, Korean, and in the languages of Tonkin and Cochin-China.[1]

It is conceivable that peoples who have never had any contact with one another each divide the ecliptic into twenty-seven or twenty-eight parts and give to each lunar day the name of the stars near which the moon appears in its progressive west-east movement. It seems quite natural that a people of hunters and herdsmen should refer to these constellations and lunar days by the names of the animals that are always either the objects of their affection or a source of fear. The heavens of the nomadic hordes could therefore be inhabited by dogs, stags, bulls, and wolves without our having to conclude from this that these hordes once formed part of the same people. Points of resemblance that are either purely coincidental or stem from a similarity of position must not be confused with those that attest either to a common origin or to ancient contact.

But the Tartar and Mexica zodiacs do not contain only animals peculiar to the climes in which these peoples currently dwell; one also finds tigers and monkeys in them. These two animals are unknown on the plateaus of central and eastern Asia, whose high elevation gives them a colder temperature than the one that predominates toward the west on the same latitude. The Tibetans, Moguls, Manchu, and Kalmyk thus inherited from a more southerly land the zodiac that is rather narrowly called the Tartar cycle. The Toltecs, Aztecs, and Tlaxcalteca surged forth from the north toward the south: we know 162 of Aztec monuments all the way to the banks of the Gila, between 33° and 34° northern latitude. History shows us the Toltec coming from even more northerly regions. Having left Aztlan, these colonists did not arrive like barbarous hordes; everything about them suggested the remnants of an ancient

1. Souciet, [*Observations astronomiques sur la Chine,*] Vol. II, p. 138.

civilization. The names they gave to the cities they built were the names of the places where their ancestors lived. Their laws, their annals, their chronology, and the order of their sacrifices were modeled on the knowledge they had acquired in their original homeland. The monkeys and tigers that figure among the hieroglyphs for the days, and in the Mexica tradition of the *four ages* or *destructions of the sun*, lived neither in the northern part of New Spain nor in the northwest coasts of the Americas. The signs *ozomatli* and *ocelotl* thus make it especially likely that the zodiacs of the Toltec, Aztecs, Moguls, Tibetans, and so many other peoples who are today separated by a vast expanse of land were probably conceived at a single location on the old continent.

The Hindu lunar mansions, in which we also find a monkey, a serpent, a dog's tail, and the head of either a gazelle or a sea monster, have other signs whose names recall those of *calli*, *acatl*, *tecpatl*, and *ollin* from the Mexica calendar.

INDIAN NAKSHATRAS	MEXICAN SIGNS
Magha, dwelling	*Calli*, dwelling
Verou, reed	*Acatl*, reed
Critica, razor	*Tecpatl*, flint, stone knife
(*Sravana*, three footprints)	(*Ollin*, movement of the sun, represented by three footprints)

We note first that the Aztec word *calli* has the same meaning as the *kuala* or *kola*[1] of the Voguls [Mansi], who dwell on the banks of the Kama and the Irtysh, just as *atl* (water) in Aztec and *itels* (river) in Vilela recall the words *atl*, *atelch*, *etel*, or *idel* (river) in the languages of the Mogul, Cheremis [Mari], and Chuvash Tartars.[2] The name *calli*, dwelling, accurately denotes a lunar mansion (in Arabic, *mendzil el kamar*), a place of rest. Similarly, among the Indian nakshatras one finds not only *dwellings* (magha and punarvasu) but also wooden bedsteads and daybeds.

The Mexica sign *acatl*, reed, is generally represented as two reeds bound together.[3] But the stone found in Mexico City in 1790, which shows the day-hieroglyphs, represents the sign *acatl* very differently. One can make out either a bundle of rushes or a sheaf of corn enclosed within a vase. We shall take this opportunity to recall that in the first thirteen-day period of the year

163

1. Vater, *[Über] Amer[ikas] Bevölker[ung]*, p. 160.

2. ▾Engel, *Ungar[ische] Gesch[ichte]*, Part I, pp. 346, 361. Georgi, *[Bemerkungen einer] Reise [im Russischen Reich]*, Vol. II, p. 904. ▾Thwrocz, *Chron[ica] Hungaror[um]*, p. 49.

3. Plate XXVII.

tochtli, the sign *acatl* is constantly *accompanied* by *Cinteotl*, who is the goddess of corn, or Ceres, the deity that presides over agriculture. Among the Western peoples, Ceres is placed in the fifth dodecatemory; one even finds very old zodiacs in which a bundle of ears of wheat[1] fills the entire place that should otherwise be occupied by Ceres, Isis, Astraea, or Erigone, in the sign of the wheat and the grape harvests. It is thus that from the earliest times and among the most distant peoples, we find the same ideas, the same symbols, and the same tendency to relate physical phenomena back to the mysterious influence of the stars.

The Mexica hieroglyph *tecpatl* shows a sharp, oval-shaped stone elongated at both ends and similar to stones used as knives or attached to the end of a pique. This sign recalls the *krrtika*, the sharp knife of the Hindus' lunar zodiac. On the large stone shown on Plate XXIII, the hieroglyph *tecpatl* is represented in a slightly different shape from the one normally given to this instrument. The flint is pierced through the center, and this hole appears to have been meant as a hand-grip for the warrior who used this dual-tipped weapon. It is known that the Americans were particularly adept at piercing through the hardest rocks and working them using the friction method. I brought back from South America and deposited in the Museum of Berlin an obsidian ring that served as a young girl's bracelet; it consists of a hollow cylinder four centimeters high and nearly three millimeters thick, with an opening of nearly seven centimeters. It is difficult to understand how this vitreous and fragile substance could have been reduced to such a thin blade. It should be noted that *tecpatl* differs from obsidian, which the Mexica called *iztli*; they used the word *tecpatl* to refer not only to jade but also to hornstone and fire-starting flint.

The sign *ollin*, or *ollin tonatiuh*, presides over the seventeenth day of the first month at the beginning of a fifty-two-year cycle. Finding an explanation for this sign proved a great challenge to the Spanish monks, who brought the Mexica calendar to light without having even the most basic knowledge of astronomy. The Indian authors translate *ollin* as *movements of the sun*. Whenever they find the number *nahui* added, they render *nahui ollin* through the words *sun* (tonatiuh) *in its four movements*. The sign ollin is represented in three ways: in some places (Plate XXXVII) as two interlaced ribbons, or better, as two curved sections that cross one another and have three visible dips at their apex; in other places (Plate XXIII) as the solar disc surrounded by four squares and containing the hieroglyphs for the numbers *one* (*ce*) and *four*

164

1. Ideler, *Sternnamen*, p. 172. Dupuis, *Origine des cultes*, Vol. II, p. 228–34. *Atlas*, number 6.

(*nahui*); and elsewhere yet as *three footprints*. As we shall explain below, the four squares may well allude to the famous tradition of the four ages, or the four destructions of the world, which occurred on the days 4 *tiger*, *nahui ocelotl*; 4 *wind*, *nahui ehecatl*; 4 *rain*, *nahui quiahuitl*; and 4 *water*, *nahui atl*; and in the years *ce acatl*, 1 reed; *ce tecpatl*, 1 *flint*; and *ce calli*, 1 *dwelling*. To these days corresponded, more or less, the solstices, the equinoxes, and the sun's passages through the zenith of the city of Tenochtitlan.

The depiction of the sign *ollin* as three *xocpalli* or *footprints*, which often appears in the manuscripts preserved at the Vatican and in the *Codex Borgianus*, folio 47, number 210, is remarkable for its apparent similarity to the *sravana*, the three footprints of Vishnu and one of the mansions of the Hindu lunar zodiac. In the Mexica calendar, the three prints indicate either the traces of the sun in its passage across the equator and in its movement toward the two tropics or the three positions of the sun at its zenith, at the equator, and at one of the solstices. It may be possible that the Hindus' lunar zodiac contained some sign like Libra that was related to the progression of the sun. We have seen that the twenty-eight-sign zodiac may have been transformed, little by little, into a zodiac of twelve mansions of the full moon, and that some nakshatras may have changed names after the *full moon zodiac* became, through the knowledge of the sun's annual movement, a veritable *solar zodiac*. Indeed, Krishna, the Apollo of the Hindus, is nothing more than Vishnu in the form of the sun,[1] which, under the name of the god Surya, is the object of special worship. Despite this similarity of ideas and signs, we think that the three prints that form the twenty-third nakshatra *sravana* bear only a coincidental resemblance to the three vestiges of feet that represent the sign *ollin*. ˙Mr. de Chézy, who combines a profound knowledge of Persian and Sanskrit languages, observes that the *sravana* of the Indian zodiac alludes to a legend that is very famous among the Hindus and is recorded in most of their sacred books, especially the *Bhagavata Purana*. Intending to punish a giant who thought himself as powerful as the gods, Vishnu appears before him in the form of a dwarf. The dwarf begs the giant to grant him, from his vast empire, the amount of space that the dwarf could cover with three of his steps. Smiling, the giant grants the wish, but at that very moment, the dwarf grows to such a prodigious size that after two steps he takes up all the space between the sky and the earth. When the dwarf asks, on his third step, where he may place his foot, the giant recognizes the god Vishnu and prostrates himself before the deity. This fable explains the figure of the nakshatra *sravana* so well that it would be difficult to accept that this sign is

1. [*Asiatick Researches*], Vol. I, p. 200.

linked to that of *ollin* in the same manner in which *cipactli* and the Mexica's Noah, *Teo-Cipactli*, are linked to the constellations of *Capricorn* and *Deucalion*, formerly placed in Aquarius.

We have just explained the connections between, on the one hand, the signs that constitute the different zodiacs of India, Tibet, and Tartary and, on the other, the day- and year-hieroglyphs of the Mexica calendar. We have found that the most striking and most numerous of these links are those presented by the twelve-animal cycle, to which we have referred as the Tartar and Tibetan zodiacs. To bring to a close a discussion whose conclusions are so important to the history of the ancient contacts between peoples, it remains for us to examine more closely the Tibetan zodiac and to demonstrate that in the system of Asian astrology with which Mexica astrology appears to share a common origin, the twelve zodiac signs preside not only over the months but also over the years, the days, the hours, and even the smallest divisions of the hours.

When one considers that the peoples of eastern Asia divide the ecliptic not only into twenty-seven or twenty-eight parts but also into twelve and twenty-four parts, and that the same solar zodiac signs have entirely different names there and often entirely different shapes, one is tempted to believe that this multiplicity of signs must create great confusion about the limits assigned to the zodiac constellations. Among the Hindus, for example, in addition to the nakshatras or lunar mansions, we find twelve *lagna* whose names are the same as those of the Greek and Egyptian zodiac. The Chinese divide the ecliptic in three different ways: into twenty-eight nakshatras, which they call *che* or *eul-che-po-sieou*;[1] into twelve *tse*, which correspond to our signs but bear names that are either mystical or borrowed from the local flora and fauna, such as *great splendor, deep emptiness, quail's tail and quail's head*;[2] and into twenty-four *tsieki*. The names of these *tsieki* or *half-tse* are related to the climate and to temperature variations.[3] In addition, the Chinese have two other twelve-sign cycles: that of the *tchi* and that of the *animals*, the names of which are identical to those of the Tibetan and Tartar cycles; seven *che* correspond to three *tse* in the same way that six *tsieki* correspond to three *celestial animals*. This cycle of twelve Chinese animals—among which we have found the monkey, the tiger, the rat (a symbol of water), the dog, the bird, the serpent, and the hare from the Mexica calendar—gives names both to the twelve-year cycle and to the short twelve-day period. These twelve ani-

166

---

1. Souciet and Gaubil, [*Observations astronomiques sur la Chine,*] Vol. III, p. 80.

2. *Ibid.*, Vol. III, p. 98.

3. *Ibid.*, Vol. III, p. 94. Bailly, *[Histoire de] astr[onomie] ind[ienne]*, p. LXXXXVI.

mals are used, Father Gaubil writes,[1] to mark the twelve moons of the year, the twelve hours of the day and the night, and the twelve celestial signs. But in eastern Asia these twelve-part divisions, designated by different names, are all either simply abstract or imaginary divisions; they are used to remind the people of the sun's transit across the ecliptic. But as Mr. Bailly has rightly observed,[2] and as the more recent studies by Mr. Jones and Mr. Colebrooke confirm, the true celestial zodiac consists of the twenty-eight lunar mansions. It is true that in China they say that the sun *enters the monkey and the hare*, just as we say that it enters Gemini or Scorpio; but the Chinese, the Hindus, and the Tartars only distribute the stars in accordance with the system of the nakshatras. Like the short seven-day period, the division of the zodiac into twenty-seven or twenty-eight parts, known from Yemen to the Turfan Basin and to Cochin-China, belongs to the earliest achievements of astronomy.

167

Wherever one encounters several coexisting ecliptic divisions that differ not in the number of catasterisms but in their names (like the *tse*, the *tchi*, and the *celestial animals* of the Chinese, Tibetans, and Tartars), this multiplicity of signs is probably due to the intermingling of several peoples who had subjugated one another. The effects of this admixture—that is, of the influence the conquerors exerted over the vanquished—are especially evident in the northeastern part of Asia, where the languages, despite the large number of Mogol and Tartar roots, are so fundamentally different[3] from one another that they seem resistant to any methodic classification. The farther one travels from Tibet and Hindustan, the more one notices the uniformity of civil institutions, knowledge systems, and worship fade away. If the hordes of eastern Siberia, which were obviously exposed to Buddhist dogma, seem nonetheless to display only weak ties to the civilized peoples of southern Asia, can we truly be surprised to find such a large number of striking dissimilarities on the new continent, aside from a few similar features within the traditions, chronology, and the style of monuments? Once the peoples of Tartar or Mongol origin, transplanted onto foreign shores and intermixed with the indigenous hordes of the Americas, had succeeded in clearing a path toward civilization, everything—their languages, mythologies, and temporal divisions—assumed an individual character that erased, so to speak, their original national physiognomy.

Indeed, instead of the sixty-year cycles, the years divided into twelve months, and the short seven-day periods common among the peoples of

1. Souciet, [*Observations astronomiques sur la Chine*,] Vol. II, pp. 156, 174.
2. [Bailly, *Histoire de l']astr[onomie] ind[ienne]*, p. V.
3. Adelung, *Mithridates*, Vol. II, pp. 533 and 560.

Asia, we find fifty-two-year cycles, eighteen-month years (of twenty days each), half-decades, and semilunar months of thirteen days among the Mexica. The system of periodic series, in which paired terms were used to designate the days and the years, is the same on both continents. A large number of   168
the signs that compose the series in the Mexica calendar are borrowed from the zodiac of the peoples of Tibet and Tartary, but neither their number nor the order in which they appear are the same as what one finds in Asia.

Unlike the Hindu zodiac, the Tartar zodiac does not begin with the dog, which corresponds to our sign Aries, but with the rat, which represents Aquarius.[1] The Tartar zodiac also displays a remarkable oddity in that the order of the *celestial animals* runs counter to that of the signs: instead of placing the latter in the order marked by the sun's movement from west to east through the ecliptic, the Tibetans, Chinese, Japanese, and Tartars list the signs as follows: *rat* (or Aquarius), *steer* (or Capricorn), *tiger* (or Sagittarius), *hare* (or Scorpio), etc. The reason for this bizarre custom lies perhaps in the circumstance that the twelve zodiac constellations preside over the different hours of the day and night during their transit across the meridian. Since these constellations participate in the general movement of the celestial sphere from east to west, people arranged them in the order in which they either rise or set, one after the other.

The day-signs of the Mexica calendar, which are identical to the signs of the Tartar cycle—the dog, the monkey, the tiger, and the hare—are placed in an order that makes it impossible to detect any similarity in their relative positions. *Cipactli*, which we have proven above to be the *gazelle-fish*, is the first catasterism, just as Capricorn appears to have been among the Egyptians.[2] Among the Mexica signs, the following order more or less dominates: *cipactli, cohuatl, tochtli, itzcuintli, ozomatli,* and *ocelotl*—or, substituting the names for our signs, Capricorn, Virgo, Scorpio, Aries, Gemini, and Sagittarius. Might this discrepancy in the distribution of signs be merely superficial, and might it have resulted from a cause similar to the one that, according to the accounts of Herodotus and ˙Cassius Dio,[3] led to the days of the week being named, among all eastern peoples, after the planets and placed in a very different order from the one established in Hindu, Egyptian, and Greek as-   169
tronomy? If, however, one considers the number of terms that make up both

1. Souciet, [*Observations astronomiques sur la Chine,*] Vol. II, p. 136. Bailly, *[Histoire de l']astr[onomie] ind[ienne,]* p. 212. Langlès, notes on the *Voyage de Thunberg*, p. 319.

2. *Fragmentum ex Gazophylacio Card[inalis] Barberini* (Kircher, *Oedipus [Aegypticus],* 1653), Vol. III, p. 160.

3. Dio Cassius, [*Historiae Romanae,*] Book XXXVII, ch. 19 (ed. Fabric[ius], 1750), Vol. I, p. 124. Herodotus, [*Histoire,*] Book II, ch. 89 (ed. Wesseling, 1763), p. 105.

the series of hours and the series of Mexica hieroglyphs, one must acknowledge that this theory is untenable.

We have explained above, while discussing the apparent similarity between the names of several of the lunar mansions and those of the solar zodiac signs, how the original order of the catasterisms may be changed when certain peoples, having regressed into barbarism, seek to reestablish their former chronological system on the basis of vague recollections. Although it is natural to assume that such changes did occur, we are not obliged to accept this theory in order to explain the divergence between the positions of the same signs within the Tartar and the Mexica zodiacs. The Hindus maintain several divisions of the ecliptic into twenty-seven or twenty-eight nakshatras, and although their names are more or less the same, they are not always placed in the same order. An ancient monument that Bianchini unveiled at the end of the last century proved that there existed solar zodiacs in the East in which the Tartar catasterisms of the horse, the dog, the hare, the dragon, and the bird appear, but in an order in which the dog corresponds to Taurus of the Greek zodiac, not to Aries, and in which the dog and the hare are separated not by four signs but by only two. Given that the nakshatras and dodecatemoria have not always followed the same order in the various lunar and solar zodiacs in Asia, one must not be surprised by the transposition of signs that we can observe in the cycle of day-hieroglyphs among the Mexica. It may well be that this transposition was merely illusory, and that it seemed real to us because we can compare the Toltec and Mexica calendar only with the cycles that we find among the Tartars and the Tibetans today. Perhaps other peoples from eastern Asia passed on their zodiac to the warlike hordes that flooded into Mexico beginning in the seventh century. Perhaps, while traversing the plateau of central Asia and examining more closely the remnants of civilizations preserved in Lesser Bukhara [in Uzbekistan], in the Turfan Basin, or near the ruins of Karakorum, the ancient capital of the Mongol empire, travelers will one day discover the same series of signs contained within the Mexica zodiac.

170 This astronomical monument, of which Bianchini sent a drawing to the Academy, is a marble fragment preserved at the Vatican and discovered in Rome in 1705. We propose to examine it with particular care here, because we believe that it has the potential to cast light on the ecliptic divisions commonly used in Mexico and eastern Asia. It shows, in five concentric zones, the figures of the planets, the ˇdecans, the catasterisms of the Greek zodiac (repeated twice), and the signs of another zodiac that bears the greatest similarity to that of the Tartar peoples. One may be surprised that ˇFontenelle, Bailly, Dupuis, and other scholars who have written about the origin of the

zodiacs have taken this bas-relief for an Egyptian work.[1] According to an observation by an illustrious scholar, Mr. Visconti, the style of the figures that
represent the planets offers clear evidence that it was sculpted in the time of
the Caesars. Among the signs in the interior section of this disfigured monument, one can make out a horse, a crab, a serpent, a somewhat wolflike dog,
a hare, two birds (one of which appears opposite the serpent), and two quadrupeds, one with a long tail, the other with goat horns. Since the catasterisms of the Greek zodiac are paired with those of the unknown zodiac, one
sees that the horse and the hare correspond to our signs Leo and Scorpio, as
they do in the Tartar dodecatemoria. The following table presents the order
in which the catasterisms are found in Bianchini's planisphere. I have added
the signs of the Tartar cycle, vestiges of which we found among the peoples
of the new continent.

Bianchini's Zodiac Signs		
EXTERIOR ZONE	INTERIOR ZONE	TARTAR CYCLE
Sagittarius	*Bird*	Tiger
Scorpio	Hare	Hare
Libra	*Goat*	Dragon
Virgo	*Long-tailed animal*	Serpent
Leo	Horse	Horse
Cancer	Cancer	Ewe
Gemini	Serpent	Monkey
Taurus	Dog or wolf	Chicken
Aries	*Bird*	Dog
Pisces	. . . .	Hog
Aquarius	. . . .	Rat
Capricorn	. . . .	Steer

Printed in *italics* are the names of those animals that are too disfigured to   171
be identified with any certainty; the catasterisms of the Greek sphere that are
completely missing but easy to supply have been marked in the same manner.
Following the custom of the Tartar peoples, I have arranged the latter *counter to the order of the signs*. It is quite remarkable that in this strange monument the planets and the decans (of which only the latter are depicted in the

1. [Fontenelle, "Planisphère céleste Égyptien et Grec 1705,"] *Hist[oire] de l'Acad[émie
Royale] des Sciences*, 1708, Vol. I, p. 110. Bailly, *Hist[oire] de l'astr[onomie] anc[ienne]*,
pp. 493 and 504. Dupuis, *Origine des cultes*, Vol. I, p. 180. Hager, *Illustraz[ione]d'uno zodiaco orientale*, 1811, p. 15.

Egyptian style, with animal heads or animal masks) are placed in opposite directions. Although in the two sections that show the Greek zodiac there are four signs repeated with the same forms, one cannot conclude from this that the others were also identical. It would have been especially desirable for Gemini and Pan (or Capricorn) to have been preserved in the two sections, for the sculptor's apparent intention was to bring together the zodiacs of different peoples and the mixed forms[1] given to the same catasterisms among the Chaldeans, the Egyptians, and the Greeks. Gemini is represented by two figures that Mr. Bailly believed to be of different sexes, one of which bears a club and the other a lyre. It is in this very form that this sign is described in Hyginus's *Astronomicon*[2] and also in Sanskrit verses by the poet ˈSrīpeti: "The couple, *Mithuna*," the Hindi author writes, "is composed of a girl who plays the veena, and a young man who brandishes a club."[3]

Like the zodiac of the Tibetans, the Chinese, and the Tartars, the inner zodiac contains only animals, true ζώδια [*zôdia*, zodiacal signs]. In the Greek sphere, half of the signs are animals found in nature; the other half is composed of human figures and fabled or allegorical beings. In some instances, the scales [Libra], ζυγός [*zygos*, scales, balances] or λίτρα [*litra*, pound, weight], are sometimes held by the claws (χηλαί [*chelai*]) of Scorpio,[4] in others by a male figure, as in Bianchini's planisphere and the Indian zodiac, and in yet others, by a virgin [Virgo] who, in this case, goes by the name of Astrea or Δίκη [Dike, goddess of justice]. The signs for the lunar mansions, or the day-hieroglyphs in the Mexica calendar, include both animals and inanimate objects. If one adopts ˈMr. Hager's inspired theory that the sacred stone that Michaux brought back from the banks of the Tigris is an ancient zodiac, one recognizes that among the Chaldeans the series of the true ζώδια [*zōdia*] was also punctuated by altars, towers, and houses.[5] This fact lends credence to the hypothesis that the dodecatemoria originated from the lunar houses or mansions. The stone itself seems to suggest yet another similarity: in the Tartar cycle, the tiger corresponds to Sagittarius, often represented by a simple arrow. In the zodiac described by Mr. Hager one recognizes, in addition to the wolf (or feral dog) and the Capricorn (or gazelle-fish), an arrow

172

1. Eratosthenes, *Cataster[ismi]* (ed. Schaubach, 1795), p. 21. ˈHyginus, *Poeticon astr[onomicon]*, Book II, ch. 28; Book III, ch. 27 (*Auctores mythographi latini*, ed. van Staveren, 1742, Vol. II, pp. 481–528).

2. Book III, ch. 21 (*Auct[ores] mythograph[i latini]*, Vol. II), p. 523). ˈDu Choul, *Discours de la religion des anciens Romains*, 1556, p. 180. Ideler, *Sternnamen*, p. 151.

3. [*Asiatick Researches*], Vol. II, p. 335.

4. ˈManil[ius], [*Astronomica*], Book I, Verse 609.

5. *Illustrazione d'une zod[iaco] orientale*, Ch. VIII, p. 39, Plate 2.

that represents the River Tigris. This similarity, however, is purely coincidental, for the name of the river is of no relation to the name that the animal "tiger" bears in the languages of the East.

When one recalls that the zodiac containing a dog, a hare, and a monkey is exclusive to eastern Asia and that it very likely spread from there into the Americas, one is surprised to learn that it was known in Rome in the first centuries of our era, the period in which Bianchini's planisphere was sculpted. The astrologers, or Chaldeans, who had settled in Greece and Italy were probably still in contact with their counterparts in Asia. These contacts must have been more frequent and more extensive the more popular astrology became among the people and at the court of the Caesars. Of the eight identifiable signs in Bianchini's planisphere, only one, Cancer, does not belong to the Tartar zodiac. The hare, which is also found among the Tibetans and the Mexica, is a bit long-legged, but its position in Scorpio makes it easily identifiable. I do not know why Mr. Bailly mistook the dog, or the wolf, for a pig. This latter animal is also found in the Tartar zodiac; it corresponds to the sign Pisces of the Greek sphere, and—what is truly remarkable—in the planispheres from the temple of Dendera, one sees two examples near this very sign of a figure holding a pig in its hand.[1] The monument Bianchini describes is all the more interesting in that there are no identifiable traces in any work of Greek or Latin astronomy—not even in the Saturnalia by Macrobius, written in the time of Theodosius—of this animal cycle which the Mongols and the other Tartar hordes that devastated Europe probably used in their chronology, but which we have come to know better only through our contacts with China and Japan. It is odd that the Academy's eloquent historian, Fontenelle, did not acknowledge that the imaginative excesses of astrology are closely connected to the foundational concepts of astronomy and that they can serve to cast light on the ancient contacts between peoples. ▼"The monument," he writes, "about which Bianchini has sought to learn more belongs to the history of human folly, and the Academy has better things to do than to occupy itself with this sort of research."

If we bring together what we have explained about the various ecliptic divisions and the signs that preside over the years, the months, the days, and the hours on both continents, we reach the following conclusions: Among those peoples who focused their attention on the starry vault of the heavens, the lunar zodiac divided into twenty-seven or twenty-eight mansions predates the twelve-part zodiac; the latter, originally only a *full-moon zodiac*, eventually became a *solar zodiac*. The names of the months either are chosen

1. Denon, *Voyage [en Égypte]*, Plates 130 and 132.

from among the lunar mansions, as with the Hindus, or are taken from the dodecatemoria, as in the Dionysian year. On the banks of the River Ganges, people still refer to the months of *Arrow*, *Dwelling*, and *Antelope Head*, just as in the time of ʼPtolemy Philadelphus people in Alexandria referred to the months of *Didymon*, *Parthenon*, and *Aegon*—the months of Gemini, Virgo, and Capricorn.[1] A close connection is apparent between the names of the dodecatemoria and and the names of the nakshatras: among several peoples, the latter passed to the lunar days. In addition to the actual division of the ecliptic, which is a region of the starry heavens, there still exist—particularly in eastern Asia—divisions of the time that it takes the sun to return, more or less, to the same stars or the same point on the horizon. Generally composed of either twelve or twenty-four parts, based on the number of elapsed lunar or semilunar months, these cycles belong to chronology rather than to astrognosy; they present only an ideal division of the ecliptic, each part of which is assigned a specific name and sign. Examples of these include the Tartar animals, and the *tse* and the *tsieki* of the Chinese. Such signs, which only mea-

174 sure time and subdivide the seasons, can be invented among peoples who do not focus their attention at all upon the stars. A genuine zodiac—composed of twelve signs that preside over the months and, through the periodic series device, over the years, days, and hours as well—might have been found in the lowlands of Peru, where a thick layer of fog blocks the inhabitants from seeing the stars while still allowing them to see the discs of the moon and the sun. The signs of the ideal zodiac, in which a complete revolution (circle, *annulus*) forms a year (*annus*, ἐνιαυτός [*eniautos*]), are easily transferred to the constellations themselves; the *division of time* thus becomes a *division of space*.

We shall not discuss whether the zodiac of the Hindus, the Chaldeans, the Egyptians, and the Greeks was not also originally a cycle[2] in which the signs indicated the climactic variations in a land subject to periodic flooding. The unequal space occupied by Virgo and Cancer and the apparent lack of connection[3] between the dodecatemory figures and the extrazodiacal constellations seem to make this a credible claim. Indeed, we see that there are peoples who use several ecliptic divisions at the same time and that the signs that are linked to constellations for one people are but divisions of time for another. Perhaps there once existed some region of Asia where the Tartar cycle of celestial animals—which Bailly regards as the most ancient of all the zodiacs,

1. Ideler, *Hist[orische] Untersuch[ungen]*, p. 264.

2. Rhode, *Versuch über das Alter des Thierkreises*, 1809, pp. 15 and 101.

3. [Schwartz,] *Recherches sur l'origine [et la signification] des constellations de la sphère grecque*, 1807, p. 63.

while Dupuis[1] strains to pass it off as one of the tables of the paranatellons [extrazodiacal constellations]—was a genuine division of the stars located on the ecliptic. In order to understand fully the relationships that since the earliest times have been forged between the peoples of the two continents, one must not lose sight of the intimate connection that exists between the imaginary zodiac and the real zodiac, between the cycles and the constellations of the ecliptic, and between the mansions and the divisions of the solar orbit.

These very reflections on the progressive development of astrognosy prevent us from deciding whether the day- and year-hieroglyphs within the Toltec and Aztec calendar merely belong, as the Chinese *tse* and *tchi* do, to an imaginary or fictitious zodiac, or whether they refer to zodiacal constellations. We have already observed above that wrapped around the great wheels that represent the fifty-two-year cycle was a serpent biting its own tail, with four folds that marked the four indictions. Since the hieroglyphs were arranged into recurring series of four terms, and since each of the intervals between the folds contained twelve years, each knot of the serpent corresponded to a different sign. I think that these four knots, designated by the catasterisms *rabbit*, *reed*, *flint*, and *dwelling*, alluded either to the points of the solstices and the equinoxes or to the intersections of the colures and the ecliptic. According to Albateginus,[2] the most ancient division of the zodiac is a four-part division. Indeed, in the first year of the long cycle of days, *matlactli tochtli* (10 rabbit), *chicuei acatl* (8 reed), *chicome calli* (7 dwelling), and *matlactli tecpactl* (11 flint), corresponded to December 22, March 22, June 20, and September 23. These days fall very close to the equinoxes and the solstices, and since the Mexica year, like the Chinese year, began on the winter solstice, it is quite natural that the first term in the recurring series of year signs would be *tochtli*, although in the series of twenty day-signs *tochtli* is still preceded by *calli*.

We also know from the concepts that Sigüenza drew from the works of ˇ[Alva] Ixtlilxochitl that both the four folds of the serpent and the four catasterisms that are linked to them referred to the four seasons, the four elements, and the cardinal points. The land was dedicated to the rabbit and the water to the reed; in our earlier discussion of the night-signs, we have seen that *Tepeyollotli*, one of the cave-dwelling deities, and *Cinteotl*, the goddess of the harvests, accompany the day-signs *rabbit* and *reed*. The meaning of these allegories is so clear that they do not require any explanation. The four signs for the equinoxes and the solstices, chosen from a series of twenty signs,

175

---

1. *Origine des cultes*, Vol. III, p. 362.
2. *De scientia stellarum*, ch. 2 (ed. Bonon[iae], 1645), p. 3.

also recall the four *royal stars*, Aldebaran, Regulus, Antares, and Fomalhaut, famous throughout Asia and presiding over the seasons.[1] On the new continent, the indictions of the fifty-two-year cycle constitute the four seasons of the *long year*, so to speak, and the Mexica astrologers liked seeing one of the four equinoctial or solsticial signs preside over each thirteen-year period.

176   Although the same signs were used and arranged in the same order in every part of the Mexica Empire, there are observable differences in the choice of the solsticial or equinoctial sign placed at the head of the *xiuhmolpilli* or *binding* of years. The inhabitants of Tetzcoco began their long year with *acatl*; those of Teotihuacan, with *calli*; the Toltec, with *tecpatl*. There has been some question whether the first day of the year always bore the sign *cipactli* among these peoples, regardless of the differences that we have just noted; but the fragments of their historical annals preserved in Boturini's museum and Father Pichardo's collection in Mexico City seem to indicate that the diversity of the dates stems from the period when the intercalation of thirteen days took place, not from differing ways of marking the beginning of the cycle.

We do not know whether the twenty Mexica day-signs are the remnants of an ancient division of the zodiac into twenty-eight lunar mansions or whether, together with the four night-signs (the names of which are not repeated among the day-signs), they once formed twenty-four catasterisms, like the *tsieki* of the Chinese zodiac. Perhaps an equal number of signs had been placed between the four equinoctial and solsticial signs; perhaps the number twenty is merely derived from a division of the visible hemisphere into ten parts. What is certain is that this very division prompted the Mexica to split the three-hundred-sixty-day year into eighteen months, and that this division became the foundation of a system of which we find no vestiges on the old continent. I am inclined to believe, however, that the division into eighteen months of twenty days each occurred after an earlier split into twelve moons of thirty days each, for the method of having a zodiac sign preside over each day, and of determining the number of months by the return of the periodic series, must have emerged later than the simpler idea of dividing the year according to the number of lunar months it contains. Although divisions of the ecliptic into twenty-four *tsiekis*[2] and into thirty-six *decans* do exist in Asia, these divisions did not give rise there to years of ten or fifteen months; and if antiquity shows us years of four, six, or twenty-four months,

1. Firmicus, [*Astronomicorum,*] Book VI, ch. 1.

2. ▾Amiot, *Mémoires concernant les Chinois*, Vol. II, p. 161. Gaubil, *Traité de l'astronomie chinoise*, p. 32.

those divisions do not result from the use of periodic series, as do the eighteen months of the Mexica year; rather, they stem from the importance given   177
to the equinoctial and solsticial points, the sixty-year cycles, and the length
of the semilunar months.

We recalled above that the Mexica year, like that of the Egyptians and the
Persians, was composed of three hundred sixty days, to which five furtive
(*musteraka*) or futile (*nemontemi*) epagomenal days were added. Had the
Mexica not been aware of the excess length of a solar revolution of three hundred sixty-five days, then the beginning of their year, like that of the Egyptians' wandering year, would have passed once through all the seasons and
through all the points on the ecliptic over a period of fourteen hundred sixty
years. By the time the Spanish arrived, four centuries had elapsed since the
reform of the Mexica calendar in 1091. At that time, all the contemporary
writers confirm, the Europeans' calendar coincided with the Aztec calendar
but for a few days. The exactness of the calculation of the solar eclipses noted
in the Mexica annals even made it likely that the difference between the two
calendars was entirely due to the fact that our calendar had not yet undergone the Gregorian adjustment. Let us now turn to the mode of intercalation
through which the Mexica succeeded in avoiding errors in their chronology.

Since the Mexica year was solar and not lunar, the mode of intercalation
could be much simpler than that used by the Greeks and the Romans before
the introduction of the ʼMercedonius. Looking broadly at the intercalations
commonly used among various peoples, we see that some allowed the hours
to accumulate until they formed a complete day, while others neglected intercalation until the extra hours formed a period that equaled one of the longer divisions of their year. The first mode of intercalation is that of the Julian Year; the second is that of the ancient Persians, who added a complete
thirty-day month to a twelve-month year every one hundred twenty years, so
that the intercalary month would traverse the entire year over the course of
12 × 120 (or fourteen hundred forty) years.[1] The Mexica obviously followed
the Persians' system: they preserved the wandering year until the surplus
hours formed a semilunar month; they therefore intercalated thirteen days
with every *binding*, or fifty-two-year cycle. The result of this, as we have observed above, was that each binding contained $^{18,993}/_{13}$ or fourteen hundred   178
sixty-one short thirteen-day periods. The Mexica year began with the first
year of *xiuhmolpilli*, on the day that corresponds to January 9 of the Gregorian calendar. In the fifth, ninth, and thirteenth years of the cycle, the first
day of the year was January 8, 7, and 6, respectively. In each year bearing the

1. Ideler, *Hist[orische] Unters[uchungen]*, p. 379.

sign *tochtli*, the Mexica lost a day, and the effect of this *retrogradation* was that the year *calli* of the fourth indiction began on December 27 and ended on the winter solstice, December 21, not counting the five futile or supplementary days. The result of this was that the last of the *nemontemi*—called *cohuatl* (serpent) and regarded as the unluckiest day because it did not belong to any of the thirteen-day periods—fell at the end of the cycle, on December 26, and that thirteen leap days returned the beginning of the year to January 9. To clarify our explanations, we add here the tableau of the final twenty-five days of the first year of a cycle [see table on p. 203].

179   The intercalation of thirteen days gave rise to the great secular festival called *xiuhmolpia* or *toxiuhmolpia* (bundling of our years), which all the historians of the conquest describe. Following an ancient prediction, the Mexica believed that the end of the world would come at the end of a fifty-two-year cycle, that the sun would not reappear on the horizon, and that men would be devoured by hideous evil spirits known as *Tzitzimimes*. This belief probably stemmed from the Toltec legend of the *four ages*, according to which the earth had already undergone four major upheavals, three of which had occurred at the end of a cycle. The people were overcome with dread for the five days that preceded the *xiuhmolpia*. On the fifth day, the sacred fire was extinguished in the temples by order of the *teoteuctli*, high priest. In the convents, which were as numerous in Tenochtitlan as they have been in Tibet and Japan since the earliest days, the monks, *tlamacazqui*, devoted themselves to prayer. As night fell, no one dared light a fire in his dwelling; people broke clay vases, tore up their clothes, and destroyed their most valuable possessions, because everything seemed useless at the terrible moment of the final day. By a bizarre superstition, pregnant women became objects of terror for men; their faces were hidden behind masks made of *agave* paper, and they were even locked up in maize storage rooms, because it was believed that if the cataclysm came to pass, the women, transformed into tigers [jaguars], would join with the evil spirits (*tzitzimimes*) in order to exact their revenge for the injustice inflicted upon them by the men.[1]

On the evening of the last day of the *nemontemi*, which is presided over by the sign of the *serpent*, the festival of the *new fire* began. The priests took the garments of their gods and, followed by an immense throng of people, they walked in solemn procession to ʼHuixachtecatl Mountain,[2] located two leagues from Mexico City, between Iztapalapa and Culhuacan. This somber

1. Torquemada, *De una fiesta grandissimja*, Book X, ch. 33–6, Vol. II, pp. 312 and 321. Acosta, [*Histoire naturelle y morale des Indes*,] Book VI, ch. 2, p. 259.
2. *Vixachtla*, according to Gómara, *Conquista*, folio 133(a).

	Metzlapohualli		
GREGORIAN CALENDAR	TONALPOHUALLI	SERIES OF THIRTEEN NUMBERS AND OF TWENTY DAY-SIGNS	SERIES OF NINE NIGHT-SIGNS
DECEMBER OF THE YEAR 1091 — 15	ATEMOZTLI OF THE YEAR 2 ACATL — 1	27TH 13-DAY PERIOD — 3 Cipactli	Tepeyollotli
16	2	4 Ehecatl	Quiahuitl
17	3	5 *Calli*	*Tletl*
18	4	6 Cuetzpalin	Tecpatl
19	5	7 Cohuatl	Xochitl
20	6	8 Miquiztli	Cinteotl
21	7	9 Mazatl	Miquiztli
22	8	10 *Tochtli*	Atl
23	9	11 Atl	Tlazolteotl
24	10	12 Itzcuintli	Tepeyollotli
25	11	13 Ozomatli	Quiahuitl
26	12	1 Malinalli — 28TH 13-DAY PERIOD	*Tletl*
27	13	2 *Acatl*	Tecpatl
28	14	3 Ocelotl	Xochitl
29	15	4 Quauhtli	Cinteotl
30	16	5 Cozcaquauhtli	Miquiztli
31	17	6 Ollin	Atl
JANUARY 1092 — 1	18	7 *Tecpactl*	Tlazolteotl
2	19	8 Quiahuitl	Tepeyollotli
3	20	9 Xochitl	Quiahuitl
4	NEMONTEMI — 1	10 Cipactli	. . . .
5	2	11 Ehecatl	. . . .
6	3	12 *Calli*	. . . .
7	4	13 Cuetzpalin	. . . .
8	5	1 Cohuatl	. . . .
9	TITITL OF THE YEAR 3 — 1	1 Cipactli — 1ST 13-DAY PERIOD	*Tletl*
10	2	2 Ehecatl	Tecpatl
11	3	3 *Calli*	Xochitl
12	4	4 Cuetzpalin	Cinteotl
13	5	5 Cohuatl	Miquiztli
14	6	6 Miquiztli	Atl
15	7	7 Mazatl	Tlazolteotl

march was called the *march of the gods*, *teonenemi*, a name that reminded the Mexica that the gods were leaving their city and they might never see them again. When they arrived at the top of the porphyritic mountain of Huixachtecatl, they waited for the moment when the Pleiades occupied the middle of the heavens before beginning the horrific sacrifice that we have described above,[1] and which is depicted on Plate XV, number 8. The victim's corpse remained stretched out on the ground, and the instrument used to light the fire through friction (πυρεῖα [*pyreīa*, flint stones] among the Greeks, *tletlaxoni* among the Mexica) was inserted in the very wound that the priest of Copulco, armed with an obsidian knife, had made in the chest of the prisoner destined for sacrifice. As soon as the wood splinters (*la harina del palillo*) broken off by the quick rubbing of the cylinder had caught fire, they lit an enormous stake that had been prepared in advance for the body of the unfortunate victim. The people shouted with joy. The glow from the stake could be seen across a large part of the Valley of Mexico due to the height of the mountain on which this bloody ceremony took place. All those unable to follow the procession stood on the flat roofs of their dwellings, on the summits of the teocalli, and on the hills that rise from the middle of the lake, their eyes fixed on the place where they expected to see the flame, a definite omen of the gods' benevolence and the preservation of humankind over the course of a new cycle. Messengers posted at equal distances and bearing torches made of highly resinous pine carried the new fire from village to village over a distance of up to fifteen to twenty leagues. In each place it was left in the temples, whence it was distributed to the individual homes. When they saw the sun rising on the horizon, their rejoicing intensified, the procession returned from Iztapalapa Mountain to the city, and the people believed that they saw their gods returning to their shrines. The women then emerged from their prison, everyone arrayed themselves in new clothes, and they used the thirteen leap days to clean the temples, whitewash the walls, and replace their furniture, dishes, and all household items.

This secular festival and the fear of witnessing the fifth sun extinguished on the winter solstice seem to constitute an additional point of commonality between the Mexica and the inhabitants of Egypt. In his commentary on Aratus, ˈAchilles Tatius[2] preserved for us the following note, which Scaliger

---

1. P. 120 [this edition].

2. Achilles Tatius, *Isag[oge] in [Arati] Phaenom[ena]*, ch. 23 (ˈPeta[vius], *De Doctr[ina] Tempor[um]*, 1703, Vol. III, p. 85). Scalig[er], *Adnot[ationes] ad Manil[ius] Astron[onmica]*, Book I, Verse 69, p. 85. See also the translation of the Letters [*americane*] of Count Carli, Vol. I, p. 398, note 1.

believes to have been borrowed from the *Oktaeteris* of Eudoxus: "When they   181
saw that the sun was descending from Cancer to Capricorn and that the days
were becoming shorter and shorter, the Egyptians had the custom of moan-
ing, fearing that the sun might completely abandon them. This period coincided with the festival of Isis, but when the sun began to show itself again,
and the days grew longer, they wore white clothes and garlands of flowers
on their heads (λευχειμονήσαντες ἐστεφανηφόρησαν [*leucheimonēsantes
estephanephoresan*, people dressed in white wore garlands])." Reading this
passage by Achilles Tatius, one has the impression of reading what Gómara
and Torquemada report about the Mexica jubilee festival; just as[1] in ˅Sextus
Empiricus's work[2] against the astrologers, one seems to find a portrait, so to
speak, of the symbolic figure[3] shown on Plate XV and reproduced from the
manuscript preserved at Velletri. Among all the peoples of the earth, superstitious ideas take the same form at the beginning and at the decline of civilization, and it is because of this general phenomenon that it is difficult to distinguish between what has been transmitted from one people to another and
what men have drawn from a strictly internal source.

   In speaking of the secular festival, Father Torquemada describes the moment of the sacrifice in a seemingly precise manner that nevertheless be-
trays a serious contradiction: "When the procession," he writes,[4] "arrived at
Huixachtecatl Mountain, the priests waited until midnight, which they determined by the position of the Pleiades that had risen to the middle of the
heavens (*estavan encumbradas en medio del cielo*) at that hour: for the time of
the jubilee, the secular festival, had come when these stars began to rise at
the onset of night, which, *for the horizon over Mexico, generally* occurs in the
month of December." The expression "when the Pleiades are in the middle
of the heavens" probably refers to the passage of these stars through the meridian or (what is more or less the same thing for the latitude of Mexico City)
their passage through the zenith. But the last secular festival was celebrated in
the sixth year of Montezuma's reign, and at that time, the culmination of the
Pleiades took place at midnight, if one takes the precession of the equinoxes
into account, not in the month of December but on November 8. On Decem-   182
ber 26, this constellation already rose 3 hours 23 minutes before sunset, and
its passage through the meridian was at 8:30 in the evening. The situation

   1. Dupuis, *Mém[oire] explicatif du zodiaque*, 1806, p. 145.
   2. Sex[tus] Empir[icus], *[Adversus] Mathem[aticus]*, Book V (ed. Stephan[us]), Vol.
III, p. 187). ˅Firmicus, *[Astronomicorum,]* Book II, ch. 27 (ed. Ald[us] Manut[ius], 1503,
folio CV). Origen, *Contra Celsum*, Book VIII, ch. 55 (ed. Delarue, 1733), Vol. I, p. 783.
   3. See p. 110 [this edition], Plate XV.
   4. Torquemada, *[Monarquía indiana,]* Vol. III, pp. 513b and 321a.

would naturally be the same in every place on earth where one might assume that the Mexica calendar was created, and if one goes back to the first sacrifice celebrated at Tlalixco in 1091, or to the Toltec migrations in the sixth century of our era, one finds that due to the effect of the precession of the equinoxes, the culmination of the Pleiades takes place closer and closer to sunset as the winter solstice approaches. The expressions "at the moment of midnight" and "in the middle of the heavens" should probably not be taken too literally. In general, Father Torquemada speaks so confusedly about Mexica chronology that one may assume that he misunderstood almost everything the Indians reported to him about astronomical phenomena. After stating categorically that the cycle—and therefore the year—ended in the month of December, he concedes that the first day of the year is February 1, and he adds that the sun over Mexico City reaches the *highest point* of its path on the winter solstice. Torquemada gathered names, legends, and isolated facts with the most scrupulous precision; but lacking any critical acumen, he contradicts himself each time he tries to link these facts together or assess their mutual relationships. Since the Mexica did not know of the use of clepsydras [water clocks], which had been used for ages[1] in Chaldea and China, they could not determine the moment of midnight exactly. Furthermore, across Asia, the cosmic setting of the Pleiades was also seen as an indication of the onset of winter.[2] One would seek in vain for a rigorous precision in popular traditions, which were perhaps conceived in the more northerly regions where the cold is felt a month before the solstice.

What we have just noted about the constellation of the Pleiades suffices, moreover, to prove how many authors were mistaken in suggesting some uncertainty as to whether the year began around the spring equinox or around the winter solstice. The further one moves away from November 5, the day of the achronic rising of the Pleiades, the less possible it becomes that the Mexica could have seen this constellation near the zenith in the middle of the night that the secular sacrifice took place.[3] Torquemada, Léon, and Betancourt nevertheless believed that the year began on February 1 or 2; Acosta and Clavijero, on the 26th day of the same month; Valadés and Alva Ixtlilxochitl, on March 1 and 20; Gemelli and ʼ[Echeverría y] Veytia, on April 10. In the sixteenth century, the culmination of the Pleiades took place on the day of the spring equinox, 3 hours and 8 minutes *before sunset*. It is true that accord-

183

1. Sex[tus] Empir[icus], [*Adversus Mathematicus,*] (ed. Stephan[us]), p. 113. Letter by ʼFather Du Croz, in Souciet, *Observat[ions]*, Vol. I, p. 245.

2. Bailly, *Astr[onomie] mod[erne]*, p. 477.

3. Gama, [*Descripción histórica y cronológica de las dos piedras,*] §35, p. 52, note.

ing to an ancient legend,[1] the disappearance of this constellation at sunrise once marked the day of the autumn equinox, which presupposes an observation conducted three thousand years before our era; but we cannot suppose that the Mexica had received their chronology from a people who began their year at the start of autumn. The correspondence of dates; several astronomical phenomena; the testimony of Spanish writers who accumulated materials without any knowledge of the true calendar system—everything corroborates Gama's system. I shall content myself to cite here only one piece of evidence. In a handwritten work[2] in the Mexica language preserved in Mexico City, the Indian historiographer Cristóbal del Castillo maintains that the five supplementary days were added to the end of the month *Atemoztli*, which, according to the unanimous accounts of Indian and Spanish authors, corresponded to our month of December. Torquemada also notes that the third festival of the god of water was celebrated on the winter solstice, which took place near the end of *Atemoztli*, and that the cycle was complete at the end of December. These circumstances conjointly place the leap days shortly after the winter solstice. The fear of seeing the day star extinguished or departed and the ideas of grief and joy expressed in the secular festival are also much more closely related to the time of the shortening of days than to that of the equinox. While it is true that the high priest in Rome took the new fire on the altar of Vesta at the start of spring and that the Persians celebrated the great festivals of Nowruz, the reasons[3] behind these festivals were different from those that motivated the Mexica and the Egyptians in their solsticial and Isis festivals, respectively.

I have explained the system of intercalation as it is described in the Mexica manuscripts and as it was adopted by Sigüenza, Clavijero, Carli, and, long before them, by ˙Boulanger and Fréret. Following this system, the length of the year is posited to be 365.25 days, the result of which is that from the reform of the calendar in 1091 to the arrival of the Spanish, the Mexica must have erred by at least three days. Yet the studies that Gama conducted on the solar eclipses of February 23, 1477, and June 7, 1481 (which are mentioned in the hieroglyphic annals), on several memorable periods of the conquest, and on the days when, according to the Mexica celebrations, the sun passes through the zenith of Tenochtitlan, seem to prove that this three-day error did not occur, and that at the beginning of the sixteenth century, as noted

184

---

1. Plin[y], *Hist[oria] nat[uralis]*, Book XVIII, ch. 25 (ed. Harduin, 1741), Vol. II, p. 129.

2. *M.S.S.*, ch. 71.

3. Dupuis, *Origine des cultes*, Vol. I, p. 156; Vol. II, Plate 2, p. 96.

above, the dates of the Aztec calendar corresponded more closely to the solstice and equinox days than did those of the Spanish calendar.

Without knowing the exact length of the year, the Mexica could have corrected their calendar from time to time, since their gnomonic observations informed them that in the first year of the cycle, the spring and autumn equinoxes moved a few days away from 7 *malinalli* and 9 *cozcaquauhtli*. The Peruvians of Cusco, who had a lunar year, based their intercalation not on the shadow cast by the gnomons, which they nevertheless measured very diligently, but on marks placed on the horizon to indicate the points where the sun rose and set on the solstice and equinox days. A periodic and exact intercalation, like the one observed by the Persians since the eleventh century, is probably preferable to the abrupt changes known as calendar *reforms*; but a people that had used a very imperfect mode of intercalation for centuries could nevertheless preserve the harmony between its calendar and that of the most civilized peoples if, prompted by direct observation of celestial phenomena, it occasionally adjusted the beginning of its year. In the annals of Mexica history there are no traces of such abrupt changes, nor of such extraordinary intercalations. The calendar had not undergone any reforms since the legendary time of the Tlalixco sacrifice; the intercalation took place uniformly at the end of each cycle, and to explain why four centuries had not sufficed to produce any observable error in their chronology, Mr. Gama suggests that the Mexica inserted only twenty-five days after each of the one-hundred-four-year cycles, *cehuehuetiliztli*, or twelve and a half days at the end of each fifty-two-year cycle, which sets the length of the year at 365.240 days. Based on the sixteenth-century historians' own accounts, he believes it is possible to conclude that the secular festival was alternatively celebrated at day and at night, and that if the years of one cycle all began at midnight, those of a another cycle all began at noon. Since I am unable to examine the works written in the Mexica language, I am in no position to pronounce on the accuracy of Mr. Gama's theories. The reasons that he invokes in his paper on the monuments discovered in 1790 no longer appear to me as conclusive as I once found them to be, before having had the opportunity to conduct an in-depth study on the Mexica calendar. Once his heirs have found the means to publish his treatise on Toltec and Aztec chronology, it will be easier to assess the true number of leap days. Gama's astronomical works, the accuracy of which we have already had occasion to verify, must moreover inspire great confidence, and it is likely that this scholar, who had the patience to calculate (using ˇMayer's Tables) a large number of solar eclipses for the parallel of the former Tenochtitlan, and then to link them to specific historical periods, would not have ventured lightly toward a new hypothesis had he not been prompted to do so by both a careful comparison of the dates and a study of hieroglyphic painting.

"The intercalation of twenty-five days into one hundred four years," writes Mr. Laplace[1] in his excellent précis of the history of astronomy, "presupposes a length of the tropical year that is more exact than that of Hipparchus and—what is quite remarkable—almost equal to the year of Al-Ma'mun's astronomers. When one considers the difficulty of reaching such an exact determination, one is tempted to think that this is not the work of the Mexica themselves and that it came to them from the old continent. But from which people, and by which means, did they receive it? If it had been transmitted to them via the north of Asia, why then do they have a division of time that differs so sharply from the ones that have been in use in that part of the world?" We cannot presume to be able to resolve these questions with the current state of our knowledge; but even if one refuses to accept the intercalation of twelve and a half days per cycle, and if one grants the Mexica only the knowledge of the ancient Persian year of 365.250 days, the day-hieroglyphs and the use of periodic series nevertheless constitute irrefutable evidence of ancient contacts with eastern Asia.

Although the Mexica cycle began with the year of the rabbit, *tochtli*, just as the Tartar cycle began with the year of the rat, *singueri*, the intercalation only occurred in the year *ome acatl*; it was this very circumstance that prompted the Mexica to use the image of a bundle of reeds to represent a *xiuhmolpilli*, or fifty-two-year cycle, in their paintings. The Mexica had left Aztlan in the year 1064, or 1 *tecpatl*; their migrations lasted for twenty-three years, until 1087, or 11 *acatl*, when they arrived in Tlalixco. Although the calendar reform took place in 1090 (the year 1 *tochtli*), the festival of the new fire was celebrated only in the following year, 2 *acatl*, "because," the Indian historian Tezozomoc writes,[2] "the people's tutelary god, *Huitzilopochtli*, had made his first appearance on the day 1 *tecpatl* of the year 2 *acatl*."

A few authors have suggested that before the calendar reform at Tlalixco, the Mexica had intercalated one day every four years; what seems to have given rise to this theory is a festival in honor of the fire god (*Xiuhteuctli*), celebrated with greater solemnity in the years that bore the symbol *tochtli*. Count Carli, whose *Lettere americane* display a peculiar mixture of precise observations, very clever ideas, and hypotheses incompatible with either the principles of good physics or the theory of celestial movements, thought he had identified the remnants of a lunar intercalation in the nine-day festivals cel-

186

1. *Exp[osition] du système du monde* (third edition), Vol. II, p. 318.
2. Gama, [*Descripción histórica y cronológica de las dos piedras,*] §7, p. 21.

ebrated every four years. He assumed that in the Mexica priests' calculations, each year contained twelve lunar months of twenty-nine days and eight hours each, and that in order to bring these three-hundred-fifty-two-day years back into harmony with true lunar years, they added nine days once every four years. This supposition is almost as hazardous as the one by which this very author attributes the error of the ancient calendars to the celestial bodies themselves, suggesting that a few thousand years before our era, the earth completed its rotation around the sun in three hundred sixty days,[1] and that one lunar month lasted only twenty-seven and a half days.

187 Since a periodic series of four terms was used to distinguish between the years contained within one cycle, the Mexica naturally found themselves compelled to observe quadrennial festivals. These included the solemn one-hundred-sixty-day fast celebrated on the spring equinox in the tiny republics of Tlaxcala, Cholula, and Huejotzingo and the horrific sacrifice that took place every four years in Quauhlitlan, in the month *itzcalli*. In this sacrifice the penitent scarified their own bodies, letting their blood stream through reed stalks that they inserted into their wounds[2] and then deposited in the temples as public displays of their devotion. These festivals, which recall the penance rituals commonly observed in Tibet and India, were repeated every time a certain sign presided over the year.

Examining in Rome the *Codex Borgianus* of Velletri, I was able to identify the curious passage[3] that led the Jesuit Fábrega to conclude that the Mexica had knowledge of the actual length of the tropical year. Spread across four pages, the codex shows twenty cycles of fifty-two years each, or one thousand forty years. At the end of this long period, the order of the day-hieroglyphs has the sign of the rabbit, *tochtli*, immediately preceding the bird, *cozquauhtli*, such that seven days are omitted, those of the *water*, the *dog*, the monkey, the grass, *malinalli*, the *reed*, the *tiger* [jaguar], and the *eagle*. In his handwritten commentary, Father Fábrega suggests that this omission stems from a periodic reform of the Julian intercalation because, through an ingenious method, the subtraction of eight days at the end of a one-thousand-forty-year cycle brings a year of 365.25 days back into harmony with a year of 356.243 days, which is only 1 minute and 26 seconds, or 0.0010 days, longer than the true average year as reported in Mr. Delambre's Tables. When one has had the opportunity to examine a large number of the

1. *Lettere americane*, Vol. II, pp. 153, 161, 167, 333, and 371.

2. Gómara, [*Historia de la conquista de Mexico,*] pp. CXXXI, CXXXII. Torquemada, [*Monarquía indiana,*] Vol. II, p. 307. Gemelli, [*Giro del mondo,*] Vol. VI, p. 75.

3. *Cod[ex] Borg[ianus]*, Folios 48–63. Fábrega, *M.S.S.*, Folio k, p. 7.

Mexica hieroglyphic paintings and has seen the extreme care with which they have been executed, down to the smallest details, one cannot accept that the omission of eight terms in a periodic series is merely accidental. Father Fábrega's observation probably deserves mention here, not because it is plausible that a people might effectively institute a calendar reform only after a long period of one thousand forty years, but because the Velletri manuscript seems to prove that their author knew the true length of the year. If at the time of the arrival of the Spanish, there existed in Mexico an intercalation of twenty-five days across one hundred four years, it must be assumed that this more perfected intercalation had been preceded by one of thirteen days spread across fifty-two years. People would have remembered this earlier method, and it may well be that the Mexica priest who composed the *ritual* manual lodged in the Borgia museum meant to demonstrate a calculation trick that could rectify the old calendar by subtracting seven days from a long period of twenty cycles. It is impossible to determine the accuracy of this theory until a larger number of Mexica paintings in Europe and the Americas has been consulted; for—I cannot repeat this often enough—everything that we have learned up to now about the former state of the peoples of the new continent will pale by comparison with the insights on this subject that will one day be reported once all the materials that are scattered throughout the two worlds, and that have survived centuries of ignorance and barbarity, are successfully gathered together in one place.

188

This precious monument, of which I commissioned an image (shown on Plate XXIII) and which had already been engraved twenty years ago in Mexico City, serves to corroborate a number of the ideas about the Mexica calendar that we have just explained. This enormous stone was found in December of 1790 in the foundations of the great temple of Mexitli in the *Plaza mayor* of Mexico City, about seventy meters to the west of the second gate of the viceroys' palace and thirty meters to the north of the flower market called *Portal de las Flores*, at a shallow depth of five decimeters. It was positioned in such a way that one could see the sculpted section only when the stone was upright. While destroying the temples, Cortés had ordered the idols and everything related to the ancient worship to be smashed to pieces. Stone pieces too large to be destroyed were buried to remove them from the sight of the vanquished people. Although the circle that contains the hieroglyphs for the days is only 3,4 meters in diameter, one can see that the stone, in its entirety, formed a rectangular parallelepiped four meters long, as many meters wide, and one meter thick.

This stone is not limestone, as Mr. Gama claims, but grayish-black trappean porphyry with a basaltic *wacke* base. By carefully examining some detached fragments, I was able to identify amphibole, several quite elongated crystals of vitreous feldspar, and, what is most remarkable, some mica particles. Cracked and filled with small cavities, this rock is lacking in quartz, as are almost all rocks of the trappean formation. Since its current weight is over four hundred eighty-two quintals (24,400 kilograms), and since none of the mountains that surround the city at a distance of eight to ten leagues could have furnished a porphyry of this grain and color, one can easily imagine the difficulties that the Mexica experienced in hauling such an enormous mass all the way to the base of the *teocalli*. The sculpture in relief displays the same finish that one finds in all Mexica works: the concentric circles and the innumerable divisions and subdivisions are etched with mathematical precision. The more one examines the detail of this sculpture, the more one discovers that taste for repeating the same forms, that spirit of order, and that sense of symmetry that among semicivilized peoples replaces the sense of beauty.

In the center of the stone is the famous sign *nahui ollin Tonatiuh* (the sun in its four movements), which we have mentioned above.[1] Eight triangular rays surround the sun; these rays are also found in the ritual calendar, *tonalamatl*, and in historical paintings wherever the sun, *Tonatiuh*, is represented.[2] The number eight alludes to the division of the day and the night into eight parts.[3] The god Tonatiuh is represented with a large open mouth armed with teeth; the open mouth and the brandished tongue recall the figure of a deity of Hindustan, that of *Kala* or *Time*. According to a passage from the *Bhagavad Gita*, "*Kala* swallows up the worlds, opening wide his flaming mouth with its horrifying teeth and an enormous tongue."[4] Tonatiuh, placed in the middle of the day-signs and measuring the year by the *four movements* of the solstices and the equinoxes, is indeed the veritable symbol of *Time*: he is *Krishna* taking the form of *Kala*; he is *Kronos* devouring his children, whom we believe to have identified among the Phoenicians under the name Moloch.

The inner circle shows the twenty day-signs. If one remembers that the first of these catasterisms is *cipactli* and that the last is *xochitl*, one sees that here, like everywhere else, the Mexica arranged the hieroglyphs from right to left. The heads of the animals point in the opposite direction, probably because an animal that turns its back to another is supposed to come before it.

1. P. 189 [this edition].

2. Plate XV, number 4. *Cod[ex] Borg[ianus] Veletr[i]*, Folio 49.

3. See p. 154 [this edition].

4. Translation by Mr. ▾[Charles] Wilkins. [*The Bhagvat-Geeta or dialogues of Kreeshna and Arjoon*. 1785.] See also Moor, *The Hindu Pantheon*, article [about] *Kala*.

Mr. Zoëga noted this same oddity among the Egyptians.[1] An exception to this rule is the death's head, *miquiztli*, which is placed next to the *serpent* and accompanies it as a *night-sign* in the third recurring series; it alone is oriented toward the last sign, while the animals face the first sign. This arrangement does not recur in the manuscripts at Velletri, Rome, and Vienna.

It is likely that the sculpted stone that Mr. Gama attempted to explain was formerly placed within the enclosure of the teocalli, in a *sacellum* [chapel] devoted to the sign *ollin Tonatiuh*. From a passage by ˙Hernández [de Toledo], which the Jesuit Nieremberg preserved for us in the eighth book of his *Historia naturae*, we know that the great teocalli contained within its walls six times thirteen (or seventy-eight) chapels, several of which were devoted to the sun, the moon, the planet Venus, called either *Ilhuicatitlan* or *Tlazolteotl*, and the signs of the zodiac.[2] The moon, which all peoples regard as a star that draws humidity, had a small temple (*teccizcalli*) made of shells. The great festivals of the sun (*Tonatiuh*) were celebrated on the winter solstice and in the sixteenth thirteen-day period, which was presided over by both the sign *nahui ollin Tonatiuh* and the Milky Way, known as *Citlalinycue* or *Citlalcueye*. During one of these festivals of the sun, the kings had the custom of retiring to a building located in the middle of the teocalli's enclosure, called *Hueyquauhxicalco*. They spent four days there, fasting and undergoing penance; afterward, a bloody sacrifice was made in honor of the eclipses (*Netonatiuhqualo, unlucky, devoured sun*). During this sacrifice, one of the two masked victims represented the sun, *Tonatiuh*, and the other the moon, *Meztli*, as if to recall that the moon is the true cause of the solar eclipse.

In addition to the catasterisms of the Mexica zodiac and the image of the sign *nahui ollin*, the stone shows the dates of ten major festivals celebrated between the spring equinox and the autumn equinox. Since several of these festivals correspond to celestial phenomena, and since the Mexica year "wanders" for the length of one cycle (intercalation taking place only every fifty-two years), the dates do not refer to the same days for four consecutive years. The winter solstice that takes place on the day 10 *tochtli* in the first year of the cycle will eight years later have retrograded by two signs and will thus fall on the day 8 *miquiztli*. The result of this is that it is necessary to add the year of the cycle to which these dates correspond in order to indicate dates by the day-signs. In effect, the sign 13, *reeds* or *matlactly omey acatl*,

191

1. Zoëga, *De [origine et usu] obel[iscorum]*, p. 464 (where the words *dextrorsum* and *sinistrorsum* are confused through a typographical error).

2. Eusebius Nieremberg. *Hist[oria] nat[uraes]*, Book VIII, ch. 22 (Antwerp, 1635), pp. 142–56. *Templi partes*, 3, 8, 9, 20, 25.

placed above the image of the sun toward the upper edge of the stone, tells us that this monument contains the celebrations of the twenty-sixth year of the cycle, from March to September.

To make it easier to understand the signs that refer to the festivals of the Mexica religion, I must note again that the dots placed near the day-hieroglyphs are terms from the first of the three periodic series, whose use we explained above. Counting from right to left, and beginning at the right of the triangle on the forehead of the god *Ollin Tonatiuh*, the point of which is oriented toward *cipactli*, one finds the following eight hieroglyphs: 4 *tiger* [jaguar]; 1 *flint*; *tletl*, fire, with no number indicated; 4 *wind*; 4 *rain*; 1 *rain*; 2 *monkey*; and 4 *water*. What follows is the explanation of the Mexica celebrations according to Mr. Gama's calendar and the order of the festivals mentioned in historians' works from the sixteenth century.

In the year 13 *acatl*, the final year in the second indiction of the cycle, the beginning of the year had retrograded by six and a half days because there had not been any intercalation for twenty-six years. Therefore, the first day of the month *tititl*, which bears the sign 1 *cipactli tletl*, corresponds not to January 9 but to January 3; and the sign that presides over the seventh thirteen-day period, 1 *quiahuitl* (1 *rain*), coincides with March 22, the spring equinox. This was the time of the great festivals of Tlaloc, the god of water, which began even as early as ten days before the equinox, on the day 4 *atl* (4 *water*), probably because on March 12, the third of the month *Tlacaxipehualiztli*, the hieroglyph for water, *atl*, was both[1] the day-sign and the night-sign. Three days after the spring equinox, on the day 4 *ehecatl* (4 *wind*) began a solemn forty-day fast held in honor of the sun. This fast ended on April 30, which corresponds to 1 *tecpatl* (1 *flint*). Since the sign for this day is *accompanied* by the *lord of the night*, *tletl*, fire, we find the hieroglyph *tletl* placed next to 1 *tecpatl*, to the left of the triangle, the point of which is oriented toward the beginning of the zodiac. To the right of the sign 1 *tecpatl* is that of 4 *ocelotl* (4 tiger [jaguar]); this day is noteworthy for the sun's transit through the zenith of Mexico City. The entire thirteen-day period during which this transit occurred, which is the eleventh of the ritual year, was also dedicated to the sun. The sign 2 *ozomatli* (2 *monkey*) corresponds to the time of the summer solstice; it is located immediately adjacent to 1 *quiahuitl* (1 *rain*), the day of the equinox.

One might be confounded[2] by the explanation of 4 *quiahuitl*, 4 *rain*: in the first year of the cycle, this day corresponds exactly to the second transit

---

1. *Nahui atl, atl, atl*; see p. 174 [this edition].
2. Gama, [*Descripción histórica y cronológica de las dos piedras,*] §75, p. 109.

of the sun through the zenith of Mexico City, but in the year 13 *acatl*, the celebrations of which are shown on this monument, the day 4 *rain* already preceded this transit by six days. Since the entire thirteen-day period in which the sun reaches the zenith is dedicated to both the sign *ollin Tonatiuh* and the Milky Way, *citlalcueye*, and since the day 4 *rain* always belongs to this same period, it is quite likely that the Mexica gave priority to the latter day, so that the image of the sun could be surrounded by four signs that all displayed the same number four, and especially in order to allude to the four destructions of the sun that legend places on the days 4 *tiger* [jaguar], 4 *wind*, 4 *water*, and 4 *rain*. The five small dots to the left of the day 2 *monkey*, immediately above the sign *malinalli*, appear to allude to the festival of the god *Macuil-Malinalli*, who had his own special altars: this festival was celebrated around September 12, called *Macuilli Malinalli*. The point of the triangle that separates the day-sign 1 *flint* from the night-sign *tletl*, fire, is oriented toward the first of the twenty catasterisms of the zodiac signs, because in the year 13 *reed*, the year 1 *cipactli* corresponds to the day of the autumn equinox. Around this time they celebrated a ten-day festival, the most solemn day of which was 10 *ollin*, 10 *sun*, which corresponds to our September 16. In Mexico City people believe that the two boxes under the tongue of the god *Ollin Tonatiuh* show the number five twice; but this interpretation seems to me just as hazardous as the one that some have attempted to give for the forty boxes or fields that surround the zodiac and for the numbers six, ten, and eighteen, which are repeated near the edge of the stone. We shall not examine either whether the holes bored into this enormous stone were made, as Mr. Gama believed, in order to place threads there to serve as gnomons. What is more plausible—and essential to Mexica chronology—is that this monument proves, contrary to the theory advanced by Gemelli and Boturini, that regardless of the year sign, the first day is always presided over by *cipactli*, a sign that corresponds to Capricorn from the Greek sphere. It is conceivable that near this stone was placed another one containing all the celebrations, from the autumn equinox to the spring equinox.

193

We have gathered into a single perspective everything we know up to now about the division of time among the Mexica, carefully distinguishing what is certain from what is merely plausible. If one considers what has been explained about the shape of the year, one sees how misguided the theories are that attribute to the Toltec and the Aztecs either lunar years or years of 286 days, divided into 22 months.[1] It would be interesting to learn about the calendar system that the northernmost peoples of the Americas and Asia follow.

1. Waddilove, in Robertson's *Hist[ory] of America*, Vol. III, p. 404 [421], note XXXV.

Among the inhabitants of Nootka we still find the Mexica 20-day months, but their year has only 14 months, to which they add a large number of leap days,[1] using very complicated methods. Whenever a people does not base the subdivisions of the year on the lunar months, it begins to view the number of months as quite arbitrary, and its choice of subdivisions seem to depend solely on a particular preference for certain numbers. The Mexica peoples preferred *double decades* because they had simple signs only for the units one, twenty, and the powers of twenty.

Both the custom of the periodic series and the day-hieroglyphs have shown us striking points of similarity between the peoples of Asia and those of the Americas. Some of these points did not escape the wisdom of Mr. Dupuis,[2] although he confused the month signs with the day-signs and had only a very imperfect knowledge of the Mexica chronology. It would be contrary to the goal that we have set for ourselves in this work for us to engage in hypotheses about the ancient civilizations of the inhabitants of northern and central Asia. Tibet and Mexico present quite remarkable parallels in their ecclesiastical hierarchy, in the number of religious congregations, the extreme austerity of their penance rituals, and the order of their processions. Indeed, it is impossible not to be astonished by this resemblance if one reads closely Cortés's report to Emperor Charles V about his ceremonial entrance into Cholula, which he calls the holy city of the Mexica.

This people, who based their festivals on the movements of the stars and who engraved their celebrations on a public monument, had likely reached a higher level of civilization than that accorded to them by ▾[de] Pauw, Raynal, and even Robertson, the most judicious of all the historians of the Americas. These authors regard as barbarous any state of humanity that diverges from the notion of culture that they have established, based on their own systematic ideas. We simply cannot accept such sharp distinctions between barbarous and civilized peoples. In examining in this work with a scrupulous impartiality, everything that we ourselves have been able to discover about the former circumstances of the indigenous peoples of the new continent, we have tried to bring together the features that distinguish them individually, as well as those that appear to link them to various groups of Asian peoples. As it is with mere individuals, so it is with entire peoples; among the former, not all the soul's faculties manage to develop simultaneously, while among the latter, the advances of civilization do not manifest themselves all at once in

194

---

1. ▾Don José Moziño, *Viaje a Noutka*, manuscript (see my *Essai politique sur [le royaume de] la Nouvelle-Espagne*, Vol. I, p. 335 [Q?]).

2. *Mémoire explicatif sur le zodiaque*, p. 99.

the tempering of public and private morals, in a sense for the arts, and in the form of institutions. Before one can classify peoples, one must study them on the basis of their specific characteristics, for external circumstances produce infinite variations in the cultural nuances that distinguish tribes of different races, especially when, settled in regions separated by vast distances, they have lived for a long time under the influence of governments and religions more or less contrary to the advancement of the mind and the preservation of individual liberty.

PLATE XXIV

The Inca's House at Callo in the Kingdom of Quito

After Tupac-Yupanqui and Huayna Capac, father of the ill-fated Atahualpa, had completed their conquest of the kingdom of Quito, not only did they have magnificent roads cleared on the ridge of the Cordilleras, but to improve the connections between the capital and the northernmost provinces of their empire, they also ordered that inns (*tambos*), storehouses, and dwellings fit to serve as residences for the prince and his coterie be built at regular intervals on the road from Cusco to Quito. These *tambos* and houses of the Inca, which other travelers describe as palaces, have existed for centuries on this segment of the great road that leads from Cusco to Cajamarca; the only buildings whose construction can be attributed to the last conquerors of Manco-Capac's race are those whose ruins we find today between the province of Cajamarca, the southern border of the ancient kingdom of Quito, and the *Los Pastos* mountains. One of the most famous and best preserved of these buildings is that of the *Callo* (or *Caïo)* described by La Condamine, Don Jorge Juan, and Ulloa in their travels to Peru. These travelers' descriptions are quite inaccurate, and the drawing that Ulloa provided of the Inca's house shows so little of the original building plan that one would almost be tempted to believe that it is completely fictitious.

When, during an excursion to Cotopaxi volcano in April of 1802, Mr. Bonpland and I visited these humble remnants of Peruvian architecture, I drew the sections shown in Plate XXIV; upon my return to Quito, I showed both my drawings and the plate contained within Ulloa's travels to the elderly friars of Saint Augustine's order. They knew the Callo ruins, which are located on a plot of land belonging to their monastery, better than anyone else. They had formerly occupied a neighboring country house, and they assured me that since 1750 and even before that time, the Inca's house had always been in the same condition as it is today. It is likely that Ulloa intended to depict a *restored* monument, and that he presumed the existence of interior walls[1] wherever he saw heaps of rubble or occasional rises in the ground. His sketch shows neither the true shape of the chambers nor the four large exterior doors, which must have existed since the building's construction.

We already noted above that the Quito plateau extends between a double crest[2] of the Cordillera of the Andes; it is separated from the plateau of Latacunga and Ambato by the heights of Chisinche and Tiopullo, which stretch crosswise like an embankment from the eastern toward the western crest or from the basaltic boulders of Rumiñahui toward the soaring pyramids of the ancient volcano of Iliniza [Ecuador]. From the top of this embankment, which divides the waters between the South Sea and the Atlantic Ocean, one discovers the *Panecillo* of Callo and the ruins of the house of the Inca Huayna Capac in an immense pumice-covered plain. The *Panecillo*, or *sugarloaf*, is a conical mound around eighty meters high and covered in small *Molina*, *Spermacoce*, and *Cactus* bushes. The natives are convinced that this bell-shaped mound, which is of a surprisingly regular shape, is a *tumulus*, one of the numerous hills that the ancient inhabitants of this land raised to serve as sepulchers for the prince or for some other dignitary. In support of this theory, they claim that the *Panecillo* is entirely composed of volcanic debris and that the very pumice stones that surround its base can also be found at its summit.

This claim might seem unconvincing to the eyes of a geologist, since the ridge of the neighboring mountain of Tiopullo, which is much higher than the *Panecillo*, also has large piles of pumice stone, probably from ancient eruptions of Cotopaxi and Iliniza. It is undeniable that in both Americas, just as in the north of Asia and on the banks of the Borysthenes [Dnieper], there are only man-made burial mounds, true *tumuli* of extraordinary height. The ones we found in the ruins of the ancient city of Mansiche in Peru are by no means dwarfed by the *sugarloaf* of Callo. It is possible, however (and 197 this seems to me a more likely theory), that the latter was an isolated volcanic mound in the vast Latacunga plain to which the natives gave a more regular shape. Ulloa, whose authority carries great weight, appears to adopt the natives' theory; he even believes that the *Panecillo* is a *military monument* and that it was used as a watchtower from which to observe what was happening in the countryside and to safeguard the prince at the first sign of an unforeseen attack. In the state of Kentucky one also finds, near ancient, oval-shaped

---

1. *Voyage historique de l'Amérique méridionale*, Vol. 1, p. 387, Plate 18.

2. See above, p. 124 [this edition] and my *Recueil d'observations astronomiques*, Vol. 1, p. 309.

fortifications, soaring *tumuli* containing human remains and covered with trees, which ˙Mr. Cutter claims to be nearly a thousand years old.[1]

The *Inca's house* is located a short distance to the southwest of the *Panecillo*, three leagues from the crater of Cotopaxi, about ten leagues south of the city of Quito. This building forms a square with thirty-meter-long sides; one can still make out four large exterior doors and eight chambers, three of which have been preserved. The walls are about five meters high and one meter thick. The doors, similar to those of Egyptian temples; the recesses, of which there are eighteen in each chamber, arranged in perfect symmetry; the cylinders used to hang up weapons; the cut of the stones, the outer face of which is convex and beveled at the edges—everything recalls the building at Cañar, which is depicted on Plate XX. I have not seen anything at Callo that suggests the opulence, grandeur, and majesty that Ulloa described; but what seems to me of the greatest interest is the uniformity of construction that all Peruvian monuments have in common. It is impossible to scrutinize a single building from the time of the Inca without noticing that the same structure is present in all the other buildings on the ridge of the Andes, over a distance of more than four hundred fifty leagues, from one thousand to four thousand meters above sea level. This mountain people was so attached to its domestic habits, its civil and religious institutions, and the form and positioning of its buildings that it almost seems as if a single architect had built this large number of monuments. With the help of the drawings contained within this work, it will one day be easy to verify whether (as the learned author of the *Noticias americanas* claims) there exist buildings in upper Canada that exhibit traces of the *Peruvian style* in the cut of their stones, the shape of their doors and small recesses, and the positioning of their chambers. Those who engage in historical research will be all the more interested in having this hypothesis confirmed, since we know from reliable witnesses that the Inca built the fortress of Cusco after the model of the older buildings at Tiahuanaco located at 17° 12' southern latitude.

The stone used in Huayna Capac's house, which Cieza[2] called *Aposentos de Mulahalo*, is a rock of volcanic origin, a spongy, burnt porphyry with a basalt base. It appears to have been spat out by the Cotopaxi volcano, for it is identical to the enormous blocks that I found in large number in the plains of Callo and Mulaló. Since this monument appears to have been built in the early years of the sixteenth century, the materials that were used in its construction belie the widely held notion that the first eruption was the one that

198

1. Carey, *[American] Pocket Atlas*, 1796, p. 101.
2. *Crónica del Perú*, ch. 41 (1554 edition), p. 108.

took place in 1533, when *Sebastián de Belalcázar conquered the kingdom of Quito. The stones at Callo are cut into parallelepipeds; they are not all of the same size, but they form terraces that are as regular as those of Roman buildings. If the illustrious author of *History of America*[1] had seen a single Peruvian building, he would probably not have written that "the natives used the stones just as they were when they had extracted them from the quarries; some were triangular, others square; some were convex and others concave; and the excessively praised art of this people consisted merely in the arrangement of shapeless materials."

During our long stay in the Cordillera of the Andes, we never came across any buildings that resembled what is called cyclopean construction. In all of the buildings that date back to the time of the Inca, the stones are cut with admirable precision on the outer face, while the inner face is uneven and often quite angular. An excellent observer, Mr. *Don Juan Larrea, has indicated that, within the walls of the Callo, or gap, between the outer and inner stones is filled with tiny pebbles cemented with clay. I myself did not see this unusual feature, but I have depicted it on Plate XXIII, following a sketch by Mr. Larrea. One does not find any vestiges of a floor or a roof, though one may assume that the latter was made of wood; nor do we know whether the building was originally only one story high; it has been damaged as much by the greed of the neighboring farmers, who have extracted stones from it for use elsewhere, as by the earthquakes that constantly plague this unfortunate country.

It is plausible that the structures to which I have heard people in Peru, in Quito, and as far as the banks of the Amazon River refer as *Inga-Pirca*, the Inca's buildings, do not date back to before the thirteenth century of our era. The structures at Viñaque and Tiwanaku are more ancient, as are the walls of unbaked brick that owe their origin to the former inhabitants of Quito, the *Puruays*, ruled by the *Conchocando* (the king of Licán) and by the *Guastays*, tributary nobles. It would be desirable for some learned traveler one day to visit the banks of the great Lake Titicaca, Collao province, and especially the Tiwanaku plateau, the center of an ancient civilization in South America. A few remnants of these buildings (which *Pedro de Cieza*[2] described with such marvelous simplicity) are still intact there: they appear never to have been completed, and when the Spanish arrived, the natives attributed their construction to a race of white, bearded men who had lived on the ridge of the Cordilleras long before the Incan empire was founded. American architecture—we cannot repeat this enough—is captivating neither for its scale nor

199

1. Robertson, *Hist[ory] of America*, Vol. III, p. 414.
2. Cieza, [*Crónica del Perú,*] ch. 105, p. 255.

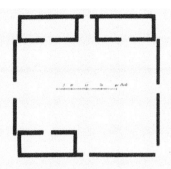

Plate XXIV

for the elegance of its shapes; but one regards it with all the more interest because it illuminates the history of the original culture of the new continent's mountain peoples.

I have drawn (1) the plan of the Inca Huayna Capac's house; (2) one segment of the inner wall of the northernmost chamber as viewed from within; (3) the same segment as viewed from the outside—that is, from the interior of the courtyard. In the outer walls, opposite the chamber doors, there are openings that look out onto the surrounding countryside, in lieu of recesses. I shall not decide whether these window openings are actually recesses through which holes (*hocos*) were bored at some point after the conquest, when this building was used as a residence for a number of Spanish families. The natives believe, on the contrary, that these openings had been made so   200 that people could observe whether some enemy conspired to launch an attack against the Inca's troops.

PLATE XXV
Chimborazo Viewed from the Tapia Plateau

The mountain has been drawn exactly as it is seen from the arid plain of Tapia, near the village of Licán, the former residence of the sovereigns of Quito before the conquest of the Inca Tupac Yupanqui. Licán is five leagues away from the summit of Chimborazo, as the crow flies. Plate XVI shows this colossal mountain with a belt of permanent snow, which near the equator occurs only at a height of four thousand eight hundred meters above sea level. Plate XXV shows Chimborazo as we saw it after one of the heaviest snowfalls in memory, on June 24, 1802, only one day after ˅our excursion to the top. I thought that it might be interesting to give an exact idea of the imposing appearance of the Cordilleras at the time of both *maximum* and *minimum* snow levels.

Only those travelers who have seen the summits of Mont Blanc and Mont Rose are able to grasp the character of this imposing, calm, and majestic scene. Chimborazo is of such enormous size that the visible area near permanent snow line is seven thousand meters wide. The extreme thinness of the air layers through which one sees the tops of the Andes greatly enhances[1] the sparkle of the snow and the magical effect of its reflections. In the tropics, at a height of five thousand meters, the azure vault of the heavens takes on a shade of indigo.[2] The contours of the mountain stand out against the background of this pure, limpid atmosphere, while the lower air layers, which rest upon a grassless plateau and reflect its radiant heat, are misty and seem to veil the distant countryside.

The Tapia plateau, which stretches to the east all the way to the foot of El Altar and Condorasto, is at an elevation of three thousand meters. Its height is more or less equal to that of Canigou, one of the highest peaks of the Pyre-

---

1. *Essai politique sur [le royaume de] la Nouvelle-Espagne*, Vol. 1, p. LXXVII.
2. See my *Géographie des Plantes*, p. 17.

Plate XXV

nees. In the arid plain grow a few stalks of Schinus molle, Cactus, Agave, and Molina. In the foreground, one sees llamas (*Camelus lacma*), drawn from life, and a few groups of Indians heading to the Licán market. The mountainside presents the very gradation of plant life that I have attempted to outline in my *Tableau de la Géographie des Plantes*, and which one can follow on the western slope of the Andes from the impenetrable palm groves to the permanent snow, lined by a thin layer of lichen.

At an absolute elevation of three thousand five hundred meters, the woody plants with their glossy, tough leaves gradually disappear. The region of shrubs is separated from that of grasses by alpine herbs, tufts of Nertera, Valerians, Saxifrages, and Lobelia, and small crucifer plants. The grasses constitute a very large belt occasionally covered in snow that lasts only for a few days. This zone, which the locals call the *pajonal*, appears from afar like a golden carpet. Its color contrasts delightfully with that of the scattered snow masses; it is the result of the stems and leaves of the grasses being burnt by the rays of the sun during great drought periods. Above the *pajonal* one finds oneself in the region of the cryptogamic plants, which here and there cover the porphyritic rocks, devoid of humus. A little farther on the line of permanent ice marks the end of organic life.

However surprising the height of Chimborazo, its summit is still four hundred fifty meters below the point at which ˅Mr. Gay-Lussac, during his memorable air [balloon] voyage, conducted experiments that were as important for meteorology as they were for understanding the laws of magnetism. The natives of Quito province have preserved a legend according to which one peak from the eastern crest of the Andes, which today is called the Altar (*el Altar*) and which partially collapsed in the fifteenth century, was once taller than Chimborazo. In Bhutan, Soomoonang, the highest mountain for which the English travelers have given us the measurements, is only 4419 meters (2,268 toises) high: but, according to ˅Colonel Crawford,[1] the tallest peak in the Cordilleras of Tibet is over twenty-five thousand English feet, or 7617 meters (3,909 toises) high. If this estimate is based on an exact measurement, then there is a mountain in central Asia that is one thousand ninety meters higher than Chimborazo. The absolute elevation of mountains is an unimportant phenomenon to the true geologist, who, being engaged in the study of rock *formations*, is accustomed to viewing nature on a large scale. He will hardly be surprised if at some point in the future and in some other part of the globe, someone discovers a peak that surpasses Chimborazo by as

202

1. Jameson, *System of Mineralogy*, Vol. III, p. 329.

much as the highest mountain in the Alps soars above the highest point in the Pyrenees.

A distinguished architect, *Mr. Thibault, who combines a knowledge of the monuments of antiquity with a deep appreciation of the beauties of nature, was gracious enough to execute the colored drawing whose engraving forms the main decorative illustration in this work. The sketch that I had made on site served no other purpose than to give an exact indication of the contours of Chimborazo. The faithfulness of the work as a whole and its details has been scrupulously maintained. To make it easier for the naked eye to perceive the gradation of the different levels and to grasp the expanse of the plateau, Mr. Thibault has enlivened the scene with figures that are grouped very intelligently. It is always a joy to acknowledge favors granted by the most selfless of friends.

PLATE XXVI

Epochs of Nature According to Aztec Mythology

The most astonishing of all the observable similarities among the monuments, customs, and traditions of the peoples of Asia and the Americas is revealed in Mexica mythology in the cosmogonic myth of the periodic destructions and regenerations of the Universe. This mythic fiction, which links the
return of the great cycles to the notion of the renewal of matter presumed to
be indestructible and which attributes to space characteristics that seem to
belong only to time,[1] dates back to the earliest period of antiquity. The sacred books of the Hindus, especially the *Bhagavata Purana*, already speak
of the four ages and the *pralayas* or cataclysms that have led to the destruc-
203   tion of the human species in various epochs.[2] A tradition of *five ages*, similar
to that of the Mexica, can also be found on the high plateau of Tibet.[3] If it is
true that this astrological fiction, which became the foundation of a particular cosmogonic system, was conceived in Hindustan, then it is also plausible
that it spread from there, via Iran and Chaldea, to the peoples in the west.
One cannot disregard a certain resemblance between the Indian tradition of
the *yuga* and the *kalpa*, the cycles of the ancient inhabitants of Etruria, and
the series of annihilated generations that ʼHesiod characterized through the
emblem of the four metals.

"The peoples of Culhua or Mexico," wrote Gómara,[4] who wrote in the
mid-sixteenth century, "believe, according to their hieroglyphic paintings,
that before the sun that now shines upon them, four suns had already existed
and had been extinguished, one after the other. These five suns constitute
the ages in which humankind was wiped out by floods, earthquakes, an all-

1. ʼHermann, *Mythology der Griechen*, Vol. II, p. 332.
2. ʼHamilton and Langlès, *Catalogue des Manuscrits sanskrits de la Bibl[ioteque]
impér[iale]*, p. 13. [*Asiatick Researches*,] Vol. II, p. 17. Moor, *Hindu Pantheon*, pp. 27
and 101.
3. Georgi, *Alphab[etum] Tibetanum*, p. 220.
4. Gómara, *Conquista*, Folio CXIX.

consuming blaze, and the effect of fierce storms. After the destruction of the fourth sun, the world was plunged into shadows for a period of twenty-five years. It was in the midst of this deep night, ten years before the fifth sun reappeared, that humanity was regenerated. At that time, the gods created one man and one woman for the fifth time. The day on which the last sun appeared bore the sign *tochtli* (rabbit), and the Mexica counted eight hundred fifty years from that time to 1552. Their annals date back all the way to the fifth sun. They used historical paintings (*escritura pintada*) in the four preceding ages as well; but these paintings were destroyed, they claim, because everything must be renewed with each new age." According to Torquemada,[1] this fable about the revolution of time and the regeneration of nature is of Toltec origin; it is a national tradition that belongs to the group of peoples whom we know as the Toltec, Chimichec, Acolhua, Nahua, Tlaxcalteca, and Aztecs and who, speaking the same language, had migrated from north to south beginning in the middle of the sixth century of our era.

While examining in 'Rome the *Cod[ex] Vaticanus*, number 3738, copied in 1566 by the Dominican friar Pedro de los Ríos,[2] I came across the Mexica drawing shown on Plate XXVI. Since it indicates the duration of each epoch using signs whose value is known to us, this historical monument is all the more intriguing. In Father Ríos's commentary, the order in which the catastrophes occurred is completely wrong: he places the final one, the flood, first. This same error is found in the works of Gómara, Clavijero,[3] and the majority of Spanish writers, who, forgetting that the Mexica arranged their hieroglyphs from right to left, starting from the bottom of the page, necessarily reversed the order of the four destructions of the world. I will adhere to the order shown in the Mexica painting from the Vatican library and described in a very curious story written in the Aztec language, a few fragments of which have been preserved for us by the Indian Fernando de Alva Ixtlilxochitl. The testimony of a native author and the copy of a Mexica painting made on site shortly after the conquest are probably more trustworthy than the accounts of the Spanish historians. Besides, this discrepancy, the reasons for which we have just indicated, applies only to the order of the destructions, for Gómara, Pedro de los Ríos, Ixtlilxochitl, Clavijero, and Gama all report the circumstances that accompanied each of them in the most consistent manner possible.

First cycle. Its duration is 13 × 400 + 6 = 5,206 years: this number is indicated on the right in the lower painting through nineteen circles, thirteen of

204

---

1. Torquemada, [*Monarquía indiana*,] Vol. I, p. 40; Vol. II, p. 83.

2. See above, pp. 97 and 106 [this edition].

3. *Storia antica del Messico*, Vol. II, p. 57.

which have a *feather* placed above them. We observed above,[1] in our discussion of the calendar, that the hieroglyph for twenty squared is a feather, and that the Mexica used simple dots to represent the number of years, in the same way that Etruscans and the Romans used nails.[2] This first age, which corresponds to the Hindus' age of justice (*Satya Yuga*), was called *Tlaltonatiuh*, epoch of the earth; it is also the age of the giants (*Qzocuilliexeque* or *Tuinametin*), for the historical legends of all peoples begin with brawls among giants. The Olmec (or Hulmec) and the Xicalanc, two peoples who preceded the Toltecs and who prided themselves on their ancient origins, claimed to have come across giants when they arrived in the Tlaxcala plains.[3] According to the sacred *Puranas*, Bacchus (or the young Rama) won his first victory over Ravana, the king of the giants of the island of Ceylon.[4]

The year presided over by the sign *ce acatl* was a year of lack, and the first generation of humans perished in the famine. This catastrophe began on the day 4 *tiger* [jaguar] (*nahui ocelotl*), and it is probably because of the hieroglyph for this day that other traditions maintain that the giants who did not die of starvation were devoured by the very tigers [jaguars] (*tequanes*) whose appearance the Mexica dreaded at the end of each cycle. The hieroglyphic painting depicts an evil spirit who descends to earth to rip out grass and flowers. Three human figures—among whom one can easily identify a woman by a hairstyle composed of two small plaits that resemble horns[5]—are each holding a sharp instrument in their right hand and fruits or cut sheaves of grain in the left. The spirit who announces the famine is wearing one of those prayer-bead chains[6] that have been used in Tibet, China, Canada, and Mexico since time immemorial and that spread from the east to the Christians in the west. Despite the fact that among all the peoples of the world, the fictional stories of the giants, the Titans, and the Cyclops appear to refer to the conflict of the elements or the state of the globe upon its emergence from chaos, it is undeniable that in both Americas the enormous fossilized skeletons of animals scattered across the surface of the earth have had a great impact on mythological history. At Punta Santa Elena to the north of Guayaquil, there are giant remains of unknown cetaceans: some Peruvian legends thus maintain that a colony of giants who slaughtered one another had once disem-

205

1. See p. 166 [this edition].
2. ▾Tit[us] Liv[ius], *Hist[oria Romana]*, Book VII, ch. 3 (ed. Gesneri, 1735), Vol. I, p. 461.
3. Torquemada, [*Monarquía indiana*,] Vol. I, p. 37.
4. Paol[ini] da Sanct. Barthol[omeo], *Syst[ema] Brahman[icum]*, pp. 24 and 143.
5. Plate XV, numbers 3–7, 3.
6. Plate XIV, number 8.

barked in this very spot. Both the kingdom of New Granada and the ridge of the Mexican Cordilleras[1] are teeming with fossilized remains of mastodons and elephants belonging to species that have disappeared from the surface of the globe. It is for this very reason that the plain that stretches from Suacha toward Santa Fé de Bogotá at an elevation of two thousand seven hundred meters bears the name *Field of Giants*. It is likely that the Hulmecs' claim that their ancestors had fought against the giants on the fertile plateau of Tlaxcala derives from findings of mastodon and elephant molars, which people across the country take for the teeth of colossally sized men.

Second cycle. Its duration is 12 × 400 + 4 = 4,804 years; it was the age of fire,   206 *Tletonatiuh*, or the red age, *Tzonchichilteque*. The god of fire, Xiuhteuctli, descends to earth in the year presided over by the sign *ce tecpatl*, on the day *nahui quiahuitl*. Since only birds could escape the all-consuming blaze, legend has it that all humans were transformed into birds, except for one man and one woman who found safety inside a cave.

Third cycle, the age of the wind or the air, *Ehecatonatiuh*. Its duration is 10 × 400 + 10 = 4,010 years. The catastrophe occurred on the day 4 wind (*nahui ehecatl*) of the year *ce tecpatl*. The drawing shows the hieroglyph for the air or the wind, *ehecatl*, repeated four times. Humans perished in the storms, and some of them were transformed into monkeys; these animals did not appear in Mexico until this third epoch. I do not know which deity it is that descends to earth, armed with a sickle—might it be Quetzalcoatl, the god of the air, and might the sickle imply that the storm uproots the trees as though they had been felled? In any case, I doubt that the yellow streaks allude to the shape of the clouds battered by the storm, as one Spanish commentator suggests. Monkeys are generally less numerous in the hot regions of Mexico than in South America. These animals undertake distant migrations whenever they are driven by hunger or inclement weather to abandon their original abode. I know lands in the mountainous parts of Peru where the people recall the time when new monkey colonies settled in one valley or another. Might the tradition of the five epochs bear some relation to the history of animals? Might it refer to a particular year in which great storms and upheavals caused by the volcanoes incited the monkeys to make forays into the mountains of Anahuac? In this *cycle of tempests*, only two humans survived the catastrophe by taking refuge inside a cave, just as at the end of the preceding age.

Fourth cycle, the age of water, *Atonatiuh*, the duration of which was 10 × 400 + 8 = 4,008 years. A great flood, which began in the year *ce calli* on

1. Cuvier, *Mém[oire] de l'Institut, classe des Sciences phys[iques] et mathém[atiques]*, Year 1, p. 14.

the day 4 water (*nahui atl*), destroyed the human species; this was the last of the great revolutions that the world suffered. Humans were transformed into fish, with the exception of one man and one woman who escaped into the trunk of an *ahahuete*, a bald cypress. The drawing shows the goddess of water, called *Matlalcueje* or *Chalchiuhcueje*, and regarded as the companion of Tlaloc, diving toward the earth. Coxcox, the Noah of the Mexica, and his wife *Xochiquetzal* are seated in a tree trunk covered in leaves and are floating in the midst of the floodwaters.

These four epochs, which are also known as *suns*, together contain eighteen thousand twenty-four years; that is, six thousand years more than the four Persian ages described in the ˅Zend-Avesta.[1] I do not see any indication anywhere of the number of years that elapsed from the flood of Coxcox to the sacrifice of Tlalixco, or until the reform of the Aztec calendar; but however short an interval one assumes between these two epochs, one always finds that the Mexica viewed the world as having been in existence for more than twenty thousand years. This duration certainly conflicts with the Hindus' great period, which is four million three hundred twenty thousand years long, and especially with the cosmogonic fiction of the Tibetans, according to which humankind has already passed through eighteen revolutions, each of which contains several *padu* represented by by sixty-two-digit numbers.[2] It is, nevertheless, remarkable to find an American people who, using the same calendar system that they employed when Cortés arrived, mark the days and the years of twenty centuries ago, when the world suffered great catastrophes.

Le Gentil, Bailly, and Dupuis[3] have offered clever explanations for duration of the great cycles of Asia. I was unable to determine any special property in the number 18,028 years: it is not a multiple of 13, 19, 52, 60, 72, 360, or 1440, which are the numbers that one finds in the cycles of the Asian peoples. If the duration of the *four Mexica suns* was three years, and if one replaced the numbers 5,206, 4,804, 4,010, and 4,008 years with the numbers 5,206, 4,807, 4,009, and 4,009, one might believe that these cycles were derived from the knowledge of the nineteen-year lunar period. Whatever their true origin, it seems clear that they are fictions of astronomical mythology, modified either by a vague recollection of some great revolution that our planet underwent or in accordance with theories about natural history and

---

1. ˅Anquetil, *Zend-Avesta*, Vol. II, p. 352.

2. [Georgi,] *Alphab[etum] Tibet[anum]*, p. 472.

3. Le Gentil, *Voyage dans les Indes*, Vol. I, p. 235. Bailly, *Astron[omie] indienne*, pp. LXXXXVIII and 212. Bailly, *Histoire de l'astronomie ancienne*, p. 76. Dupuis, *Origine des cultes*, Vol. III, p. 164.

geology that the physical appearance of marine petrifactions and fossilized remains inspires, even among those peoples furthest removed from civilization.

Examining the paintings shown on Plate XXVI, one finds in the four destructions the emblem of four elements: *earth, fire, air,* and *water.* These same elements were also represented by the four year-hieroglyphs:[1] *rabbit, dwelling, flint,* and *reed. Calli,* or *dwelling,* seen as the symbol of fire, recalls the customs of a northern people whom the inclement weather forced to heat their huts, as well as the idea of Vesta (Ἑστία [Hestia]), which, in the most ancient system of Greek mythology, represents not only the *house* but also the *hearth* and the *fire* in the hearth. The sign *tecpatl, flint,* was dedicated to the god of the air, *Quetzalcohuatl,* a mysterious figure who belongs to the heroic period of Mexica history, and of whom we have had occasion to speak at many points in this work. According to the Mexica calendar, *tecpatl* is the *night-sign* that accompanies the hieroglyph for the day called *ehecatl, wind,* at the beginning of the cycle. Perhaps the story of an aerolite that fell from the heavens onto the top of the Cholula pyramid, dedicated to Quetzalcohuatl, inspired the Mexica to conceive this bizarre relationship between fire-starting flint (*tecpatl*) and the god of the winds.

We have seen that the Mexica astrologers gave a historical character to the tradition of the destructions and the regenerations of the world by citing the days and years of the great catastrophes in accordance with the calendar they used in the sixteenth century. Only a very simple calculation was needed for them to find the hieroglyph for a year that came 5,206 or 4,804 years before a given epoch. It was in this manner that, according to Macrobius and ˙Nonnus, the Chaldean and Egyptian astrologers determined even the position of the planets both at the time of the creation of the world and at that of the great flood. Using the system of the periodic series, I recalculated the signs that presided over the years several centuries before the sacrifice at Tlalixco (in the year *ome acatl,* 2 *reed,* which corresponds to the year 1091 of the Christian era) and found that the dates and signs did not correspond exactly to the duration of each Mexica epoch. These dates are also not indicated in the Vatican paintings; I took them from a fragment of Mexica history preserved by Alva Ixtlilxochitl, who sets the length of the four ages not at 18,028 but only at 1,417 years. This discrepancy in astrological calculations should not surprise us, for the first number contains almost as many indictions as the second does years. Likewise, in the Hindus' mystical chronology, substituting days for *divine years*[2] reduces the four ages from 4,320,000 years to 12,000. 209

---

1. See p. 171 [this edition], and Sigüenza, in Gemelli, *Giro del Mondo,* Vol. VI, p. 65.
2. Bailly, *Astr[onomie] ind[ienne],* p. CI.

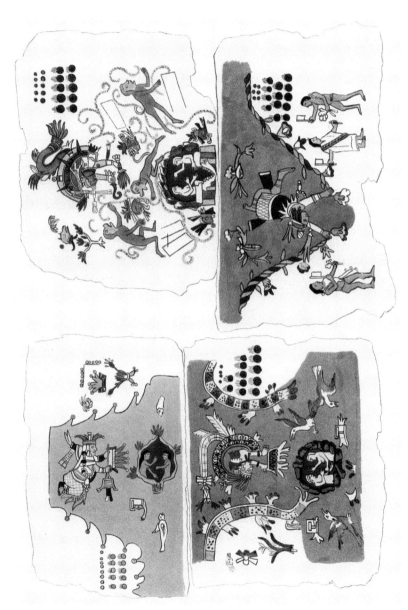

Plate XXVI

SYSTEM OF THE CODEX VATICANUS, No. 3738	SYSTEM OF THE TRADITION PRESERVED BY IXTILXOCHITL	
Duration of the *first age* . . . . 100 × 52 + 6 = 5,206 years	13 × 52 + 676 years	
Period of the first destruction	. . . .	1 *Acatlt*
Duration of the catastrophe	. . . . 13 years	
Duration of the *second age* . . . . 92 × 52 + 20 = 4,804 years	7 × 52 = 364 years	
Period of the second destruction	. . . .	1 *Tecpatl*
Duration of the *third age* . . . . 77 × 52 + 6 = 4,010 years	6 × 52 = 312 years	
Period of the third destruction	. . . .	1 *Tecpatl*
Duration of the *fourth age* . . . . 76 × 52 + 4 = 4,008 years	1 × 52 = 52 years	
Period of the fourth destruction	. . . .	1 *Calli*
346 cycles of 52 years + 36 = 18,028 years	109 indictions of 13 years, or 1,417 years	

Examining the numbers in this table in accordance with the Mexica calendar system, one sees that two ages separated by an interval cannot have different signs if the number of years in the interval is a multiple of 52. It is impossible for the fourth destruction to have taken place in the year *calli* if the third occurred in the year *tecpatl*. I cannot imagine what might have caused this error; it may be, however, that it merely appears to be an error, and that in the historical monuments that have been handed down to us, no mention was made of the small number of years that nature required for each regeneration. The Hindus make a distinction between the interval separating two cataclysms and the amount of time that each of them actually lasted. Similarly, in the fragment from Alva Ixtlilxochitl, we read that the first catastrophe is separated from the second by seven hundred seventy-six years but that the famine that killed the people lasted for thirteen years, or one-quarter of a cycle. In the two chronological systems that we have just described, the epoch of the creation of the world, or better, the starting point for the great periods, is the year presided over by *tochtli*; this sign was for the Mexica what the catasterism 210 *aries* [the ram] was for the Persians. The astrological traditions of all peoples note the position of the sun at the moment when the stars begin their course; and in our discussion above[1] of the observable connections between the fiction of the epochs and the meaning of the hieroglyph *ollin*, we have suggested the likelihood that *tochtli* corresponds to one of the solsticial points.

In the Mexica system, the four great revolutions of nature are caused by the four elements: the first catastrophe is the devastation of the productive force of the earth, while the other three are due to the effect of fire, air, and

1. Pp. 190 and 214 [this edition].

water. The human species regenerates after each destruction, and any members of the previous race who did not perish are transformed into birds, monkeys, or fish. These transformations also recall the legends of the East; yet in the Hindus' system the ages, *yugas*, all end in floods, while in that of the Egyptians[1] the cataclysms alternate with conflagrations, and men find refuge either on the mountains or, at other times, in the valleys. We would digress from our topic were we to explain here the small local revolutions that occurred on several occasions in the mountainous part of Greece[2] and discuss the famous passage from the second book of Herodotus, which has so challenged commentators' wisdom. It seems quite clear that it is not a question of *apocatastases* in this passage but, rather, of four (visible) changes that affected the moments of sunset and sunrise[3] and that were brought about by the precession of the equinoxes.[4]

Since one might be surprised to find five ages, *suns*, among the peoples of Mexico, while the Hindus and the Greeks recognize only four, it is useful to observe here that Mexica cosmogony aligns closely with that of the Tibetans, who also regard the present age as the fifth one. By examining closely the lovely passage of Hesiod[5] in which he explains the Eastern system of the renewal of nature, one sees that this poet actually counts five generations within four epochs. He divides the Bronze Age into two parts that cover the fourth and fifth generations,[6] and one may be surprised that such a clear passage has occasionally been misinterpreted.[7] We do not know the number of ages mentioned in the books of the Sibyl,[8] but we believe that the similarities we just highlighted are not accidental, and that it is not unimportant for the philosophical history of humanity to note that the same fictional stories are widespread, from Etruria to Latium to Tibet and from there all the way to the ridge of the Cordilleras of Mexico.

In addition to the tradition of the four suns and the costumes that we de-

---

1. ▾Timaeus, ch. 5 (Plato, *Oper[a omni]*, 1578, ed. Serran[us]), Vol. III, p. 22. *De legib[us]*, Book III (*Op[era] omn[i]*, Vol. II, pp. 678–9). ▾Origenes, *Contra Celsum*, Book I, ch. 20; Book IV, ch. 20 (ed. Delarue, 1733), pp. 332 and 514.

2. Arist[otle], *Meteor[ologia]*, Book I, ch. 14 (*Op[era] omn[i]* (ed. Duval, 1639), p. 770).

3. Herodotus, [*Histoire d'Hérodote*,] Book II, ch. 142 (ed. Larcher, 1802), Vol. II, p. 482.

4. Dupuis, *Mémoire explicatif du zodiaque*, pp. 37 and 59.

5. Hesiod, *Oper[um] et die[rum]*, verse 174 (*Op[era] omn[ia]* (ed. Cleric[us], 1701), p. 224.

6. Hesiod, [ibid.,] verses 143 and 155.

7. ▾Fabrici[us], *Bibl[iotheca] graeca, Hamb[urg]*, 1790, Vol. I, p. 246.

8. ▾Virgil, *Bucol[ica]*, IV, verse 4 (ed. Heyne, Lond[on,] 1793), Vol. I, pp. 74 and 81.

scribed above,[1] the *Cod[ex] Vatican[us] Anon[ymous]* Number 3738 contains several other peculiar images, among which we cite the following: Folio 4, the *chichiuhalquehuitl, milk tree* or *celestial tree,* which secretes milk from the tips of its branches and around which are seated the children who died a few days after their birth; Folio 5, a three-pound molar, perhaps from a mastodon, which Father Ríos gave to the viceroy ʼDon Luis de Velasco in 1564; Folio 8, the volcano *Catcitepetl,* the *speaking mountain,* famous for the penance exercises of Quetzalcohuatl and designated by a mouth and a tongue, which are the hieroglyphs for speech; Folio 10, the pyramid of Cholula; and *Folio 67,* the seven chiefs of the seven Mexica tribes, dressed in rabbit skins and emerging from the seven caves of Chicomoztoc. From folio 68 to 93, this manuscript contains copies of hieroglyphic paintings made after the conquest: here one sees natives hung from trees, holding crosses in their hands; a number of Cortés's soldiers on horseback setting fire to a village; friars who are baptizing unfortunate Indians at the very moment when the latter are put to death by being cast into the water. In these images, one recognizes the arrival of the Europeans in the new world.

---

1. Plate XIV, p. 108 [this edition].

PLATE XXVII

Hieroglyphic Painting from the Borgia Manuscript of Velletri and Day-Signs from the Mexica Almanac

The twenty day-signs have been chosen from the first pages of the Velletri manuscript, each of which shows five rows of thirteen hieroglyphs and in total, 5 × 13 × 4 = 260 days, or one year of twenty *semilunar months* of the ritual almanac. These two hundred sixty signs are arranged such that four double pages serve to convert the thirteen-day periods into half-decades of the civil almanac, fifty-two of which form a ritual year. It is also noteworthy that in order to make these tables easier to read, the author has repeated the last sign from the preceding row at the beginning of each row. Mr. Zoëga observed this same feature in the arrangement of the Egyptian hieroglyphs, and it was by virtue of such observations that he was able to judge whether the hieroglyphs were read from right to left or from left to right. In the *Codex Borgianus* one occasionally finds a footprint, the sign for movement, added to a day-sign. I do not know the reason for this odd combination.

The first of the five rows of day-hieroglyphs (Plate XXVII, number 1)—which, according to the Mexica writing system, is the lowest series—shows, from right to left, *cipactli, ehecatl, calli, cuetzpalin,* and *cohuatl*; the second one shows *miquiztli, mazatl, tochtli, atl,* and *itzcuintli*; the third one shows *ozomatli, malinalli, acatl, ocelotl, quauhtli,* and *cozcaquauhtli*; and the fourth or upper series shows *ollin, tecpatl, quiahuitl,* and *xochitl*. We have explained the meaning of these hieroglyphs above.[1] If one compares the images from Plate XXVII with those published by Valadés, Gemelli, Clavijero, and Cardinal Lorenzana, one sees how inaccurate the theories proposed about the Mexica calendar signs have been to this point.

The painting that shows a figure that would appear to have four hands (Plate XXVII, number 2) comes from the *Codex Borgianus, Folio* 58. I had an entire page of it copied to give a clearer idea of the *layout* of this strange manuscript. Just as one finds no indications of lingam (φαλλός [*phallos*, phal-

1. Pp. 163, 171, and 129–91 [this edition].

I.

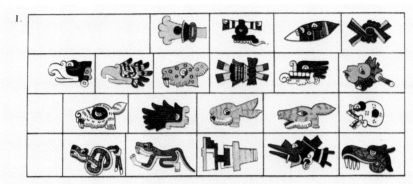

II.

Plate XXVII

lus]) worship in the Mexica hieroglyphs, neither does one see any of those figures with several heads and hands that also characterize, so to speak, the mystic paintings of the Hindus. The man placed on the right-hand side of the upper box is a priest dressed in the skin of a freshly sacrificed human victim. The painter has shown the drops of blood that cover this skin; since the skin of the hands is hanging down from the arm of the sacrificer, the latter appears to have four hands. Both this costume and the horrific and revolting ceremonies that it recalls are described in Torquemada.[1] A chapel, known as a *yopico*, was built above the cave where the human skins were stored. We have seen above that the fourth Mexica month, *tlacaxipehualitzli*, which corresponds to our month of March, took its name from these bloody festivals. In the *Codex Borgianus*, which is a ritual calendar, one actually finds the figure of a priest wrapped in a human skin under the day-sign that indicates the spring equinox.[2] The sacrificer's head is capped by one of those pointed hats used in China and on the northwest coasts of America. Across from this figure sits the god of fire, Xiuhteuctli Tletl; at his feet is a sacred vessel. In the first year of the Mexica cycle, Tletl is the *night-sign* for the day on which the spring equinox falls.

The lower box (Plate XXVII, number 2) shows the god *Tonacateuctli* holding in his right hand a knife, agave leaves, and a bag of incense. We have absolutely no idea what is meant by the two children who are holding one another's hand, and of whom one commentator has written that "they seem to speak the same language." The serpent placed below the temple might lead one to suspect that they are the twin children of *Cihuacohuatl*, the famous *wife of the serpent*, the Aztec Eve. But the small figures in the *Codex Borgianus, Folio 61*, are both female, as their hair arrangement clearly indicates, while those shown in the Vatican manuscript[3] are male.

---

1. *Mon[arquía] ind[iana]*, Book 10, ch. 12, Vol. II, p. 271.

2. *Cod[ex] Borg[ianus]*, folio 25 (Fábr[ega], MSS, numbers 105, 275, and 299). See also p. 160 [this edition].

3. See Plate XXIII in this atlas.

PLATE XXVIII 214

Aztec Ax

This ax, of a compact feldspar that borders on true saussurite, is covered in hieroglyphs. I am indebted to the kindness of 'Don Andrés Manuel del Río, professor of mineralogy at the School of Mines in Mexico City and author of an excellent Treatise on Oryctognosy [mineralogy], for this ax, which I have deposited in the cabinets of the king of Prussia, in Berlin. Jade, compact feldspar (*dichter feldspath*), Lydian-stone, and a few varieties of basalt are mineral substances that, in both continents and in the South Sea islands, have furnished savage and semicivilized peoples with the raw material for their axes and various defensive weapons. Just as the Greeks and the Romans maintained the use of bronze for a long time after the introduction of iron, the Mexica and the Peruvians still used stone axes even when copper and bronze were already quite common among them. Despite our long and frequent journeys throughout the Cordilleras of both Americas, we have never been able to find jade on site, and the scarcer this rock appears to be, the more astonished one is by the large number of jade axes one finds almost everywhere that one digs into the earth in formerly inhabited places, from Ohio to the mountains of Chile.

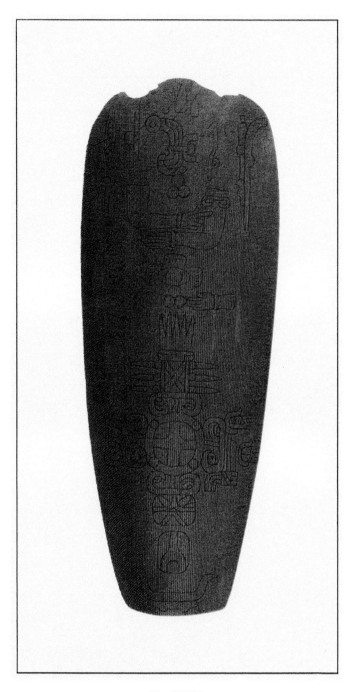

Plate XXVIII

PLATE XXIX

Aztec Idol Made of Basaltic Porphyry, Found under the Cobblestones of the Main Square of Mexico City

With the sole exception of the group of figures shown on Plate XI, all the remnants of Mexica painting and sculpture that we have examined thus far demonstrate a complete ignorance of the proportions of the human body, a great deal of coarseness and inaccuracy in the drawing, but also an effort toward meticulous fidelity in rendering the details of the accessories. One may be surprised to find the arts of imitation present in such an unrefined state among a people whose political existence had for centuries suggested a certain degree of civilization, yet among whom idolatry, astrological superstitions, and the desire to preserve the memory of events magnified the number of idols, sculpted stones, and historical paintings. One must not forget, however, that several peoples who played a role on the world stage (particularly the peoples of central and eastern Asia, whom the inhabitants of Mexico seem to resemble through quite close ties) exhibit this very contrast between social perfection and artistic infancy. One might be tempted to apply to both the inhabitants of Tartary and the mountain peoples of Mexico what a great historian of antiquity[1] said about the Arcadians: "The somber, cold climate of Arcadia lends a hard, austere character to its inhabitants, since it is natural for men, in their customs, their figure, their color, and their institutions, to resemble the climate in which they live." But the more one examines the state of our species in various regions and becomes accustomed to comparing the physiognomy of countries with that of the peoples who settled there, the less faith one has in the specious theory that attributes exclusively to the climate the effects of a combination of numerous moral and physical circumstances.

Among the Mexica, the cruelty of their customs, sanctioned by a blood-thirsty religion, the tyranny exercised by nobles and priests, the fanciful concepts of astrology, and the frequent use of symbolic script all appear to have greatly contributed to perpetuating a lack of artistic refinement and a predi-

215

---

1. ▾Polyb[ius], *Hist[oriae]*, Book IV,§ 80 (ed. Casaub[on], 1609), p. 290, D.

lection for inaccurate, hideous forms. These idols, before which the blood of human victims flowed daily, "these original deities spawned by fear," had features that combined the strangest aspects of nature. The character of the human figure disappeared under the weight of the garments, the helmets shaped like the heads of carnivorous animals, and the serpents wound around bodies. A religious respect for signs resulted in each idol's having its individual type, from which it was forbidden to diverge. It was in this manner that the religion perpetuated the inaccuracy of forms, and the people became accustomed to these assemblages of monstrous parts, which were nevertheless arranged in accordance with systematic ideas. Both astrology and the complicated manner of graphically representing temporal divisions were largely responsible for these lapses in imagination. Each event appeared to be influenced not only by the hieroglyph that presided over that day but also by those that ruled the half-decade and the year; hence the idea of pairing signs and creating those purely fantastical creatures we find repeated so many times in the astrological paintings that have been transmitted to us. The genius of the American languages, which, like Sanskrit, Greek, and the languages of Germanic origin, enables one to convey a large number of ideas in a single word, probably facilitated these wondrous creations of mythology and the mimetic arts.

216

Faithful to their original habits, all peoples—regardless of their intellectual culture—pursue for centuries the course that they once set for themselves. In regard to the imposing simplicity of the Egyptian hieroglyphs, one shrewd writer[1] remarked that "these hieroglyphs show a lack of imitation rather than an excess of it." It is, on the contrary, precisely this excess of imitation, this predilection for the minutest details, and the repetition of the most common forms that characterizes Mexica historical paintings. We have already cautioned above[2] that one must not confuse images in which almost everything is individualized with simple hieroglyphs capable of representing abstract ideas. Whereas the Greeks drew from the latter a sense of the ideal style,[3] the Mexica peoples encountered insurmountable obstacles to the progress of the mimetic arts in their frequent use of historical and astrological paintings and their respect for forms that were frequently bizarre and always inaccurate. It was in Greece that religion became the wellspring of those arts

1. Quatremère de Quincy, "Sur l'idéal dans les arts du dessin," in *Archives littéraires [de l'Europe]*, 1805, number 21, pp. 300 and 310.

2. P. 198–99 [this edition].

3. Quatremère de Quincy, "Sur l'idéal dans les arts du dessin," pp. 303–7.

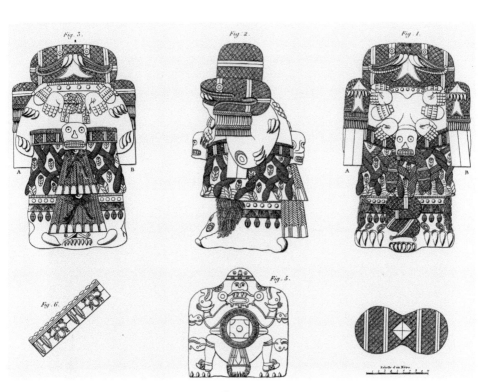

Plate XXIX

to which it gave birth. The imagination of the Greeks was able to endow even the gloomiest of subjects with sweetness and charm. But among a people who bear the yoke of a bloodthirsty religion, death manifests itself everywhere in the most terrifying emblems: it is engraved onto each stone; one finds it inscribed on each page of their books; the religious monuments have no other purpose than to incite terror and dread.

I thought it necessary to recall these ideas before focusing the reader's attention on the monstrous idol on Plate XXIX. This rock, sculpted on all sides, is over three meters high and two meters wide. It was found under the cobblestones of the *Plaza Mayor* in Mexico City, within the walls of the great temple, in August of 1790, and thus only a few months before[1] people discovered the enormous stone that shows both the celebrations and the day-hieroglyphs of the Aztec calendar. Workers carrying out excavations for the construction of an underground aqueduct discovered it in a horizontal position, thirty-seven meters to the west of the viceroy's palace and five meters to the north of the *Acequia de San José*. Since it is hardly likely that while burying the idols to remove them from the natives' sight, Cortés's soldiers had ordered masses of considerable weight to be transported quite far from the *sacellum* [chapel] where they had been originally placed, it is important to give an exact indication of those places where each remnant of Mexica sculpture was found. These ideas will be of particular interest should some government, eager to cast light upon the ancient civilization of the Americans, decide to dig around the cathedral in the main square of the former Tenochtitlan and in the market of Tlatelolco,[2] where, in the final days of the siege, the Mexica had retreated with their household gods (*Tepitotan*), their sacred books (*Teoamoxtli*), and all of their most precious possessions.

Looking at the idol represented in Plate XXIX as it appears when seen from the front (*Figure* 1), from behind (*Figure* 3), from the side (*Figure* 2), from above (*Figure* 4), and from below (*Figure* 5), one might at first be tempted to believe that this monument is a *teotetl*, a *divine stone*, a kind of betyl [omphalos][3] adorned with sculptures, a rock on which hieroglyphic signs have been engraved. Yet when one examines this shapeless mass more closely, one can make out in the upper part the heads of two monsters positioned side by side, and one finds on each face (*Figures* 1 *and* 3) two eyes and a wide mouth armed with four teeth. Perhaps these monstrous figures represent only masks, for it was a custom among the Mexica to mask the idols

217

1. See p. 211 [this edition].
2. Gama, *Descripción [histórica y cronológica] de las [dos] piedras*, p. 2.
3. Zoëga, *De [origine et usu] obel[iscorum]*, p. 208.

whenever the king was ill[1] and during any other public calamity. The arms and feet are hidden under a drapery ringed by enormous serpents that the Mexica called *cohuatlicuye, garment of serpents.* All of these accessories, especially the feather-shaped fringes, are sculpted with great care. Mr. Gama suggested in a separate paper that this idol (*Figure* 3) very likely represents the god of war, *Huitzilopochtli*, or *Tlacahuepancuexcotzin*, and his wife (*Figure* 1), called *Teoyamiqui*[2] (from *miqui*, to die, and *teoyao*, divine war), because she led the souls of the warriors who died defending the gods to the *house of the Sun*, the paradise of the Mexica,[3] where she transformed them into hummingbirds. The skulls and the severed hands, four of which surround the goddess's breast, recall the horrific sacrifices (*teoquauhquetzoliztli*) celebrated in the fifteenth thirteen-day period after the summer solstice in honor of the god of war and his companion *Teoyamiqui*. The severed hands alternate with the images of particular vessels in which incense was burned. These vessels were called *top-xicalli, calabash-shaped bags* (from *toptli*, purse woven from century plant fiber, and *xicalli*, calabash).

Since all of the faces of this idol were sculpted, even the underside (*Figure* 5) where one finds an image of *Mictlanteuhtli, lord of the realm of the dead*, it was undoubtedly held up in the air by means of two columns that supported parts A and B, in figures 1 and 3. This odd positioning meant that the head of the idol was actually raised five to six meters above the floor of the temple, such that when the priests (*Teopixqui*) dragged the unfortunate victims to the altar, they made them pass under the figure of *Mictlanteuhtli*.

The viceroy, the count of Revillagigedo, had this monument brought to the building of the University of Mexico, which he regarded as "the most appropriate place in which to preserve one of the most peculiar remnants of American antiquity."[4] But the professors of that University, friars of the Dominican order, did not want the Mexica youth to gaze upon this idol; they buried it anew, half a meter deep in the corridors of the college. I would not have been so fortunate as to examine it if the bishop of Monterrey, 'Don Feliciano Marín, who passed through Mexico City on his way to his diocese, had not, at my request, petitioned the rector of the University to have it disinterred. I found Mr. Gama's drawing, which I have had copied on Plate XXIX, to be very accurate. The stone used in this monument is bluish-gray basaltic *wacke*, cracked and filled with vitreous feldspar.

1. Gómara, *Conquista de México*, p. 123.
2. Boturini, *Idea de una nueva historia general*, pp. 27 and 66.
3. Torquemada, [*Monarquía indiana*,] Book XII, ch. 48, Vol. II, p. 569.
4. *Officio del 5 sept[iembre] 1790.*

In January of 1791, the same excavations to which we owe the sculptures shown on Plates XXI, XXIII, and XXIX also led to the discovery of a tomb two meters long and one meter wide, filled with very fine sand and containing the well-preserved skeleton of a carnivorous quadruped. The tomb was square and formed of slabs of a porous amygdaloid called *tezontli*. The animal appeared to be a *coyote* or Mexica wolf. Clay vessels and small, well-molten bronze bells had been placed next to the remains. This tomb was probably that of some sacred animal; for sixteenth-century writers inform us that the Mexica erected small chapels dedicated to the wolf (*chantico*), the tiger ([jaguar] *tlatocaocelotl*), the eagle (*quetzalhuexoloquauhtli*), and the grass snake. The *cu*, or *sacellum* [chapel], of the *chantico* was called *tetlanman*; and what is more, the priests of the sacred wolf formed a special congregation, whose monastery bore the name *Tetlacmancalmecac*.[1]

One can easily imagine how the zodiac divisions and the names of the signs that preside over the days, the semilunar months, and the years might lead men to the worship of animals. The nomadic peoples count by lunar months; they distinguish the moon of the rabbits from that of the tigers and of the goats, et cetera, depending on the different periods of the year in which wild or domestic animals bring them joy or inspire fear among them. As the temporal divisions gradually become spatial divisions[2] and peoples form the dodecatemoria of the *full-moon zodiac*, the names of the wild and domestic animals pass over to the constellations themselves. It is thus that the Tartar zodiac, which contains only true ζώδια [*zōdia*], can be considered to be the *zodiac of the hunting and herding peoples*; the tiger, unknown in Africa, lends it an exclusively Asian character. This animal is no longer found within the Chaldean, Egyptian, and Greek zodiacs, in which the tiger, the hare, the horse, and the dog were replaced by the lion of Africa, Thrace, and western Asia; the scales; the twins; and, what is quite remarkable, agricultural symbols. The Egyptian zodiac is a *farming people's zodiac*. As peoples became civilized and the mass of their ideas grew, the names of the zodiac constellations lost their original uniformity and the number of *celestial animals* decreased; this number has nevertheless remained considerable enough to exert a noticeable influence on religions. The imaginative excesses of astrology led men to attach a great importance to the signs that preside over the various divisions of time. In Mexico City, each day-sign had its own altar. In the main *teocalli* (θεοῦ καλια [*theoū kalia*, dwelling of the god), near the column that

220

1. Nieremberg, *Hist[oria] nat[uraes]*, Book VII, ch. 22, p. 144. Torquemada, [*Monarquía indiana*,] Book II, ch. 58; Book VIII, ch. 13 (Vol. I, p. 194; Vol. II, p. 29).

2. See p. 199 [this edition].

supported the image of the planet Venus (*Ilhuicatitlan*), one could see small chapels for the catasterisms *macuilcalli* (5 dwelling), *ome tochtli* (2 rabbit), *chicome atl* (7 water), and *nahui ocelotl* (4 jaguar). Since the majority of the day-hieroglyphs were composed of animals, the worship of the latter was closely tied to the calendar system.

PLATE XXX

Río Vinagre Falls near the Puracé Volcano

The city of Popayán, the administrative center of a province in the kingdom of New Granada, is located in the beautiful Río Cauca valley, at the foot of the great volcanoes of Puracé and Sotará. Since its elevation above sea level is only eighteen hundred meters, it enjoys, at latitude 2° 26' 17", a delightful climate, much cooler than that of Quito and Santa Fé de Bogotá. Going up from Popayán toward the peak of the Puracé volcano, one of the high peaks of the Andes, one finds, at an elevation of two thousand six hundred fifty meters, a small plain (*Llano del Corazón*) inhabited by Indians and cultivated with the greatest care. This lovely plain is bordered by two extremely deep ravines, and it is on the edges of the precipices that the houses of the village of Puracé are built. Springs burst forth everywhere from the porphyritic rock; each garden is ringed by a live hedge of euphorbia (*lechero*) with thin leaves of the palest green. There is nothing more pleasant than the contrast between this gorgeous greenery and the wall of black, arid mountains that surround the volcano and are rent by the effect of earthquakes.

The small village of Puracé, which we visited in November 1801, is famous in the country because of the beautiful waterfalls of the *Pusambio* River, whose water is acidic, which is why the Spanish call it *Río Vinagre*. This small stream runs hot near its source: it probably owes its origin to the daily thawing of snow water and to the sulfur that burns inside the volcano. Near the plain of the *Corazón*, it forms three cataracts, of which the upper two are quite considerable. The second of these falls (*chorreras*) is shown on Plate XXX; I have drawn it just as it is seen from the garden of an Indian, the neighbor of the missionary of Puracé, who is a Franciscan friar. The water, which carves a path through a cave, plunges to a depth of more than one hundred twenty meters. The waterfall is quite delightful, and it attracts the attention of travelers; but the inhabitants of Popayán would prefer it if the river rushed into some crevice instead of mixing with the Río Cauca, for the Río Cauca is deprived of fish for over four leagues because of the blending of its waters

Plate XXX

with those of the Río Vinagre, which are laden not only with iron oxide but also with sulfuric and muriatic acid.

The foreground of the drawing shows a group of *Pourretia pyramidata*, a plant closely related to *Pitcairnia*, known in the Cordilleras by the name *achupallas*. The stem of this plant is filled with a starchy pith that serves as food for the great black bear of the Andes and sometimes, during food shortages, even for humans.

PLATE XXXI

Mail Service in the Province of
Jaén de Bracamoros

To hasten communications between the South Sea coasts and the province of
Jaén de Bracamoros to the east of the Andes, the Peruvian mail carrier swims
for two days, first down the Guancabamba (or Chamaya) river, then down
the Amazon, from Pomahuaca and Ingatambo all the way to Tompenda. He
wraps the few letters with which he is entrusted every month either in a ker-
chief or in a kind of loincloth called a *guayucu*, which he ties around his head
in the shape of a turban. This turban also contains the large knife (*machete*)
with which every Indian is armed, less for his defense than to cut his way
through the forests.

The Río de Chamaya is unnavigable due to a large number of small cas-
cades: I determined[1] its descent to be five hundred forty-two meters over a
short distance of eighteen leagues, from the ford of Pucara to its confluence
with the Amazon River below the village of Choros over a short distance of
eighteen leagues. The locals call the mail carrier of Trujillo *the swimming
postman* (*el correo que nada*). Plate XXXI depicts him just as we encountered
him in the village of Chamaya, about to jump into the water. To tire less during his swim downriver, he holds on tightly to the trunk of a Bombax or an
Ochroma (*palo de balsa*), which are trees with very light wood. Whenever the
riverbed is obstructed by the ledge of rock, he climbs onto the bank above
the falls, goes around it through the forests, and jumps back into the water as
soon as he sees that there is no longer any danger. He does not need to take
any supplies with him, for he finds hospitality in the many huts surrounded
by banana plantations and situated along the shore between Las Huertas de
Pucara, Cavico, Sonanga, and Tomependa. Sometimes he invites another
Indian to join him so as to make his journey more enjoyable. Fortunately, the
rivers that mingle their waters with those of the Marañón above Pongo de
Mayasi are not infested with crocodiles; for this reason, almost all the wild

---

1. See my *Recueil d'observ[ations] astr[onomiques]*, Vol. I, p. 314.

Plate XXXI

hordes travel in the same manner as the Peruvian mail carrier. It is quite rare for this postman to lose letters or get them wet on his way from Ingatambo to the residence of the governor of Jaén. After resting for a few days in Tompenda, he returns either through the *Páramo del Paretón* or by way of the terrible path that leads through the villages of San Félipe and Sagique, whose forests are teeming with cinchona of the highest quality.

PLATE XXXII

Hieroglyphic History of the Aztecs, from the Great Flood to the Founding of Mexico City

This historical painting was already published at the end of the seventeenth century in Gemelli Careri's travel narrative. Although the *Giro del Mondo* by that author is a widely known work, we found it necessary to reproduce this piece, whose authenticity has been the subject of some rather baseless conjectures that warrant scrupulous examination. It is only by bringing together a large number of monuments that one can hope to shed some light on the history, the customs, and the civilization of those peoples of the Americas who were not acquainted with the admirable art of breaking down sounds and representing them either as isolated or as grouped signs. Comparing monuments to one another not only makes it easier to explain them; it also provides reliable information about the degree of trustworthiness of the Aztec traditions recorded in the writings of the first Spanish missionaries. I think that these powerful motives will provide ample justification for our having chosen a few monuments scattered in printed works to supplement the numerous previously unpublished ones in this collection.

The hieroglyphic drawing shown in Plate XXXII has been all the more neglected until now because of its inclusion in a book that, due to the most extraordinary skepticism, has been considered a mass of frauds and lies. "I did not dare speak of Gemelli Careri," writes the illustrious author of the *History of America*, "because it appears to be a generally accepted opinion now that this traveler never left Italy, and that his *Tour of the World* [*Giro del mondo*] is the account of a fictitious voyage." It is true that even though he voices this opinion, Robertson appears not to share it, for he adds prudently that the grounds for this accusation of deceit do not appear very plausible to him.[1] I shall not judge whether or not Gemelli ever went to China or Persia; but having traveled a large part of the itinerary in Mexico that the Italian traveler so meticulously described, I can say that it is just as undeniable

1. Robertson, *History of America*, 1803, Vol. III, p. 401.

that Gemelli was in Mexico City, Acapulco, and the small villages of Mazatlán and San Augustin de las Cuevas as it is certain that ʼPallas was on the Crimea and Mr. Salt in Abyssinia. Gemelli's descriptions have that local color that constitutes the main charm of travel narratives, even when they are written by the least enlightened of men, and that only those who have had the advantage of seeing with their own eyes can provide. A respectable cleric, the Abbot Clavijero,[1] who had traveled across Mexico nearly half a century before me, already raised his voice in defense of the author of the *Giro del Mondo*. He very rightly observed that had Gemelli never left Italy, he would not be able to speak with such great precision, characteristic of the people of his time, about the monasteries of Mexico City and the churches of several villages whose names were unknown in Europe. The same truthfulness—and we must insist on this point—is not, however, evident in the ideas that the author claims to have drawn from his friends' accounts. Gemelli Careri's work, like that of a famous traveler who has been subjected to such harsh treatment in our own day, seems to show an inextricable mixture of errors and precisely reported facts.

This drawing of the Aztecs' migration was once part of the famous collection of Dr. Sigüenza, who had inherited the hieroglyphic paintings from a noble Indian, Juan de Alva Ixtlilxochitl. As the Abbot Clavijero assures us, this collection was preserved intact in the Jesuits' college in Mexico City until 1759. It is not known what became of it after the destruction of the order; I have looked in vain through the Aztec paintings preserved in the library of the university without being able to find the original of the drawing that is shown in Plate XXXII. But there exist in Mexico City several old copies that were certainly not made from Gemelli Careri's engraving. If one compares all the symbolic and chronological content in the painting of the migrations to the hieroglyphs contained in the manuscripts at Rome and Velletri and in the collections of Mendoza and Gama, one will certainly not be inclined to give any credence to the theory that Gemelli's drawing is the fictitious creation of some Spanish monk who attempted to prove, through apocryphal testimony, that the legends of the Hebrews are also found among the indigenous peoples of the Americas. Everything we know about the history, the 225 religion, the astrology, and the cosmogonic fables of the Mexica forms a system with tightly interconnected components. The paintings, the bas-reliefs, the decorations of the idols, and the *divine stones* (*teotetl* among the Aztecs, θεοῦ πέτρα [*theoū petra*], stone of god, among the Greeks)—everything has the same character, the same physiognomy. The cataclysm with which the

---

1. *Storia antica del Messico*, Vol. I, p. 24.

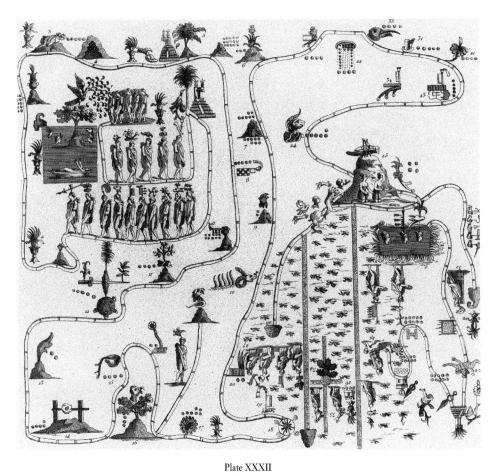

Plate XXXII

history of the Aztecs begins, and from which Coxcox escapes in a boat, is shown with the very same circumstances in the drawing that depicts the destructions and the regenerations of the world.[1] The four indictions (*tlalpilli*) that are related[2] to these catastrophes or the subdivisions of the *long year* are sculpted on a stone discovered in 1790 in the foundations of the teocalli of Mexico City.

In the most recent edition of his work, Robertson, whose pursuit of facts demonstrates a most exacting critical perspective, has also acknowledged the authenticity of the paintings in Sigüenza's museum. It is undeniable, this great historian writes, that these paintings were produced by the natives of Mexico, and the accuracy of the drawing seems to prove only that this copy was either made or retouched by some European artist. This latter observation does not, however, appear to be entirely corroborated by the large number of hieroglyphic paintings preserved in the archives of the viceroyalty in Mexico City. One detects in these paintings a noticeable improvement in the standard of drawing since the conquest, and especially since the year 1540. In Boturini's collection I have seen only cotton canvases or rolls of agave paper that show, with quite faithful contours, bishops mounted on mules, Spanish lancers on horseback, oxen pulling a plow, ships landing at Veracruz, and several other objects unknown to the Mexica prior to Cortés's arrival. These paintings were made not by Europeans but by Indians and Mestizos. Leafing through hieroglyphic manuscripts from different periods, one follows with interest the progressive advance of the arts toward a higher level of perfection. The once stocky figures become slimmer; the limbs become more distinct from the torso; when the heads are seen in profile, the eyes do not gaze directly at the viewer; the horses that in Aztec paintings resembled Mexican stags gradually assume their true form. The figures are no longer grouped in *procession style*; their interactions become more frequent; one sees them in action; and symbolic painting, which involves alluding to or recalling events more than expressing them, is thereby subtly transformed into an animated style that uses only a few phonetic hieroglyphs[3] for the names of persons and places. I am inclined to believe that this painting, which Sigüenza passed on to Gemelli, is a copy made after the conquest by either a native or a Mexican Mestizo. The painter probably did not want to reproduce the inaccurate forms of the original; he imitated the hieroglyphs of the names and the cycles with scrupulous precision, but he changed the proportions of the human fig-

226

1. Plate XXVI.
2. See pp. 199 and 233 [this edition].
3. See p. 84 [this edition].

ures, which he clothed in a manner similar to that which we have acknowl-edged[1] in other Mexica paintings.

Here, then, are the principal events depicted on Plate XXXII, according to Sigüenza's explanation, to which we shall add some concepts taken from the Mexica's historical annals.

History begins with the great flood of Coxcox, or the fourth destruction of the world, which in Aztec cosmogony concludes the fourth great cycle, *atonatiuh*, the *age of water*.[2] According to the two accepted chronological systems, this cataclysm occurred either one thousand four hundred seven-teen or eighteen thousand twenty-eight years after the beginning of the *age of the earth*, *tlatonatiuh*. The huge difference between these numbers should surprise us, especially if we recall the theories on the duration of the Hindus' four *yugas* that Bailly, William Jones, and 'Bentley[3] have recently advanced. Among the various peoples who lived in Mexico, the Aztecs, the Mixteca, the Zapoteca, the Tlaxcalteca, and the Michoacans were all found to have paintings that represented Coxcox's flood. The Noah, Xisuthros, or Manu of these peoples is called Coxcox, Teo-Cipactli, or Tezpi. Together with his wife, Xochiquetzal, he saved himself in either a small boat or, according to other legends, a raft made of Ahuahete (*Cupressus distichia*). The painting shows Coxcox adrift on the water, stretched out in a boat.

227   The mountain whose tree-crowned summit soars above the waters is the Ararat of the Mexica, the Peak of Culhuacan. The horn on the left is the phonetic hieroglyph for Culhuacan. At the base of the mountain, one sees the heads of Coxcox and his wife; the latter is identifiable by the two horn-shaped plaits that specify the female gender, as we have already mentioned several times. The humans born after the flood were mute; a dove perched on top of a tree distributes to them tongues imaged as small commas.[4] One must not confuse this dove with the bird that brought to Coxcox the news that the waters had receded. According to a legend preserved among the peoples of Michoacan, Coxcox, whom they call Tezpi, boarded a spacious *acalli* with his wife, children, several animals, and seeds whose conservation was of the utmost importance to the human species. When the great Tezcatlipoca com-manded the waters to retreat, Tezpi ordered a vulture, zopilote (*Vultur aura*), to leave the boat. This bird, which feeds on dead flesh, did not return because

1. Plate XIV, numbers 5 and 7.

2. See p. 231 [this edition].

3. ["VI. On the Hindu Systems of Astronomy, and their connection with History in an-cient and modern times,"] *Asiat[ick] Research[es]*, Vol. VIII, p. 195.

4. See the "Trial Document" above; Plate XII.

of the large number of corpses strewn across the recently dried land. Tezpi sent out other birds, among which only the hummingbird returned, holding in its beak a leaf-covered branch. Then, seeing that the ground was being covered anew with greenery, Tezpi left his boat near Culhuacan Mountain.

These legends, we repeat here, recall others of a high and venerable antiquity. Among peoples who had no contacts with each other, finding the fossils of sea creatures on the highest mountain tops might well have inspired the idea of great floods that had temporarily extinguished organic life on earth. But is it not imperative to acknowledge the traces of a common origin wherever the cosmogonic ideas and the earliest legends of peoples exhibit striking analogies, even in very minor details? Does not Tezpi's hummingbird recall Noah's dove, that of Deucalion, as well as the birds that, according to ʼBerosus, Xisutrus ordered to leave his ark to determine whether the waters had ebbed, and whether he could begin building altars to the patron gods of Chaldea?

Since the tongues that the dove had distributed to the peoples of the Americas (Number 1) varied infinitely from one another, these peoples scattered, and only fifteen heads of family, who spoke the same language and from whom the Toltecs, the Aztecs, and the Acolhua descended, banded together and arrived in ʼAztlan (land of herons or flamingos). The bird in the hieroglyph for water, *atl*, refers to Aztlan. The terraced pyramidal monument is a *teocalli*. I am surprised to find a palm tree near this teocalli: this plant surely does not suggest a northerly region, yet it is almost certain that the original homeland of the Mexica peoples—*Aztlan, Huehuetlapallan,* and *Amaquemecan*—must have been north of at least the 42nd degree of latitude. Perhaps the Mexica painter, an inhabitant of the Torrid Zone, placed a palm tree next to the temple of Aztlan merely because he was unaware that this tree does not grow in the lands of the North. The simple hieroglyphs for the names of the fifteen chiefs are placed above their heads.

The images placed along the road from the Aztlan teocalli to Chapultepec mark the places where the Aztecs stayed for some time and the cities they built: *Tocolco* and *Oztotlan* (numbers 3 and 4), *humiliation* and *place of caves*; *Mizquiahuala* (number 5), identified by a fruit-bearing mimosa next to a teocalli; *Teotzapotlan* (number 11), *place of the divine fruits*; *Ilhuicatepec* (number 12); *Papantla* (number 13), *wide-leaved herb*; *Tzompango* (number 14), *place of human remains*; *Apazco* (number 15), *clay pot*; *Atlicalaguian* (slightly above the preceding hieroglyph), crevice into which a stream plunges; *Quauhtitlan* (number 16), the grove where the eagle dwells; *Atzcapozalco* (number 17), *anthill*; *Chalco* (number 18), *place of precious stones*; *Pantitlan* (number 19), *the place of spinning*; *Tolpetlac* (number 20), *mats made of*

228

rushes; *Quauhtepec* (number 9), *the Eagle's mountain*, from Quauhtli, eagle, and tepec (in Turkish, tepe), mountain; *Tetepanco* (number 8), *wall composed of many small stones*; *Chicomoztoc* (number 7), *the seven caves*; *Huitzquilocan* (number 6), *place of thistles*; *Xaltepozauhcan* (number 22), *the place where sand originates*; *Cozcaquauhco* (number 33), name of a vulture; *Techcatitlan* (number 31), place of obsidian mirrors; *Azcaxochitl* (no. 21), *ant flower*; *Tepetlapan* (number 23), the spot where one finds *tepetate*, a clayey breccia that contains amphibole, vitreous feldspar, and pumice stone; *Apan* (number 32), *place of water*; *Teozomaco* (number 24), place of the divine monkey; *Chapultepec* (number 25), *mountain of the grasshoppers*, a site shaded by ancient cypresses and famous for the magnificent view that one enjoys from the top of the hill;[1] *Coxcox*, king of Culhuacan (number 30), identified by the same phonetic hieroglyphs found in the square that depicts both Coxcox's flood and Culhuacan mountain; *Mixiuhcan* (number 29), *birthing place*; the city of *Temazcatitlan* (number 26); the city of *Tenochtitlan* (number 34), identified both by the causeways that cross its swampy terrain and by the Indian fig (*cactus*) on which the eagle reposes, which the oracle had designated as the spot where the Aztecs were to build their city and end their migrations; the founders of *Tenochtitlan* (number 35); those of *Tlatelulco* (number 27); the city of *Tlatelulco* (number 28), which today is but a suburb of Mexico City.

We shall not enter into historical detail about the events to which the simple and composed hieroglyphs of Sigüenza's painting refer. These events are reported in Torquemada and the ancient history of Mexico published by the Abbot Clavijero. This painting is less interesting as a historical monument than for the method that the artist employed to link these events together. Suffice it to mention here that the ribbon-tied bunches of rushes (number 2) do not represent four-hundred-year periods, as Gemelli claimed, but rather cycles or bindings of years, Xiuhmolpilli, of fifty-two years.[2] The painting as a whole shows only eight of these bindings, or four hundred sixteen years. If one recalls that the city of Tenochtitlan was founded in the twenty-seventh year of a Xiumolpilli, one finds that according to the chronology of the painting (Plate XXXII), the departure of the Mexica peoples from Aztlan took place five cycles before the year 1298, or in the year 1038 of the Christian era. Gama's estimate for this departure, based on other information, is 1064. The dots accompanying the hieroglyph of a binding of years had been tied since the famous sacrifice of Tlalixco. Yet in the painting that we are examining, the hieroglyph of the cycle is followed by four nails, or units, placed

---

1. See my *Essai politique sur [le royaume de] la Nouvelle-Espagne*, Vol. I, p. 179.
2. See p. 158 [this edition].

near the hieroglyph of the city of Culhuacan (number 30). It was therefore in the year 208 of their era that the Aztecs emerged from their enslavement under the kings of Culhuacan, and this timing conforms to Chimalpahin's annals. The dots placed next to the hieroglyphs for the cities (numbers 14 and 17) mark the number of years that the Aztec people remained in each place before resuming their migrations. I think that binding number 2 represents the cycle that ended at Tlalixco; for according to Chimalpahin, the festival of the second cycle was celebrated in Cohuatepetl, and that of the third cycle in Apuzco, while the fourth- and fifth-cycle festivals took place in Culhuacan and in Tenochtitlan, respectively.

The odd idea of recording on such a small sheet what in other Mexica paintings often fills canvases or skins ten to twelve feet long made this historical summary quite incomplete. It deals only with the Aztecs' migration, not with that of the Toltec, who preceded the Aztecs in the land of Anahuac by more than five centuries, and who differed from the latter by the same love for the arts and the same religious and peace-loving character that distinguish the Etruscans from the original inhabitants of Rome. The heroic period of Aztec history extended to the eleventh century of the Christian era. Up to then, the deities had been involved in the actions of men; at that time appeared, on the coasts of Panuco, Quetzalcoatl, the Buddha of the Mexica, a white, bearded man, both priest and legislator, devoted to strict penance, and the founder of monasteries and congregations similar to those of Tibet and western Asia. Everything prior to the departure from Aztlan is mixed with childish fables. Among the barbarous peoples, who lack the means to preserve the memory of their deeds, self-awareness is a relatively recent phenomenon; there is a point in their existence beyond which they no longer measure the intervals between events. Distant objects approach one another both in time and in space and become confused. The very cataclysm that the Hindus, the Chinese, and all the peoples of the Semitic race place thousands of years before the perfecting of their social state is believed by the Americans, a people no less ancient perhaps, but whose awakening occurred later, to have taken place only two cycles before their departure from Aztlan.

230

PLATE XXXIII

Rope Bridge near Penipe

The small river of Chambo, which begins in Lake Coley, separates the pretty village of Guanando from that of Penipe. It flows through a ravine whose bottom is at two thousand four hundred meters above sea level and which is famous for *cochineal harvesting,[1] to which the natives have devoted themselves since the earliest times. While traveling through this country en route from Riobamba to the western slope of the volcano Tungurahua, we stopped to examine the plots of land destroyed by the memorable earthquake of February 7, 1797, which killed thirty to forty thousand Indians in the space of a few minutes. In June 1802, we crossed the Chambo River on the bridge of Penipe. This is one of those rope bridges that the Spanish call *puente de ma-roma* or *de hamaca*; the Peruvian Indians, in the Quechua or Incan language, call it *cimppachaca*, from *cimppa* or *cimpasca*, ropes or braids, and *chaca*, bridge. Between three and four inches in diameter, the ropes are made from the fibrous part of the roots of the *Agave americana*. On both sides of the river, they are attached to a crude scaffolding that consists of several trunks of *Schinus molle*. Because their weight makes them sag toward the middle of the river and it would be unwise to stretch them too tightly, it is necessary to build steps or ladders at either end of the *hammock* or *suspension bridge* wherever the riverbank is not particularly high. The Penipe bridge is twenty feet long and seven to eight feet wide; but there are bridges of even more impressive dimensions. The thick ropes made of century plant are covered crosswise with small cylindrical pieces of bamboo. These constructions, which the peoples of southern America had used long before the Europeans' arrival, recall the *chain bridges* that one encounters in Bhutan and the interior of Africa. In his fascinating travels in Tibet, *Mr. Turner[2] gave us the plan of the bridge

231

---

1. See my *Essai politique sur [le royaume de] la Nouvelle-Espagne*, Vol. II, p. 465.
2. *Account of an Embassy to the Court of the Teshoo Lama in Tibet*, 1800, p. 53.

of Chinchu near the fort of Chukha (latitude 27°14'), which is one hundred forty feet long and can be crossed on horseback. This bridge in Bhutan (*chain bridge*) consists of five chains covered in bamboo pieces.

Every traveler has spoken of the extreme danger involved in crossing these rope bridges, which look like ribbons stretched precariously above either a ravine or a raging torrent. This peril is not particularly great when a single person, thrusting his body forward, crosses the bridge as quickly as possible; but the swinging of the ropes becomes very strong when the traveler is led by an Indian who walks much more quickly than he does, or when, frightened by the sight of the water that he glimpses through the gaps in the bamboo, the traveler is imprudent enough to stop in the middle of the bridge and grab on to the ropes that serve as a handrail. A hammock bridge generally remains in good condition for only twenty to twenty-five years; even so, it is necessary to replace a few ropes every eight to ten years. But in these lands, law enforcement is so lax that it is not uncommon to see bridges with bamboo pieces that are mostly broken. On these old bridges one must tread very cautiously to avoid holes so wide that one's entire body could fall through them. A few years before my stay in Penipe, the hammock bridge over the Río Chambo completely collapsed. This event was sparked by very dry wind that had followed on the heels of long rains, causing all the ropes to snap at once. On this occasion, four Indians drowned in the river, which is very deep and has an extraordinarily swift current.

The ancient Peruvians also built wooden bridges whose framework rested on stone piers, but they usually contented themselves with rope bridges. The latter are extremely useful in a mountainous country where the depth of the crevices and the raging torrents are an obstacle to building piers. By attaching lateral cords to the middle of the bridge and stretching them diagonally toward the riverbank, the swinging of the ropes can be lessened. It was by using an extremely long rope bridge that travelers can cross with pack mules that people succeeded in establishing a permanent link between the cities of Quito and Lima a few years ago, after a million francs had been needlessly expended to build a stone bridge near Santa, across a mountain torrent that gushes down from the Cordillera of the Andes.

232

Plate XXXIII

PLATE XXXIV

Cofre de Perote

This mountain of basaltic porphyry is less remarkable for its height than for the bizarre shape of a small boulder positioned on the eastern side of its summit. This boulder, which resembles a square tower, inspired the natives of the Aztec race to name the mountain *Nauhcampatepetl*, from *nauhcampa*, four parts, and *tepetl*, mountain; the Spanish called it Cofre de Perote [Perote's Coffer]. From the top of this mountain, one enjoys a magnificent view of both the Puebla plateau and the western slope of the Mexican Cordilleras covered with thick forests of liquidambar [sweetgum], tree-ferns, and mimosas. One can make out the port of Veracruz, the castle of San Juan de Ulúa, and the Ocean coasts. The Cofre does not at all rise above the line of the perpetual snows; using a barometer, I found the elevation of its summit to be 4088 meters (2,097 toises) above sea level. This height surpasses that of the Peak of Tenerife by 400 meters. I sketched the mountain near the large town of Perote, in the arid, pumice-covered plain that one crosses while traveling from Veracruz to Mexico City. At its crest, the Cofre is only a naked boulder surrounded by pine forest. While climbing to the top, I noticed that the oaks disappeared at an elevation of 3165 meters (1,619 toises), but the pines, whose needles make them resemble *Pinus strobes*, did not completely cease to grow until an absolute elevation of 3942 meters (2,022 toises) was reached. In each zone, the temperature and the barometric pressure set growth limits for the plants which it is impossible for them to cross.

Plate XXXIV

PLATE XXXV

Mount Iliniza

Among the colossal peaks that one discovers around the city of Quito, that of
Iliniza is one of the most majestic and most picturesque. The summit of this
mountain is divided into two pyramidal points; these points are very likely   234
the debris of a collapsed volcano. They have an absolute elevation of 2,717
toises. Mount Iliniza is positioned in the western range of the Andes, at the
same parallel as the Cotopaxi volcano. It is linked to the summit of Rumiñahui by *Alto de Tiopullo*, which forms a transverse secondary mountain chain
from which the waters flow down toward both the South Sea and the Atlantic
Ocean.[1] The pyramids of Iliniza are visible from a vast distance in the plains
that form part of the province of *Las Esmeraldas*. Bouguer conducted trigonometric measurements of their elevation in relation both to the Quito plateau and the Ocean coasts. Using the difference between the heights obtained
by these two measurements, the French academicians determined the absolute elevation of the city of Quito and the approximate value of the barometric coefficient. Physicists interested in the history of scientific progress will
place the name of Iliniza next to that of Puy-de-Dôme, where 'Périer, guided
by Pascal's counsel, first attempted to measure the height of the mountains
with the help of a barometer.

1. See p. 219 [this edition].

Plate XXXV

PLATE XXXVI

Fragments of Aztec Hieroglyphic Paintings from the Royal Library of Berlin

These fragments are taken from ancient manuscripts that I acquired during my stay in Mexico City. They are most definitely rolls drawn up by tribute collectors, *tlacalaquiltecani*, but it is not easy to determine the objects to which these rolls refer.

Number I is part of a *Codex Mexicanus* on agave paper three to four meters long. Here one seems to recognize corn, gold bars, and other products that composed the tribute, *tequitl*. I have absolutely no idea what the painter meant to depict through the large number of small, symmetrically arranged squares. In the second row, counting from right to left, one finds four hieroglyphs repeated in a periodic series. The days marked here and there indicate when the tribute was to be paid.

Numbers II–IV. How to explain these female heads placed near the sign for twenty? The cocks and turkeys that appear in number III might lead one to believe that these two birds had already been familiar to the Mexica before the conquest, were it sufficiently established that the paintings from which these images are taken date back to before the fifteenth century. I have explained in another work[1] that the Indian cock, quite common in the South Sea islands, was transplanted to America by the Europeans. The *tlamama*, hauler (number V), appears to be holding in his hand stalks of either corn or sugarcane. I shall not endeavor here to determine the species of animal represented below the *tlamama*, which bears a slight resemblance to the *tochtli* or Mexica rabbit. Number VII indicates the type of punishment that was meted out to the unfortunate natives when they did not pay the tribute at the prescribed times. Three Indians, whose hands are tied behind their backs, appear to be condemned to the strappado. In each community the tribute rolls were displayed to all the *tequitqui*, or tributaries, and the collectors were in the habit of adding to the bottom of the roll the type of punishment in store for those who did not obey the law.

235

1. *Essai pol[itique sur le royaume de la Nouvelle-Espagne]*, p. 452.

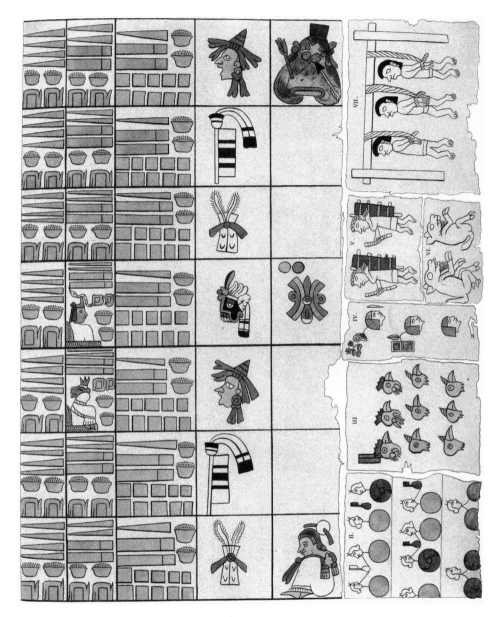

Plate XXXVI

PLATE XXXVII

Hieroglyphic Paintings from the Borgia Museum in Velletri

We have already explained above[1] the layout of the *Codex Mexicanus* from the Borgia museum. Since there is little hope of seeing this Mexica *ritual book* published in its entirety anytime soon, I have gathered in a single plate a large number of images that are remarkable for their shapes and for the ways in which they relate to the customs of a people at once fierce and superstitious.

Number I (*Codex Borgianus*, folio II, Mss. Fábrega, number 18). The mother of humanity, the *wife of the serpent, Cihuacohuatl*, whom the first missionaries referred to as *Señora de nuestra carne*, or *Tonacacihua* (from *tonacayo*, our flesh, and *cihua*, woman). Compare the *Codex Vaticanus*, Plate XII, number 2.

Number II. The same wife of the serpent, the Mexica's Eve. The rabbit, *tochtli*, on the right indicates the first year of the world, as each cycle begins with the sign of the *rabbit*. In his commentary Father Fábrega claims that the mother of humankind is depicted in a state of humiliation, eating *cuitlatl* (κόπροσ [*kopros*, feces]).

Number III (*Codex Borgianus*, folio 58, Mss. number 275). The lord of the dead, *Mictlanteuhtli*,[2] devouring a child.

Number IV (*Codex Borgianus*, folio 24, Mss. number 98). An already aged Noah, his chin bearing a long beard, *Huehuetonacateocipactli*, from *huehue*, old, *tonacayo*, our flesh, *teotl*, god, and *cipactli*. See the explanations given above, pages 183–84 and 232. This same figure appears in the *Codex Borgianus*, folio 60.

Number V (*Codex Borgianus*, folio 56, Mss. number 265). The same deities that we have seen assembled in the monstrous group depicted on Plate XXIX, namely, the god of war, *Huitzilopochtli*, club in hand, and the goddess *Teoyamiqui*. They are shown seated on a human skull. I had only the

236

---

1. Plate XXVII, p. 239 [this edition].
2. Plate XXIX, Figure 5, p. 245 [this edition].

Plate XXXVII

goddess copied, holding in her left hand a kind of scepter with a hand at the tip. This scepter was called *Maquahuitl*, from *maitl*, hand, and *quahuitl*, wood. It is surely remarkable that one finds in Aztec paintings a hand of justice similar to the one that figures on the seal of ʼHugues Capet[1] and that recalls the *manus erecta* of the Roman cohorts.[2]

Number VI. *Teocipactli*, the same figure shown in Number IV. I chose it because of the extraordinary shape of the forehead. Typically, the natives of Mexico and Peru have an oddly indented forehead, and painters make an effort to exaggerate this trait when depicting heroic characters.

Number VII (*Codex Borgianus*, folio 33, Mss. number 150). Five little devils, which recall the famous painting of the temptation of Saint Anthony. On the same page is the image of a temple of Quetzalcoatl with a serpent wrapped around its triangular roof. The idol, placed in a recess, receives the offering of a human heart. Next to the temple one sees the goddess of the underworld, *Mictlanteuhcihua*, stretching her arms toward the victim's body.

Number VIII (*Codex Borgianus*, folio 47, Mss. number 210). The astrological sign *nahui Ollin tonatiuh*, the *Sun in its four movements*, seems to recall, through footprints, *xocpalli*, the positions of the sun at its zenith, at the equator, and during the solstices.[3] Next to this are the dates of the days that are ruled by the catasterisms *ozomatli*, monkey; *calli*, dwelling; and *quiahuitl*, rain. If these dates were 8 *rain*, 5 *dwelling*, and 5 *monkey*, then they would correspond, with the help of the periodic series device, to the days when the sun is found at one of the tropics, the equator, or the zenith of Mexico City; but the numbers added to the hieroglyphs deviate by several digits from those we have just mentioned. The sign *ollin* is placed at the end of a cylindrical insect that appears to be a *millipede* or scolopendra. I do not know the meaning of the astrological symbol that resembles a cross.

Number IX (*Codex Borgianus*, folio 59). A man and a woman holding children in their arms and raising one hand toward the heavens.

No. X (*Codex Borgianus*, folio 23, Mss. number 94). The drinking Devil, *Tlacatecolutl motlatlaperiani*, holding a heart in one hand and drinking the blood from another heart; a third one hangs from his neck. This hideous figure confirms what we have suggested above[4] about the savagery of the Mexica people.

237

1. ʼMontfaucon, *Monumens de la monarchie françoise*, Vol. I, p. 36. ʼMénestrier, *Nouvelle méthode raisonnée du Blason* (Lyon, 1750), p. 52. *Dictionnaire de Trévoux*, Vol. III, p. 127. ʼGilbert De Varennes [*Le Roy d'Armes*] (Paris, 1635), p. 184.

2. ʼAugustinus, *Antiquit[atum] Romanor[um] Hispanarumque in nummis veterum Dialogi* (Antwerp, 1654), p. 18. ʼLipsius, *De militia romana*, p. 41.

3. See pp. 189–99 and 212 [this edition].

4. P. 244 [this edition].

PLATE XXXVIII

Migration of the Aztec Peoples; Hieroglyphic Painting from the Royal Library of Berlin

This poorly preserved fragment appears to have been part of a larger painting that formerly belonged to Chevalier Boturini's collection. The figures are very crudely painted on *amatl* or *maguey* (*Agave americana*) paper. On the left one sees a swampy land, indicated by the hieroglyph for water, *atl*; footprints (*xocpal-machiotl*) representing the migrations of a warlike people; arrows shot from one bank to the other; clashes between two peoples, one of which is armed with shields, the other naked and without any means of defense. It is likely that these conflicts are among those that took place in the sixth century of our era, during the Aztecs' wars against the Otomi and other hunting peoples who lived to the north and west of the Valley of Mexico. The figures placed near the hieroglyph perhaps indicate the founding of a few cities. The shields of each Aztec tribe are adorned with the coat of arms typical of each tribe and bear those leather and cotton-cloth appendices that are meant to deaden the blows of the spears and that one finds on some Etruscan vases.[1] The figures are arranged in a symmetrical order; one might be surprised to see them using their left hand rather than the right; but as we have had occasion to observe above, the two hands are often confused in Mexica paintings, just as they are in some Egyptian bas-reliefs.

1. See Plate XIV, number 2.

Plate XXXVIII

PLATE XXXIX

Granite Vase Found on the Coast of Honduras

These granite vases, four times larger than the drawing on Plate XXXIX, are preserved in England in the ˅collections of Lord Hillsborough and Mr. Brander. They were excavated on the Mosquito Coast in a country inhabited today by a barbarous people that does not think of sculpting stones. Images and descriptions of these vases are provided by ˅Mr. Thomas Pownall in the interesting Papers published by the London Society of Antiquaries.[1] I thought that I should reprint the drawings here to demonstrate the similarity between the decorative details with which they are covered and the details on the ruins of Mitla. This similarity completely belies the notion that these details were added after the conquest by Indians attempting to imitate the form of some Spanish vases. It is known that the Toltecs, passing through the province of Oaxaca, migrated all the way beyond Lake Nicaragua. One can therefore speculate that these vases adorned with bird and tortoise heads are the work of some tribe of the Toltec race. If one reflects for a moment on the style of the furnishings that the Spanish used in the sixteenth century, it is impossible to accept that Cortés's soldiers might have introduced to Mexico vases similar to those that Mr. Pownall brought to our attention.

239

1. ["Observations Arising from an Enquiry into the Nature of the Vases found on the Mosquito Shore in South America,"] *Archaeologia: or Miscellaneous Tracts Relating to Antiquity. Published by the Soc[iety] of Antiquaries of London*, Vol. V, Plate XXVI, p. 318.

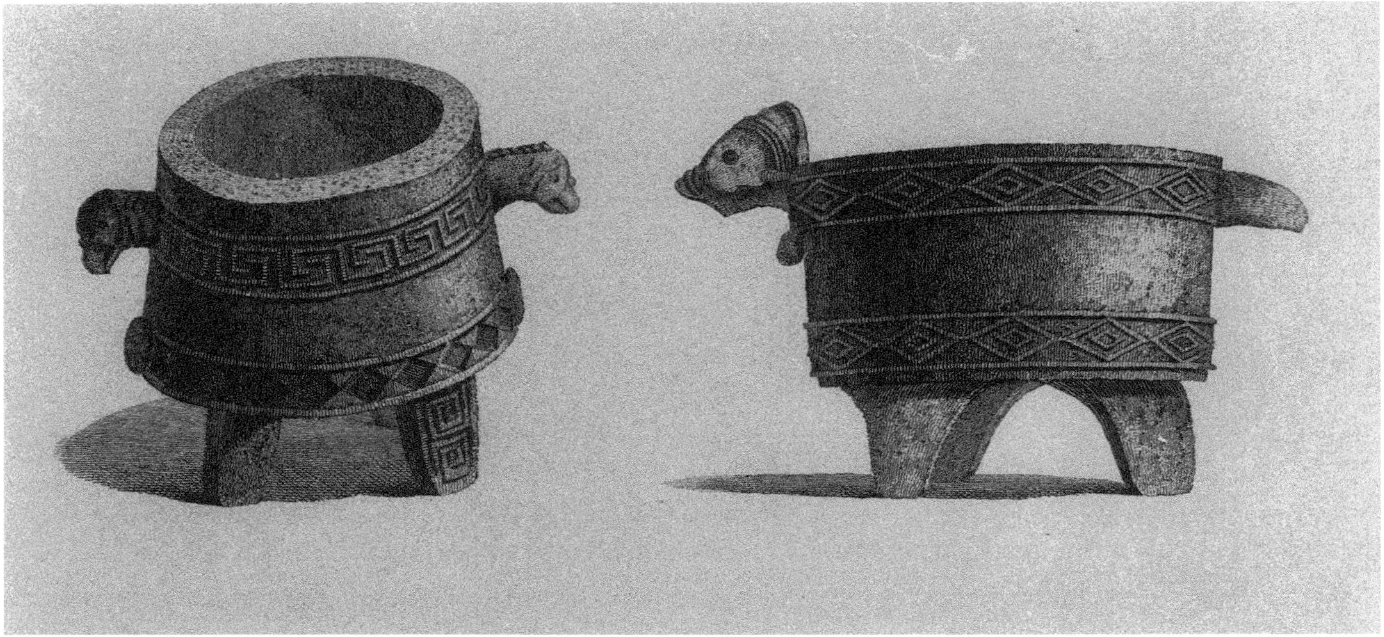

Plate XXXIX

PLATE XL

Aztec Idol in Basalt, Found in the Valley of Mexico

This small idol made of porous basalt, which I deposited in the cabinet of the King of Prussia in Berlin, recalls the bust of the priestess placed at the head of this work.[1] Here one recognizes the same headdress that resembles the *calantica* of the Isis heads, the pearls from California that line the forehead, and the bag attached with a bow, with two appendages hanging down from the bottom of the bag to the middle of the body. The circular hole visible in the chest appears to have been used to hold the incense (*copalli* or *xochitle-namactli*) that was burned on the idols. I do not know what the figure is holding in its left hand: the shapes are extremely inaccurate, and everything here is suggestive of artistic infancy.

1. Plates I and II, pp. 18 and 19 [this edition].

Plate XL

PLATE XLI
Air Volcano of Turbaco

To escape the summertime heat and diseases in Cartagena de Indias and on the arid coasts of Barú and Tierra Bomba, nonacclimated Europeans take refuge in the interior, in the village of Turbaco. This small Indian village is perched on a hill at the edge of a majestic forest that stretches to the south and to the east, all the way to the Mahates Canal and the Magdalena River. The houses are mostly made of bamboo and are covered with palm fronds. Here and there limpid springs bubble forth from the calcareous rock that contains a large amount of petrified coral debris; they are shaded by the glossy foliage of the *Anacardium caracoli*, a tree of colossal size that, according to the natives, has the property of attracting from afar the vapor spread throughout the atmosphere. Since the terrain of Turbaco is at over three hundred meters above sea level, it is delightfully cool there, especially at night. We stayed in this charming place in April 1801 after a difficult crossing from the island of Cuba to Cartagena de Indias, and while preparing for a long journey to Santa Fé de Bogotá and the Quito plateau.

240

The Indians of Turbaco, who accompanied us on our plant-collecting trips, often spoke to us of a swampy land that lay in the heart of a palm forest and that the Creoles called the Small Volcanoes, *Los Volcancitos*. They recounted that according to a legend preserved among them, this terrain had once been covered in flames, but the village curate, a good monk known for his great piety, had succeeded in extinguishing the underground fire through frequent aspersions of holy water; they added that the fire volcano had since become a water volcano, *volcán de agua*. Having lived in the Spanish colonies for quite some time, we were reasonably well acquainted with the bizarre and fabulous tales through which the natives enjoy drawing travelers' attention to natural phenomena. We know that these tales are typically less a product of the Indians' superstition than of that of Whites, Mestizos, and African slaves, and that the flights of fancy of a few individuals, who speculate on the progressive changes of the surface of the globe, take on the character of historic

Plate XLI

legends over time. Somewhat skeptical about the existence of a land formerly covered in flames, we nevertheless had the Indians lead us to the *Volcancitos de Turbaco*, and this excursion had phenomena in store for us that were much more significant than we had expected.

The *Volcancitos* are located six thousand meters to the east of the village of Turbaco, in a thick forest that abounds in *balsam of Tolu*, *gustavia* with waterlily-like flowers, and *Cavanillesia mocundo*, whose membranous, transparent fruit give the impression of lanterns dangling from the ends of its branches. The terrain rises gradually to a height of forty to fifty meters above the village of Turbaco; but since the ground is entirely covered in vegetation, one cannot determine the nature of the rocks superimposed on the shelly limestone. Plate XLI shows the southernmost part of the plain where the *Volcancitos* are found. This engraving was based on a sketch made by one of our friends, ꞌMr. Louis de Rieux. This young draftsman, with whom we traveled up the Río Grande de la Magdalena, was then accompanying his father, who, under the ministry of ꞌMr. de Urquijo, was responsible for inspecting the cinchona trees of Santa Fé.

At the heart of a vast plain bordered by *Bromelia karatas* rise eighteen to twenty small cones with a height of only seven to eight meters. These cones are of blackish-gray clay, and at their tips is a water-filled aperture. As one comes closer to these small craters, one hears an intermittent sound, dull yet quite loud, which occurs 15 to 18 seconds before a large quantity of air is released. The force with which this air rises to the surface of the water might lead one to assume that it is subjected to high pressure in the earth's interior. I counted about five explosions over two minutes. This phenomenon is often accompanied by a muddy ejection. The Indians assured us that the cones do not noticeably change shape over a large number of years; but both the upward force of the gas and the frequency of the explosions appear to vary in accordance with the seasons. Through analyses conducted with the help of nitrous gas and phosphorus, I found that the air released does not contain even half a percent of oxygen. It is a nitrogen gas that is purer than what we generally prepare in our laboratories. The physical cause of this phenomenon is discussed in the *Relation historique* of our voyage to the interior of the new continent.

241

PLATE XLII

Volcano of Cayambe

Of all the peaks of the Cordilleras whose height has been determined with some degree of exactitude, Cayambe is the second tallest after Chimborazo. Bouguer and La Condamine found its elevation to be 5901 meters (3,028 toises), and this result is confirmed by a number of angles that I took in the Ejido of Quito to observe the progress of terrestrial refractions at different hours of the day. The French academicians[1] named this colossal mountain *Cayambur* instead of Cayambe-Urcu, which is its real name: in Quechua the word *urcu* means mountain, like *tepetl* in Mexica and *gua* in Muisca. This error has been reproduced throughout all the works that present the most important elevations of the globe.

I have drawn Cayambe as it is seen from the Ejido of Quito, which is thirty-four thousand toises away. Its shape is that of a truncated cone; it recalls the contour of the *Nevado del Tolima* shown on Plate V. Among the mountains blanketed in permanent snow that surround the city of Quito, Cayambe is the most beautiful and the most majestic. One does not tire of admiring it at sunset, when the volcano of Guagua Pichincha, located to the west, toward the South Sea, casts its shadow on the vast plain in the foreground of this landscape drawing. This grass-covered plain is completely treeless. One sees only a few Barnadesia, Duranta, and Berberis shrubs, and those beautiful Calceolaria that belong almost exclusively to the southern hemisphere and the western part of the Americas.

A number of distinguished artists from the North have recently brought attention to the falls of the Kyrö River near the hamlet of Yervenkyle in Lapland, through which (as 'Maupertuis and Mr. Swanberg have observed) passes the polar circle. The equator crosses the top of Cayambe. This colossal mountain can be considered one of those eternal monuments through which nature has marked the great divisions of the terrestrial globe.

1. La Condamine, *[Journal d'un] Voyage à l'Équateur*, p. 163.

Plate XLII

PLATE XLIII

Volcano of Jorullo

The Plate on which I shall comment here brings to mind one of the most noteworthy catastrophes in the history of our planet. Despite the active exchanges between the two continents, this catastrophe has remained almost entirely unknown to the geologists of Europe. I have described it in my *Essai politique sur le royaume de la Nouvelle-Espagne*.[1]

According to my measurements, the Jorullo volcano is located at 19° 9' latitude and 103° 51' 48» longitude, in the intendancy of Valladolid to the west of Mexico City, at a distance of 36 leagues from the Ocean. It rises above the neighboring plains to an elevation of 513 meters (263 toises). Its height is thus three times that of Monte Nuovo di Pozzuoli, which emerged from the earth in 1538. My drawing shows Jorullo (Xorullo or Juruyo) surrounded by several thousand small basaltic cones, just as one sees it while descending from Areo and the hills of Aguasarco toward the Indian huts of the *Playas*. In the foreground one sees a part of the savannah where that enormous upheaval occurred on the night of September 29, 1759. The former level of the devastated land is now known as the *Malpaís*. The fractured layers, seen from a frontal perspective, separate the still-intact plain from the *Malpaís*. The latter, spiked with small, two- to three-meter high cones, covers an area of four square miles. At the place where the warm waters of the Cuitimba and the San Pedro descend toward the savannahs of the *Playas*, the elevation of the cracked layers is only twelve meters; but the raised land is shaped like a bladder whose arch becomes more convex toward the center, such that the ground at the foot of the great volcano is already at an elevation of 160 meters above the Indian huts at the *Playas de Jorullo* in which we stayed. Its profile, attached to the *Atlas géographique et physique* that will accompany my *Relation historique*, will make these differences in elevation easier to grasp.

1. Vol. I, p. 248. See also my *Receuil d'observ[ations] astr[ologiques]*, Vol. I, p. 327, and Vol. II, p. 521.

243

The cones are all *fumaroles* that emit thick fumes and make the surrounding air unbearably hot. In this extremely insalubrious country, they call these cones small ovens, *hornitos*. They contain clumps of basalt set in a mass of hardened clay. The slope of the great volcano is constantly in flames and covered in ashes. We went all the way into the interior of the crater by scaling the hill of scoriaceous branched lava that is shown on the left side of the engraving and that rises to a considerable height. We mention here as a noteworthy fact[1] that all of the volcanoes of Mexico are arranged on a single east-west line, which thereby forms a *parallel of the great heights*. When one considers this fact and compares it to what has been observed about the *bocche nuove* of Vesuvius, one is tempted to believe that the subterranean fire came to light through an enormous cleft that stretches in the earth's interior from the South Sea to the Atlantic Ocean at 18° 59' and 19° 12' latitude.

244

1. *Essai politique [sur le royaume de la Nouvelle Espagne]*, Vol. I, p. 47.

Plate XLIII

PLATE XLIV

Calendar of the Muisca Indians, the Ancient Inhabitants of the Bogotá Plateau

This stone, covered in hieroglyphic signs from the lunar calendar and show-ing the order of the intercalation that returns the start of the year to the right season, is a particularly remarkable monument since it is the work of a people whose name is almost entirely unknown in Europe and who have heretofore been confused with the roaming hordes of savages of South America. ˙Don José Domingo Duquesne de la Madrid, the canon of the archbishopric of Santa Fé de Bogotá, can be credited with the discovery of this monument. For a long time this cleric, a native of the kingdom of New Granada and a mem-ber of a French family that settled in Spain, had been the curate of an Indian village located on the plateau of the former Cundinamarca. His position en-abled him to win the trust of the natives, descendants of the Muisca Indians, and he attempted to collect everything that their legends had preserved about the state of these regions over the course of three centuries before the arrival of the Spanish in the new continent. He succeeded in obtaining one of the sculpted stones that the Muisca priests used to regulate the division of time; he became familiar with the simple hieroglyphs that referred to both the num-bers and the lunar days; and he displayed the full extent of his knowledge, 245 the fruits of his long and arduous studies, in a paper that bears the title *Dis-ertación sobre el calendario de los Muyscas, Indios naturales del Nuevo Reyno de Granada*. This handwritten paper was given to me in Santa Fé in 1801 by the famous botanist ˙Don José Celestino Mutis. Mr.Duquesne granted me permission to commission a drawing of the pentagonal stone, which he had attempted to explain, and it is this drawing that has been engraved on Plate XLIV. In offering here a few scattered ideas about the Muisca Indians' cal-endar, I shall make use of materials from the Spanish paper I have just men-tioned and to which I have added some reflections on the observable similari-ties between this calendar and the cycles of the Asian peoples.

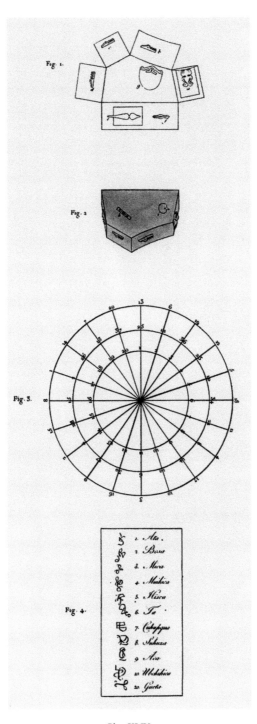

Fig. 1.

Fig. 2.

Fig. 3.

Fig. 4.

1. *Ata*.
2. *Bosa*.
3. *Mica*.
4. *Muihica*.
5. *Hisca*.
6. *Ta*.
7. *Cuhupqua*.
8. *Suhuza*.
9. *Aca*.
10. *Ubchihica*.
11. *Gueta*.

Plate XLIV

When, in 1537, the *Adelantado* Gonzalo Jiménez de Quesada, nicknamed The Conqueror, reached the high savannahs of Bogotá from the banks of the Magdalena, he was astounded by the contrast between the civilization of the mountain peoples and the savage state of the scattered hordes that lived in the hot regions of Tolu, Mahates, and Santa Marta. On this plateau, around four and five degrees latitude, where the centigrade thermometer constantly hovers between 17 and 20 degrees during the day and between 8 and 10 degrees at night, Quesada found the Muisca Indians, the Guanes, the Muzos, and the Colimas organized into communities, devoted to agriculture, and dressed in cotton cloth, while the tribes that wandered in the neighboring plains, at only a slight elevation above the Ocean surface, appeared to be dim-witted, deprived of clothing, without industry and arts.[1] The Spanish were stunned to find themselves transported to a land where, despite the relatively infertile soil, the fields everywhere provided abundant harvests of corn, Chenopodium quinoa, and *turmas* or potatoes. I shall not examine here whether, despite the introduction of grains and horned cattle, the Bogotá plateau is presently less populated than it had been before the conquest. I shall merely observe that when I visited the rock salt mines of Zipaquirá, I was shown undeniable traces of an ancient culture in plots of land to the north of the Indian village of Suba, which have not been reclaimed.

246   Among the different peoples of Cundinamarca, those whom the Spanish called the Muisca, or Mosca, appear to have been the most numerous. The mythical stories of this people date back to the distant time when the moon did not yet accompany the earth and when, during the flooding of the Funzha River, the Bogotá plateau formed a lake of considerable size. In the description of the Tequendama waterfall above,[2] we have spoken of this fabled man, known in American mythology by the name Bochica or Idacanzas, who opened up a passageway for the waters of Funzha lake, gathered the scattered people into communities, introduced the worship of the sun, and, similar to the Peruvian Manco Capac and the Mexica Quetzalcoatl, became the legislator of the Muisca. These same legends maintain that Bochica, son and symbol of the sun, high priest of Sogamozo or Iraca, advised the leaders of the various Indian tribes, who struggled over supreme authority, to choose as *zaque*, or sovereign, one of their own named Hunca-

1. *Historia general de las conquistas del Nuevo Reyno de Granada por el Doctor D. Lucas Fernández Piedrahita*, p. 15. (The author, who was bishop of Panama when he died, had drafted this history based on the manuscripts of Quesada the Conqueror; Juan de Castellanos, curate of Tunja; and the Franciscan monks Friar Antonio Medrano and Fr. Pedro Aguado.)

2. See p. 38 [this edition].

hua, revered for his fairness and his great wisdom. The high priest's counsel was universally followed, and Huncahua, who reigned for two hundred
fifty years, succeeded in bringing to heel the entire country that stretches
from the savannahs of San Juan de los Llanos to the Opon Mountains. Devoted to austere penance, Bochica lived for one hundred Muisca cycles,
or two thousand years. He disappeared mysteriously in Iraca, to the east
of Tunja. The latter city, which was then the most populous of all, had
been founded by Huncahua, the first of the dynasty of zaques of Cundinamarca. It took the name of its founder, Hunca, which the Spanish changed
to Tunca or Tunja.

The form of government that Bochica gave to the inhabitants of Bogotá is
quite remarkable for its similarity with the governments of Japan and Tibet.
In Peru, the Incas combined in their persons both secular and ecclesiastical powers. The sons of the sun were at once sovereigns and priests, as it
were. In Cundinamarca, at a time probably before Manco Capac, Bochica
had appointed as electors the four tribal chiefs, Gameza, Busbanca, Pesca,
and Toca. He had ordered that after his death, these electors and their descendants should have the right to select the high priest of Iraca. The pontiffs
or lamas, successors of Bochica, were supposed to inherit his virtues and his   247
saintliness. What Cholula was for the Aztecs during the time of Moctezuma,
Iraca became for the Muisca. The people gathered there in crowds to offer
gifts to the high priest. They visited the places that Bochica's miracles had
made famous; and in the midst of the bloodiest wars, the pilgrims enjoyed the
protection of the nobles whose territory they had to cross on their way to the
shrine (*chunsua*) and to the feet of the lama who resided there. The secular
leader called the *zaque* of Tunja, to whom the *zippa*, nobles, of Bogotá paid
tribute, and the pontiffs of Iraca were therefore two distinct powers, much
like the dairi and the secular emperor are in Japan. It seemed to me important
to record here these two historical concepts, virtually unknown in Europe,
in order to generate some interest in a people whose calendar we shall presently introduce.

Bochica was not merely regarded as the founder of a new religion and the
legislator of the Muisca: symbol of the sun, he also ruled time, and the invention of the calendar was attributed to him. He had also prescribed the order
of the sacrifices that were to be performed at the end of the short cycles on
the occasion of the fifth lunar intercalation. In the zaque's empire, day (*sua*)
and night (*za*) were divided into four parts, namely *sua-mena*, from sunrise
to noon; *sua-meca*, from noon to sunset; *zasca*, from sunset to midnight; and
cagui, from midnight to sunrise. In the Muisca language, the word *sua* or *zuhe*
refers both to the day and the sun. From *sua*, which is one of Bochica's nick-

names, comes *sue, European* or *white man*,[1] a strange name that derives from the fact that when Quesada arrived, the people regarded the Spanish as the sons of the sun, *sua*.

The smallest division of time among the Muisca was a period of three days. The seven-day week was unknown in the Americas, as it was in a part of Eastern Asia. The first day of the short period was reserved for a great market held in Turmequé.

The year (*zocam*) was divided by moons: there were twenty moons in a 248 *civil year*, which regulated daily life. The *priestly year* contained thirty-seven moons, and twenty of these long years formed a *Muisca cycle*. To distinguish between the lunar days, the moons, and the years, they used periodic series, the ten units of which were all numbers. Since the words used to refer to these units have several very remarkable features, we must discuss the language of Bogotá in some detail.

This language, which has almost disappeared from use since the end of the last century, had become predominant due to the victories of both the zaque Huncahua and the Zippa—and to the influence of the high lama of Iraca—over a huge area of the country, from the plains of the Ariari river and the Río Meta all the way to the north of Sogamoso. Just as the language of the Inca is called qquichua [*Quechua*] in Peru, that of the Mosca or Muisca is known in the country by the name *Chibcha*. The word *muisca*, of which *mosca* appears to be a corruption, means *man* or *person*; but the natives generally apply it only to themselves. The phenomenon illustrated by this expression is also exhibited by the Qquichua word *runa*, which refers to an Indian of the coppery race and not to a white person or a descendant of European colonists. The Chibcha or Muisca language, which at the time of the discovery of the new continent was one of the most widespread tongues of South America, along with the Incan and Carib languages, contrasts sharply with the Aztec language so remarkable for its reduplication of the syllables *tetl, tli,* and *itl*. The Indians of Bogotá or *Bacata* (farthest *edge of the fields* or *of the plowed ground*) know neither *l* nor *d*. Their language is characterized by the frequent repetition of the syllables *cha, che,* and *chu*, for example, in *chu chi,* us; *hycha chamique*, myself; *chigua chiguitynynga*, we must fight; *muisca cha chro guy*, a worthy man; here, the particle *cha*, added to *muysca*, which indicates the male sex.

The numbers in the Chibcha language—the first ten of which were selected as the terms of the periodic series appropriate for indicating the large

---

1. ▾Friar Bernardo de Lugo (professor of Chibcha in Santa-Fé de Bogotá), *Gramática de la lengua general del Nuevo Reyno llamada Mosca*, Madrid, 1619, p. 7.

and small divisions of time—are as follows: one, *ata*; two, *bozha* or *bosa*; three, *mica*; four, *mhuyca* or *muyhica*; five, *hicsca* or *hisca*; six, *ta*; seven, *qhupqa* or *cuhupqua*; eight, *shuzha* or *suhuza*; nine, *aca*; ten, *hubchibica* or *ubchihica*. Above ten, the Muisca Indians add the word *quihicha* or *qhicha*, which means *foot*. To indicate eleven, twelve, and thirteen, they say *foot one, foot two, foot three, quihicha ata, quihicha bosa, quihicha mica*, etc. These simplistic expressions suggest that after having counted on the fingers of both   249 hands, they continued with their toes. As we have seen above in our discussion of the calendar of the peoples of the Mexica race, the number twenty, which corresponds to the total number of toes and fingers, plays an important role in the American numerical system. In the Chibcha language, twenty is indicated either by *foot ten, quihicha ubchihica*, or by the word *gueta*, which derives from *gue, dwelling*. They continued with twenty-one, *guetas asaqui ata*; twenty-two, *guetas asaqui bosa*; twenty-three, *guetas asaqui mica*, etc., up to thirty, or *twenty plus (asaqui) ten, guetas asaqui ubchihica*; forty or two twenties, *gue-bosa*; sixty or three twenties, *gue-mica*; eighty or four twenties, *gue-muyhica*; one hundred or five twenties, *gue-hisca*. We shall recall here that beyond their units, which resembled the Etruscans' nails, the Aztecs only had a simple number or hieroglyph for twenty, for twenty squared or four hundred, and for twenty cubed or eight thousand. I would like to insist on the uniformity exhibited by the peoples of both Americas in the first stages of the development of their most basic ideas and in the methods through which they could graphically express numerical quantities above ten. This uniformity is deserving of our attention all the more in that it suggests a numerical system very different from what we find on the old continent, from the Greeks, whose notation was already less inaccurate than that of the Romans, to the Tibetans, the Hindus, and the Chinese, each of whom claims the honor of having invented numerals whose value changes depending on their position.

Among the large number of erroneous ideas that have been spread about the languages of peoples barely advanced in civilization, none is more extravagant than the assertion by [de] Pauw and other equally system-bound writers, who claim that none of the indigenous peoples of the new continent is able to count beyond three in its own language.[1] Today, we know the numerical systems of forty American languages, and the work of ˅Abbé Hervás, *L'Arithmétique de toutes les nations*, presents nearly thirty of them. In studying these various languages, one observes that when peoples emerge from

1. [de Pauw,] *Recherches philosophiques sur les Américains*, Vol. II, Part 5, section 1, p. 162 (1769 edition).

their original state of mindlessness, their subsequent advances no longer establish any significant difference in their manner of expressing quantities. The

250 Peruvians were at least as capable as the Greeks and the Romans of expressing numbers in the millions in their own language; they even had a noncompound word (*hunu*) for one million, of which there is no equivalent in any of the idioms of the old world. *Huc*, one; *iscay*, two; *qimça*, three . . . *chuncha*, ten ; *chuc huniyoc*, eleven ; *chunca iscayniyoc*, twelve . . . *iscaychunca*, twenty; *qimça chunca*, thirty; *tahuachunca*, forty . . . *pachac*, one hundred; *iscaypachac*, two hundred . . . *huaranca*, one thousand; *iscayhuaranca*, two thousand . . . *chunca-huaranca*, ten thousand; *iscay-chunca-huaranca*, twenty thousand; *pachachuaranca*, one hundred thousand; *hunu*, one million; *iscayhunu*, two million; *qimça hunu*, three million. . . . This same simple and regular progression manifests itself in several other American languages in which the numerical expressions have no other flaw than being extremely long and difficult to pronounce for the mouths of Europeans. The urge to count is clearly felt in a state of society that is still far away from what we so vaguely call the state of civilization.

According to the missionaries, a few of the peoples of the new continent whose numerical system is known to us are unable to count higher than twenty or thirty and refer to everything in excess of these numbers as *many*. At the same time, however, the missionaries assure us that to express one hundred, these peoples nevertheless make small piles of corn[1] of twenty kernels each; this obviously proves that the Yaruros of the Orinoco and the Guarani of Paraguay do count in *twenties*, like the Mexica and the Muisca, and that either stupidity or the extreme intellectual laziness inherent in even the most intelligent savages leads them to make *three twenties* or *four twenties* easier to enumerate by counting as children do, either with their toes and hands or by piling up kernels of corn. When travelers report that entire peoples in the Americas cannot count higher than five, one must not lend any more credence to this assertion than one would to that of a Chinese person who might boast that the Europeans cannot count higher than ten because "dix-sept" and "dix-huit" are compounds of ten and the basic units. One must not confuse the alleged impossibility of expressing large quantities with the limits

251 that the genius of different languages prescribes to the number of noncompound numerical signs. These limits are reached at five, ten, or twenty, depending on whether the peoples, while counting out the units, prefer to stop at the fingers of one hand, those of both hands, or all fingers and toes.

1. Hervás, *Idea del Universo: Aritmetica delle nazioni e divisione del tempo fra l'Orientali*, Vol. XIX, pp. 96, 97, and 106.

In the tongues of the American peoples, the furthest removed from the full development of their faculties, six is expressed as *four with two*, seven as *four with three*, and eight as *five with three*. The languages of the Guarani and the Lulos function in this manner. Other tribes that are already slightly more advanced—for example, the Omagua, and in Africa, the Wolof and the Fula—use words that mean at once *hand* and *five* just as we use the word ten; they express seven as *hand and two*, and fifteen as *three hands*. In Persian, *péndj* means five and *péntcha* hand. The Roman numerals also reveal a few traces of a quinary numerical system: the units accumulate until one arrives at five, which has a specific sign, as do fifty and five hundred.[1] Among both the Zamuca and the Muisca, *eleven* is expressed as *foot one*, twelve as *foot two*; but the remaining numerals are of tiresome length because instead of simple words they use childish circumlocutions; for example, they say *an entire hand* for five, *one from the other hand* for six, *both whole hands* for ten, and *the entire feet* for twenty. At times, this latter number is identical to the word for *man* or *person*, indicating that two hands and two feet form a complete person. Thus, among the Yaruro, *noenipume*, which derives from *noeni*, two, and *canipume*, man, means either *two men* or *forty*. The Sapibocono do not have a simple expression for one hundred or one thousand; for ten, they say *tunca*; for one hundred, *tunca-tunca*; and for one thousand, *tunca-tunca-tunca*. They form squares and cubes through reduplication, just as the Chinese sometimes form their plural and the Basques their superlative. Finally, the twenty-unit groups of the Muisca, the Mexica, and many other peoples and nations of the Americas are also found in the old world, among the Basques and the inhabitants of the Armorican peninsula. The latter two peoples count: one, *bat* or *unan*; two, *bi* or *daou*; three, *iru* or *tri*; twenty, *oguei* or *hugent*; forty, *berroguei* or *daouhgent*; sixty, *iruroguei* or *trihugent*. It is interesting to consider the small groups of five, ten, or twenty of these   252
numerical systems, which display such a wide range of nuances, but which nevertheless exhibit the same uniformity of features that characterizes all the inventions of humanity at the first stage of its social existence.

Mr. Duquesne has conducted several etymological studies on the number words in the Chibcha language. He insists that "all these words are meaningful, that they all have root words that are related either to the waxing or waning phases of the moon or to agricultural and religious objects." As there is no Chibcha dictionary, we are unable to verify the accuracy of this claim. One cannot be wary enough when it comes to etymological research, and we

---

1. Hervás, [*Idea del Unvierso,*] pp. 28, 96, 102, 105, and 127. ▾Mungo Park, *Voyage [dans l'intérieur de l'Afrique,]* Vol. I, pp. 25 and 95.

shall content ourselves with presenting here the meanings of the numbers one through twenty, just as they are shown in the manuscript that I brought back from Santa Fé. We shall merely add that Father Lugo, without entering into any other discussions on the numbers, reports in his *Grammar of the Chibcha language* that the word *gue* refers to a *dwelling* and that it is also found in its entirety in *gue-ata* (by elision, *gueta*), twenty, a dwelling; in *gue-bosa*, two-twenties, forty, or two dwellings; in *gue-hisca*, five twenties, one hundred, or five dwellings.

1. *Ata*, etymology uncertain: perhaps this word derives from an ancient root word that meant water, like the Mexica *atl*. Hieroglyph: a frog. The call of these animals, very common on the Bogotá plateau, announces that the time for sowing maize and quinoa is approaching. The Chinese designated their first *tse*, *water*, not by a frog but by a *water-rat*.

2. *Bosa*, around [environs]. The same word refers to a kind of enclosure intended to protect the fields from ravaging animals. Hieroglyph: a nose with flared nostrils, part of the lunar disk represented as a face.

3. *Mica*, variable; according to a different etymology, that which is chosen. Hieroglyph: two eyes open, also part of the lunar disk.

4. *Muyhica*, everything black, menacing storm cloud. Hieroglyph: two closed eyes.

5. *Hisca*, to rest. Hieroglyph: two joined figures, the marriage of the sun and the moon. Conjunction.

6. *Ta*, harvest. Hieroglyph: a stake with a rope, alluding to the sacrifice of the *Guesa*, who was tied to a column that was perhaps used as a gnomon.

7. *Cuhupqua*, deaf [or secret]. Hieroglyph: two ears.

8. *Suhuza*, tail. Mr. Duquesne does not know the meaning of this numeral, nor that of the following word.

9. *Aca*. Hieroglyph: two mating frogs.

10. *Ubchihica*, brilliantly shining moon. Hieroglyph: an ear.

20. *Gueta*, dwelling. Hieroglyph: a frog stretched out.

253   The numerical hieroglyphs are found engraved on Plate XLIV, figure 4; the explanations of them that we have just given are those preserved among a small number of Indians whom Mr. Duquesne found to be well acquainted with the calendar of their ancestors. Those who have studied the Chinese keys and the little that is known about their origin will not find the explanations of the American numerals completely fanciful. The extended use of the signs results in a gradual fading of their characteristic features. Who could recognize in the shapes of today's Hebraic and Samaritan letters those of the simple animal, dwelling, and weapon hieroglyphs that appear to have given rise to them? Our Tibetan and Hindu numerals, incorrectly called Arab numerals, likely

also contain a mysterious meaning. Among the Indians of Bogotá, *bosa*, *hisca*, *ubchihica*, and *gueta* undoubtedly retain a few image features. The latter hieroglyph is almost identical to the [East] Indian sign for four.[1]

It is interesting to find numerals among a semibarbarous people who had no knowledge either of the art of preparing paper or of writing itself. *Maguey* (Agave americana) is indigenous to both the Americas, and yet it is only among the peoples of the Toltec and Aztec race that the use of paper was as widely known as it has been, from the earliest times, in China and in Japan. When one recalls how many obstacles the Greeks and Romans faced in obtaining papyrus, even in a period when their literature had already attained its greatest brilliance, one feels a strange regret upon realizing that the raw material for paper was so common among the American peoples, who had no knowledge of symbolic script and only fanciful astrological concepts and the memories of an inhuman religion to bequeath to posterity, through their coarse paintings.

If it were true, as Mr. Duquesne claims, that in the Chibcha language the number words share roots with other words referring to the phases of the moon or items related to rural life, this would be one of the most remarkable items in the philosophical history of languages. One can imagine an occasional coincidental similitude in sounds between number words and objects that bear no relation to numbers, as in the French "neuf" (*novem*, or *nava*, in Sanskrit), nine, and "neuf" (*novus*, or *nava*, in Sanskrit), new; *acht*, an example from German, is *acht* [eight] and *achtung* [respect]; ἕξ [*hex*, six] and ἐξ [*ex*, out of], the preposition *de* [of]; *bosa*, which means *two* in Chibcha, and *bosa*, the preposition *for*; similarly, one can conceive of how in languages rich in figurative expressions, the words *two*, *three*, and *seven* might be connected with notions of the couple or pair (*jugum* [yoke]), omnipotence (the Hindus' *trimurti*), and magic and misfortune; but can one possibly accept that when the uncultivated man feels the initial urge to count, he calls four a *black thing* (*muyhica*); six, *harvest* (*ta*), and twenty, *dwelling* (*gue* or *gueta*), because according to the arrangement of a lunar almanac and the repetition of the ten terms of a periodic series, the *four* comes one day before the conjunction of the moon or because the harvest occurs *six* months after the winter solstice? In all languages there is a certain independence between the roots that express numbers and those that refer to other objects from the physical world, and we must assume that wherever this independence disappears, two numerical systems exist, one of which emerged later than the other, or, rather, that the etymological affinities one believed to have discovered are

254

1. Hager, "Memoria sulle cifre de la Cina," *Fundgruben des Orients*, Vol. II, p. 73.

only so in appearance because they are based on figurative meanings. Indeed, Father Lugo, who wrote in 1618, explains to us that the Muisca had two ways of expressing the number twenty and that they said either *gueta*, dwelling, or *quihica-ubchihica*, meaning *foot ten*; but we shall not enter here into discussions that are extraneous to the goal of this work. What we know for certain about the Muisca's lunar calendar and the origin of their numerical hieroglyphs does not require the support of arguments taken from the grammar of a language that can almost be regarded as a dead language.

We have seen above that the Muisca had neither the *decades* of the Chinese and the Greeks, nor the half-decades of the Mexica and the peoples of Benin,[1] nor the Peruvians' short periods of nine days, the Romans' *ogdoades*, or the Hebrews' eight-day weeks (*schebua*), which we also find in Egypt and in India, but which were unknown to the inhabitants of Latium and Etruria, as well as to the Persians and the Japanese. The Muisca week was distinct from all those presented by the history of chronology: it was only three days long. Ten of these groups formed a lunar month called *suna*, *great road*, *paved road*, *embankment*, because of the sacrifice that was performed every month at the time of the full moon in a public space to which led, in each village, a great road (*sina*) that started at the dwelling (*tithua*) of the tribal chief.

The *suna* did not begin at the new moon, as it did among the majority of the peoples of the old world, but rather on the first day after the full moon, whose hieroglyph as represented on the *intercalary stone* was a frog (Plate XLIV, figure Ia). The words *ata*, *bosa*, *mica*, and their graphic signs, arranged into three periodic series, were used to identify the thirty days of a lunar month, such that *mica* was at once the fourth, the fourteenth, and the twenty-fourth day of the month, like *quartidi* from the French Republican calendar. The same custom was found among the Greeks, who, however, added a few words to specify that the number belonged to either the *month's beginning*, μηνὸς ἀρχομένου [*menos archomenou*], the *middle of the month*, μηνὸς μεσοῦντος [*menos mesountos*] or the *month's end*, μηνὸς φθίνοντος [*menos phthinontos*]. Since the lesser festivals (*feriae*) or market days recurred every three days, each of them was ruled by a different sign throughout the course of a Muisca month, because the two periodic series of three and ten terms—those of the weeks and of the *suna*—do not have a common divisor and can coincide only after three times ten days. According to the following table, in which the lesser festivals are shown in italics, *cuhupqua* (both ears) falls in the last quarter; *muyhica* (both eyes closed) and *hisca* (the joining of

255

---

1. ▾Palin, *De l'étude des hiéroglyphes*, Vol. I, p. 52.

two figures, the marriage of the moon, *chia*, and the sun, *sua*) correspond to the period of the conjunction [new moon]; *mica* (both eyes open) refers to the first quarter; and *ubchihica* (one ear) to the full moon. The relationship we find here between the thing and the hieroglyph, between the phases of the moon and the lunar day-signs, manifestly demonstrates that these signs, which also served as actual numerals, were invented at a time when the artifice of the periodic series had already been applied to the calendar. Among the Egyptians, the number-hieroglyphs appear to have been independent from those for the lunar phases. According to ʿHorapollon, the image of a star indicated the number five, either because of the divergent rays that the naked eye perceives in stars of the first and the second magnitude or because of a mystical allusion to the world being governed by five stars. Ten was represented by a horizontal line placed on a perpendicular line. ʿMr. Jomard, a scholar fortunate enough to examine on site the monuments of Upper and Lower Egypt, who drew and described them carefully and who, through his position, was able to compare more hieroglyphs than any other antiquarian of our time, is currently engaged in a fascinating work on the Egyptians' numerical system.

256

The Lunar Days of the Suna of the Muisca Indians,
Divided into 10 Short Three-day Periods

	Ata
	Bosa
	Mica
	Muyhica
	Hisca
FIRST SERIES........	*Ta*
	Cuhupqua.* Last quarter
	Suhuza
	Aca
	Ubchihica
	Ata
	Bosca
	Mica
	Muyhica
	Hisca. Conjunction
SECOND SERIES......	Ta
	Cuhupqua
	Suhuza
	Aca
	Ubchihica

	⎧	*Ata*
	⎪	Bosa
	⎪	Mica.* First quarter
	⎪	*Muyhica*
	⎪	Hica
THIRD SERIES	⎨	Ta
	⎪	*Cuhupqua*
	⎪	Suhuza
	⎪	Aca
	⎩	*Ubchihica.** Full moon

Since twenty moons or *suna* formed the Muisca common year called *zocam*, one understands that the *zocam* was only a short lunar cycle, not a year in the true sense of the words *annus*, *annulus* [year, circle], and ἐνιαυτός [*eniautos*, year], both of which assume the return of a star to its point of 257 origin. The *zocam* and the long cycle of twenty leap years most likely originated in the preference for the year twenty, *gueta*. In addition to the *zocam*, the Muisca had an astronomical cycle, the *priestly year*, commonly used in religious festivals and containing thirty-seven moons, as well as a *rural year*, which was reckoned from one rainy season to the next.

The *suna* did not have any specific names, such as those we find among the Egyptians, Persians, Hindus, and Mexica; they were identified only by their number. This custom seems to me to have been observed the longest in eastern Asia; it has been preserved to the present day among the Chinese, and the Jews observed it until the period of Babylonian rule. But in their rural, civil, and religious calendars, the inhabitants of Cundinamarca did not count up to twelve, twenty, or thirty-seven; for the suna, as for the days within the same moon, they used only the first ten numbers and their hieroglyphs. The first month of the second agricultural year was ruled by the sign *mica*, three; the third month of the third year, by the sign *cuhupqua*, seven, and so forth. This predilection for periodic series and the existence of a sixty-year cycle, which is equal to the seven hundred forty *suna* contained within the cycle of twenty *priestly years*, appear to reveal the Tartar origin of the peoples of the new continent.

Since the rural year was supposed to be composed of twelve *suna*, the *xeques*, unbeknownst to the people, added a thirteenth month, similar to the *jun* of the Chinese,[1] to the end of the third year. The table of the Muisca moons that we shall provide demonstrates that through the use of the peri-

1. Souciet and Gaubil, *Observ[ations] mathém[atiques]*, Vol. I, p. 183.

odic series, this leap *suna* was, in the first intercalation, presided over by *cuh-upqua*. This sign was called the *deaf* [or secret] moon because it did not factor into the fourth series, which, without the use of a *supplementary term*, would necessarily have begun not with *suhuza* but with *cuhupqua*. This mode of intercalation, which is also found in the north of India and according to which two common lunar years of three hundred fifty-four days and eight hours are succeeded by an embolismic lunar year of three hundred eighty-three days and twenty-one hours, is the very one that the Athenians followed before Meton: it is the Dieteride [two-year cycle] in which they intercalated a Ποσειδεὼν δεύτερος [*Poseideōn deūteros*, second Poseideon] after the month   258
Poseidon. In his praise of Egyptians' solar calendar, Herodotus[1] very clearly expresses his opinion of this simple but quite imperfect practice: ὅσῳ῎Ελληες μὲν διὰ τρίτου ἔτεος ἐμβόλιμον ἐπεμβάλλουσι, τῶν ὡρέων εἵνεκεν [*hoso Hellenes men tritou eteos embolimon epemballousi, tōn hōreon heineken* (I believe the calculation of the Egyptians is cleverer than that of the Greeks), in that the Greeks added a leap month every third year because of the seasons].

We have seen above [in the table on p. 304] that the Mexica system of in-   259
tercalation was much more exact and highly regular, whereas the Peruvians adjusted their lunar year only from time to time, through observations of solstices and equinoxes carried out from cylindrical towers on the mountain of Carmenga near Cusco,[2] which they used to measure azimuths.

Among the Muisca the odd use of numbers, the series of which contains two fewer terms than the rural year has moons, explains the imperfection of a calendar in which, over a six-year period and despite the intercalation of the thirty-seventh month, *cuhupqua*, the harvest each year fell in a month bearing a different name. The *xeques*, therefore, announced every year which sign would preside over the *month of the ears of corn*, which corresponds to the *Abib* or *Nisan* of the Hebrew calendar. Since the power of one class of society is founded upon the ignorance of the other classes, the lamas of Iraca preferred to have a bizarre calendar in which the eighth month (October) was sometimes called the third month, at other times the fifth, and in which the differences in season (which, despite the proximity of the equator, are still quite noticeable on the Bogotá plateau) did not coincide with the *suna* of the same name. The priests of Tibet and Hindustan know how to take advantage of this multiplicity of the catasterisms that preside over the years, the months,

---

1. Herodotus, [*Histoire,*] Book II, ch. 4 (ed. Wesseling, 1763), p. 105. ᵛCensorin[us], *De die natali*, ch. 18. Ideler, *Histor[ische] Untersuchungen*, p. 176.

2. Nieremberg, [*Historia naturaes,*] p. 139. Cieza, [*La crónica del Perú,*] p. 230.

Three Types of Zocam in the Muisca Calendar

Rural Years of 12 and 13 moons		Priestly Years of 37 moons			Common Years of 20 moons		
I. *Ata*	1	I.	*Ata*	1	I.	*Ata*	1
	2		Bosa	2			2
	3		Mica	3			3
	4		Muyhica	4			4
	5		Hisca	5			5
	6		Ta	6		Harvest	6
COMMON YEAR	7		Cuhupqua	7			7
	8		Suhuza	8			8
	9		Aca	9			9
	10		Ubchihica	10			10
	11		*Ata*	11			11
	12		Bosa	12			12
II. *Mica*	1		Mica	13			13
	2		Muhica	14			14
	3		Hisca	15			15
	4		Ta	16			16
	5		Cuhupqua	17			17
	6		Suhuza	18		Harvest	18
COMMON YEAR	7		Aca	19			19
	8		Ubchihica	20			20
	9		*Ata*	21	II.	*Ata*	1
	10		Bosa	22			2
	11		Mica	23			3
	12		Muyhica	24			4
III. *Hisca*	1		Hisca	25			5
	2		Ta	26			6
	3		Cuhupqua	27			7
	4		Suhuza	28			8
	5		Aca	29			9
	6		Ubchihica	30		Harvest	10
▾EMBOLISMIC YEAR	7		*Ata*	31			11
	8		Bosa	32			12
	9		Mica	33			13
	10		Muyhica	34			14
	11		Hisca	35			15
	12		Ta	36		Embolismic month	16
Secret /deaf month	13		Cuhupqua*	37			17
IV. *Suhuza*	1	II.	Suhuza	1			18
	2		Aca	2			19
	3		Ubchihica	3			20
	4		*Ata*	4	III.	*Ata*	1

the lunar days, and the hours: they announce them to the people in order to levy a tax on their gullibility.[1]

The goal of the Muisca's intercalation was to ensure that both the beginning of the rural year and the festivals that were celebrated in the sixth month (the name of which was, consecutively, *suna ta*, *suna suhuza*, and *suna ubchihica*) always occurred in the same season. Mr. Duquesne thinks that the beginning of the *zocam* was (as among the Mexica, the Peruvians, the Hindus, and the Chinese) the first full moon after the winter solstice, but this tradition is uncertain. The first numeral, *ata*, represents water, symbolized by a frog. Among the Chinese, the first catasterism in the cycle of the *tse* is also that of *water*, and it corresponds to our sign Aquarius.[2]

Just as among the peoples of the Tartar race,[3] the sixty-year cycle, presided over by twelve animals, was divided into five parts, the Muisca cycle of twenty years of thirty-seven *suna* was divided into four short cycles, the first of which ended in *hisca*, the second in *ubchihica*, the third in *quihicha hisca*, and the fourth in *gueta*. These short cycles represented the four seasons of the long year. They each contained one hundred eighty-five moons, which corresponded to fifteen Chinese and Tibetan years and therefore to the genuine *indictions* commonly used in the time of ˅Constantine. In this division by sixty and by fifteen, the Muisca calendar more closely approximates that of the peoples of eastern Asia than does the Mexica calendar, which had cycles of four times thirteen (or fifty-two) years. Since each rural year of twelve or thirteen *suna* was identified by one of the ten hieroglyphs shown in the fourth figure, and since the series of ten and of twenty terms have a common divisor, the indictions consistently ended in the two signs of *conjunction* and *opposition*. We shall not pause here to demonstrate how citing the hieroglyph for the year, as well as the particular sixty-year cycle to which this year belongs, could have served to regulate the chronology; we have explained these methods while introducing the relationships among the Mexica, the Tibetan, and the Japanese calendars.

The beginning of each *indiction* was marked by a sacrifice whose barbaric ceremonies, according to the little we know about them, all appear to have been related to astrological concepts. The human victim was called *guesa*, *wanderer*, *homeless*, and *quihica*, *gate*, because his death announced, as it were, the opening of a new cycle of one hundred eighty-five moons.

260

1. Le Gentil, *Voyage dans les Indes*, Vol. I, p. 207.

2. See pp. 183–84 [this edition].

3. See pp. 175–76 and 199 [this edition]. Dupuis, *Origine des cultes*, Vol. III, Plate I, p. 44. Bailly, *Astronomie indienne et orientale*, 1787, p. 29.

This name recalls the Romans' *Janus*, stationed at the *gates* of heaven, and to whom Numa dedicated the first month of the year, *tanquam bicipitis dei mensem* [as the month of the two-headed deity].[1] The *guesa* was a child torn from his father's home. It was customary that he be taken from a specific village located in the plains that we call today the *Llanos de San Juan* and that stretch from the eastern slope of the Cordillera to the banks of the Guaviare.

261 It was from this very land in the *East* that *Bochica*, symbol of the *sun*, came when he first appeared among the Muisca. The *guesa* was raised with great care in the temple of the sun at Sogamoso, until the age of *ten*: at that point, he was taken out to walk the paths that Bochica had followed at the time when, traveling through these very places to teach the people, he had made these sites famous through his miracles. At *fifteen* years of age, when the victim had reached a number of *suna* equal to that contained within an *indiction* of the Muisca cycle, he was immolated in one of those circular public spaces with a tall column at its center. The Peruvians were familiar with gnomonic observations. They especially venerated the columns erected in the city of Quito because the sun, as they said, "positioned itself directly on their apex, and because the gnomon cast shadows there that were shorter than in the rest of the Inca empire." Were not the Muisca's stakes and columns, represented in some of their sculptures, also used to observe the length of the equinoctial and solsticial shadows? This supposition is all the more plausible since we find two instances of a rope attached to a stake among the ten *month-signs*, in the numerals *ta* and *suhuza*, and since the Mexica were acquainted with the use of the *wire gnomon*.[2]

During the celebration of the sacrifice that marked the *opening* of a new indiction or fifteen-year cycle the victim, *guesa*, was led in procession along the *suna* that gave the lunar month its name. He was led to the column that appears to have been used to measure the solsticial and equinoctial shadows and the sun's transits through the zenith. The priests, *xeques*, followed the victim; they were masked, like Egyptian priests were. Some of them portrayed Bochica, who is the Osiris or the Mithras of Bogotá and whom they believed to have three heads because, like the Hindus' *Trimurti*, he combined within him three persons who formed a single deity. Others bore the emblems of *Chia*, the wife of Bochica, Isis, or the moon. One group was covered in frog-like masks, an allusion to the first sign of the year, *ata*, while a final group portrayed *Fomagata*, the symbol of evil, with one eye, four ears, and a long

262 tail. This Fomagata, whose name in *Chibcha* means *fire* or *boiling molten*

1. Macrobius, [*Saturnalia*,] Book I, ch. 13.

2. On a sculpted stone found at Chapultepec. See Gama, *Descripción [histórica] de [las] dos piedras*, p. 100.

mass, was regarded as an evil spirit. He traveled by air between Tunja and Sogamoso and transformed humans into snakes, lizards, and tigers [jaguars]. According to other legends, Fomagata was originally a cruel prince. To ensure his brother *Tusatua*'s inheritance, *Bochica* had him dealt with on his wedding night just as Saturn dealt with Uranus. We do not know which constellation bore the name of this specter; but Mr. Duquesne believes that in the Indians' minds it was linked to the vague memory of the appearance of a comet. When the procession, which recalls both the *astrological processions*[1] of the Chinese and that of the festival of Isis, had reached the end of the *suna*, the victim was tied to the column, as we mentioned above. He was covered in a swarm of arrows, and his heart was ripped out as an offering to the *Sun King*, Bochica. The *guesa*'s blood was collected in sacred vessels. This barbaric ceremony bears a striking relation to the one the Mexica performed at the end of their long cycle of fifty-two years, which is represented on Plate XV.[2]

The Muisca Indians engraved onto stones the signs that presided over the years, the moons, and the lunar days. These stones, as we have noted above, reminded the priests, *xeques*, in which *zocam* or Muisca year a particular moon became intercalary. The petrosilex [felsite] stone, shown both in orthographic projection (in figure 1) and perspectivally in its true dimensions (in figure 2), appears to list the embolismic months of the first *indiction* of the cycle. It is pentagonal because this *indiction* contains *five* ecclesiastical *years* of thirty-seven moons each. It shows *nine* signs because *nine* Muisca years contain five times thirty-seven moons. In order to grasp fully the explanation that Mr. Duquesne gives of these signs, one must first recall that in an indiction of nine Muisca years and five Muisca months, the intercalary months, through the use of periodic series, fall on *cuhupqua, muyhica, ata, suhuza,* and *hisca*, respectively, and that no intercalation can occur in the first, third, seventh, or ninth year. What make these coincidences particularly evident are the three concentric circles shown in the third figure. The first circle, which is the innermost one, indicates the signs of the moons, *suna*; the second circle, in the middle, recalls in which Muisca year of twenty *suna* one of the signs contained within the series of ten terms became intercalary; finally, the outer circle determines the number of intercalations that take place over thirty-seven years. For example, if one asks in which *zocam* the sign *bosa* is intercalated, one finds that this is the sixth intercalation and that it happens in the twelfth year of the cycle.

Guided by Indians who still know the Muisca calendar signs, Mr. Duquesne believes he has identified the intercalations *ata, suhuza,* and *hisca*

1. Souciet, [*Observations mathématiques,*] Vol. III, p. 33.
2. See above, pp. 120–21 and 204 [this edition], Plate XV, number 8.

on three faces of the stone; these are the ones that occur in the nine years of twelve and thirteen *suna*, which correspond to the sixth, eighth, and tenth Muisca year of twenty *suna*. I do not know why the first two intercalations, *cuhupqua* and *muyhica*, are not marked there. What follows is an at times somewhat arbitrary interpretation of figures 1 and 2.

The headless frog, *a*, recalls that the indiction begins with the sign *ata*, the emblem of water. In *b*, *c*, and *d* are three small pieces of wood, each marked with three cross-cut lines. The one in the middle is not of the same height as the others, an indication that only six Muisca years are represented, after which the intercalation falls on *quihichata*, *e*, a tadpole frog provided with a long tail but without legs, *resting frog*. This emblem indicates that the month over which the animal presides is *useless*, or *unnecessary*, and does not factor into the twelve *suna* that elapse from one harvest to the next. The two frog figures, *a* and *e*, are placed on a kind of quadrangular platform. One might be skeptical about the interpretation of hieroglyph *e*, but Mr. Duquesne claims to have seen the same astrological symbol of a leap moon in several jade idols. In those idols, the legless animal was draped in the Indian tunic (*capisayo*) that is still commonly used among the lower classes. One recalls that among the Aztecs the *day-signs* even had their own altars.[1] Figures *f* and *h* indicate, through the eight cross-cut lines arranged in two groups of five and three, that in the eighth Muisca year they intercalated the moon ruled by *suhuza*. This sign is represented in *i* by a circle traced around a column with a rope. The Indians maintain that *f* and *h* represent serpents, which are emblems of time among all peoples. In *g*, the underside of the stone shows the sign *hisca*, which alludes to the marriage of Bochica and Chia,[2] a sign of lunar conjunction represented as a *closed temple*. This is the end of the first revolution of the cycle. The sacrifice of the *guesa* will shortly *reopen* the temple and begin the second indiction. The intercalation of *hisca* occurs after nine Muisca years, which are indicated by nine lines in *b*, *c*, and *d*. The lock with which the temple is shut is, by the way, the same as the one that the natives use today. It has holes on both sides, meant for two cylindrical pieces of wood. In comparing this lock to that of the Egyptians, sculpted on the walls of Karnak and widely used for thousands of years on the banks of the Nile,[3] one sees the same difference between the works of a crude people and those of an ingenious people that was both artistically and technologically advanced.

264

---

1. See above, p. 248 [this edition].
2. Plate XLIV, figure 4, number 5.
3. Denon, *Voyage en Égypte*, Plate CXXXIX, figure 14.

The Indians claim that four of these pentagonal stones explained the twenty intercalations of the *secret moon* that, according to the imperfect calendar of the Muiscas, took place in a cycle of seven hundred forty *suna*. This cycle contained twenty *priestly years* of thirty-seven years each, or sixty *rural years*; it is known to all the peoples that live to the east of the Indus, and it appears to be linked to the visible movement of Jupiter in the ecliptic. We have demonstrated above[1] that among the Hindus, the dodecatemoria of the solar zodiac originated in the nakshatras, the lunar zodiac, with each month taking the name of the lunar mansion in which the full moon occurred. We have also observed that both the twelve-year indictions and the nakshatra names given to these years are related to the heliacal rising of Jupiter. It is conceivable that at that remote time, when the first astronomical ideas were developed, people were stunned to see a planet crossing through the twenty-eight lunar mansions in almost the same number of years as, according to their observations, the lunar revolutions from one winter solstice to the next. To group together these *long years* of twelve solar years, they were obliged to use one of the numbers that, among all peoples, serve as resting points during enumeration; namely, 5, 10, or 20. Perhaps they gave preference to the smallest of these numbers, because 5 × 12 (or 60) is contained six times within the number 360, which was used in the division of the circle because of the three hundred sixty days that the most ancient peoples of the East attributed to the year represented by the emblem of the ring. Among the American peoples, for example the Mexica and the Muisca, we find four indictions instead of five; and this unusual preference for the number four is due to their interest in the solsticial and equinoctial points that designate the four seasons or *long weeks* of the *long year*.[2] The number of five intercalations, however, led the Muiscas to establish groups of fifteen rural years, four of which form the Asian cycle of sixty years.

265

If one follows the vague ideas that have reached us concerning the lunar signs carried in the procession of the *guesa* and the relationship between the constellation of the frog, *ata*, and the sign of *water* or the *water rat*, which among the Chinese and the peoples of the Tartar race opens the cycle of the catasterisms, one can speculate that, like the Mexica day-signs,[3] the ten hieroglyphs[4] *ata, bosa, mica*, etc. originally marked the division of a zodiac into ten parts. Among the Chinese we also find—and this is very important—

1. P. 182–83 [this edition].
2. See above, p. 199 [this edition].
3. See above, p. 198–99 [this edition].
4. Plate XLIV, figure 4.

a cycle of ten *can*, to which the Manchu give the names of ten colors.[1] It is likely that the Muisca's *can* also had specific names, and one may suspect that the numerals that Mr. Duquesne passed on to us alluded to those very names. All this leads me to assume that the numerical words *ata*, *bosa*, *mica*, etc. were only substituted for the names of the signs to indicate the *first sign* of the zodiac, the *second sign*, the *third sign*, etc., and that this substitution imperceptibly gave rise to the strange notion that the numbers themselves had meanings. This subject, which is not without interest to the history of the migrations of peoples, can be clarified only once a larger number of American monuments have been compared.

1. Souciet and Gaubil, [*Observations mathématiques*,] Vol. II, p. 135.

PLATE XLV 266

Fragment of a Hieroglyphic Manuscript Preserved at the Royal Library of Dresden

Following this very principle, namely, that an explanation for one monument can be found in another, and that in order to enter into the history of a people in greater depth, one must have in front of one all of the works infused with the character of that people, I decided to commission engravings (on Plates XLV–XLVIII) of fragments taken from the ▾Mexica manuscripts of Dresden and Vienna. The first of these manuscripts was entirely unknown to me when the printing of these pages began. It is not easy to provide a complete explanation for the hieroglyphic paintings that escaped the destruction with which they were threatened during the discovery of the Americas by the monks' fanaticism and the foolish insouciance of the first conquerors.[1] ▾Mr. Böttiger, an antiquarian who has conducted scholarly studies on the arts, mythology, and domestic life of the Greeks and the Romans, first brought to my attention the *Codex mexicanus* of the royal library at Dresden. He mentioned it quite recently in a work that offers the most advanced theories on the painting of the barbarous peoples, as well as on that of the Hindus, the Persians, the Chinese, the Egyptians, and the Greeks.[2] It is to the generosity of this scholar and to the particular kindness of ▾Count Marcolini that I owe the copy of the fragment shown on Plate XLV.

According to the information that Mr. Böttiger was good enough to pass on to me, this Aztec manuscript appears to have been purchased in Vienna by the librarian ▾Götze[3] during his literary travels in Italy in 1739. It is of *Metl* (Agave Mexicana) paper or pasteboard, like those that I brought back from New Spain. It forms a *tabella plicatilis* [accordion-folded plate] nearly six meters long and containing forty sheets covered in paintings on both sides. Each page is 0,295 meters (7 inches 3 lines) long and 0,085 meters (3 inches

---

1. P. 94 [this edition].

2. Böttiger, *Ideen sur Archäologie der Malerei*, Vol. I, pp. 17–21.

3. Götze, [*Merckwürdigkeiten*] *der Dresdner Bibliothek, erste Sammlung*, 1744, p. 4.

2 lines) wide. This format, similar to that of the ancient *Diptyches*, distinguishes the Dresden manuscript from those of Vienna, Velletri, and the Vatican; but what makes it especially remarkable is the arrangement of the simple
hieroglyphs, several of which are ordered into columns, as in true symbolic
script. In comparing Plate XLV with Plates XII and XXVII, one sees that
the *Codex mexicanus* of Dresden does not resemble any of the *ritual books*
in which the image of the astrological sign that presides over the *semilunar
month*, a short period of thirteen days, is surrounded by the catasterisms of
the lunar days. Here a large number of simple unconnected hieroglyphs follow each other, as in the Egyptian hieroglyphs and the Chinese keys.

In my view there is nothing that takes the lofty character of the works of
the Mexica to newer heights than the coarse paintings of sacred animals,
lying on the ground and pierced with arrows, that one sees at the bottom of
the first three pages of the manuscript. This similarity even extends to the
linear signs: these signs recall the *kouas* that, two thousand nine hundred
forty-one years before our era,[1] Emperor Tai-hao-fo-hi substituted for the
thin cords, or *quippu*, which we also find on the inscriptions of Rosetta, in
the interior of Africa, in Tartary, Canada, Mexico, and Peru. The *kouas* and
especially the *Ho-tous* are perhaps only a linear imitation[2] of the thin cords,
for the first of the eight trigrams also contains unbroken lines, like the hieroglyphs of the Dresden manuscript. We shall not decide whether the latter,
which contain dots intermixed with lines parallel to one another, express numerical quantities, as, for example, in a list of tributes, or whether they are
true cursive [continuous] characters.

1. ▼Julius Klaproth, [Preface to Hager, "Über die vor kurzem entdeckten Babylo-
nischen Inschriften,"] *Asiatisches Magazin*, 1802, Book I, pp. 91, 521 and 545.

2. Palin, *De l'étude des hiéroglyphes*, 1812, Vol. I, pp. 38, 107, 114, 120; Vol. V, pp. 19,
31, and 112. Souciet and Gaubil, *Observ[ations mathématiques]*, Vol. II, pp. 88 and 187;
Vol. III, p. 4, figure 7.

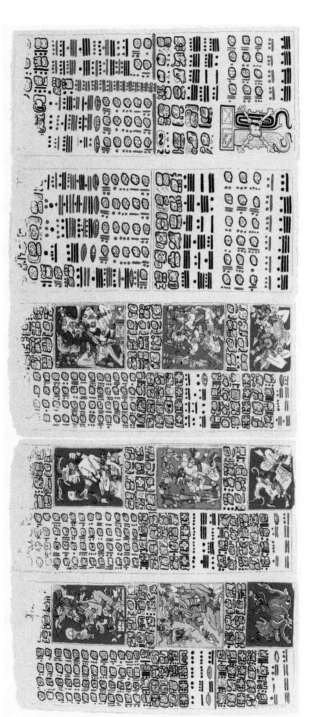

Plate XLV

Hieroglyphic Paintings from the Mexica Manuscript Preserved in the Imperial Library of Vienna, Numbers 1, 2, and 3

Of all the Mexica manuscripts that exist in the various libraries of Europe, the one in Vienna has been known the longest [to the European public]. It is the one that Lambeck and 'Nessel[1] described in their catalogs and of which Robertson commissioned an outline engraving. I had occasion to examine it during my most recent stay in Vienna, in 1811, and I owe the three-page colored copy shown on Plates XLVI, XLVII, and XLVIII to the good offices of a distinguished scholar, 'Mr. von Hammer [Purgstall], whose various works, especially the *Fundgruben des Orients* [Treasures of the Orient], have greatly contributed to facilitating the study of the relationships between the peoples of Central Asia and those of the Americas.

The *Codex Mexicanus* of the imperial library of Vienna is quite remarkable for its excellent state of preservation and for the strikingly vivid colors that distinguish its allegorical figures. Its outward shape makes it resemble the Vatican and Velletri manuscripts, which are folded in the same manner. It has fifty-two pages, and each page is 0,272 meters (10 inches 1 line) long and 0,220 meters (8 inches 2 lines) wide. The skin covering these hieroglyphic paintings is certainly not human skin, as has been erroneously suggested; it is likely that it is the skin of the Mazatl, which naturalists call the Louisiana Deer, common in the north of Mexico. The pages are shiny, as though they were varnished: this is the effect of a white, earthy coating on the skin. A similar coating is found on the Dresden manuscript, although the latter is made not of parchment but of *metl* paper. The *Codex Mexicanus vindobonensis* contains over a thousand human figures arranged in the most varied manner; one sees no trace of that uniform arrangement that one finds in the *Ritual books* of Velletri and the Vatican. Occasionally two figures are shown interacting with one another, but for the most part each figure is isolated and

1. Nessel, *Catal[ogus] Biblioth[ecae] Cæsaræ [Vindobonensis]*, Vol. VI, p. 163. See also above, pp. 95–96 [this edition].

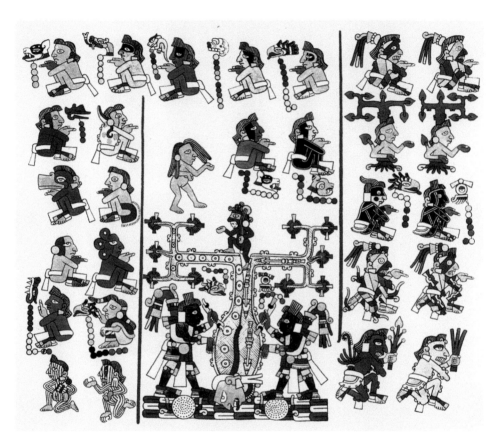

Plate XLVI

Plate XLVII

Plate XLVIII

appears to be pointing toward something with a finger. The thirteenth page is quite remarkable: divided by three horizontal lines, it manifestly shows that the Mexica read from right to left and from bottom to top, βουστροφηδόν [*boustrophedon*, as the ox plows, that is, having alternate lines written in the opposite direction]. Although the number of pages is equal to the number of years in a Mexica cycle, I was unable to find anything related to the recurrence of the four hieroglyphs that distinguish the years from one another. On nearly every page there is a representation of—in addition to the solstitial and equinoctial signs, *rabbit*, *reed*, *flint*, and *dwelling*—the catasterisms of the Jaguar, *Ocelotl*; the monkey, *Ozomatli*; and the *richly feathered eagle*, *Cozcaquauhtli*. These signs preside over the days and not the year. When one examines the remaining pages in groups of thirteen, one finds nothing periodic about them, and—what is especially striking—the dates, of which I counted 373 on the first twenty-two pages of the manuscript, are arranged in a manner that bears no relation to the order in which they are placed in the Mexica calendar. One finds *ome ehecatl* (1 *wind*) just before *matlactli calli* (10 *dwelling*), and *ce miquiztli* (1 *skull*) right next to *chicome miquiztli* (7 *skull*), although the days over which these signs preside are quite far apart from each other. If this manuscript deals with astrological matters, as is very likely, there is reason to be surprised that there are entire pages, for example the first and the twenty-second, that offer no indication of dates. If the latter were present, they would be easily recognizable by the circles that express the different terms of the periodic cycle of thirteen numerals.

Plate XLVI depicts a very strange symbolic figure, a man whose foot is trapped in the crack of either a tree trunk or a rock; on Plate XLVII, a woman spinning cotton; an isolated, bearded head; shells; a large bird, perhaps an *alcatras* [pelican] drinking water; a priest lighting the sacred fire through friction;[1] a man with a bushy beard, holding in his hand a kind of *vexillum* [a small piece of linen or silk attached to the upper part of a crozier], etc. Surrounded by ten other hieroglyphs, these same characters are repeated on Plate XLVII.

In looking at the Mexica's rough script, one cannot but observe that the sciences would stand to gain very little if anyone ever succeeded in deciphering what a people barely advanced in civilization recorded in its books. Despite the respect we owe to the Egyptians, who exerted such a powerful influence on the progress of enlightenment, one must also suspect that the numerous inscriptions drawn on their obelisks and on the friezes of their temples do not hold truths of great importance. However accurate they may

269

---

1. See above, p. 120 [this edition] and Plate XV, number 8.

be, such considerations should not, I think, lead us to neglect the study of symbolic and sacred script. The knowledge of such script is intimately linked to mythology, customs, and the particular genius of a people. It sheds light on the history of the ancient migrations of our species and is of great interest to the philosopher in that it offers him an image of the initial development of the faculties of humans on the farthest points of the earth in the consistent advance of the language of signs.

270

Ruins of Miguitlan, or Mitla, in the Province of Oaxaca; Survey and Elevation

After having described in this work so many barbarous monuments of purely historical interest, I feel some satisfaction in bringing attention to a building constructed by the Zapoteca, the ancient inhabitants of Oaxaca, and covered in decorative details of the most remarkable elegance. This building is known in the country as the *Palace of Mitla*. It is located in a granitic landscape ten leagues southeast of the city of Oaxaca, or Guaxaca, on the Tehuantepec road. *Mitla* is merely a contraction of the word *Miguitlan*, which means, in Mexica, *place of devastation, place of sadness*. This name seems well chosen for such a wild and gloomy site since, according to travelers' accounts, one almost never hears birdsong there. The Zapoteca Indians call these ruins *Leoba* or *Luiva, sepulcher*, an allusion to the caverns that are found below the arabesque-covered walls. I had occasion to discuss this monument in my *Political Essay on the Kingdom of New Spain*.[1]

According to the legends that have been preserved to this day, the main purpose of these structures was to mark the resting place of the ashes of the Zapoteca nobles. Upon the death of his son or his brother, the sovereign retired inside one of these dwellings, which are positioned above the tombs, to devote himself to grief and religious ceremonies. Other legends maintain that a family of priests, responsible for the expiatory sacrifices that were made for the rest of the dead, lived in this solitary place.

The floor plan of the *Palace*,[2] drawn by a very distinguished Mexica architect, ˅Don Luis Martín, shows that there were originally five separate buildings at Mitla, all arranged with great regularity. A very wide gate (6), some vestiges of which can still be seen, led to a spacious courtyard fifty square meters in size. Piles of dirt and the remnants of subterranean structures suggest that four small oblong-shaped buildings (8 and 9) surrounded the courtyard. The one on the right is still quite well preserved; one can even see the remnants of two columns.

271

1. Vol. I, p. 263.
2. Plate XLIX.

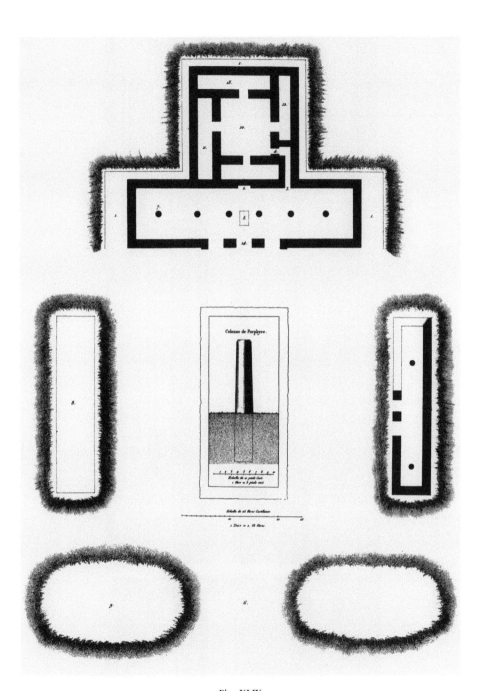

Colonne de Porphyre.

Echelle de 10 pieds Cast.
1 Pieu = 3 pieds cast.

Echelle de 28 Varas Castellanas.

1 Toise = 1, 38 Varas.

Plate XLIX

In the main building one can make out these features:

1. A terrace raised one to two meters above the level of the courtyard and surrounding the walls for which it also serves as a base, as one can see more clearly on Plate L;

2. A recess built into the wall, at a height of one and a half meters above the level of the *Column Room*. This recess, the width of which exceeds its height, likely contained an idol. The main door to the room is covered in a stone that is 4,3 meters long, 1,7 meters wide, and 0,8 meters high;

3 and 4. Entrance to the interior courtyard;

5 and 6. Shaft or tomb opening. A very wide staircase leads to a cross-shaped cavern held up by columns. The two galleries intersect at a right angle and are both twenty-seven meters long and eight meters wide. The walls are covered in Greek frets and in arabesques;

7. Six columns designed to support the *Sabino* beams that formed the ceiling. Three of these beams are very well preserved. The roofing was made of very large slabs. The columns, which testify to the formative stages of Zapoteca art and which are the only ones that have thus far been found in the Americas, do not have capitals. Their shaft is made of a single piece of stone. Some individuals very well versed in mineralogy told me that the stone is an exquisite amphibolic porphyry; others have insisted that it is porphyritic granite. The columns' total height is 5,8 meters, but they are buried up to a third of their height. I have had one column drawn separately and on a larger scale;

10. The interior courtyard;

11, 12, and 13. Three small chambers surrounding the courtyard, which are unconnected to a fourth chamber found behind the recess. The various parts of this building exhibit the most remarkable irregularities and flaws in symmetry. In the interior of the chambers one notices paintings of weapons, trophies, and sacrifices. Nothing suggests that there were once windows.

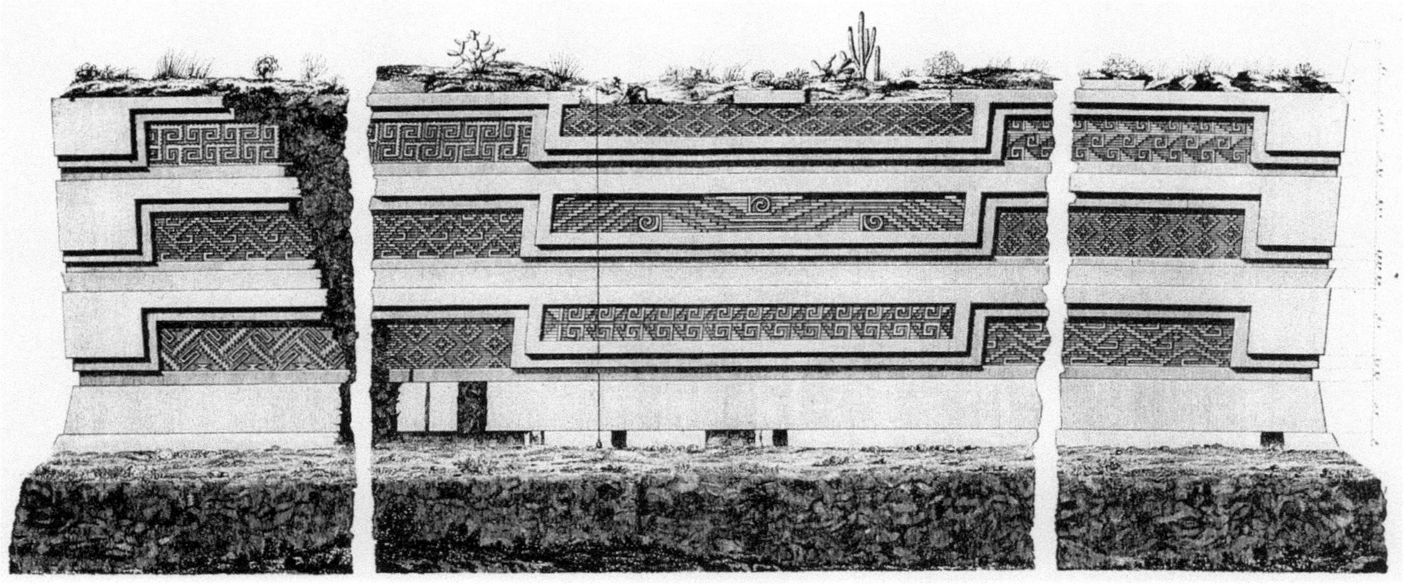

Plate L

Don Luis Martín and Colonel de la Laguna have very precisely drawn the *Greek frets*, *labyrinths*, and *meanders* that cover the outside of the palace walls of Mitla. These drawings, which are worthy of being engraved in their entirety, are in the hands of the Marquis of Branciforte, one of the most recent viceroys of New Spain. It was Mr. Martín, with whom I had the pleasure of conducting several geological excavations in the vicinity of Mexico City, who passed on to me the section I display on the fiftieth plate. It brings together three wall fragments and shows that the decorative details abutting one another are never alike. These arabesques[1] form a kind of mosaic composed of small square stones placed next to one another in a very artistic manner. The mosaic is affixed to a clay mass that fills the interior of the walls, as one also observes in some Peruvian buildings. The walls at Mitla do not stretch more than around forty meters in a single direction; their height likely never exceeded five to six meters. Even though it was fairly small, this building was nevertheless able to impress through the layout of its sections and the elegant form of its decorative details. Many temples in Egypt, near Syene, Philae, Elethyia, and Latopolis, or Esna,[2] are even smaller in scale.

273   Near Mitla are the remnants of a great pyramid and a few other structures that closely resemble those we have just described. Farther to the south, near Guatemala, in a place called *El Palenque*, the ruins of an entire city prove the predilection of the peoples of the Toltec and the Aztec race for architectural detail. We have no idea how old these buildings are; it is rather unlikely that they date back before the thirteenth or fourteenth century of our era.

The *Greek frets* of the Mitla palace surely bear an astonishing similarity to those of the vases of Magna Graecia and other decorative details that one finds across almost the entire old continent; but as I have already observed elsewhere, similarities of this kind prove very little about the former contacts between peoples, as humans in all regions have delighted in the *rhythmic repetition* of the same shapes, a repetition that forms the primary character of what we vaguely call *Greek frets*, *meanders*, and *arabesques*. What is more, the perfection of these details does not even suggest a highly advanced civilization among the people who used them. The interesting travels of the ˙Chevalier Krusenstern[3] introduced to us impressively elegant arabesques *tattooed* onto the skin of the fiercest inhabitants of the Washington islands.

1. Compare above, Plate XXXVIII, p. 277 [this edition].

2. [*Description de l'Égypte, monumens anciens,*] Vol. I, Plate XXXVIII, figures 5 and 6; Plate LXXI, figures 1 and 2; Plate LXXIII and Plate LXXXV.

3. Krusenstern, *Reise um die Welt* (Petersburg, 1810), Vol. I, p. 168. *Atlas*, Plates 8, 10, and 16.

PLATE LI

View of Corazón

Covered in perpetual snows, the mountain of Corazón took its name from the heartlike shape of its summit. I have drawn it just as it is seen from the *Alto de Poingasi* near the city of Quito. This *Nevado* is located in the western Cordillera between the peaks of the Pichincha and the Iliniza. One of the pyramids of the latter mountain[1] appears on the left-hand side, above the eastern slope of the Corazón. The seeming proximity of these two summits and the contrast between their shapes make for a unique vantage point.

It was on the peak of Corazón, before our trip to the Americas, that the mercury had been observed at the lowest point of the barometer. "Mr. Bouguer and I had departed," writes Mr. La Condamine in his historical introduction,[2] "in reasonably good weather; those whom we had left behind in our tents soon lost sight of us in the clouds, which, after we had plunged into them, were no more than fog for us. A cold, biting wind quickly covered us in ice; in several places we were forced to struggle up the rock with the help of our feet and hands; finally, we reached the summit. When we beheld each other there, with one entire side of our clothes, an eyebrow, and half of our beards spiked with small ice crystals, we were a strange spectacle. The mercury did not go above fifteen inches ten lines. No one has ever seen the barometer so low in the open air, and probably no one has ever climbed to such a great height: we were at 2,470 toises above sea level, and we can vouch for the accuracy of this determination, within four or five toises." 274

Now that we know the effect that temperature and the decrease in heat have on calculations made with a barometer, we are permitted to be somewhat skeptical about the precision of a measurement in which the margin of error would not even rise to 1/490 of the total height, although the calculation was made through the simple subtraction of logarithms. Mr. La Condamine

---

1. Plate XXXV.

2. *[Journal de] Voyage [. . .] à l'équateur*, p. 58. This excursion took place in July 1738.

Plate LI

did not have any instruments when he visited Rucu-Pichincha's crater. If at that point this famous astronomer had reached an elevation equal to that of a boulder of which I shall speak elsewhere, and on which I nearly perished, together with the Indian ˈFelipe Aldas, on May 26, 1802, then without knowing it, he found himself higher[1] than he had been on the peak of Corazón. According to Mr. Laplace's formula, the absolute elevation of that rock is 4858 meters (2,490 toises); it therefore surpasses the elevation of the point that the French Academicians measured in 1758 by nearly forty meters. Those scholars' determinations, moreover, are all affected by the uncertainty over the elevation of Caraburn's signal, to which Bouguer ascribes 2366 meters (1,214 toises) and Ulloa 1270 meters (1,268 toises).

1. See my *Recueil d'observations astronomiques*, Vol. I, p. 308.

PLATES LII AND LIII
Dress of the Indians of Michoacan

The Indians of the province of Valladolid, the former kingdom of Michoacan, are the most industrious in all of New Spain. They have a remarkable talent for carving small wooden figures and dressing them in garments made from the pith of an aquatic plant. This highly porous pith soaks up the most brilliant colors, and, cut in a spiral, it provides pieces of cloth of considerable size. I had brought back for ˟Her Majesty the Queen of Prussia a group of these Indian figures arranged intricately and with great care. This princess, who combined an enlightened appreciation for the arts with a very noble character, commissioned drawings of those of the figures that had suffered the least in shipment. These drawings are shown on Plates LII and LIII. In examining them, one is struck by the peculiar combination of the ancient Indian attire with the garb introduced by the Spanish colonists.

Plate LII

Plate LIII

PLATE LIV

View of the Interior of the Crater
of the Peak of Tenerife

Since the *Views of the Cordilleras* are also the *Picturesque Atlas* of the account of my voyage to the Tropics, one thought it fitting to add this plate, even though it bears no relation at all to the new continent. It shows the summit of the *Piton* or *Sugarloaf* that contains the *Caldera* of the Peak of Tenerife. One can make out the steep slope of the cone, covered in volcanic ashes; a circular lava wall surrounding the crater, which is now no more than a fumarole; and a wide gap in the wall on the western side. I sketched this drawing from a purely geological perspective: the lithoid lava is eroded by the constant action of the sulfuric acid fumes and stacked up in layers, like the beds in mountains of the secondary formation.

Similar to those on the rim of the ancient crater of Vesuvius, at the *Somma* [collapsed Mount Somma], these layers appear to be the result of successive eruptions. They are a combination of vitrified lava, an obsidian-based porphyry, and pechstein. For centuries the Peak of Tenerife, whose perpendicular height is over nineteen hundred toises, has produced only lateral eruptions. The most recent of these was that of Chahorra, which occurred in 1798. When one sees the enormous amount of ejecta from the Peak in the plain, which is covered in *Spartium nubigenum*, one is surprised at the small size of the crater from which supposedly so much ash, pumice, and chunks of volcanic glass were spewed out; but *Mr. Cordier, who of all the mineralogists has spent the most time on the island of Tenerife, has made the important observation that the current crater, the *Caldera* of *Pitón*, is not the main opening of the volcano. That learned traveler has found on the Peak's southern slope a funnel-shaped crater of enormous size that appears to have played the primary role in the past eruptions of the volcano of Tenerife.

276

Plate LIV

SUPPLEMENT

PLATES LV AND LVI

Fragments of Paintings from the Codex Telleriano-Remensis

The Library of Paris does not have an original of a Mexica manuscript, but a very precious volume is preserved there in which a Spaniard, who lived in New Spain, copied a large number of hieroglyphic paintings, either toward the end of the sixteenth century or at the beginning of the seventeenth. These copies were typically made with great care: they bear the character of the original drawings, as can be judged from the symbolic figures that also appear in the manuscripts of Vienna, Velletri, and Rome. This little-known volume,[1] from which we have excerpted the fragments shown on Plates LV and LVI, once belonged to Le Tellier, the archbishop of Reims; it is not known through which channels it came into his possession. On the outside it resembles the manuscript preserved in the Vatican Library under the number 3738. Each hieroglyphic figure is accompanied by several written explanations from different periods, it appears, and in both Mexica [Nahuatl] and Spanish. It is likely that these notes, which shed light on the history, chronology, and religion of the Aztecs, were composed in Mexico by some Spanish cleric both from his own observations and as told by the natives. They are more informative than those found in the *Raccolta di Mendoza*, and the Mexica names in them are reproduced much more correctly.

The *'Codex Mexicanus Tellerianus* contains a copy of three different works, the first of which is a ritual almanac, the second a book of astrology, and the third a history of Mexico from the year 5 *tochtli*, or 1197, to the year 4 *calli*, or 1561. We shall give here a brief overview of these three manuscripts. 280

1. *Ritual book.* Here are found the images of twelve Toltec and Aztec deities, the main festivals that gave their name to the eighteen months of the year; for example, the festivals of Tecuilhuitontl, *all the lords*; of Micaylhuitl, *all*

---

1. Manuscript of 96 pp. in-folio, bearing the title *Geroglyficos de que usavan los Mexicanos* (Codex Telleriano-Remensis 14, Reg. 1616).

Plate LV

Plate LVI

the dead; of Quecholi, etc. The hieroglyph for the five supplementary days[1] ends the series of festivals. In his notes the owner of the manuscript followed the misguided system, which maintains that the Mexica year began eighteen days before the spring equinox.

2. *Astrological part*. Here one finds the days that should be considered neutral, lucky, or unlucky. Among the latter are eleven that the Mexica believed to be very perilous to domestic peace. Husbands had to fear women born in this period, and one may assume that the latter took great pains to hide either the astrological almanac or the date of their birth. Although regarded as the effect of an impartial destiny, infidelity was nonetheless harshly punished by the law. A rope was tied around the adulterous woman's neck, and she was dragged into a public square, where she was stoned to death in front of her husband. This punishment is represented on the ninth sheet[2] of the manuscript.

3. *Annals of the Mexica Empire*. These cover three hundred sixty-four years. This part of the work, unknown to Boturini, Clavijero, and Gama, and apparently of the utmost authenticity, merits consideration by anyone who wishes to undertake a classical history of the Mexica peoples. These annals report only a very small number of events from the year 1197 to the mid-sixteenth century, often barely one or two in a thirteen-year interval. After 1454, the narration becomes more detailed, and from 1472 to 1549, it recounts in detail, and almost year by year, everything noteworthy about the physical and political state of the country. Missing are the pages covering the periods from 1274 to 1385, from 1496 to 1502, and from 1518 to 1529. The Spaniards' entry into Mexico City fell into this last interval. The paintings are coarse but often of a great naiveté. Among the items worthy of attention, we shall mention the image of King Huitzilihuitl, who, having had no legitimate children with his wife, took as a mistress a female painter[3] who died[4] in the year 13 *tochtli*, or 1414; the snowfalls[5] that occurred in 1447 and 1503, which caused widespread death among the natives by destroying their seeds; the earthquakes of 1460,[6] 1462, 1468, 1480, 1495, 1507, 1533, and 1542; the solar eclipses[7] of

281

1. Plate LV, figure 1.
2. Same Plate, figure 2.
3. Plate LV, figure 3.
4. Same Plate, figure 4.
5. Same Plate, figures 5 and 6.
6. Same Plate, figure 7, and Plate LVI, figure 2.
7. Plate LVI, figure 7.

1476, 1496, 1507, 1510, 1531; the first human sacrifice;[1] the appearance of two comets in 1490[2] and in 1529; the arrival[3] and death[4] of the first bishop of Mexico City, Fray Juan Zumárraga, in 1532 and 1549; the departure of ᵛNúño de Guzmán[5] for the conquest of Jalisco; the death of the famous Pedro [de] Alvarado, whom the natives called Tonatiuh, the *sun*, because of his blond hair;[6] the baptism of an Indian by a monk;[7] an epidemic that depopulated[8] Mexico under the viceroy Mendoza in 1544 and 1545; the uprising and punishment[9] of the Blacks of Mexico City in 1537; a storm that devastated the forests;[10] the ravages that smallpox[11] made among the Indians in 1538, etc.

Although the annals of the Le Tellier Manuscript are in agreement with the chronology the Abbot Clavijero adopted in his treatise in the fourth volume of his history of ancient Mexico, the system of correspondence that the former presents between the Aztec and the Christian years nevertheless differs quite sharply from the one Boturini and Acosta followed. The annals begin in the year 5 *tochtli*, or 1197, at the time of the Mexica's arrival in Tula, which is the northern border of the valley of Tenochtitlan. The great comet, whose appearance is indicated next to the hieroglyph for the year 11 *tochtli*, or 1490, is the one that was seen as an omen of the Spaniards' arrival in the Americas. On this occasion, Montezuma, annoyed with the court astrologer, had him put to death.[12] The sinister omens continued until 1509, when, according to the Le Tellier Manuscript, a bright light was seen toward the east for forty nights. This light, which appeared to radiate from the earth itself, was perhaps the zodiacal light, the brightness of which is both very strong and very uneven in the Tropics. The people regard the most common phenomena as novel whenever superstition decides to attach a mysterious meaning to them.

The comets of 1490 and 1529 were either comets that appeared near the South Pole or comets that, as ᵛFather Pingré[13] reports, were also seen in Eu-

282

1. See above, p. 116 [this edition].
2. Plate LV, figure 8.
3. Plate LVI, figure 1.
4. Same Plate, figure 6.
5. Plate LV, figure 9.
6. Plate LVI, figure 4.
7. *Ibid.*
8. Plate LVI, figure 5.
9. Same Plate, figure 2.
10. Same Plate, figure 5.
11. Same Plate, figure 3.
12. Clavijero, [*Storia antica del Messico*,] Vol. I, p. 288.
13. [Pingré,] *Cométographie*, Vol. I, pp. 478 and 486.

rope and China. It is noteworthy that the hieroglyph for a solar eclipse[1] is composed of the disks of the moon and the sun, one of which moves in front of the other. This symbol demonstrates a precise understanding of the cause of eclipses; it recalls the Mexica priests' allegorical dance that enacted the devouring of the sun by the moon. The eclipses of the latter star, which correspond to the years *Matlactli Tecpatl*, *Nahui Tecpatl*, and *Ome Acatl*, are those of February 25, 1476, August 8, 1496, January 13, 1507, and May 8, 1510; these are all fixed points of reference for Mexica chronology. *L'Art de vérifier les dates* makes no mention of any solar eclipse in the year 1531, while our annals indicate one for *Matlactli Ome Acatl*, which corresponds to that year in our era. Mexica historians used the eclipse of 1476 to establish the time of king Axayacatl's victory over the Matlatzinca; it is about this very eclipse that Mr. Gama made such a large number of calculations.[2]

283 I do not know to which phenomenon[3] the commentary often refers with the phrase "in this year, the star gave off smoke." The Orizaba volcano was called Citlaltepetl, *mountain of the Star*, and one may well believe that the Annals of the Empire recorded the various eruption periods of this volcano. Nevertheless, on page 86 of the Le Tellier Manuscript, it is expressly stated that "the star that smoked, *la estrella que humeaba*, was *Sitlal choloha*, which the Spanish call Venus, and which was the subject of a thousand fantastic stories." Yet I wonder: what optical illusion might have lent Venus the appearance of a smoking star? Could this be a reference to a kind of halo formed around the planet? Since the Orizaba volcano is located to the east of the city of Cholula and its inflamed crater at night resembles a rising star, there was perhaps some confusion in the symbolic language between the volcano and the morning star. The name that Venus bears today among the natives of the Aztec race is *Tlazolteotl*.

1. Plate LVI, figure 7. See p. 212–13 [this edition].

2. Gama, *Descripción [histórica] de [las] dos piedras*, pp. 85–9. Torquemada, [*Monarquía indiana*,] Vol. I, book II, ch. 59. Boturini, [*Idea de una nueva historia*,] § 8, n. 13.

3. Plate LVI, figure 2.

PLATE LVII

Fragment of a Christian Calendar from the Aztec Manuscripts Preserved in the Royal Library of Berlin

This is a hieroglyphic calendar made after the arrival of the Spanish that we mentioned at the beginning of this work.[1] The paper is made of *metl*; the figures are in outline and are lacking colors, as in some strips from Egyptian mummies; this is script rather than painting. The festival days are indicated by the dots that indicate the units. The Holy Spirit is represented in the form of the Mexica eagle, *cozcaquauhtli*. "At the time when this calendar was composed, Christianity was blending with Mexica mythology; the missionaries not only tolerated but even, to a certain extent, encouraged this mixture of ideas, symbols, and worship. They persuaded the natives that the Gospel had already been preached in the Americas in very distant times; they searched for evidence of this in the Aztec rite with the same ardor with which present-day scholars devoted to the study of Sanskrit discuss the similarity between Greek mythology and mythologies from the banks of the Ganges and Brahmaputra."[2]

1. Pp. 90–100 [this edition].
2. *Essai politique sur [le royaume de] la Nouvelle-Espagne*, Vol. I, p. 95.

Plate LVII

Hieroglyphic Paintings from the
Raccolta di Mendoza

These plates serve to cast some light on what we have said above about the rite and the customs of the ancient Mexica.[1] We know of no better way to introduce the interesting manuscript known as the *Raccolta di Mendoza* than to quote here Mr. Palin's explanation of it from his *study of hieroglyphs*. We are far from accepting without reservation the comparisons that clever author made; but we think that it is a beautiful and fruitful idea to consider all the peoples of the earth as belonging to the same family, and to identify in the Chinese, Egyptian, Persian, and American symbols examples of a language of signs that is common, as it were, to the entire species, and that is the natural product of the intellectual faculties of humankind.

"The collection, preserved by Purchas and Thévenot, presents in three parts the founding of the city and its growth from the conquests of its nobles; the support it received from the tributes that the conquered cities pay; its institutions; and a glimpse into the life of its citizens. All of this is evident at first glance: one first makes out the ten leaders of the founding colony of the empire, with the symbols for their names above their heads. They arrive near the objects that form the arms of Mexico City; the stone surmounted by an Indian fig cactus [*Opuntia*], on which sits an eagle,[2] recalls the eagle perched on a tree and the cup that the god Astrochiton gave as the identifying sign for the site where Tyre[3] was to be built. A house or dwelling represents the new city,[4] a shield with arrows, the armed occupation.[5] The symbols next to two other houses surrounded by combatants reveal to us the names of the first   285
two conquered cities. The rest of the history is composed in the same spirit

1. Pp. 97–98 [this edition].
2. Plate LVIII, figure I.
3. Nonnus, [*Dionysiaca*,] XL, verse 4773.
4. [Nils Palin,] *Monum[entum trouvé à] Rosette*, and Denon, [*Voyage*,] Plate CXXXIII.
5. Horapollon, [*Hieroglyphica*,] II, 5, 12.

and consists of similar components: everywhere one sees arms, the instrument of the conquest, between the figures of the conquering nobleman and the subjugated cities, with the symbols for their names and the years. The latter are arranged next to the representation of each event in a kind of frame that surrounds each tableau, and which contains the hieroglyphs of a chronological cycle of fifty-two years.

"The notes of the tributes form the second part of the *Mendoza Collection*, which is composed of the names of the tributary cities and the items that each was required to deliver in kind to the treasury and to the temples designated at the top of this list by the symbol for *calli*. These items consist of all the useful products of nature and art: gold,[1] silver, and precious stones; arms, mats, mantles, and coverings;[2] animals and birds, feathers; cocoa beans, corn, and vegetables; colored paper, borax, salt, etc. They were represented either by showing the container in place of the content (such as vases,[3] baskets, sacks, crates, and bundles of certain shapes) or through a representation of their own shapes. The quantity is expressed by means of number signs that refer to the units with points and balls, to multiples of twenty[4] with a character also found among the hieroglyphs; four hundred, or twenty times twenty, with an ear of corn,[5] a pineapple, or a feather replete with grains of gold; twenty times four hundred, or eight thousand, with a bag,[6] a value that presumably goes back to the custom of storing this many thousand cocoa beans in a sack. This is the same way in which a sum of money was designated in the Byzantine Empire and is still designated in the Ottoman states.

"This method and these names indicate the origin of the number symbols in the Mexica book. One can see how many similarities this painting, which depicts a state of a primitive society, has with the historical inscriptions in the ruins of Thebes, which Tacitus mentions, and in which a long list of conquests was also followed by that of the tributes paid in kind by the subjugated peoples.[7] Like the religious precepts concerning the mysteries, the laws were

286

1. Plate LVIII, figure 5.
2. Plate LVIII, figure 9.
3. Plate LVIII, figure 6.
4. Plate LVIII, figure 5.
5. Plate LVIII, figure 10.
6. Plate LVIII, figure 16.

7. *Legebantur et indicta gentibus tributa, pondus argenti et auri, numerus armorum equorumque, et dona templis, ebur atque odors, quasque copias frumenti et omnium utensilium quæque natio penderet.* [Also read were the tributes imposed upon the people, the weight of silver and gold, the number of arms and horses, and, as offerings for the temple, ivory and incense, furthermore what quantities of grain and items of practical use of all sorts each people had to render.]

exhibited inside the temples and on mummy caskets, much like the tableaus of the Eleusinian mysteries, copied from those of Egypt, which traced life from the cradle to the verge of death.[1]

"A number of Mexica laws form the third part of the manuscript that we are examining, and they cover the entire life of the citizens, placing before their eyes a tableau of all the actions prescribed by the law, thereby presenting them with models. Just as the hieroglyphs for amulets presuppose the optative mode, one has to read this entire passage as an imperative: that the mother teach the child in the cradle by addressing him with speech, is represented by a tongue; that the child be placed in the cradle from the first day after his birth, marked by a flower attached to the cradle and accompanied by three others; that the midwife, after having dedicated the child to the gods,[2] bathe him on the fifth day in the courtyard in the midst of arms and the tools necessary for the work done by his gender. This ceremony is performed in front of three children (who represent children in general); they name the newborn and celebrate his birth by eating corn.[3] In the Rosetta inscription a decree ordains this very thing and through a similar image, in which the three celebrants are brought together with the three flowers to form the character for the celebration of the day of birth, which is also represented by the sunrise.[4] All of the details of this tableau, or this table of Mexica law, recall the baptism of the proselytes of Judaism in the presence of three witnesses, as well as the αμφιδρομια [amphidromia, celebration at the naming of a child] of the Greeks, in which the child, on the fifth day after his birth, was dedicated to the gods and given a name after expiatory ceremonies. In this first part the law still ordains that the parents present the child in the cradle to the high priest and the master of arms, and that they think about his future purpose. His education is prescribed within the subsequent tableaus, which explain oral instruction and the rations of a half and a whole [unleavened] cake (with the hermetic mark of seven[5]) that parents are supposed to give their children at three and four years of age, respectively. The number of years is marked by circles, as in both the hieroglyphs and the language of the Romans. At age five the boy is hauling loads, and the girl is watching her mother spin; at age six she herself spins and, like the boy, receives a cake and a half at each meal. At eight years the instruments of punishment are shown to the

287

1. Themistius in ᵛStobaeus, Serm[ones] 119, p. 104.

2. With five prayers to the gods, masters of the heavens and the waters, to all the gods, to the moon and the sun.

3. Plate LIX, figure 1.

4. [Palin,] Analyse de l'inscr[iption] en hiéroglyphes du monument trouvé à] Rosette, p. 145.

5. Plate LIX, figure 2.

disobedient and lazy children; they are threatened; but it is not until ten years that they are actually punished.[1] At thirteen and fourteen years, the children of both sexes share their parents' work; they row, they fish, or they cook and make textiles.[2] At fifteen years, the father presents two sons to two different instructors at the temple and the military college; this is the age for choosing a trade; the girl receives one by marrying. Henceforth, the years are no longer counted: one sees the young man follow and serve the priests and the warriors, receiving both instruction and punishments in this dual career. He achieves the honors of the positions, the blazoned shields that are the marks of great deeds, the red ribbon tied around the head of the initiated knight, and the other distinctions that the sovereign grants for valor, according to the number of prisoners taken; all of these ranks are depicted, from the simple soldier to the top leaders and the army generals, even to the cacique who rebelled and was punished. The story of this cacique brings onto the scene state administrators, spies, sergeants, judges, the high courts of the empire, and finally the sovereign himself, seated on his throne.

"These tableaus are followed by depictions of several trades that are given regulations, and several offenses along with their punishment. At the end of the whole work are the man and the woman, both seventy years old, enjoying, at the edge of the grave and surrounded by their descendants, the Persian royal privilege of becoming intoxicated—that is, of escaping the law—in order to forget their troubles.[3] The circle indicating the year is repeated here several times, but it is divided by a Greek double cross and topped by the numeral for twenty, to mark each multiple of twenty. Particularly noteworthy among the other inscriptions in this part of the work is that of an astronomer-priest observing the night sky.[4] This section of the circle, this arc covered with small circles with eyes, recalls the Egyptian hieroglyph for the sky and its images covered with eyes."[5]

In what follows we record the notes that, following the Mexica text, are added to the *Mendoza Collection* in the two editions by Purchas[6] and by Thévenot.[7]

288

1. Plate LIX, figures 3 and 4.
2. Plate LVIII, figure 12.
3. Plate LIX, figure 7.
4. Palin, [*Analyse de l'inscription en hiéroglyphes du monument trouvé à Rosette,*] Vol. I, pp. 88–97. Since the text of the original is riddled with typographical errors, some slight changes have been made, without which several sentences would have been unintelligible.
5. Plate LVIII, figure 8.
6. *Pilgrim, in Five Books*, Vol. III, pp. 1068, 1071, 1087, 1089, 1091, and 1097.
7. Melchisédech Thévenot, *Relation de divers voyages curieux,* Vol. II, p. 47.

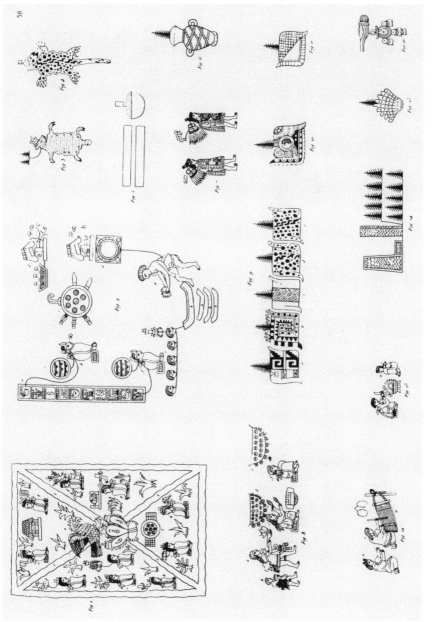

Plate LVIII

Plate LIX

Plate LVIII, Figure I. The ten founders of Tenochtitlan: *a*, Acacitli; *b*, Quapan; *c*, Ocelopan; *d*, Aguexotl; *e*, Tecineuh; *f*, Tenuch; *g*, Xominitl; *h*, Xocoyol; *i*, Xiuhcaqui; and *k*, Acotl. The city of Tenochtitlan, or Mexico City, is indicated by the weapons that were used to conquer the land on which it was built. Below these weapons is the tuna or Indian fig cactus, *m*, atop a rock, and the eagle, *n*, perched on the cactus. (An ancient prophecy foretold that the Aztecs' migrations would not come to an end until the leaders of the peoples came across an eagle resting on a cactus. The location of this miracle was to be the site of the new city.) The lines *t*, which form a cross, indicate either causeways or canals that crossed the swampy country inhabited by the founders of Tenochtitlan.

Figure 2. *a*, ten years of the reign of Chimalpupuva *b*; a shield *c*, and spears, to represent the conquest of Tequixquiac *d* and Chalco *e*. Death of Chimapupuca *f*. Uprising of the inhabitants of Chalco *g*. They destroy four enemy boats *h*, and kill five Mexica *i*. (It is quite surprising that the memory of such a minor event should be preserved for centuries.)

Figure 3. Tribute of eight hundred tiger [jaguar] skins.

Figure 4. Tribute of twenty tiger [jaguar] skins.

Figure 5. Tribute of gold bars and gold dust.

Figure 6. Tribute of four hundred pots of honey from the Maguey, Agave americana.

Figure 7. Soldiers of the order of the priests.

Figure 8. "One of the principal priests, *a*, goes at night, *d*, to the mountains to do penance there; he carries fire with him and a small bag filled with copal resin; he is followed by a novice, *b*. Another priest, *c*, plays an instrument called *teponatztli* at night. A third priest, *f*, determines what hour it is by observing the stars, *e*." 289

Figure 9. Tribute of fabrics for clothes. Each bale (*a*, *b*, *c*, *d*, and *e*) has four hundred pieces, as the inscribed numeral shows.

Figure 10 and 11. *Idem.*

Figure 12. A mother, *n*, teaching her daughter, *o*, to weave, *q*.

Figure 13. A goldsmith teaching his son.

Figure 14. Tribute: ten times four hundred, or four thousand, mats and the same number of chairs made of rushes.

Figure 15. Tribute: four thousand ocean shells from the coasts of Colima.

Figure 16. Tribute: eight thousand bundles of copal.

Plate. LIX. Figure 1. "The figure, *a*, is of a woman who has just given birth. Her child has been placed in the cradle, *c*, and, after four days, marked with the four circles, *b*, the midwife, *d*, carries the completely naked child into the courtyard of the new mother's house and places it on rushes, called

Tule, i, spread out on the ground. Three young boys, *f, g, h,* seated close to the rushes, eat *ixicue,* or roasted corn mixed with cooked fava beans represented by the vase that sits in front of them. Having washed the child, the midwife asks the boys to call him out loud by the name that would be given to him. When they take the child to be washed, if it is a boy they place in his hand the tools, *e,* that his father uses in his trade: a shield and spears, for example, if the father is in the military; and if it is a girl, a distaff and a spindle, *l,* a basket, *m,* a broom, *k.* After this ceremony (ablution and baptism) is finished, the midwife carries the child back to the mother. If the boy is the son of a warrior, they bury the shield and the spear near the place where he will likely fight one day against his enemies; they bury the tools used by the girls under a *metate,* the stone on which they knead corn cakes. If the father, *q,* and the mother, *r,* of the child, *o,* want him to devote himself to the ecclesiastical profession, they carry him to the temple on the twentieth day after the ablution. While presenting him to the altar, they add offerings of rich fabrics and food. When the child is of age, they place him in the hands of the high priest, *n,* to teach him the order of the sacrifices. If the parents want the child to bear arms, they offer him to the Teachauch, *p,* whose role is to instruct young men in the art of war."

Figure 2. "Ration, or the quantity of food to which each child is entitled at each meal: the father, *a,* teaches precepts to his son, *c,* whose age, three years, is marked by the three dots, *b.* A boy of this age receives half a corn cake, *d,* at every meal. The mother, *e,* instructs her three-year-old daughter, *g;* the girl also receives the ration of a half-cake, *f.*"

Figure 3 and 4. Punishment for children: they are pricked with maguey leaves and exposed to the smoke from burning hot peppers.

Figure 5. An adulterous woman and her lover, tied together to be stoned to death. See the Le Tellier manuscript in the Library of Paris, Plate IV, figure 2.

Figure 6. "The father, *a,* places one of his sons, *b,* fifteen years old, in the hands of the *Tlamacazqui, c,* or high priest, of the Calmacac temple, *d,* so that he can train him and make a priest of him. Another son, *e,* of the same age, *h,* is sent by his father to school, *g,* to be instructed there by the teacher responsible for children."

"When a girl was married, the *Amanteza,* or matchmaker, *i,* carries her toward evening on his back, *w,* to the home of the boy who is to wed her; his path is lighted for him by four women, *x, z,* each of whom holds in her hand a kind of torch made of pine wood, marked with the numerals 1, 2, 3, and 4. The boy's parents come to greet the girl at the entrance to the courtyard of their house and escort her into a room where the boy awaits her. They sit down on chairs arranged on a mat, *o,* and the entire marriage ceremony consists of tying

together one of the lower corners of the boy's garment, *l*, with one corner of 291
the girl's garment, *m*. They offer their gods copal resin, *q*, as a sacrifice, which
they burn on a dish containing fire. Two old men, *i*, *r*, and two old women,
n, *v*, serve as witnesses. Afterward, the newlyweds eat the served meat and
drink pulque from cups, *t*, represented by the pot, *s*. The old men and women
eat as well, and after the meal each of them exhorts the newlyweds individually to live together in harmony as a couple."

Figure 7. "The law permits an old man aged seventy years, *f*, to get drunk
in public and in private. His wife, *g*, enjoys the same privilege, provided that
she is a grandmother."

PLATE LX

Fragments of Aztec Paintings
from a Manuscript Preserved at
the Vatican Library

These symbolic figures were chosen from among those in the manuscript that we discussed at the beginning of this work, page 106 [this edition].

Plate LX

PLATE LXI

Pichincha Volcano

This view is from Chillo, from the country house of the ⟨Marquis of Sel-valegre, whose son accompanied us on our journey to Mexico and the Amazon River. The volcano is seen rising above the Cachapamba savannah. In my drawing one can make out (1) Rucu Pichincha, the snow-capped summits that surround the crater; (2) the cone of Tablahuma; (3) Picacho de los Ladrillos; (4) the rocky peak of Guagua Pichincha, which is the *cacumen lapideum* [stony summit] of the French academicians; and finally, (5) the peak on which was placed the famous cross that served as a marker for the measurement of the meridian. According to my observations, the absolute elevations of these peaks are from two thousand three hundred to two thousand five hundred toises; but since the Chillo plain is already at an elevation of one thousand three hundred forty toises above sea level, the sight of the Pichincha volcano is less impressive from the eastern side than from the western side, where the vast forests of the Esmeraldas begin. Both the distances and several of the height angles that were used in the execution of this drawing were determined by means of a Ramsden sextant.

1. *Rucu-Pichincha.*
2. *Tablahuma.*
3. *Picacho de los Ladrillos.*

4. *Guagua-Pichincha.*
5. *La Cruz.*
6. *Le Pic (3) vu de près.*

Plate LXI

PLATE LXII

Plan of a Fortified House of the Inca on the Ridge of the Cordillera of Azuay; Ruins of a Part of the Ancient Peruvian City of Chulucanas

I. The plan of the fortified house of Cañar was drawn up by Mr. La Condamine in 1739; using the notations that I made in 1803, attempts were made to correct the drawing found in the archives of the Bureau des Longitudes in Paris and used for the plate included in the *Mémoires de l'Académie de Berlin*.[1]

A. B. A terreplein made by hand at a height of five to six meters above the former ground level.

C. D. A square dwelling, of which we have provided a drawing on Plate XX. In the chamber on the western side, one can make out cylindrical stones that project a half a meter outside the wall at a right angle and that seem to be designed to hang weapons.

L. F. A terrace that supports the terreplein, and that rests on a second terrace, G H, two meters wide and five meters high. The platform at the end of the terreplein is shaped like an elongated oval, the main axis of which forms a right angle with the magnetic meridian, angle N. 6° W, assuming that the declination of the needle is 8° to the northeast.

I K. and L M. Two ramps by which one climbs to the esplanade to the south and the north of the fortress, the first one leading to the middle and the second to a point one-quarter of the length of the platform. At the end of the northern ramp, M, begins the lower terrace, G H.

293   N O. A wall sloping from one gable to the other and dividing the square building into two chambers.

P and Q. The two doors facing the two semi-circular edges, A D, at the ends of the platform.

R S. A stone-covered terrace four meters lower than the oval platform. This terrace originates at the western edge of the terreplein. It begins by jutting out a few feet to the north, R, as if to block and bring an end to the faussebraie [first rampart], G H; from there it turns at a right angle toward

---

1. *Mém[oires] de l'Académie de Berlin*, 1746, pp. 448–54.

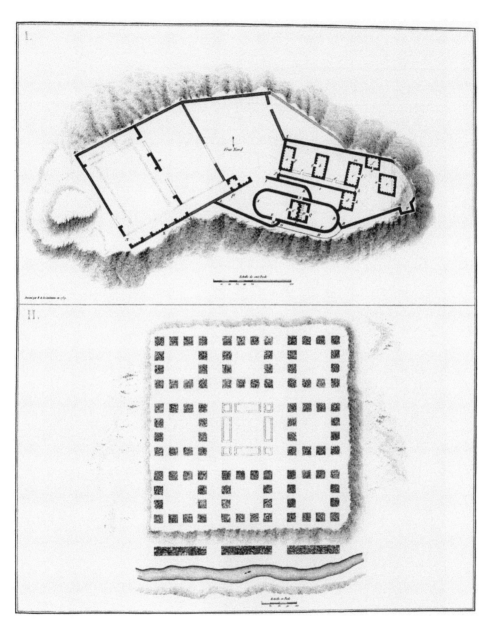

Plate LXVII

the west and extends over a length of twenty-eight meters, forming a curtain wall whose western edge is supported by a kind of square bastion, T, consisting of two side sections and a front. Beyond this bastion there are only the vestiges of a simple wall with no apparent fortification. This wall always followed the most elevated part of the land, which gradually flattens out, and turned back toward the east through the south, making a half-circle, T V, to run parallel once more to the length of the terreplein. The part V X of the wall is well preserved.

X Y Z W L. A quite irregular compound divided into four courtyards; the first, of which some vestiges remain on the eastern side (in w and Δ Γ), is a square eighty to one hundred feet long. It appears to have been surrounded by small, isolated corps de logis, longer than they are wide, the foundations of which are still distinguishable in some places.

Γ z μ Δ. The second courtyard, slightly smaller than the first and without any vestiges of buildings.

X Y Z μ s g. The third courtyard, the largest of all, but very irregularly shaped. The walls of this part of the compound are of modern construction, and it may well be that the small, square building, the ruins of which, μ, are visible, was originally located outside the fortress.

a b c d e f. Six rooms of the fourth courtyard contained within the irregular compound R S T V X to the south and west of the fortress.

294   r and s. Vestiges of two doors cut into a wall that was once parallel to the wall g i h.

g h. Narrow gallery through which one reaches the bastion S T; it is next to the inner ramp, I K, by which one climbs to the platform of the fortress on the southern side.

k and l. The doors of the two buildings d and c.

n and o. Doors open to the east and the north, leading to the small buildings e and f. These buildings, intended to lodge the Inca's guard, appear to be much less carefully built than the previous ones and without the support of a flat angle bracket. Mr. La Condamine thinks that the prince and his wife lived in the buildings indicated by the letters a and b. The gates p, q, g, and h are tall enough for a man seated in a litter and carried on the shoulders of his domestic servants. The recesses[1] hollowed into the interior walls are indicated in the plan.

Since the primary goal of this work is to give an exact idea of the state of the arts among the civilized peoples of the Americas, we have preferred to present the ruins of the house of the Inca at Cañar as one could see them in

---

1. See pp. 128 and 141 [this edition].

1739. Since then many walls have been knocked down, and I had difficulty recognizing all of the divisions included in Mr. La Condamine's plan.

II. The ruins of the ancient city of Chulucanas are quite amazing for the extreme regularity of both the streets and the alignment of the buildings. They are found at an elevation of fourteen hundred toises on the ridge of the Cordilleras in the Páramo de Chulucanas, between the Indian villages of Ayavaca and Guancabamba. The Inca's great road, one of the most useful and at the same time most gigantic works ever carried out by humans, is still quite well preserved between Chulucanas, Guamani, and Sagique. On the crest of the Andes, in extremely cold places that could appeal only to the inhabitants of Cusco, one sees the remnants of great buildings everywhere: I counted nine of them between the Páramo de Chulucanas and the village of Guancabamba; the locals refer to them by the rather pompous name of house or palace of the Inca, but it is likely that the majority of them were caravanserais built to facilitate military communications between Peru and the kingdom of Quito. 295

The city of Chulucanas appears to have been placed on the slope of a hill, on the bank of a small river from which it was separated by a great wall. Two openings bored into this wall led to the two main streets. The houses, built of porphyry, are spread across eight quarters formed by streets that intersect at right angles. Each quarter contains twelve small dwellings, such that there are ninety-six of the latter in the part of the city shown in the plan that we are providing in the sixty-second Plate. I prefer the word dwelling to house, because the latter gives the impression of several interconnected chambers in the same compound, while the dwellings of Chulucanas, like those of Herculaneum, have only one room that probably opened onto an interior courtyard. At the center of the eight quarters that we have just mentioned are the remnants of four large, oblong-shaped buildings separated by four small square structures occupying the four corners. To the right of the river that skirts the city one finds some rather bizarre structures that together form a kind of amphitheater: the hill is divided into six terraces, each step of which is covered in cut stone. Farther on are the *Inca's baths*, which I will describe in greater detail in the *Personal Narrative* of my journey. It is surprising to find baths on a plateau where the natural springs have a temperature of barely ten to twelve degrees on the centigrade thermometer, and where the air temperature drops down to six to eight degrees.

PLATE LXIII

Raft from the Guayaquil River

This drawing has the dual function of presenting a group of fruits from the equinoctial region and of introducing the shape of the great rafts (*balzas*) that the Peruvians have used since the earliest times on the coasts of the South Sea and at the mouth of the Guayaquil River. The raft, loaded with fruit, is depicted at the moment of its being docked at the riverbank. Toward the prow one can make out pineapples, the pear-shaped drupes of the Avocado tree, the berries of the Theophrasta longifolia, whole stalks of bananas, and Passiflore and Lecythis flowers shaded by Heliconia and Coconut Palm fronds. Used for fishing and transporting goods, these rafts are sixteen to twenty-five meters long. They are composed of eight to nine balks of a very light wood.[1] ᵛDon Jorge Juan[2] published some fascinating observations on the maneuvers of these boats, which, though heavy in appearance, tack quite close to the wind.

1. Bombax and Ochroma.
2. *Voyage historique de l'Amérique Méridionale*, Vol. I, p. 168.

Plate LXIII

PLATE LXIV

Summit of Los Órganos Mountain at Actopan

The porphyritic mountain of Mamanchota, famous in Mexico under the name Los Órganos, is located to the northeast of the Indian village of Actopan. The slender part of this rock is one hundred meters high, but the absolute elevation of the mountain's summit at the point where *Los Órganos* begins to break away is 1,385 toises. It is on the road from Mexico City to the mines of Guanajuato that one can make out from afar—and standing out against the horizon—the rock of Mamanchota; it soars from the middle of an oak forest[1] and offers a most picturesque sight.

---

1. *Essai politique sur [le royaume de] la Nouvelle-Espagne*, Vol. I, p. 289.

Plate LXIV

PLATE LXV

Columnar Porphyry Mountains of El Jacal

This view was drawn in the plain of Copallinchiche, which is a part of the great Mexica plateau and at an elevation of thirteen hundred toises (2530 meters) above sea level. Composed of enormous columns of trappean porphyry, the mountains of the Oyamel and El Jacal are crowned with pines and oaks. Between the tenanted farm of Zembo and the Indian village of Omit-lan are the famous *iztli* (obsidian) *mines* exploited by the ancient Mexica. In the country itself, this area is called *the mountain of the Knives, el Cerro de las Navajas*. The peak of El Jacal has an absolute elevation of sixteen hundred three toises (3124 meters). My drawing shows the contours of the Cerro de Santo Domingo (1) and Mocaxetillo (2), along with Los Orcones (3) and El Jacal, or Cerro Gordo (4).

297

Plate LXV

PLATE LXVI

Head Engraved in Hard Stone by the Muisca Indians; Obsidian Bracelet

This sculpted head is the work of the ancient inhabitants of the kingdom of New Granada. The stone, which some mineralogists regard as smaragdite, is no doubt merely a green quartz verging on hornstone. This extremely hard quartz is perhaps dyed by nickel oxide, like chrysoprase; it is perforated in such a way that the openings of the cylindrical hole are positioned in planes that intersect at right angles. One may assume that this perforation was carried out using tools made of copper mixed with tin, for iron was not in use among the Muisca and the Peruvians. The obsidian bracelet was found in an Indian tomb in the province of Michoacan in Mexico. It is extremely difficult to conceive of how they succeeded in working such a fragile substance. Completely transparent, the volcanic glass has been whittled down to a cylindrically curved band less than one millimeter thick.

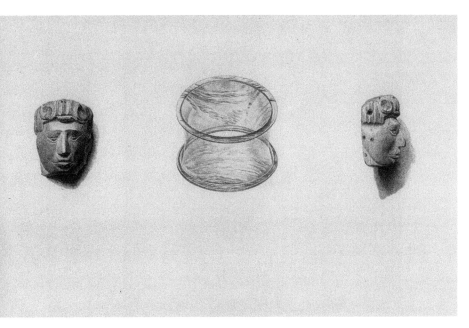

Plate LXVI

PLATE LXVII
View of Lake Guatavita

This lake is located to the north of the city of Santa Fé de Bogotá at an absolute elevation of over fourteen hundred toises on the ridge of the Zipaquirá Mountains, in a wild and lonely place. In this drawing one sees the remnants of a staircase used in the ablution ceremony and an incision in the mountains. Shortly after the conquest, attempts had been made to cut this breach in order to dry out the lake and remove the treasures that, according to legend, the natives had hidden there when Quesada and his cavalry appeared on the plateau of New Granada.

Plate LXVII

PLATE LXVIII
View of the Silla de Caracas

This granitic mountain, which is very difficult to scale because its slope is covered in a thick turf, has an absolute height of over thirteen hundred fifty toises. From the coast of Paria to the Sierra Nevada de Santa Marta, there is no other peak with an elevation that equals that of the Silla de Caracas, also called *Montaña de Ávila*. The two rounded summits bear the name *Saddle* (*Silla*); they serve as landmarks for identifying the port of La Guaira. I have drawn this mountain from the southern side, as it is seen from the coffee plantation of ˙Don Andrés Ibarra.

Plate LXVIII

PLATE LXIX

Dragon Tree of La Orotava

This Plate shows the colossal trunk of the Dracæna Draco on the island of Tenerife, of which all travelers have spoken but of which there had been no published image until now. Its height is between 50 and 60 feet; its circumference, near the roots, is 45 feet; it had already reached this thickness when the Spanish first landed at Tenerife in the fifteenth century. Given that this plant, of the Monocotyledon family, grows extremely slowly, it is likely that the dragon tree of La Orotava is more ancient than most of the monuments we have described in this work.

Plate LXIX

LETTER FROM MR. VISCONTI,
MEMBER OF THE INSTITUT DE FRANCE,
TO MR. VON HUMBOLDT,
ON SOME MONUMENTS OF THE
AMERICAN PEOPLES

While reading through the sections of your works that are devoted to the monuments of the peoples of the Americas, and in which you were gracious enough to make such a precious gesture of your friendship, I came across, among the large number of heretofore unknown facts and novel observations contained within this volume, a few points in which my opinion differs from yours. Frankly, this difference applies only to details of little importance, and my remarks may appear persnickety; but since we are speaking of an entirely new branch of archaeology—if I may use this term to refer to studies of the monuments of the new world—I felt it necessary to pass on to you a few observations on the subject. If accurate, they may contribute to the understanding and explanation of some very remarkable monuments; if you do not find them to be so, then the trust I have in your wisdom will dispel my doubts.

The first item that drew my attention is the ronde-bosse figure of an Aztec priestess, or, if one prefers, princess (Plates I and II). Your thought was that the sculptor's ignorance led him to omit the arms from this figure and that he clumsily attached the feet to the sides. I do not have any more insight than you do into the sculptor's skill, but it seems to me that this figure, although completely out of proportion, is nevertheless neither mutilated nor crippled. In the extremities that you take for feet, I believe I recognize the hands of the statue. To me, the figure seems to kneel and is seated on its lower legs and heels, ὀκλὰξ καθημένη [*oklax kathemene*, to sit with bent knees, squat], as ʽLucian would say.[1] Suggested to humanity by nature herself, this resting posture is carefully described by the Greek lexicographers and in artistic monuments is specifically linked to female figures. ʽHesychius (see ὀκκύλαι [*okkylai*, kneeling, sitting on the ankles] and ὀκλάζειν [*oklazein*, kneel]) and ʽErotianus in his Lexicon Hippocratis (see ὄκλασις [*oklasis*, kneeling, sitting on the ankles]) describe this posture by way of periphra-

---

1. In *Lexiphane*.

ses that refer to the position in which one sits on one's legs and heels: ἐπὶ
τῶν πτερνῶν καθέζεσθαι: ἐπὶ τὰς κνήμας καὶ τὰς πτέρνας κάμψαντα τὰ
γόνατα καθίσαι [epi tōn pternōn kathezesthai: epi tas knemas kai tas pternas
kampsanta ta gonata kathisai; to sit on one's heels: to sit down with bent
knees on one's calves and heels]. The learned ʼHemsterhuis speculates that
the original verb that expressed this state of repose was ὄκειν [hokein; or alternatively ὀκνεῖν, to shrink from, hesitate] and that it was the root of a large
number of Greek words that subsequently passed into other languages.[1] Suffice it to mention the nouns ὄκνος [oknos], hesitation, fear, and οἶκος [oikos],
house or dwelling; such was the familiarity of this pose to tired people in original and nearly savage societies in the quiet moments that they spent inside
their rustic retreats.

On the monuments of Egypt one sees a large number of women represented in this position, either nursing their children or praying at the feet
of their idols, playing instruments or displaying grief at the funerals of their
parents or fellow countrymen.[2] On the same monuments[3] one also finds men
represented in this pose, but this is much rarer. One might even go so far as
to think that the Pythagoreans' precept of praying while seated was, in distant times, merely a reference to this posture commonly used in the rites of
the Egyptians. This pose is so natural, especially to women, because of the
suppleness of their limbs, that the country women in several rural parts of
Italy generally assume this position even in church. We should, therefore, not
be surprised that the Aztec women used it as well. It is also found in some of
this people's symbolic paintings: on Plate XXVI the goddess of water who
descends to earth in order unleash the floods is shown seated on her heels;
and several other figures in other Mexica paintings sit more or less in the same 301
pose, except that they have only one knee on the ground. And in regard to the
statue about which I have had the honor of writing you, it seems to me that
the rear of this figure (Plate II) offers undeniable proof of what I have just suggested: one can make out distinctly the feet with clearly indicated toes; they are
placed right next to one another, and the chiaroscuro in the drawing (Plate I)
enhances the projection of the knees hidden underneath the stiff, uniform
cloth that covers the entire figure.

1. In the Hesychius edition by Alberti, see the note about the word Οἰκίδδειν [oikíi-
dein, to sit].

2. See, in the superb work *Description de l'Égypte*, Vol. I, Plates XII, number 2; LXII,
number 2; LXIX, number 1; LXX, number 2; LXXXI, XCVI, and elsewhere; and in
Denon, *Voyage de la Basse et [la] Haute Égypte*, Plates CXXVI, CXXXI, and CXXXV.

3. *Sculture de la ʼvilla Borghese*, St. VIII, number 4; Winckelmann, [*Geschichte der
Kunst des Alterthums*] (Rome edition), Vol. I, Plate VI.

So as not to dwell any longer on this odd trace of the arts of a people who have nearly disappeared, I will limit myself to observing that excessively sized heads are a common flaw in most of the works of this people. This same flaw is quite noticeable in the sculpted figures mounted on the lids of Etruscan funerary urns. It seems that for ignorant artists, the attempt at expressing the features of this most important body part with greater precision and exactness became an occasion for enlarging it to the point of exaggeration. I now move on to another observation that was suggested to me by the examination and explanation of one of the hieroglyphic paintings that I have just mentioned, and on the subject of which you read a paper to our class: the representation of the four destructions of the world (Plate XXVI). You compare these periods to the four ages of Greek mythology; and since you find five ages of the world in the Aztec traditions, you attempt to make this difference disappear by proving that the Bronze Age in Hesiod can easily be divided in two because of the two generations that the poet describes (pp. 235 and 236 [this edition]). I note that Hesiod, like the Aztecs, counted five ages, also including the one that was as yet incomplete and in which he himself lived. He states it expressly (*Opera et dies*, verse 174):

Μηκέτ' ἔπειτ' ὤφελλον ἐγὼ πέμπτοισι μετεῖναι.
[*Meket' epeit' opheilin ego pemptoisi meteinai*. If only I did not have to live among the men of the fifth age!]

302 This tradition of the five ages would have been known to the Chaldeans, if it is permitted to speculate, following Dante,[1] that Nebuchadnezzar's dream vision of the Colossus[2] bears some relation to this theory. The Colossus was composed of five different and separate materials: gold, silver, bronze, iron, and clay.

Finally, it remains for me to share with you another observation, which is as unimportant as the previous ones. It relates to the manner in which the Aztecs drew their hieroglyphs. You note (p. 238 [this edition]) that to make them easier to read and understand, the Aztecs occasionally placed at the end of one line the first signs or, as it were, the first characters of the hieroglyphic phrase of the subsequent line, as a result of which the first signs are repeated. Following Mr. Zoëga's account, you compare this method to that of the Egyptians, who, according to him, dealt with the characters in their hieroglyphic script in the same manner. I cannot conceal from you that my own studies have not convinced me of this similarity. If you have no other author-

1. *Inferno*, ch. 14.
2. Daniel, ch. 2.

ity than the passage of the profound work of the Danish antiquarian on the obelisks (p. 464), then I must confess to you that I understand his comments in a completely different manner, and I shall add that my view of them seems to be confirmed by an examination of the monuments themselves. To prove that in hieroglyphic script the direction in which the figures of men and animals are turned determines whether the hieroglyphic line should be read from right to left or from left to right, Mr. Zoëga uses certain sets of signs, which are repeated in the same monument and are sometimes found drawn in their entirety on the same line, at other times half on one line, half on the next; for example, in the Sallustian obelisk[1] one of these sets shows the figure of a dove, followed by those of a scarab and a knife, all on the same line. This set is repeated on the same column, but the hieroglyphs are distributed across two lines. Following the rule proposed by the learned antiquarian, the figures are found again in the same order, in such a way that the scarab and the knife appear after the dove.

This is what Mr. Zoëga says, though somewhat less clearly.[2] But if, as a consequence of this remark, I am depriving you of a felicitous analogy, I shall compensate you forthwith by offering you a similar analogy in the method the Hebrews followed in their manuscripts. Whenever they cannot fit an entire word on one line, they include the first characters on that line and write out the word in its entirety on the next line, so that the initial characters appear twice, exactly as you have noted in the manuscripts or, better, the paintings of the Aztecs. This method was used in several printed editions of the Hebrew Bible, demonstrating that the human mind, regardless of differences in century or climate, tends to act in the same manner in similar circumstances, without having any need for tradition or example.

It is to this very principle that I attribute the invention of tools for making fire by rubbing together two pieces of wood.[3] It was surely not Mercury

303

---

1. See, in Mr. Zoëga's work, *De origine et usu obeliscorum*, the Plate *Obeliscus Sallustianus lat[ero] septentrionale*.

2. *Nam præter quod hac ratione antecedens figura sequenti dorsum obvertere et eam post se reliquere agnoscitur, etiam in repetitis inscriptionibus, dum propter loci angustiam nota aliqua ex superiore spatio ad inferius sit removenda, hoc in ea fieri videmus quae ex illa nostra sententia ultima erat superioris spatii.* [For regardless of the fact that one can see clearly in this way that the prior sign turns its back on the preceding one and leaves it behind, we also find that in cases of several inscriptions following each other when, due to lack of space, a sign is at times moved from the upper field to the lower one, what occurs is precisely that which occurs in the inscription that, we believe, was the last one of the upper field.] (Zoëga, loco citato).

3. Pp. 119 and 120 [this edition].

who taught the custom of the *pyreïa* or *igniaria* to the Indians of the Orinoco. There is no Greek monument that shows us this custom from heroic
times, while you provide two examples of its representation in the Aztecs' hieroglyphic paintings.[1] It was nevertheless familiar to the ancient inhabitants
of Greece, and the figures that you publish here prove the accuracy of the
description of these fire tools that 'the scholiast Apollonius bequeathed to
us.[2] He says that the upper piece of wood, the one that is turned, resembles
a brace-and-bit, παραπλήσιον τρυπάνω [*paraplesion trypano*, similar to an
auger, borer]. This is precisely the sense one gets from your paintings. No
philologist has drawn any attention to the allusion that Apollonius makes here
to the passage in the Homeric hymn to Mercury. This allusion nevertheless
seems to me grounds for debunking the doubts that the scholar 'Rhunken has
raised about the interpolation of this passage.[3]

304   The resemblance between the *pyreïa* [fire tools] and the brace-and-bit
should make the invention of the latter date back to very remote periods, and
one would rightfully be surprised to see it attributed to 'Daedalus, a contemporary of Theseus,[4] were not the Athenian artist's invention more precisely
related to the sculptors' trepan, an instrument much more advanced than the
simple brace-and-bit when it comes to the increase in speed from the cord
and the moving strut. This relationship between the *pyreïon* [fire tool] and
the brace-and-bit did not escape those ancient authors who wrote on the
subject of tree farming.[5] They complain that the drilling of the gimlet often
leads to burns in the wood, fatal to the success of the operation. It was to
avoid this drawback that the Gauls invented another kind of drill (*terebra gal-
lica*), which was a genuine auger, whose more even and less rapid movements
did not run the risk of combustion. It seems to me that Pliny's commentators
have not, to this point, given a sufficiently accurate sense of either Daedalus's
invention or the Gallic drill.

These, my dear colleague, were some observations that I wished to submit
to your judgment. Your friendship will, I trust, consider them as proof of my
own and of the keen interest that your works hold for me.

E. Q. VISCONTI
Paris, December 12, 1812.

1. Plate XV, number 8, and Plate XLVII.
2. Book I, verse 1184.
3. *Ep[istola] Crit[ica] I, ad Hymn[um] in Mercurium*, verse 25.
4. Pliny, [*Historia naturalis*,] Book VII, § 57.
5. Pliny, [*Historia naturalis*,] Book XVII, § 25; 'Columella, [*Scriptorvm*, Vol. I,] Book
IV, verse 29.

NOTES

Page 24. The pyramid of Cholula is also called *Toltecatl*, *Ecaticpac*, and *Tlachi-huatepetl*. I assume that the latter name derives from the Mexica verb *tlachiani*, to look around, and from *tepetl*, mountain, because the Teocalli served as lookouts from which to watch the enemy's approach during the frequent wars between the Cholulans and the inhabitants of Tlaxcala. On the important question whether the temple, or, rather, the gradated pyramid of Jupiter Belus, served as prototype for the pyramids in the Sahara and for those in India and China, see Julius von Klaproth, *Asiatisches Magazin*, Vol. I, p. 486.

Page 72. Some people have recently questioned whether the Peruvians knew symbolic painting other than the *Quippu*. A passage from the *Origen de los Indios del Nuevo Mundo* (Valencia, 1610), p. 91, leaves no doubt that they did. After having spoken about Mexica hieroglyphs, Father García adds: "At the beginning of the conquest, the Peruvian Indians made confessions with the help of paintings and characters that showed the Ten Commandments and the sins that had been committed in violation of these commandments. One could conclude from this that the Peruvians made use of symbolic painting, but that those paintings were much cruder than the hieroglyphs of the Mexica and that the people generally used knots or quippu." See also Acosta, *Historia natural y moral de las Indias*, Book V, ch. 8, p. 267.

Page 125. The word *atl* or *atel* can also be found in the east of Europe. According to Friedrich Schlegel's observations, before the conquest of Hungary, the country inhabited by the Magyars was called *Atelkusu*. The name encompassed Moldavia, Bessarabia, and Wallachia, three provinces around the mouth of the Danube, a river that, like the Volga, was called *great water*, *atl* (see also pp. 187–88 above). The Mexica hieroglyph for water, *atl*, shows the movement of the waves through the undulations of several parallel lines and recalls the Phoenician sign for water, *mem*, which became part of the Greek alphabet and that of all western peoples. See the ingenious work by ▾Mr. Hug, *Die Erfindung der Buchstabenschrift*, 1801, p. 30.

Chevalier Boturini left us the names of the twenty days of the Toltec month, following the calendar of the inhabitants of Chiapas and Soconusco. Here are the signs together with their correspondences in the Aztec calendar:

Mox	Cipactli
Igh	Ehecatl
Votan	*Calli*
Ghanan	Cuetzpalin
Abagh	Cohuatl
Tox	Miquiztli

Moxic	Mazatl
Lambat	*Tochtli*
Mulu	Atl
Elab	Itzcuintli
Baz	Ozomatli
Enob	Malinalli
Been	*Acatl*
Hix	Ocelotl
Tziquin	Quauhtli
Chahin	Cozcaquauhtli
Chic	Ollin
Chinax	*Tecpatl*
Cahogh	Quiahuitl
Aghual	Xochitl

308   It is surprising to find such different names among peoples of the same race. The denominations Mox, Igh, Tox, Baz, Hix, and Chic do not seem to belong to the Americas but to that part of eastern Asia that is home to peoples whose languages are monosyllabic (see above p. 171 and Boturini, *Idea de una historia general de la América septentrional*, p. 118). We want to note on this occasion that the Chinese suffix *tzin* also appears in a large number of proper Mexica names, for instance, in *Tonantzin, Acamapitzin, Coanacotzin, Cuitlahuatzin*, and *Tzilacatzin*.

According to the scholarly studies of Mr. Klaproth, the Uygur, or Uyghur, never lived on the banks of the Selenga River, as Mr. Langlès assumes, but in the Ulugh Muz Tagh Mountains, on the banks of the Ssir [Syrdarja River], which the ancients called Iaxartes, and in the Steppes of Karakum east of the Aral Sea. (See above p. 166 [this edition], and Hammer, *Fundgruben des Orients*, Vol. II, p. 194.)

Page 194. To cast more light on the research that was the object of my writings on the Mexica calendar, I want to report here some very sensible observations that I received from Mr. Jomard. This scholar is very highly thought of by all those who study Ancient Egypt,[1] and I hasten to provide here an excerpt from a letter that he kindly wrote me.

"In your paper on the division of time among the Mexica peoples in relation to that among the Asian peoples, I have found most astonishing correspondences between the Toltec calendar and the institutions have been observed on the banks of the Nile. One of these correspondences is particularly noteworthy: the use of a wandering year of 365 days composed of equal months and 5 epagomenal days. This year is used in both Thebes and Mexico City, at a distance of three thousand leagues. It is

1. See the interesting paper by Mr. Jomard about Lake Moeris compared to Lake Faiyum, about Syene and the cataracts [of the Upper Nile River], about Elephantine Island [in the Nile], about Ombos and its environs, as well as about the antiquities of Edfu and Hermonthis [Armant], all in his *Description de l' Égypte ancienne et moderne*, which we owe to the generosity of the French Government.

true that the Egyptians did not use intercalation, whereas the Mexica insert an extra 13 days into the calendar every 52 years. What is more: in Egypt, intercalation was held in such ill repute that the kings swore at the crowning that they would never allow this practice during their reign. Despite this difference, one can see a basic similarity in the length of a solar year. In effect, since the Mexica intercalation consisted of adding 13 days each cycle of 52 years, it corresponded to that of the Julian calendar, that is, an addition of one day to every four years, and thus presupposed a year length of 365 days and 6 hours. This was, in fact, also the length of the year among the Egyptians, since the Sothic cycle contained at once 1,460 solar years and 1,461 wandering years; this meant that an entire year of 365 days was added every 1,460 years. The property of the Sothic cycle of returning the seasons and the holidays to the same point in the year, after they had run through all the points, is probably one of the reasons why they prohibited intercalation, another being the resistance Egyptians had toward foreign institutions. It is remarkable, then, that the same solar year of 365 days and 6 hours, adopted by such different peoples whose whose respective states of civilization perhaps separated them further than did their geographical distance, relates to an actual astronomical epoch and really belonged to the Egyptians. This is a point that ˈMr. Fourier will place above all doubt in his fine work about the Egyptian zodiac. No one is as well equipped as he to examine this question from an astronomical perspective, and he alone can shed light on the fortunate discoveries that he has made. I would like to add here that the Persians, who intercalated 30 days every 120 years, the Chaldeans, who used the era of ˈNabonassar, the Romans, who inserted an extra day every four years, and finally the Syrians and almost all the peoples who have regulated their calendar in accordance with the solar year also appear to me to have adopted from the Egyptians the concept of a solar year of 365 ¼ days, and the use of months of equal length and of five epagomenal days. As for the Mexica, it would be unnecessary to research how they arrived at this knowledge; it is unlikely that such a question would be answered in the near future. But the practice of intercalating 13 days per cycle, that is, the use of a 365 ¼ year, necessarily testifies either to Egyptian influence or to a common origin. "Let us add that the Peruvian year is not solar but is regulated by the trajectory of the moon, as it is among the Jews, Greeks, Macedonians, and Turks. The division of the year into 18 months of 20 days each, instead of 12 months of 30 days each, also makes a very significant difference. The Mexica are the only people to have divided the year in this way.

"A second correspondence that I noticed between Mexico and Egypt is that the number of weeks, or semilunar months of 13 days which composed a Mexica cycle, consists is that same as that of the Sothic cycle, that is, 1,461. You regarded such a correspondence as either accidental or fortuitous; but perhaps it has the same origin as the knowledge of the duration of the year. For if, in effect, the year did not have 365 days and 6 hours, that is, $146\frac{1}{4}$ days, the 52-year cycle would encompass $52 \times 146\frac{1}{4}$ or 13 times 1,461 days, which results in 1,461 periods of 13 days each. One must, however, grant that the 13-day weeks, the 13-year tlalpilli, the intercalation of 13 days at the end of a cycle, and finally the cycles of four times 13 years are all based on a prime number that is completely foreign to the Egyptian system.

309

"You have highlighted a fact that is all the more important for being connected with peoples' customs: the winter solstice festival, which both Egyptians and Aztecs celebrate. If one were to believe Achilles Tatius, the Egyptians fell into mourning upon seeing the sun set in the direction of Capricorn and the days become shorter; but when the sun rose again in the direction of Cancer, they dressed in white and put on wreaths. The Mexica custom that you have described unquestionably corresponds to that of the Egyptians; one could contest this similarity only if one placed the beginning of the Mexica year at a different time, as several authors have done. But you have eliminated any doubt that with each new cycle, the beginning of the year fell on January 9; consequently, if one counted the 13 leap days and the epagomenal days with which the festival began, the new fire would have been lit on the day of the winter solstice. It remains to be explained why the shortening of the days was a phenomenon that frightened the Mexica only once every 52 years, as if, at the end of a cycle, the sun fell farther than it normally did![1] Was it the case that in the absence of celebration of some sort they would not even have noticed the shorter appearance of the sun and that they waited for a sign before falling into mourning and fright? I can imagine that had the celebration been held on the same day each year, they would have lamented the setting of the sun at a time when it was already visibly rising again. But to ensure that their laments did not occur at the wrong time, it was easy enough to advance the celebration one day every four years, so that over the course of 52 years it would have fallen on 13 different days. A similar difficulty confounds me about the custom attributed to the Egyptians. Achilles Tatius does not mention at which point in time this custom was observed: he only uses the vague expression *one day*, ποτε [*pote*, once, at one point] (*Uranologion*, p. 146) and adds that it was at the time of the Isis festivals, without clarifying whether this practice occurred each year. If this had been the case, one would have seen during the course of a Sothic cycle that the Egyptians, in their fear of being abandoned by the sun, gave themselves over to the pain, tore at their hair and their clothes precisely at the time when this star was at its zenith and sent forth its most fiery rays. You must admit, Monsieur, that this is highly unlikely. Achilles Tatius has told us too little for us to be able to understand this alleged Egyptian custom. If the festival took place every year on the same day, then it was absurd for fourteen and a half centuries of a Sothic period; if it took place only in the year of the renewal of the period, then why in that particular year? And, finally, if they moved the celebration up by one day every four years, one has to concede that the Egyptians mourned the imminent departure of the sun without good reason, because at the winter solstice the sun in Thebes was at an angle of about forty degrees.

310

"You have compared the names of the Mexica years and days with the names of the signs of the Tartar zodiac and various other zodiacs of the old continent. You have demonstrated that the Mexica spoke of the *rabbit*, *tiger*, or *monkey day*, etc., much as in Asia people spoke of the *rabbit month*, the *tiger month*, the *monkey month*, etc.;

1. Contrary to the notion of the Greeks, Geminus claims that the celebration did not only occur on the days of the solstice and that it ran through all days of the year during the course of one Sothic cycle.

you have also shown that several of these animals are equally foreign to Tartary and to Mexico, and this latter remark leads one to think that the custom of using periodic series for calculating time (which, like the denominations themselves, is shared by the Mexica and the Asians) might have come from another, very distant land. These questions are of great interest; but I will limit myself here to analyzing the resemblance of one the Aztec signs, Cipactli, to Capricorn from the zodiac of the Greeks, or, rather, the Egyptians. This is the only one of the twenty Mexica day names to exhibit such a resemblance. Is it not noteworthy that Cipactli is the first day-sign, just as Capricorn is at the head of the catasterisms? Whatever divergences there may be in the order of the signs in the various zodiacs, this similarity of position of the first sign of all these systems is a fact; in this I seem to see a confirmation of the primacy of the Egyptian zodiac. The question of observing the colure of the summer solstice in the first point of Capricorn is immaterial, for it is now certain that the zodiac we use, following the Romans and the Greeks, and which the latter adopted from the Egyptians, belongs essentially to Egypt and nowhere else, and that one can explain the zodiac only by tracing the summer solstice back to Capricorn. The rural year of the Egyptians began on the day of the summer solstice. One must not, therefore, be surprised that Capricorn formerly occupied the first place among the dodecatemoria. If one knew at what point the year formerly began in Tartary, Tibet, or Japan, one could deduce something similar from the position of Aquarius at the head of the zodiac among these different peoples. Indeed, the first sign there is the *rat*, which corresponds to Aquarius. *Mahara*, the sea monster of the Hindu zodiac, which corresponds to Capricorn, is in the second position, which presupposes Aquarius to be once more in the first position. In this way, then, the successive positions of the colure of the summer solstice in Aquarius, Capricorn, and later in Virgo, Leo, and Cancer, would be indicated by the oldest and most authentic monuments, that is, the zodiacs of peoples. But I will not dwell on this idea, which I cannot yet support with proof. Suffice it to say that Capricorn's placement at the head of the signs in both Egypt and Mexico constitutes another correspondence between the two countries.

"You have also observed that the Pisces of the Egyptian zodiac are accompanied by a hog, an animal that in the Tibetan zodiac replaces the catasterism of the fish, and that Libra corresponds to the dragon of the Tartar zodiac, whose name is equivalent to the word *cohuatl* or serpent, one of the Mexica day-signs. The sign of Libra, whose ancientness people have doubted for no reason, can be found among the dodecatemoria of the [East] Indians and in their lunar mansions and in the Egyptian zodiac as well. Clearly, those who object that this is not a ζώδιον [*zōdion*, zodiac sign] are unaware that the Libra's scales are always held by a human figure, as are Virgo's ear [of wheat] and Aquarius's jug. If Libra is a sign that the Romans added, who then hewed in stone in Elephanta? It is true that before Augustus, Scorpio, because of its expanse, took up the space of two signs in the Greek and Roman zodiac. Vitruvius is the first author in whose writings one finds the word *libra* [scale]. Aratus, Eudoxus, and Hipparchus had used the word χηλαί [*chelai*, claws, talons, hoofs] for the sign of Libra; it refers to the *claws* of the scorpion. But since the conquest by Julius Caesar, the Romans frequently traveled to Egypt, and they probably noticed the scales

311   depicted on the monuments and adopted the sign's use. Germanicus, who, according to Tacitus, studied Ancient Egypt, translated the poem by Aratus, as Cicero had
done before him, but unlike Cicero, he did not render the word χηλαί [chelai] as chelae [claws]. He used the word libra, and one can see that Virgil, ˅Manutius, Vitruvius, Hyginus, Macrobius, and Festus Avienus, all of whom wrote after the conquest
of Egypt, also used this word. The same can be said about Ptolemy and Achilles Tatius. It would have been more likely that the Chaldeans, rather than the Egyptians,
were not aware of Libra, because, as ˅Servius remarks in his commentary on the
well-known verse—Anne novum sidus tardis te mensibus addas, etc. [whether you
will join the slow moons as a new star]—the Chaldeans divided the zodiac into eleven
signs, while the Egyptians used twelve. Germanicus's commentary illuminates the
questions perfectly by showing that the Libra of the Egyptians was what the Greeks
called chelae; I find the same remark in ˅Eratosthenes: χηλαί ὁ ἐστι ζυγός [chelai ho
esti zygos, balance or scales] corresponds to the scale. How could he have made this
connection if Libra had not yet existed in his day? Eudoxus was Greek; because he
was addressing the Greeks, he had to use the word chelae, which was familiar to them.
Eratosthenes, however, was writing in Egypt, explaining the Greek heavens, and was
able to say to which of the Egyptian signs this name corresponded. We also know
from the Zend Avesta that the ancient Persians knew of the astronomical Libra; and
˅Saint Epiphanius says the same about the Pharisees. Finally, what is more convinc-
ing than the following passage from Achilles Tatius: "The chelae that the Egyptians
call Libra" (Uranologion, p. 168)? There is no end to the writers I could cite here. As
for the monuments, few are known to us, and with the exception of Egypt and India,
they are so new that they do not teach us anything about the age of this celestial sign.
But everything points to its antiquity. In Rome itself, people knew the name of Libra
even before it had been identified in the skies. Cicero employs the word jugum [yoke],
as does Varro; Geminus uses the word ζυγός [zygos, balance, scales]. Even the school
of Alexandria was not unaware of this sign; but the fall of Egypt first had to occur in
order to open up, as it were, the temples, make known the Egyptian celestial map,
and make available the sign of Libra for the Romans to adopt and pass on.

   "I have gone into such length about the ancientness of Libra, which has already
been proved by others, because this point is closely connected with the Egyptian zodiac system; this does not appear to be your impression, Monsieur, since you seem
to accept the ancientness of this celestial sign in Egypt more readily than the idea of
the movement of the fixed stars. What may be somewhat precarious in the attribu-
tion of an origin to the monuments of Thebes is the precise determination of a year,
rather than an approximation of a date, which provides some flexibility. One does not
need a vast knowledge of astronomy to recognize the point in the sky or the constellation where the sun is at its zenith; and because this point changes all the time, it is
nearly impossible to paint it in the same spot for twenty or forty successive centuries.
Is there anything astonishing about the fact that a people for whom this point marked
the beginning of a year would have represented this point successively as Virgo, Leo,
and Cancer, and probably as other signs before that? I do not, therefore, intend to
take away from the Egyptians the merit of this and all those other discoveries that the

Greeks, so skilled in the appropriation of knowledge, passed on to us. I only want to suggest that it was entirely natural and easy for the Egyptians to mark the beginning of their year there where they saw it begin.

"You have drawn the attention of scholars to the monument of Bianchini. This planisphere makes me think that we have seen a similar zodiac in Panopolis, composed of concentric cycles divided into twelve fields; Pococke saw it when he was passing through. Time did not allow for the excavations that would have been necessary for making a copy. I have seen there the figure of a bird, like the one that you notice on Bianchini's planisphere, where it corresponds to Aries, while in the Tartar and Japanese zodiacs, the bird corresponds to Taurus. It is possible that this marble sculpture, like the Isis plaque, was made either in Egypt or after an Egyptian model, but it was clearly sculpted by foreign and not very faithful hands."

The observations in Mr. Jomard's letter touch upon several very important points 312 about ancient astronomy: the use of a wandering year of 365 days and 6 hours; festivities that are tied to material phenomena; and the catasterisms of the solar zodiac. There is probably a sort of elementary astronomy that one might call natural, and that came to light among peoples who had no direct contact with each other at the same stage of their civilization. It was to these theories that belonged the first ideas about the number of full moons that correspond to one solar revolution, about the amount of time over 365 days contained within this revolution; about the 27 to 28 equal parts of the heavens that the moon traverses during the course of a lunar month; about the stars that disappear at the first sun rays; about the length of the gnomon's shadows; and about how one draws a meridian line by way of corresponding heights or equal lengths of shadows. To develop the bases of natural astronomy, it is enough to choose a marker on the horizon, either a tree or the top of a boulder, with which to relate the rising or setting sun, and to be attentive to the phenomena that recur in small intervals. (Fréret, *Oeuvres completes*, Vol. XII, p. 78.) The dodecatemoria of the ecliptic, the lunar mansions, the intercalations of one day every four years, or of multiples of these numbers, the tested means for bringing the lunar almanac into correspondence with the solar and for making the same elements of the periodic series coincide with the same seasons, the use of gnomons, the importance given to the points in time when the shadows are either longest or shortest, the fears at the end of a great year, the idea of regeneration at the beginning of a cycle—all these originated from both the observation of the simplest phenomena and the individual nature of human beings.

We believe it necessary to repeat here that it is extremely difficult to distinguish what peoples drew, so to speak, from inside themselves and from the objects around them from what was passed on to them by other peoples more advanced in their arts. The hieroglyphs and symbolic writing emerge from the need to express ideas through figures. Either a *tumulus* or pyramids come into existence when people pile up earth and stones in order to mark a grave. Meanders, labyrinth, and Greek frets are found everywhere, either because humans take pleasure in the rhythmic repetition of the same shapes or because they took as models the regularly traced figures on the skin of large water snakes or on the carapaces of turtles. A half-savage people, the ▾Araucanians of Chile, used a year (*sipantu*) that resembles that of the

Egyptians more than that of the Aztecs. Three hundred sixty days are divided into twelve months of equal duration, to which are added five epagomenal days at the end of the year, at the winter solstice (*huamatihpantu*). The nycthemeral [daily] oscillations are divided into twelve hours (*llagantu*), as they are among the Japanese. It is possible that the Araucanians had received this temporal division from eastern Asia, from the same source from which the Asian cycle of 20 by 37 suna, or 60 years, came to the Muisca of Cundinamarca; at the same time, there is nothing to say that the Araucanian calendar might not have been created on the new continent. At first, many peoples only had 360-day years, not because the solar revolutions were once shorter but, as Count Carli—a noteworthy author—seriously claims, because they had stopped at a round number upon first calculating the length of the year. Twelve of the full moons that were observed during this period of 360 days led to thirty-day months, and the five supplementary days were added in when people noticed that confusion resulted from using years that were too short. Peoples' mores and customs are like the similarities between their languages; there are certain characteristics that point directly to the same origin or to contacts among peoples. One can, for instance, imagine that the signs of our solar zodiac might have derived their names from Egypt or India, or from other regions that are bathed by large rivers and are located on the same parallel; but once the names are fixed, one can no longer doubt that the peoples who use the same catasterisms have passed these names on to each other. In this way, one distinguishes shared linguistic roots, which are (so to speak) the arbitrary signs of things or capricious grammatical forms from all that is a matter of imitative harmony, the construction of our speech organs, and the nature of our intelligence.

313

The priests of Heliopolis, whom Herodotus consulted, prided themselves that the Egyptians were the first humans to have invented the division of the year into twelve parts: "ἔλεγον ὁμολογέοντες σφίσι, πρώτους Αἰγυπτίοίυς ἀνθρώπων ἀπάντων ἐξευρέειν τὸν ἐνιαυτὸν, δυώδεκα μέρεα δασαμένους τῶν ὡρέων ἐς αὐτόν" [*Elegon homologeontes sphisi, prōtous Aigyptias anthrōpon hapanton exeureein ton eniaouton, dyōkdeka merea dasamenous tōn horeon hes auton*. One told me in unison that the Egyptians were the first among the humans to have invented the year and to divide it into twelve time periods] (Herodotus, [*Herodoti Halicarnassei Historiarum*,] Book II, ed. Wesseling, p. 104). We think that this invention is not that of the Egyptians alone, any more than the ways of numbering in groups of five, ten, or twenties belong to a single people who then passed it on to other peoples in faraway countries.

In our day, the Egyptian calendar, after having been the object of learned studies by Fréret, ˅de la Nauze, and Bainbridge, has been further clarified by the work of Mr. Ideler, who has brought together a profound knowledge of ancient languages with that of astronomical calculations. We will not discuss any further whether different calendars and modes of intercalation were simultaneously used on the banks of the Nile, as some distinguished scholars have done based on passages from Theon, Strabo, ˅Vettius Valens, and Horapollon (De la Nauze, *Mém[oire] de L'Acad[émie] des Inscr[iptions]*, Vol. XIV, p. 352; Fréret, *Œuvres [completes]*, Vol. X, p. 86; Bainbridge, *Canicularia*, p. 26; Scaliger, *De emendat[ione] tempor[um]*, Book III, p. 195; ˅Gatterer, *Abriss der Chronologie*, p. 233; *Weltgeschichte bis Cyrus*, pp. 211, 507, and

567; Ideler, *Histor[ische] Untersuchungen*, p. 100; Rhode, *Über Dendera*, p. 43). We shall limit ourselves here to some observations about the mobility of the festivals.

In Egypt and in Persia, where they used the wandering year, and in Greece and Italy, where imperfect intercalations often disrupted the calendar, the festivals that were related to natural phenomena would certainly have lost the people's interest if they were celebrated sometimes in one season, at other times in another. On the banks of the Nile, as on the banks of the Tiber, one distinguished between festivals that were attached to a specific date of the month (*feriae stativa*) and those that the priests announced and whose dates were determined by the priestly institutions. The Romans called the latter *feria conceptiva* [movable feasts], and among them they distinguished between *sementivae* [celebrations of sowing], *paganalia* [rural holidays], and *compitalia* [*Lares Compitales*: household deities of the crossroads] (Marini, *Atti [e] monumenti de' Fratelli Arvali*, Vol. I, p. 126). In Egypt, the festival of Thoth, which ran through the eponymous month during all seasons in the Sothic cycle, did probably not coincide with the festival in honor of the heliacal ascent of Sirius. Is it plausible that the processions, in which the emblems of water were carried, took place during the season of the greatest droughts? Yet the following passage from Geminus is unequivocal: "*βούλονται γὰρ (οἱ Αἰγύπτιοι) τὰς θυσίας τοῖς θεῖς μὴ κατὰ τὸν αὐτὸν καιρὸν τοῦ ἐνιαυτοῦ γίνεσθαι, ἀλλὰ διὰ πασῶν τῶν τοῦ ἐνιαυτοῦ ὡρῶν διελθεῖν, καὶ γίνεσθαι τὴν θερινὴν ἑορτὴν καὶ χειμερνὴν, καὶ φθινοπωινὴν, καὶ ἐαρινὴν*" [*Boulontai gar (hoi Aigyptioi) tas thusias toîs theoîs me kata ton auton kairon toū eniautoū ginesthai, alla pasōn tōn toū eniautoū horōn dieltheīn, ka ginesthai ten therinen heorten kai cheimerinen, kai phthinoporinen, kai earinen.* For they (the Egyptians) desire that these feast for the gods not be limited to a single point in the year but range across all seasons, so that there is a celebration in the summer, the winter, the autumn, and the spring] (*Elem[entorum] astronom[iae]*, ch. 6). Geminus of Rhodes, who lived during the times of ˇSulla and Cicero, generally chides Eudoxus and the Greeks for their assumption that the feast of Isis had always coincided with the winter solstice whereas, in keeping with the wandering year, it in fact had to pass through thirty days within one hundred twenty years. If one assumes that all festivals related to the seasons and to astronomical phenomena remained tied to the fixed dates of the months Phamenoth, Pachon, or Mechir, what happens to Plutarch's clever explications in his treatise *De Iside et Osiride* about the reasons why the Egyptians celebrated the festival in the spring and the other on the day of the summer (Plut[arch,] *Opera omnia*, ed. Reiske, Vol. VII, pp. 446, 452, and 484)? Should these relations between natural phenomena, this intimate connection between symbol and object, only have existed in the first year of each Sothic cycle? Mr. Jomard's very fitting remarks about a passage from Achilles Tatius applies to all *stativa* festivals. The festival of Isis, cited by Geminus and Plutarch, was a mournful celebration; and even though it did not belong to the mobile festivals, it sometimes fell on dates when the days were already getting longer (*Uranol[ogion]*, p. 19, note 35). Is not the oath to preserve the wandering year, which the Egyptian priests made the king swear ([Hyginus,] *Comm[entarii] in Ger[manici] interpret[tationem] Arati: [signum] Capricorni*; ed. Basiliensis, 1535, p. 174), evidence of a privileged caste that, through a ruse, re-

314

serves the right for itself to announce festivals bound to astronomical phenomena in order to make itself indispensable to the people and to preserve its authority?

Plutarch, who lived during the reign of Trajan, already used the fixed days of the Alexandrians, in which the first Thoth corresponds to August 29 of the Julian calendar (Ideler, *Hist[orische] Unt[ersuchungen]*, p. 127); and he relates the names of the months and the festivals to the unchanging dates of the solstices and equinoxes. Achilles Tatius, who was a Christian and perhaps even a bishop, lived several centuries after Plutarch; one does not have to follow de la Nauze in assuming that the Ptolemaists had a fixed year in order to explain why Achilles Tatius talks about the Egyptians' laments at the Isis festival as a custom immutably tied to the time of the winter solstice. Besides, if among the Mexica we see the recurrence of this fear at the disappearance of the sun only after 5 wandering years, this is probably due to significance that all peoples attached to the end of a great cycle. Even today, we observe that the last day of the year is something of a solemn occasion even among those peoples with no superstitions whatsoever (Boulanger, *Œuvres*, 1794, Vol. II, p. 61.)

In Mexico, as in Thebes, the sun is still rather high up in the sky when its southern declination begins to wane, and one should think that the fear of the complete disappearance of this star would have occurred earlier in those regions of Asia, where Mr. Bailly places the origins of astronomy, than among the peoples who live near the tropics. And yet one understands how, in a culture whose symbols relate to the state of the sky, ideas about the progressive sinking of the sun and the shortening of the length of days, however noticeable these phenomena may be, lead to mournful ceremonies and to proclamations of pain and fear.

The catasterism to which different peoples have, at different times, assigned the premier place in the zodiac is a very interesting object of study for the history of astronomy. Since the years begin either with solstices or with equinoxes, the order of the signs, or better, the preference given to one over the others in order to open the sequence of the catasterisms, lends insight into the zodiac's origin. In this regard, the simple series of signs, through the precession of the equinoxes, offers unequivocal historical evidence, provided, however, that (1) the people among whom one finds this monument did not use a wandering year, and that (2) they did not intend to retrace, using systematic ideas, the earlier state of things, the point of origin, or the beginning point of a cycle. The peoples of eastern Asia calculated the positions of planets in remote times with the help of rather inexact tables; their books speak of a conjunction of all planets, which seems more of a result of their calculations than of observations. Can one not imagine that one day a monument will be discovered in India on which this conjunction is depicted, without having to place this monument in high antiquity?

The ancients do not say anything that directly proves that the Egyptians knew about the precession of the equinoxes. Hipparchus made this discovery by comparing his observations with those of ˅Timocharis; as Mr. Delambre has recently proved, it is almost certain that he made few, if any, field observations in Alexandria. Although Hipparchus did not owe the Egyptian priests anything, it is more than probable that those priests focused their attention on the relationship between the heliacal rise of Sirius and the day of the summer solstice. Within a period of 1,400 years, the difference between them varied by thirteen days.[1] We know too little about Egyptian astronomy to devalue it only because the Greeks and ˇManetho say nothing about it, especially since the latter knew as little about the exact sciences as he did about the rules of versification. Mr. Fourier will soon inquire further into this topic, which is important for the history of the advance of the human mind; his learned comments will be published in the anxiously awaited *Description des Monumens anciens de l'Égypte.*

The very ancient origin of Libra, claimed in the middle of the last century by ˇAbbé Pluche but challenged recently by two distinguished antiquarians, Mr. Testa and Mr. Hager, has been proved by the work of Mr. Ideler and Mr. Buttmann.[2] I think that scholars of ancient astronomy will be pleased to find assembled here passages that relate to Libra, which I have examined carefully: Hipparchus, *Comm[entarii] in Arat[um]*, Book III, ch. 2 (Petavius, *Uranolog[ion]*, ed. 1703, p. 134); Geminus, *Elem[entorum] Astron[omiae]*, chs. 1 and 16 (*Uranol[ogion]*, p. 139); Varro, *De lingua latina*, Book VI, ch. 2 (*Auctores lat[inae] linguae*, ed. Gothofred, 1585, p. 48); Cicero, *De div[inatione]*, Book II, ch. 146 (ed. Jos[ephus] Olivetus, 1740, Vol. III, pp. 81 and 478); German[icus] C[aesar], *In Arati Phae[nomena]*, verse 89 (Hygin[ius], *Opera*, Bas[el], 1535, pp. 164 and 187); Vitruv[ius], *De architect[ura]*, Book IX, ch. 4 (ed. Joannes de Laet., Amsterdam 1649, p. 190); Manilius, *Astro[nomica]*, Book I, verse 609, and Book IV, verse 203 (ed. Mich[ael] Fayus, Vol. I, pp. 77 and 313); Virgil. *Georg[ica]*, Book I, verse 34; Servius, *Commenta[rii] in Virg[ilium]*, Book V, p. 208 (ed. Pancrat[ius] Mascivius, Vol. I, p. 131); Plin[y], *Hist[oria] nat[uralis]*, Book XVIII, ch. 25, section 59 (ed. Harduin, 1723, Vol. II, p. 130); Ptolem[y], Book IX, ch. 7; Plut[arch], *De placi[ta] phil[osophorum]*, Book I, ch. 6 (ed. Reiske, Vol. IX, p. 486); Manetho, *Apotoles[matica]*, Book II, verse 137 (ed. Gronov[ius], 1698, p. 23); Macrob[ius], *Comment[arii] in Somnium Sci[pionis]*, Book I, ch. 19, and *Saturn[alia]*, Book I, chs. 12 and 22 (*Opera omnia*, ed. Gronov[ius], 1670, verses

1. "Over time, Sirius rose at even different points of the horizon and, at the same time, ever later during the solar year. In 2782 BC, it appeared on the second day after the summer solstice; in 1322 BC, on the thirteenth day; and in 139 BC only on the twenty-sixth day. . . . That, regardless of the precession of the equinox, it rose on the same date in the Julian calendar for 3,000 years was taken to be a consequence of its accidental position in relation to the latitudes and longitudes during this period of time" (Ideler, [*Historische Untersuchungen*,] pp. 88–90). [Translated from the German passage, which is slightly longer than the one Humboldt quotes.]

2. Ideler, *Hist[orische] Untersuchungen*, 1806, p. 371; *Sternnamen*, p. 175. Pluche, *Hist[oire] du ciel* (1740 ed.), Vol. I, p. 21. ˇMontucla, *Hist[oire] des mat[hématiques]*, Part I, Book II, § 7, p. 79. Bailly, *Hist[oire] de l'astr[onomie] ancienne]*, Vol. I, pp. 499 and 501. ˇSchmidt, *De zodiac[i] nostri] origine [aegyptia]*, p. 54. *Asiat[ick] Researches*, Vol. II, p. 302, and Vol. IX, p. 347. Dupuis, ["Observations sur le Zodiaque de Dendra,"] *Revue philos[ophique]*, May 1806, p. [321]. ˇSchwarz, *Rech[erches] sur l'origine de la sphère [grecque]*, p. 99. ˇSchaubach, *Gesch[ichte] der Griech[ischen] Astron[omie]*, pp. 242, 296, and 370. Hager, *Illustraz[ione] d'uno zodiaco [orientale]*, pp. 25–35. Anquetil, *Zend-Avesta*, Vol. II, p. 549. Testa, *Dissertaz[ione] sopra due Zodiaci dell' Egitto]*, 1802, pp. 20, 39, and 42. Delambre, *Astronomie*, Vol. I, p. 478.

90, 44, and 306); Achilles Tatius, *Isagoge [et Arati Phaenomena]*, ch. 23 and fragments (*Uranol[ogion]*, pp. 85 and 96); Theon, *Comment[arii] in Pto[lemeos]* (ed. Bas[ilensis], 1538, p. 386); ˅Martianus Capella, *De nupt[iis] philologiae et Mercurii*, Book VIII (ed. Princeps, 1498, Folio R. III); ˅Luc[ius] Ampelius, *Liber mem[orialis]*, Ch. 2 (ed. Bipontina ad calcem Flori, p. 158); Kircher, *Oedip[us] Aegypt[iacus]*, 1653, Vol. II, p. 906.

Among the writers of antiquity who mention the sign of Libra (ζυγός [*zygos*], τὰ ζυγά [*ta zyga*], λίτραι [*litrai*, pounds], *iugum, libra*), Hipparchus is the only one who lived prior to Julius Caesar's calendar reform. In his scholarly works, Abbé Testa missed the passage in Hipparchus's commentary on Aratus; Testa claims that Greek astronomers did not know the word ζυγός [*zygos*], and he adds: "Ne tre libri del commentario d'Ipparco sopra Arato, la libra non comparisce e *non si nomina mai, come ognuno può assicurarsene da per se*" [Libra does not appear in the three books of Hipparchus's commentary on Aratus, and it is never named, as anyone can see for themselves] (Testa, *Del Zodiaco*, pp. 21 and 46). I should point out here that the passage from Hipparchus to which I am referring is from the three-book commentary and not from the apparently apocryphal fragment that some attribute to Hipparchus, others to Eratosthenes. The words ζυγός [*zygos*] and *iugum* may well describe a couple, everything that is double or paired; but the prosaists prefer the use of ζεῦγος [zeūgos] for this, and Ptolemy opposes τὰ ζυγά [*ta zyga*] to χηλαί [*chelaí*, claws]; he would not do this if ζυγός [*zygos*] and ζυγά [*zyga*] explained χηλαί [*chelai*]. "The star," he writes, "that according to them (the Chaldeans) is found in Libra's basin and that according to our principles (our way of dividing up the zodiac) is located in the claws of Scorpio."[1]

Page 196. *Man-made hills.* People in both Americas ask themselves what the natives intended when they built so many artificial hills, of which some seem to have been used neither as tombs nor as overlooks nor even as foundation of a temple. An established custom in eastern Asia might shed some light on this important question. Two thousand three hundred years before our era, the Chinese made sacrifices to the Supreme Being, Chan-ty, on four high mountains called the *Four Yo*. The rulers found it inconvenient to travel there themselves, and they had artificial hills that

316

---

1. Ptolem[y], [*Opera*,] (ed. Bas[iliensis]), p. 232. In his commentary, Theon uses the singular ζυγός [*zygos*] and the plural τὰ ζυγά [*ta zyga*], often also the word λίτραι [*litrai*, pounds], a substitution that eliminates all doubts about the meaning of ζυγός [*zygos*]. Manetho talks about the "claws of Scorpio, which the holy men call the beam of Libra." This passage would be very remarkable were it proven that the astronomer Manetho were identical with the author of Αἰγυπτιακα [*Aigyptiaka*, work about Egypt] and had thus lived in the reign of ˅Ptolemy II Philadelphus (Fabricius, *Bib[lioteka] graeca*, 1795, Vol. IV, pp. 135–9). The word ζυγός [*zygos*] does not appear in the catasterisms of Erastothenes (ed. Schaubach, ch. 3, p. 6), but it does appear in the commentary about Aratus (*Uran[ologion]*, p. 142) that erroneously bears the name of this old astronomer but seems to have been written by Achilles Tatius.

represented these mountains built in proximity to their residences. ˟Lord Macartney, *[Voyage en Chine et en Tartarie,]* Vol. I, p. LVIII. Hager, *Monument de Yu*, 1802, p. 10.

Page 200. *Plain of Tapia, near Licán.* In order not to promote any false ideas about the garb of the Indians in the province of Quito, I must observe here that their clothes are typically black; but people of some greater means, for instance the Mestizos [Métis], wear *ruanas* made of striped woolen cloth (*listado*), which cover the Indian tunic called *capisayo*. These *ruanas* are depicted on Plate XXV, so that the human figures contrast with the background landscape and, at the same time, make it more varied. The style of the clothes is very faithful, but the colors of the *listado* are too vivid in some cases.

Page 210. *System of the Hindus.* Putting my faith in some of the *Sastras*, I erroneously claimed that among the Hindus all the Yugas [epoch within a cycle of four ages] ended with floods. ˟Mr. Majer, in his interesting work *on peoples' religious ideas*, notes that according to the teachings of the Bania [merchants], the first generation died in the floods, and the second perished in hurricanes; in the third age, the earth opened up and devoured the humans; and the fourth age will end in fire. Friedrich Majer. *Mythologisches Taschenbuch*, Vol. II, p. 299; and *Allgemeines Mythol[ogisches] Lexicon*, Vol. II, p. 471. These teachings are astonishingly similar to the Mexica tradition, down to the order of the destructions.

Page 218. *Tlacahuepancuexcotzin.* For a European, there is nothing more striking about the Aztec, Mexica, or Nahuatl language than the excessive length of the words. Contrary to what some scholars have claimed, this length is not always due to the circumstance that the words are compounds, as they are in Greek, German, and Sanskrit; rather, it is a result of the way in which the noun, the plural, and the superlative are formed. A kiss is called *tetennamiquiliztli*, which consists of the verb *tennamiqui*, to kiss, and the added particles *te* and *liztli*. Similarly: *tlatolana*, to request, and *tetlatolaniliztli*, a request or plea; tlayhiouiltia, to torment, and *tetlayhiouiltiliztli*, torment. To form the plural, the Aztecs doubled the first syllable of many words, such as *miztli*, cat, *mimiztin*, cats; *tochtli*, rabbit, *totochtin*, rabbits. *Tin* is the ending that indicates the plural. Sometimes, reduplication can also occur in the middle of the word, for instance, *ichpochtli*, girl; *ichpopochtin*, girls; *telpochtli*, boys; *telpopochtin*, boys. The most astonishing actual compound that I know is the word *amatlacuilolitquit-catlaxtlahuilli*, which signifies the reward given to a messenger who delivers a paper with some piece of news either in symbolic signs or as painting. This word, which by itself forms an Alexandrine verse, contains *amatl*, paper made from the Agave Americana; *cuiloa*, to paint or draw meaningful characters; and *tlaxtlahuilli*, the payment or wage of a laborer. The Aztec language lacks the letters B, D, F, G, and R. (˟Carlos de Tapia Zenteno, priest of Tampamolon, *Arte novissima de Lingua Mexicana*, 1753, p. 7.) Similarly, the Basque language lacks the letter F, and not a single one of their words starts with the letter R. However isolated some languages may appear at first glance, however extraordinary their moods and idiotisms [idiomatic or common expressions], there are still similarities between them all, and these manifold correspondences will come to light more and more with the perfection of the study of peoples'

317

philosophical history and of languages that are both products of intelligence and expressions of human individuality.

Page 226. *First age of the earth.* The Franciscan monk Andrés de Olmos, who knows much about the different languages of Mexico whose grammars he compiled, has left us a very intriguing note about the cosmogony of Anahuac. (˘Marieta, third part of the *Historia Eclesiástica*, 1596, p. 48.) The god *Citlalatonac* was bethrothed to the goddess *Citlalicue*; the fruit of their union was a flint stone, *tecpatl*, that fell to earth near a place known as the Seven Caverns, or *Chicomoztotl*. This betyl [sacred stone] can be found again among the hieroglyphs of the years and the days; it is an aerolite, a divine stone, a *teotetl*, which when it splintered produced 1,600 minor deities of the earth. These found themselves without slaves to serve them and asked their mother for permission to create humans. *Citlalicue* ordered *Xolotl*, one of the gods of the earth, to descend to the underworld in search of a bone, and this bone, upon splintering like the aerolite or tecpatl, created the human species (Torquemada, [*Monarquía indiana*,] Vol. II, p. 82). The same tradition has it that the first human, *Itzacmixcuatl* or *Iztacmixcohuatl*, remained in *Chicomoztotl*, where he reached a very old age. His woman, *Ilancueitl*, bore him six sons, from whom descended all the people of Anahuac. *Xelhua*, his oldest son, peopled Quauhyuechola, Tzoca, Epatlan, Teopantla, Tehuacan, Cozcatla, and Totctlan. *Tenuch*, the second son, became the father of the Tenochca [Toltecs], or Mexica in the proper sense. *Ulmecatl* and *Xicalancatl*, from whom descended the Olmecs and the Xicalanca, peopled the regions around Tlaxcala, Cuatzacualco, and Totomihuacan. *Mixtecatl* and *Otomitl* became the heads of the Mixteca and the Otomi (Torquemada, [*Monarquía indiana*,] Vol. I, pp. 34 and 35). This genealogy of peoples recalls the ethnographic tablet of Moses; it is all the more remarkable because the Toltecs and the Aztecs, among whom we find this tradition, saw themselves as belonging a privileged race and as very different from the Otomi and the Olmecs. The purpose of this genealogy was to reduce the diversity of languages to a single unifying principle and to explain this diversity through the common origin of all peoples.

Page 227. *Departure from Aztlan.* To facilitate the reading of this work about the monuments of the ancient peoples of Mexico, I will take up here a fragment from the Sketch of the history of Anahuac, which I began to write during my stay in Mexico City. This fragment will be of use to all those who do not have the leisure to go back to the sources and who have to limit themselves to Robertson's *History of America*, which is admirable for the wisdom of its organization but is too abridged in the part about the Toltecs and the Aztecs. I have been careful to cite all the authors on which I have based my dates.

Chronological Tableau of the History of Mexico

318

Like the Caucasus, the mountainous region of Mexico had been inhabited by a great many peoples of different races since the earliest times. Some of these peoples can perhaps be considered the remnants of the numerous tribes that passed through the land of Anahuac during their migrations from north to south. There some families separated from the rest of the group, held in Anahuac by the love of the land, which they cultivated, and they maintained their language, their customs, and the original form of their government.

The most ancient peoples of Mexico, those who regarded themselves as autochthonous, are the Olmecs, or Hulmecs, who went as far in their wanderings as the Gulf of Nicoya and Léon in Nicaragua; the Xicalanca, the Cora, the Tepaneca, the Tarasca, the Mixteca, the Zapoteca, and the Otomi. The Olmecs and the Xicalanca, who lived on the Tlaxcala plateau, prided themselves on having conquered or destroyed, on their arrival, the giants, or *quinametin*; this belief was probably founded on the findings of fossilized Elephant bones in the high regions of the mountains of Anahuac (Tor[quemada], [*Monarquía indiana*,] Vol. I, pp. 37 and 364). Boturini states that the Olmecs were chased away by the Tlaxcalteca and proceeded to people the Antilles and South America.

The Toltecs, who left their motherland Huehuetlapallan, or Tlapallan, in the year 544 of our era, arrived in Tollantzinco in the land of Anahuac in 648 and in Tula in 670. During the reign of the Toltec king Ixtlicuechahuac, the astrologer Huematzin composed, in 708, his famous divine book, the Teoamoxtli, which contained the history, mythology, calendar, and laws of the nation. The Toltecs are also the ones who appear to have built the pyramid of Cholula, after the model of the pyramids of Teotihuacan. The latter pyramids are the oldest of them all, and Sigüenza believes them to be the work of the Olmecs (Clav[ijero], [*Storia antica del Messico*,] Vol. I, pp. 126 and 129; Vol. IV, p. 46).

The Mexica Buddha appears during the time of the Toltec monarchy, or even during earlier centuries: Quetzalcohuatl, a white, bearded man accompanied by other strangers who were wearing black garb that looked like cassocks. The people used these Quetzalcohuatl clothes until the sixteenth century to dress up in disguise during festivals. In the Yucatan, this holy man was called Cuculca; in Tlaxcala, he was known as Camaxtli (Torq[uemada], [*Monarquía indiana*,] Vol. II, pp. 55 and 307). His robe was covered with red crosses. As High Priest of Tula, he founded religious congregations. "He ordered the sacrifice of flowers and fruits and covered his ears when there was talk of war." His fellow traveler Huemac was in possession of worldly power, while he himself enjoyed spiritual powers. This form of government was similar to the ones in Japan and Cundinamarca (Torq[uemada], [*Monarquía indiana*,] Vol. II, p. 237). But the first monks, Spanish missionaries, seriously debated the question whether Quetzalcohuatl was from Carthage or from Ireland. From Cholula, he sent colonies to Mixteca, Huaxaycac, Tabasco, and Campeche. It is assumed that the palace of Mitla had been built on the orders of this unknown person. At the time of the Spaniards' arrival, some green stones were kept in Cholula as if they were precious relics; they were said to have belonged to Quetzalcohuatl. And Father Toribio de Motolinía even saw a sacrifice in honor of the saint on the summit of the mountain of Matlalcuye near Tlaxcala. In Cholula, Father Toribio also witnessed penitential practices ordered by Quetzalcohuatl, during which the penitents had to pierce their tongues, ears, and lips. The high priest of Tula made his first appearance in Panuco; he left Mexico intending to return to Tlalpallan; on this voyage, he disappeared but not, as one should have assumed, in the north but in the east, on the banks of the Río Huasacualco (Torq[uemada, *Monarquía indiana,*] Vol. II, pp. 307–11). The people hoped for his return for many centuries. "When, upon my way to Tenochtitlan, I passed through Xochimilco," recounts the monk Bernardino de Sahagún, "everyone wanted to know if I was coming from Tlalpallan. At the time, I did not understand the meaning of this question, but I found out later that the Indians believed us to be descendants of Quetzalcohuatl" (Torq[uemada, *Monarquía indiana,*] Vol. II, p. 53.) It would be fascinating to trace the life of this mysterious figure—who, belonging to heroic times, probably came before the Toltec—down to its minutest details.

319

Plague and the destruction of the Toltec in the year 1051. They move farther south during their migrations. Two children of the last king and some Toltec families remain in Anahuac.

After they left their homeland Amaquemecan, the Chichimecs arrived in Mexico in 1170.

Migration of the Nahuatlaca (Anahuatlaca) in 1178. This people included the seven tribes of the Xochimilca, the Chalca, the Tepaneca, the Acolhua, the Tlahuica, the Tlaxcalteca, or Teochichimecs, and the Aztecs, or Mexica, all of whom, like the Chichimecs, used the Toltec language (Clav[ijero, *Storia antica del Messico,*] Vol. I, p. 151; Vol. IV, p. 48). These tribes called their homeland Aztlan, or Teo-Acolhuacan, and they said that it bordered on Amaquemecan (ʾGarcía, *Origen de los Indios*, pp. 182 and 502). According to Gama, the Aztecs had left Aztlan in 1064; according to Clavijero in 1160. The Mexica properly speaking separated from the Tlaxcaltecas and the Chalcas in the mountains of Zacatecas (Clav[ijero, *Storia antica del Messico,*] Vol. I, p. 156 ; Torq[uemada, *Monarquía indiana,*] Vol. I, p. 87; Gama, *Descripción de [las] dos piedras*, p. 21).

Arrival of the Aztecs in Tlalixco or Acahualtzinco in 1087; reform of the calendar and first festival of the new fire since the departure from Aztlan.

The Aztecs arrive in Tula in 1196, in Tzompanco in 1216, and in Chapultepec in 1245. "During the reign of Nopaltzin, the king of the Chichimecs, a Toltec named Xiuhtlato, master of Quaultepec, instructed the people in the cultivation of maize and cotton and taught them how to bake bread from cornmeal. The few Toltec families who lived on the banks of the lake of Tenochtitlan had completely neglected the cultivation of this graminaceous plant, and American wheat would have been lost forever had Xiuhtlato not preserved some grains of it from his youth" (Torq[uemada, *Monarquía indiana,*] Vol. I, p. 74).

The alliance of the Chichimecs, the Acolhua, and the Toltecs. Nopaltzin, son of king Xolotl, weds Azcaxochitl, daughter of a Toltec prince, Pochotl, and the three sisters of Nopaltzin enter liaisons with the heads of the Acolhua. There are few nations whose annals show such a multiplicity of family and place names as the hieroglyphic annals of Anahuac.

In 1314, the Mexica are enslaved by the Acolhua, but due to their valor, they soon succeed in freeing themselves from slavery.

Founding of Tenochtitlan in 1325.

Mexica kings: I. Acamapitzin, 1352 to 1389; II. Huitzilihuitl, 1389 to 1410; III. Chimalpopoca, 1410 to 1422; IV. Itzcoatl, 1423–1436; V. Moctezuma Ilhiucamina or Moctezuma the First, 1436–1464; VI. Axayacatl, 1464–1477; VII. Tizoc, 1477–1480; VIII. Ahuitzotl, 1480–1502; IX. Moctezuma-Xocoyotzin or Moctezuma the Second, 1502–1520; X. Cuitlahuatzin, who ruled for only three months; XI. Quauhtemotzin, who ruled for nine months of the year 1521 (Clav[ijero, *Storia antica del Messico,*] Vol. IV, pp. 55–61).

Nezahualcoyotl, king of Acolhuacan or Tezcoco, died during the reign of Axayacatl; he is as memorable for the culture of his mind as he is for the wisdom of his laws. This king of Tetzoco wrote sixty hymns in honor of the Supreme Being in the Aztec language, along with an elegy for the destruction of the town of Azcapotzalco and another on the fickleness of human power, as was proved by the example of the tyrant Tezozomoc. The great nephew of Nezahualcoyotl, baptized as Fernando Alva Ixtlilxochitl, translated some of these verses into Spanish, and the Chevalier Boturini owned the original of two of his hymns; they had been composed fifty years before the conquest and in Cortés's day were written down in Roman letters on *metl* paper. I have searched in vain for these hymns amidst the remains of Boturini's collection in the palace of the viceroy in Mexico City. It is also quite noteworthy that the famous botanist Hernández [de Toledo] made use of many of the drawings of plants and animals with which king Nezahualcoyotl had decorated his house in Tetzcoco and which had been created by Aztec painters. 320

Cortés lands on the beach of Chalchicuecan in 1519.

Conquest of the city of Tenochtitlan in 1521.

The Counts of Moctezuma and of Tula, who lived in Spain, are descended from Ihuitemotzin, the grandson of King Moctezuma-Xocoyotzin, who had married Doña Francisca de la Cueva. The origin of the illustrious houses of Cano Moctezuma, of Andrade Moctezuma, and of the Count of Miravalle (in Mexico City) goes back to Tecuichpotzin, the daughter of King Moctezuma-Xocoyotzin. This princess, baptized Elisabeth, survived five spouses, among them the last two kings of Mexico, Cuitlahuitzin and Quauhtemotzin, as well as three Spanish officers.

Page 235. *Cihuacohuatl.* Mr. Majer thinks that the figure of the mother of humanity, like the one on Plate XIII, refers to the history of Ata-Entsik and to his two small children, Juskeka and Tahuitzaron, both of whom are famous among the Huron and the Iroquois (*Mytholog[isches] Taschenb[uch]*, Vol. II, p. 241, and Vol. II, p. 294; Du Creux, *Hist[oriae] Canad[ensis], seu Novae Franciae*, 1664, Book I, p. 79).

Page 236. *Shape of the forehead.* The head of Teocipactli, Plate. XXXXVII, number 6, strangely resembles the one shown on Plate XI. According to information from Mexico, which I have received since the publication of the first part of this work, this remarkable sculpture was not found in Oaxaca, which I claimed in error (pp. 67–71 [this edition]), but farther to the south, near Guatemala, in the old *Quauhtemallan.* This circumstance further eliminates any doubts that might have been raised about the origin of such a strange monument. Besides, the ancient inhabitants of Guatemala were not a particularly cultivated lot, as the ruins of the large city prove, which can be found in a place that the Spaniards called *El Palenque.*

Page 255. *The number hieroglyphs*. In his sketch of the history of the world, Mr. Gatterer credits the Phoenicians and the Egyptians with the admirable invention of expressing tens by the position of the numbers. He expressly insists that in Egyptian manuscripts written in cursive characters, one can recognize nine letters of the alphabet, which indicate nine units, as well as a tenth sign that the Hindus and Tibetan used as a zero. The same scholar claims that Cecrops and Pythagoras had known this Egyptian system of numbering and that it originated in linear hieroglyphic arithmetic, in which the vertical lines have a positional value, whereas several rows of horizontal bars signify the tens and multiples of ten (Gatterer, *Weltgeschichte bis Cyrus*, p. 586). According to this theory, the Hindus' notation would have been introduced to Europe for the second time by the Arabs; but these claims do not appear to rest on very solid foundations (Kircher, *Obel[iscus] Pamph[ilius]*, p. 461). It is known that among the Romans, whose numerical system was infinitely more imperfect than that of the Greeks, units changed their value depending on whether they were placed before or after the signs for five or ten. An actual positional value lies in the notation, which, ˅Pappus reports, Apollonius used for myriads (Delambre, "[Sur l']arith[étique] des Grecs dans les Œuvres d'Archimède," 1807, pp. 5–8). But none of the peoples of whom we have more concrete knowledge appears to have advanced to the simple and uniform method that the Hindus, Tibetans, and Chinese have followed since high antiquity.

321   Page 257. *Twelve Suna*. The inhabitants of Otahiti divided the year not into twelve but into thirteen months or moons, to which they gave the names of the sons of the sun (˅Wilson, *[A] Missionary Voyage to the Pacific Ocean*, 1799, pp. 341–4). This division into thirteen parts is clearly rather extraordinary; but we know that even peoples who are highly advanced in their civilization kept using in the calendars, for a long time, numbers that were not particularly suited to temporal division. See ˅Mr. Niebuhr's fine study of the Roman and Etruscan year (*Römische Geschichte*, Vol. I, pp. 91 and 192).

Page 266. *Complete list of the paintings*. It is remarkable that the Franciscan Torquemada had already accused Bishop Zumárraga of barbarism—the latter is infamous for the destruction of the historical paintings of the Aztecs ([Torquemada,] *Mon[arquía] ind[iana]*, Vol. I, p. 276). One of the editors of the *Gazette littéraire* in Göttingen (1811, p. 1553) recalls that there are five Mexica manuscripts in the Bodleian Library at Oxford (*Monthly Mag[azine]*, Vol. [XI], p. 337). In his review of my studies of the monuments the indigenous peoples of America, this same scholar compares the bust represented on Plates I and II with the engraving of a head in ˅Tassie, *[A Descriptive] Cat[alogue]*, Vol. VII, p. 948.

ANNOTATIONS

Weights and Measures

ARROBAS: one Spanish arroba = 25 lbs; but one Portuguese arroba = about 32 pounds.

CROWNS (*couronnes*): the French *couronne du soleil* (literally: sun crown) was a gold coin used in the sixteenth century.

LEAGUE: any of a number of European units of measurement ranging from 2.4 to 4.6 statute miles (3,9 to 7,4 km). In English-speaking countries, the land league is typically 3 statute miles (4,83 km), although variations can range from 2,29 to 4,57 km. The Normans introduced this ancient Gallic unit to England, and the Romans estimated it at 1,500 paces (a pace, or *passus*, was nearly 5 feet = 1,5 meters). The Spanish used land leagues of about 2.63 miles (4,23 km). The league was also occasionally used as a unit of area measurement. The square league of old California surveys equals 4,439 acres (1,796 hectares). In the late eighteenth century, *league* also referred to the distance from which cannon shots could be fired at menacing ships offshore, which resulted in the three-mile offshore territorial limit. Nautical or geographical leagues come in a variety of different lengths. In the eighteenth and nineteenth centuries, one nautical league equaled 3 nautical miles. The international nautical league is equivalent to 5556 kilometers, while the British nautical league measures 18,240 feet (about 5559 kilometers).

LIVRE TOURNOIS: (literally, pound from Tours) a French gold coin introduced in the seventeenth century and subdivided into 20 sols. It was often referred to as *franc*, an older French monetary unit from the fourteenth century. After the French Revolution in 1795, *livres tournois* were replaced with the new franc.

MINUTE OF ARC: also arcminute or MOA; a unit of angular distance equal to one-sixtieth of a degree.

PALM: equal to the breadth or length of a hand.

PALMI ROMANI: according to Humboldt, 326 centimeters = 11 *palmi romani*. One centimeter (cm) = 0,01 meters (m) in the metric system; it is also the equivalent to 0.3937 inches. According to James Riddick Partington, one *palmus romani* (Roman palm) = 7.39 cm. He notes that 12 palms equaled about 34.8 inches, which suggests that either the conversion tables for *palmi romani* were very much at variance or that Humboldt missed a comma, writing 326 cm instead of 32,6 cm. As is noted by Donald W. Kurtz (2005), however, 1 Roman palm could also equal 22,3322 cm.

PIASTER: a unit of currency originally equal to one silver dollar or peso. Original French word for the US dollar; modern French uses *dollar* for this unit of currency as well. Slang for US dollar in the Francophone Caribbean, especially in Haiti.

POUCE (INCH): *Pouce* is French for thumb (among others). It is also used to translate the British and US-American inch into French. Since 1959, an inch has been determined to be 2.54 cm. In 1811, Humboldt could still choose from among varying lengths of inches or pouces. An inch or *pouce* is the twelvth part of a foot.

QUINTAL: also hundredweight, which initially corresponded to about 100 lb (or 45.4 kg) and later to 112 lb (c. 50.8 kg). It is still used in German-speaking countries.

SQUARE MILE: an area of 640 acres.

STADIUM (*stade*): A Greek unit of length denoting 600 feet (180 m)—the distance covered in a footrace. Also used in Rome, it survives today in the stadium structure of a sports arena.

TOISE: an old French measure equaling 1.949 m.

VARA: (stick or pole) a traditional Spanish and Portuguese unit of distance whose length varied: in Spanish Latin America, typically about 33 inches. Often used in land measurement in Texas as equal to 33⅓ inches (84.667 cm), 33 inches (83.82 cm) in California, but only 32.993 inches (83.802 cm) in Mexico. About 34 inches (86.4 cm) in the Southern Cone. The Spanish *vara* equaled 32.908 inches (or 83.587 cm), whereas the Portuguese *vara* equaled 5 *palmos* (palms), about 110 cm (43.3 in).

Annotations

Except where context makes it more sensible to do otherwise, we have annotated a reference or allusion at its first occurrence. Boldface signals related annotations, and readers should consult the index for page numbers. For detailed archaeological and anthropological information, we refer the reader to the notes in Eduardo Matos Moctezuma's edition, *Vistas de las cordilleras y monumentos de los pueblos indígenas de América* from 1995 (see also our editorial note).

ii. The Jesuit Francisco Javier CLAVIJERO Echegaray, also known as Francesco Saverio Clavigero (1731–87), was a historian from the Viceroyalty of New Spain (which includes modern-day Mexico). A polyglot who knew Nahuatl, Otomi, Mixteca, Latin, and ancient Greek, among other languages, he analyzed the documents on Aztec history and early conquest narratives that **Carlos de Sigüenza y Góngora** had donated to the Colegio Máximo de San Pedro y San Pablo (1576–1767) in Mexico City. Exiled to Italy after Spain expelled the Jesuits in 1767, Clavijero wrote in Spanish and published in Italian the four-volume *Storia antica del Messico* (*History of Mexico*, 1780–81), in which he publicized the existence of the **Codex Cospi** and refuted arguments by **de Pauw**, Count Georges-Louis Leclerc de Buffon (1707–88), **Raynal**, and **Robertson** (1721–93) about Europeans' superiority to Native Americans and Creoles (Americans of European descent). Shortly after its Italian publication, the book was translated into English (1787), German (1789–90), and Spanish (1826). Clavijero's Spanish manuscript was published in Mexico in 1945. In addition to numerous essays and letters, Clavijero authored the *Historia de antigüa Baja California* (History of ancient Lower California, 1852).

An accomplished poet and dramatist, in 1661 Antonio de SOLÍS y Rivadeneyra (1610–86) was appointed Spain's *cronista mayor*, an office in which he succeeded Antonio de León Pinelo (c. 1590–1660). A native of Alcalá, where he studied classics, philosophy, and law, Solís authored the *Historia de la conquista de México* (History of the conquest of Mexico, 1684), defending Spain's conquest of the Mexica empire to improve his country's image in the international community. Solís corrected earlier accounts of the conquest and argued against what he believed to be other historians' biases in order to defend the Spaniards' actions. Although he spent twenty years preparing the manuscript for the *Historia*, a literary elaboration of known historical events written in epic style, Solís completed only the first part, which covers the conquest throught the fall of Tenochtitlan in 1521. The most popular chronicle of the late seventeenth century, the *Historia* was translated into all the major European languages. It served as the reference book on Cortés and the Conquest until the nineteenth century.

iii. The ASIATICK SOCIETY of Bengal, or Asiatick Society of Calcutta (now Asiatic Society Kolkata) was founded in 1784 by **Sir William Jones** (1746–94), a judge in the employ of the East India Company. Its journal, *Asiatick Researches, or Transactions of the Society instituted in Bengal, for inquiring into the history and antiquities, the arts, sciences, and literature of Asia*, was distributed and pirated throughout Europe. The journal was originally published in Calcutta, then reprinted in London by John Murray between 1801 and 1818 under the title *Asiatick Researches*, later modernized to *Asiatic Researches*. A pirated edition appeared in London between 1798 and 1799 under a slightly changed title.

A distinctive trait of ancient Mesoamerican cultures is the use of books of HIEROGLYPHIC PAINTING. Only sixteen preconquest pictorial manuscripts still exist, among them the ones to which Humboldt refers throughout: the Codices Borgia, Cospi, Dresden, Vaticanus B, Vienna, and Borbonicus. The remaining codices are Codex Féjérváry-Mayer, Codex Laud, Aubin Manuscript no. 20, Codex Becker No. 1, Codex Bodley, Codex Colombino, Codex Nuttall, Codex Madrid, Codex Paris, and the Tonalamatl Aubin, the date of which is in dispute. Pictorial manuscripts continued and developed during the sixteenth century under European influence. Humboldt worked with several pictorial Mesoamerican codices that were produced under Spanish patronage: the Codices Mendoza, Ríos, Telleriano-Remensis, and Vaticanus A (we will discuss them in greater detail below). The numerous fragments he collected in Mexico, known as the **Humboldt Fragments**, also date from after the conquest.

v. *ORIENTAL SCENERYS* is the title of a six-volume work of 144 hand-colored aquatint engravings based on a selection of drawings by John Frederic Daniell (1749–1840) and his nephew William (1769–1837). These volumes introduced many of India's most famous buildings and sites to the European public. While the Daniells, who arrived in Calcutta in 1786, would probably have chosen to be remembered for their oil paintings, it is the volumes of *Oriental Scenerys* that secured their artistic reputation.

The oil paintings are scattered in various galleries today, such as the Fine Arts Museums of San Francisco and the Royal Academy of Arts in London.

viii. On June 5, 1799, Humboldt left Europe from the Spanish port of La Coruña on the *Pizarro*. After a passage of thirteen days, the ship reached the Canary Islands, where Humboldt, who had requested a stopover there, first set foot on non-European soil. He remained on the Spanish archipelago for eight days, mostly in Santa Cruz de Tenerife. On June 21, Humboldt and Bonpland ascended the highest mountain on the Canary Islands, the active volcano MOUNT TEIDE. They were accompanied by local mountaineers and the French vice consul Louis Le Gros. Le Gros, who was also a passionate naturalist, had shown Humboldt the rich collection of tropical flora at the island's botanical garden in Orotava. It was here that Humboldt scrutinized the famous Dragon Tree (see Plate LXIX). In the morning of June 22, the expedition reached Teide's peak, where Humboldt—engulfed by sulfuric vapor—sketched the crater's caldera and tried to descend as deeply as possible into its basin (see Plate LIV). The Mount Teide experience would served Humboldt as an important point of reference on his American journey, and he frequently compares the geological formations of the Cordilleras with this impressive volcano on the Canaries. The climb also inspired Humboldt's future works on plant geography, as he had noticed how the uncommonly diverse flora of Tenerife followed different temperature zones on the mountain's steep slope.

The German theologian and philologist Johann Severin VATER (1771–1826) is best remembered for the fact that he edited volumes 2–4 of **Johann Christoph Adelung**'s *Mithridates* after Adelung's death in 1806. Vater was on the theological faculty of the universities in Halle, Jena, and Königsberg (Kaliningrad), where he also taught Oriental languages. *Mithridates* assembled samples from all languages that were then known, organizing them geographically. Alexander's brother Wilhelm von Humboldt contributed to the fourth volume additions concerning the Basque language.

The German explorer, scholar, and naturalist Ulrich Jasper SEETZEN (1767–1811) graduated in medicine from Humboldt's alma mater, the University of Göttingen. He made a pilgrimage to Mecca in 1809 and traveled the Arabian Peninsula after that. He was found dead in what is now Yemen in 1811; it is believed that he had been poisoned.

x. Benjamin Smith BARTON (1766–1815) was a physician and botanist from Pennsylvania educated at the University of Edinburgh (1786–88). He had a lifelong interest in the early history and language of Native Americans, even as he was working on the faculty of the medical school at the College of Philadelphia, soon to be renamed the University of Pennsylvania. In 1803, President **Thomas Jefferson** asked Barton to help train **Meriwether Lewis** in scientific observation to prepare him for the Lewis and Clark expedition to the Pacific. Barton was also recruited to prepare that expedition's natural history reports but was unable to complete the task, partly on account of ill health. In 1797, Barton published a treatise on the antiquity and the languages of Native Americans titled *New Views of the Origin of the Tribes and Nations of America*, which he dedicated to Jefferson. In this work Barton also compared Native American languages to Asiatic languages in an attempt to trace the origins of Native American peoples, who, Barton argued, came from Asia. Humboldt met Barton during his brief visit to the United States in the spring of 1804.

xiv. Humboldt is referring to a chronicle in **Francisco Clavijero's** *Storia antica del Messico* (*History of Mexico*, 2.86–90) about the expansion of the Chichimecs through peaceful union with tribes that had migrated to the valley of Mexico from the north. According to Clavijero, three princes of the CITIN FAMILY—Acolhuatzin, Chiconquauhtli, and Tzontecomatl—led the Acolhua tribe to Tetzcuco or Teztcoco (Texcoco), the capital of the Chichimec kingdom. The Chichimec King Xolotl married his two daughters to the newcomers and administered his expanded kingdom together with his son and sons-in-law. Clavijero's source is the colonial historian **Fernando de Alva Cortés Ixtlilxochitl.**

MANCO CAPAC, or Ayar Manco, is the mythical founder of the Inca empire in Cusco. Two of the three different versions of the myth start with Manco Capac and his siblings' journey to Cusco. First, they emerged from a mythical cave located at Pacariqtambo. The siblings traveled northward to Cusco, where they established a new dynastic order after defeating the indigenous people of the region. The second myth refers to the decision of the deity of the sun, Inti or Apu-Punchau, to send his children, among them Manco Capac and Mama Ocllo, to found a civilization at Cusco. The subsequent Sapa Inca (sole rulers) were believed to be direct descendants of Manco Capac and the Sun. The third version, popular among Spanish historians, speaks about a ruler named Manco Capac who impressed his guilible subjects with splendid attire and elaborate customs.

1. The legendary king and conqueror Alexander III of Macedonia, known as Alexander the Great (356–323 BCE), overthrew the Achaemenid Persian empire in 331 BCE in the battle of Arbela, in which he defeated Darius III of Persia. Historians and archaeologists have long been debating the exact location of this battle, which is assumed to have taken place not in ARBELA (today's Arbil in northern Iraq) but close to Gaugamela (today's Mosul in northern Iraq). The hard-won battle gave Alexander domination over the vast Persian territories—at the time the largest empire of the ancient world—and opened the way for Hellenistic forays into Asia. In 1779, José Nicolás de Azara (1730–1804), a Spanish ambassador to the Holy See (Vatican) and later France, found a BUST OF ALEXANDER THE GREAT in the villa of Pisons near Tivoli, Italy. This famous bust is actually a herm, that is, a squared stone pillar topped with a bust made of fine marble from Mount Pentelicus near Athens. Herms have appeared in Greek art since the sixth century BCE, and early examples typically feature the head of the god Hermes (from which it derives its name) with a phallus carved at the front of the pillar. Known as the Azara Herm, this particular sculpture is a Greco-Roman copy of an original from around 330 BCE and attributed to **Lysippus**; it is believed to be the closest representation of Alexander. Azara gave the herm to Napoleon Bonaparte, who, in 1803, donated it to the Louvre, where it is still housed.

2. Humboldt uses the phrase "esprit de système," which signifies here a unified theory or belief system, as an allusion to the schematic thinking of **Cornelius de Pauw** and other eighteenth-century cabinet philosophers, notably their belief in the inferiority of the New World.

LYSIPPUS and his pupil PRAXITELES were famous sculptors in ancient Greece during the fourth century BCE. They shaped a period in Greek art that can be considered decisive for the emergence of the classical style in sculpture. In the eighteenth century this style, representative of ancient Hellenic aesthetics, became a model for European classicism. Humboldt's aesthetics—the harmony and perfection of mimetic form—centered on this classicist ideal. It is from this point of view that he contemplates American cultural artifacts, comparing them to what he—in the tradition of the German art historian Winckelmann—considers the timeless achievements of ancient Greek culture. Johann Joachim WINCKELMANN (1717–68) is considered the father of modern art history. He shifted interest away from the focus on Rome, which was prevalent until the eighteenth century, and placed Athens and Greek art at the center of attention. Winckelmann involuntarily proved that scholarly research can be a matter of life and death, being almost killed by a falling statue while examining a collection.

4. Humboldt and Guillaume DUPAIX (1750–1819) met in the Mexican capital, where the Prussian naturalist saw Dupaix's private collection of Mexica sculptures, from which the "Aztec priestess" (Plates I and II) is taken. The small sculpture is of the water goddess and patroness of birth Chalchihutlicue (She of the Jade Skirt). It was later acquired by Henry Christy (1810–65), who visited Mexico in 1856, and is now at the British Museum. Dupaix was a former French naval captain of Flemish descent and a collector of Mexican antiquities. Having lived in the kingdom of New Spain for most of his life, Dupaix was sent on three archaeological expeditions (in 1805, 1806, and 1807) in search of relics of pre-Columbian art by the Spanish king **Charles IV**. He was accompanied by the Mexican illustrator José Luciano Castañeda (1774–183?). Dupaix's observations about Xochicalco, Monte Albán, Mitla, and Palenque were among the first European efforts to distinguish the various architectural and art styles of ancient Middle America. A letter from Humboldt to the Louisiana collector François Latour-Allard (fl. 1827), dated July 28, 1826, attested to the accuracy of Castañeda's work, some 177 sketches and paintings, which Latour-Allard sold in England and France. Castañeda's drawings and Dupaix's travel account resulted in a voluminous manuscript that remained unpublished until it was incorporated into the highly acclaimed *Antiquities of Mexico*, edited by Edward King, Lord Kingsborough (1795–1837) in 1831. In 1834, years after Dupaix's death in 1819, it was released in Paris as *Antiquités mexicaines*. Humboldt's comments, alongside other favorable reviews of Dupaix's travelogue, spurred enormous interest among French archaeologists in the Mexica.

5. The CALANTICA (or *calautica/calvatica*) was a headdress similar to a miter that women wore in antiquity. Humboldt uses the Spanish word.

The many statues, busts, and faces preserved of ANTINOUS present him as the ideal type of youthful beauty in the ancient Roman empire. Antinous was the lover and confidant of the Roman emperor Hadrian (76–138). After Antinous's mysterious death, Hadrian decreed his deification and promoted acts of worship all across his empire. Antinous's image became ubiquitous in the ancient world and was associated with, and indeed depicted as, the Egyptian god Osiris and the Roman fertility god Bacchus.

APIS (or Hape) is the Egyptian bull-deity embodying the eternal soul of the god of creation and fire, Ptah. Worshiped mainly at Memphis, the ancient capital of Egypt, as a living animal-god and oracle, Apis was identified with Osiris, the Greco-Egyptian deity Serapis (or Sarapis), and Jupiter. In 1851, Auguste Ferdinand François Mariette (1821–81) rediscovered a vast underground necropolis for mummified Apis-Bulls at Saqqara, the cemetery of Memphis. Humboldt is likely describing the Apis bull statue currently at the Vatican Museo Gregoriano Egizio, founded in Rome in 1839. That statue was found in Emperor Hadrian's famous villa at Tivoli, Italy (built c. 118–34). Sculpted of dark, reveined granite, the statue is the torso of a human figure with an Apis head; it is wearing a sun disc between the horns, a *klaft*-like (folded linen) headdress, and a broad collar.

The Capitoline Museum is composed of a group of art and archaeology museums around the Piazza del Campidoglio at the Capitoline Hill in Rome. In the sixteenth century, Michelangelo Buonarroti (1475–1564) designed the piazza (and the buildings' remodeling) that now forms the museum. Administered by the municipality of Rome, it is the oldest public museum in the world. Its creation dates from 1471, when Pope Sixtus IV (1414–84) donated a group of bronze statues, including the famous She-Wolf (fifth century BCE), to the people of Rome. After its inauguration by Pope Clement XII (1652–1740) in 1734, the museum opened to the public.

The ancient town of TENTYRIS or Tentyra (today's Dendera) lies on the west bank of the Nile in central Egypt. It is best known for the famous, well-preserved temple complex of Hathor, whose construction lasted for over two hundred years, from the reign of the Greek emperor Ptolemy VI (180–145 BCE) to that of the Roman emperor Nero (37–68, r. 54–68). The composition and design of the temple's columns and its colonnaded street added new ornamentation and detail, which would become typical of Egyptian architecture in the Greco-Roman period. Some of these ornamentations were first sketched by Dominique Vivant DENON (1747–1825), a French writer, diplomat, and antiquarian. Director-general of the Imperial Museums (today's Louvre) during the Napoleonic empire (1804–14), he had taken part in Bonaparte's Egyptian campaign (1798–1801) as his chronicler and courtier, a position that Denon had already held at the court of Louis XV (1710–74). Denon's own impressions of the Egyptian expedition were woven into his epic account *Voyage dans la Basse et la Haute Égypte pendant les campagnes du Général Bonaparte* (Voyage in Lower and Upper Egypt during the campaigns of General Bonaparte), which made him well-known among European archaeologists and other scholars.

While visiting his brother Wilhelm in Rome in March 1805, Humboldt became acquainted with the Danish archaeologist and consul general Johann Georg ZOËGA

(1755–1809), who showed him the city's archaeological collections. With his works on Roman bas-relief and sculptures, Zoëga became one of the founding fathers of European archaeology in the spirit of **Johann Joachim Winckelmann.** See also Lysippus and Praxiteles.

Upon becoming a member of the Etruscan Academy of Cortona in 1750, the Italian theologian STEFANO BORGIA (1731–1804) began to assemble one of the richest private collections of antiquities. It became kown as the Borgia Museum of Antiquities in Velletri. Humboldt's **Codex Borgia**, which is now also known as the Codex Yoalli Ehecatl, is named after him. Borgia held various high positions in the Vatican administration. Benedict XIV appointed him governor of Benevento in 1759; Pope Pius VI made him a cardinal in 1789; and Pope Pius VII chose him as rector of the Collegium Romanum in 1801. These and other positions allowed Borgia to gather an impressive collection from missionaries all over the world. It included fine arts, coins, statues, idols, rare Mexica manuscripts, codices from the East, and a many documents in Coptic. Borgia wrote several books, among them *Monumento di papa Giovanni XVI* (Monument of Pope Giovanni XVI, 1750) and *Breve istoria dell antica città di Benevento* (Brief history of the antique city of Benevento, 1763–69). **Johann Georg Zoëga** (1755–1809) described Borgia's collection in his posthumous *Catologus codicum Copticorum manuscriptorum qui in Museo Borgiano Velitris adservantur* (1810). The collection is now housed at the Museo Nazionale in Naples and at the Vatican Library.

6. EPAGOMENAL DAYS are one of five (or six) days used to match the Egyptian civil, or wandering, calendar to the solar year.

7. The "great geometrician" to whom Humboldt refers here is Pierre Simon, marquis of LAPLACE (1749–1827), whom he first met in Paris in 1798. This influential French mathematician and astronomer was the author of the five-volume *Traité de mécanique céleste* (Treatise on celestial mechanics, 1799–1825), a groundbreaking study of the scientific achievements in astronomy and astrophysics at the time. The *Exposition du système du monde* (The system of the world) can be considered its successor. Humboldt and Laplace extended their scientific relationship during Humboldt's years in Paris (1804–27). Laplace's writings were an important inspiration for Humboldt's *Kosmos*.

AL-MA'MŪN, whose full name is Abū Al-'abbās 'abd Allāh Al-ma'mūn Ibn Ar-rashīd (786–833), was a caliph during the 'Abbāsid dynasty, the Islamic dynasty that founded Baghdad and ruled in Iraq between 749 and 1258. This philosopher and astronomer was known for his efforts to end sectarian rivalry in Islam and to impose a rationalist Muslim creed. Al-Ma'mūn encouraged the translation of Greek philosophical and scientific works and imported manuscripts from Byzantium. In Baghdad in 830, he founded a translation academy and research center called Bayt al-Hikmah (House of Wisdom), which survived until the thirteenth century. Al-Ma'mūn also established observatories at which Muslim scholars could verify the astronomic knowledge handed down from antiquity.

Juan Vicente Güemes Pacheco de Padilla, COUNT OF REVILLAGIGEDO (1740–99), was born in Havana and, like his father before him, served as viceroy of New Spain (1789–94). He was an effective administrator, improving living conditions in Mexico City and introducing the system of intendancies. He also founded the General Archives and, in 1793, the Museum of National History.

The "great temple of MEXITLI," known today as the Templo Mayor, was called Hueyteocalli in Nahuatl. *Mexitli* is another name for Huitzilopochtli, god of war and the sun, especially in the context of the emigration of the Mexica from their homeland in today's northern Mexico. Constructed in seven stages, the temple was placed at the dead center of the great plaza of Tenochtitlan to represent the center of the universe. It also signified access to the celestial levels and the underworld. Dedicated to both Huitzilopochtli and Tlaloc, the god of rain, water, and fertility, this temple was oriented to the west (the direction of the setting sun); it had two shrines at the top that could be accessed by dual stairways that represented two sacred mountains. The shrine on the northern side (Hill of Sustenance) was for Tlaloc, the one on the southern side (Hill of Coatepec) for Huitzilopochitli. The temple reflected Tenochtitlan's two means of support: agriculture and military conquest. Between 1978 and 1982, archaeologists were able to excavate the Templo Mayor, discovering hundreds of offering niches with thousands of artifacts, such as stone masks, ceramic vessels, turtles, crocodiles, and other animal remains. The excavations illustrate the evolution of the building from 1390 until 1521. At its zenith, the temple measured 131 feet in height and about 263 feet in width on each side. The Spaniards and their native allies destroyed the building after the Conquest.

Teocalli is Nahuatl for "god-house" (from *teo*, god, and *calli*, house or dwelling). In modern contexts, the word is also used by the Chicano congregations of the Native American Church, who refer to the organization as a *teocalli*.

The temple or tower of JUPITER BELUS in the Sumerian city of Etemenanki is believed to belong to the Akkadian god Bel, whom **Herodotus** hellenized to Zeus Belus. In 440 BCE, Herodotus described the sacred area as consisting of a "square enclosure two furlongs [402 m] each way, with gates of solid brass. . . . In the middle of the precinct there was a tower of solid masonry, a furlong [201 m] in length and breadth, upon which was raised a second tower, and on that a third, and so on up to eight. The ascent to the top is on the outside, by a path that winds around all the towers."

Hernán (or Fernando) CORTÉS de Monroy y Pizarro (1485–1547) was a Spanish *hidalgo* (low gentry) and adventurer. Sent to the New World in 1519 by the colonizer and first governor of Cuba, Diego Velázquez de Cuéllar (c. 1461–1524), to pillage the eastern coast of Mexico, Cortés instead organized a small expedition of six hundred men to conquer the new land. With the help of his native interpreter and mistress, Malintzín (also known as Marina or Malinche), and the allied Tlaxcalteca, Cortés and his men rapidly gained control of the Mexican heartland and, in 1521, overthrew the Mexica capital of Tenochtitlan, where they captured and finally killed **Moctezuma** and his successor, Cuauhtemoc (c. 1495–1522), the last kings of the Mexica empire. Cortés's first marriage, in Cuba, was to Catalina Xuárez Marcaida (d.1522). Because she was rumored to have been killed by him, an investigation of Cortés was launched, but it never led to any charges. He did not have children from this marriage. In 1528, Cortés traveled to Spain to request in person the governorship of Mexico from **Charles V** (1500–1558), who instead confirmed him as a captain-general of New Spain and named him marquis of Valle de Oaxaca. Cortés remarried, this time to Juana Ramírez de Arellano Zúñiga, with whom he had several children, including his heir Martín and three daughters named María, Catalina, and Juana. He also had five children out of wedlock, notably Martín (from Malintzín) and Leonor (from Isabel Moctezuma, a close relative of Moctezuma II). Cortés's female descendants were eventually married to the dukes of Monteleone from Italy.

MONTEZUMA II (c. 1466–1520), also known as Motecuhzoma Xocoyotzin or Montezuma the Younger—alternate spellings are Motēuczōmah and MOCTEZUMA—is the most widely described and documented of all Mexica emperors. He was the last *tlatoani* (great speaker, by extension king or great lord) of Tenochtitlan (1502–20). The nephew of **Ahuizotl** (r. 1486–1502), this eighth ruler of the Mexica empire conquered various towns in Oaxaca and Tlaxcala and suppressed rebellions inside the empire's borders. Following Mexica political traditions of announcing and establishing the terms of war before attacking, Montezuma met and welcomed Cortés in 1519. Once inside Tenochtitlan, Cortés captured Montezuma, who died shortly thereafter under historically contested circumstances, after fleeing and spreading smallpox and other infectious diseases among his people.

8. CHARLES V (1500–1558) was Holy Roman emperor (1519–56), king of Spain (as Charles I, 1516–56), and archduke of Austria (as Charles I, 1519–21). The Spanish and Habsburg empire he inherited comprised large parts of Europe—Spain and the Netherlands, Austria and the kingdom of Naples—as well as the Spanish overseas colonies.

Humboldt dedicated his *Political Essay on the Kingdom of New Spain* to Charles IV (1748–1819), king of Spain (1788–1808), who had approved an all-access royal passport for Humboldt and Bonpland's visit to Spanish territory in the Americas. According to a letter Humboldt wrote to **Thomas Jefferson**, he did this as a way to pacify the attitude of the Spanish crown toward certain individuals in New Spain who had given Humboldt more information than the court would have liked. Because of the Napoleonic upheavals in Europe, Charles IV was forced to abdicate power to his son Ferdinand VII (1784–1833), who was forced to do the same for Napoleon's brother, Joseph Bonaparte.

Miguel de la Grúa Talamanca de Carini y BRANCIFORTE (1755–1812), the first marquis of Branciforte, was a Spanish military officer of Italian origin and captain-general of the Canary Islands from 1784 to 1790. Charles IV named Grúa viceroy of the kingdom of New Spain, where he stayed in power from 1794 to 1798. Considered one of the most corrupt rulers in Spanish colonial history, Grúa was responsible for *El Caballito*, the famous equestrian statue of Charles IV, which he commissioned in 1796 from Valencian architect and sculptor Manuel TOLSÁ (1757–1816). Since 1791 Tolsá had been director of sculpting in the Academy of San Carlos in Mexico City, where he made important architectural contributions to the cathedral and the Palace of Mines. The statue of the ruler features a neoclassical style, giving the Spanish king, usually characterized as politically weak, the air of a regal commander. In 1803, the completed statue was transported to the Plaza Mayor in Humboldt's presence; it remained in its original location until 1822, when it was removed due to postrevolutionary ill-feelings against the Spanish colonial heritage. Subsequently stored in the patio of the city's university, *El Caballito* was not returned to the public until 1852, when President Mariano Arista (1802–55) was in power. It can still be admired today in front of the National Museum of Arts in Mexico City.

9. The Iberian artist RAFAEL XIMENO Y PLANES (1759–1825) was professor of painting at the Academy of San Carlos in Mexico City (founded in 1783). Together with **Manuel Tolsá**, he introduced the neoclassical style to New Spain, revived mural painting in Mexico, and painted the dome of the cathedral that burned in 1967. He also painted two pictures for the ceiling of the chapel in the Palace of Mines, one of which includes groups of Mexica. Particularly famous are his portraits of Tolsá and Gerónimo Antonio Gil (1732–98), both at the Museo Nacional de Arte, Mexico City. An unsigned 1803 portrait of Humboldt in blue uniform, which is currently housed at the Palacio de Minería in Mexico City, is also attributed to Ximeno y Planes.

The French geologist and botanist Louis-François Elisabeth RAMOND (1755–1827), baron of Carbonnières, was the first explorer to study the Pyrenees. Starting in 1787, his expeditions led him to climb many of the mountains' peaks and passes. In 1802, Ramond became the driving force behind the first ascent of the Pyrenees' Monte Perdido (11,007 ft), next to the enormous glacial valley of Ordesa in today's Ordesa y Monte Perdido National Park. He wrote extensively about this experience in his 1801 travelogue *Voyages au Mont-Perdu et dans la Partie adjacente des Hautes-Pyrénées* (Travels to Mont-Perdu and in the neighboring parts of the High Pyrenees). Inspired by German Romanticism, Ramond's works on landscape descriptions, with their unique fusion of science and poetry focusing less on precise landscape description than on the act of perception itself, became an important inspiration for European naturalists and explorers at the turn of the century.

The botanist and physician Aimé Goujaud BONPLAND (1773–1858) accompanied Humboldt during his entire American journey. Although not much of a writer, Bonpland, who much preferred fieldwork to desk work, is named as coauthor of Humboldt's thirty-volume *Voyage aux régions équinoxiales du Nouveau Continent fait en 1799, 1800, 1801, 1802, 1803 et 1804* (*Travels to the Equinoctial Regions of the New Continent during the Years 1799–1804*, in Helen Williams's translation). After returning to Europe, Bonpland was named chief gardener of Malmaison, Napoleon's residence near Paris. After Napoleon's fall, in 1816 Bonpland emigrated to South America, and he died alone and forgotten in what is now Santa Ana, Argentina.

11. The BERTHOUD CHRONOMETER, a marine chronometer, is a timepiece sufficiently accurate to determine longitude at sea by means of celestial navigation. Humboldt's chronometer was crafted by Pierre-Louis Berthoud (1754–1813), a nephew of Ferdinand Berthoud (1727–1807), who helped develop and perfected marine chronometers. Humboldt's chronometer (Berthoud No. 27) once belonged to the famous French mathematician and naval general Jean-Charles Borda (1733–99).

JORGE TADEO LOZANO (1771–1816), a native of Bogotá, studied chemistry, mineralogy, and botany in Spain. In 1801, as the cofounder and coeditor of the *Correo Curioso, Erudito, Económico y Mercantil de la Ciudad de Santa Fé de Bogota* (Interesting and learned economic and commercial review of the city of Santa Fé de Bogotá), the first independent periodical in New Granada, he published a series of short essays addressing problems of development. In 1803, he was officially appointed a zoologist of the Mutis's famous botanical expedition. An active thinker and writer, he contributed to the *Semanario del Nuevo Reino de Granada* (1808–10), edited by Francisco José de Caldas (1768–1816). Among other articles he contributed to the *Semanario*, Lonzano published a Spanish translation of an early manuscript of Humboldt's *Essay on the Geography of Plants* as "Geografía de las plantas o quadro físico de los Andes Equinoxiales y de los países vecinos, levantado sobre las observaciones y medidas, hechas sobre los mismos lugares desde 1799 hasta 1803 y dedicado con los sentimientos del mas profundo renacimiento al ilustre patriarca de los Botánicos D. José Celestino Mutis; Por Federico Alexandro Barón de Humboldt" (1809). Lozano was also the first governor of the state of Cundinamarca, where he promulgated Cundinamarca's first constitution.

12. THOMAS JEFFERSON (1743–1826) was not only author of the US Declaration of Independence and the third president of the United States. He was also an architect and an accomplished scholar and naturalist with wide-ranging intellectual interests that cumulated in 1825 with his founding the University of Virginia in Charlottesville, an institution that reflected the ideal of the modern research university in the United States long before the idea took root there in the late nineteenth century. Humboldt first met Jefferson during a short visit to the United States near the end of his travels in the New World. Jefferson invited him to Washington, DC, where Humboldt supplied the government with the latest statistical and geographical findings about New Spain. Humboldt referred to Jefferson's *Notes on the State of Virginia* (1782) in a letter to Jefferson from Philadelphia dated May 24, 1804 (available in the Library of Congress Jefferson Papers online). On April 14, 1811, Jefferson sent Humboldt an inscribed copy as a token of their "constant friendship & respect."

13. Wilhelm Friedrich GMELIN (1745–1821) was an engraver and painter from Germany who studied under Christian von Mechel (1737–1817) in Basel, Switzerland. His best works were inspired by Claude Lorrain (c. 1605–82) and Nicolas Poussin (1594–1665). In 1788 Gmelin moved to Rome, where he worked and died. LOUIS BOUQUET (fl. 1800–1817) was an engraver who focused on landscapes. He and Gmelin collaborated in the production of Humboldt's plates (see "Notes on the Plates").

14. Pierre BOUGUER (1698–1758) and Charles-Marie de LA CONDAMINE (1701–74), famous French explorers, mathematicians, and astronomers, were members of the scientific expedition led by **Godin** to the Audiencia of Quito in the viceroyalty of Peru (present-day Ecuador). In 1735, the French Academy of Sciences dispatched two expeditions, one to South America, another to Lapland, to settle the question of earth's shape. Once the scientists, along with their Spanish companions **Ulloa** and **Juan**, reached Quito, they split the expedition in two and carried out geodesic measurements, surveying Ecuador's mountain range. During his explorations, La Condamine became fascinated with rubber trees, curare poison and its antidote, and quinine (made from cinchona bark). Bouguer conducted pendulum measurements in the mountains and discovered the influence of the density of local rocks on the earth's gravity (called the "Bouguer anomaly"). Bouguer returned to Paris in 1743 via Cartagena. La Condamine reached Paris two years later via Pará, after traveling down the Amazon River. Once back in Paris, both scientists resumed a rivalry that had begun during their expedition, feuding over their publications within the Academy of Sciences. Bouguer was the first to present to the Academy of Sciences the expedition's report—it appeared in his *Mémoires* in 1744—and he published it separately as *La Figure de la Terre* (1749). Bouguer also authored many other works of science, including *Traité de navigation* (1746), the first modern text on naval architecture written by a scientist. La Condamine reported his results to the academy in 1745, including his adventure down the Amazon and his efforts to map its course and tributaries; he published them as *Journal du voyage fait par ordre du Roi, a l'Équateur* (1751). In addition to several other scientific writings based on his travels, La Condamine authored *Mesures des trois premiers degrés du Méridien dans l'hémisphére austral* (Measures of the three first degrees of the meridian in the Southern Hemisphere, 1751) and *Histoire des Pyramides de Quito* (History of the pyramids of Quito, 1751).

The Spaniard Juan de LA CRUZ Cano y Olmedilla (1734-90) studied engraving and mapmaking in Paris under the patronage of Ferdinand VI. He is famous for his *Mapa geográfico de América Meridional* (1775), a large, detailed map of South America; Humboldt owned an original copper engraving of it. Cano y Olmedilla based his *Mapa* on earlier maps and manuscript reports from various sources, such as the Archives of the Indies. He also made a map of the Straits of Magellan, which appeared in Casimiro Gómez Ortega's *Resumen histórico del primer viage hecho al rededor del mundo* (Historical summary of the first voyage around the world, 1769). The map's 1799 version can be found in the David Rumsey Collection.

20. Gonzalo Jiménez de QUESADA (c. 1500-1579) was a Spanish jurist who, at the age of thirty-six, began the conquest of New Granada for Spain. Quesada founded Santa Fé de Bogotá, now the capital of Colombia, on August 6, 1538, as Nueva Ciudad de Granada. The Spanish crown named him marshal (*mariscal*) of the New Kingdom of Granada and later, in 1565, *adelantado*, that is, a representative of the king with judicial and administrative powers over specific districts. Jiménez de Quesada tried obsessively to find El Dorado, the mythological land of gold and emeralds. After returning from a failed expedition in search of El Dorado in the eastern plains of Colombia (today's Meta), he died in Mariquita. Jiménez de Quesada narrates his inland expedition from the Caribbean coast of South America and the founding of Santa Fé de Bogotá in his *Epítome de la conquista del Nuevo Reino de Granada* (1550). He also wrote three unpublished (and now lost) books.

20n. LUCAS FERNÁNDEZ DE PIEDRAHITA (1624–88), a Jesuit historian from New Granada, had humble origins but was related to Inca royalty. His *Historia general de las conquistas del Nuevo Reino de Granada* (General history of the conquests of the New Kingdom of Granada) was written in the 1640s; parts of it were published in 1668, but most of it did not see print until 1881. Describing the cultural and physical geography of the kingdom of New Granada, Piedrahita's five volumes were inspired by the writings of earlier chroniclers such as **Herrera y Tordesillas**, **Jiménez de Quesada**, Antonio de la Calancha (1584–1654), and El Inca **Garcilaso de la Vega**. Like Garcilaso, Piedrahita was a Mestizo, his mother, Catalina de Collantes, being a granddaughter of Inca princess Francisca Coya. Drawing on the work of the Franciscan historians **Pedro de Aguado** and Pedro Simón (1574–c. 1626), Piedrahita wrote extensively about the Muisca Indians, whom he considered the most civilized indigenous group in all of South America.

22n. After completing studies with the Company of Jesus in 1749, ANTONIO JULIÁN (1722–90) performed missionary work at New Granada. He was professor of theology at the Universidad Javeriana in Bogotá, Colombia, until 1763. After the Jesuit expulsion from the colonies in 1767, Julián lived in Italy, where he wrote a trilogy about South America. *La Perla de la América: Provincia de Santa Marta* (The pearl of America: the province of Santa Marta, 1787) is the only volume ever to be published. Julián also wrote *Monarquía del diablo en la gentilidad del nuevo mundo americano* (The monarchy of the devil among the nobles of the New American World), a manuscript finally published in 1994 by the Instituto Caro y Cuervo in Bogotá, Colombia.

24. This list of ethnic groups reflects accounts written by the Spanish chroniclers. Preconquest pictorial manuscripts, such as the Codex Aubin, mention other tribes, such as the Tepanecs; the Codex Vaticanus A refers to seven tribes that come from seven caves. There has been much debate among Mesoamericanists about the degree to which terms such as *Toltec* and *Chichimec* can be regarded as ethnonyms, as well as about the degree of historical accuracy of the Spanish accounts, which contain much that is legendary. These groups are all thought to have spoken languages belonging to the Nahuatl language family, the southernmost of the Uto-Aztecan language families. Today's Mexico has over a million and a half Nahuatl speakers. The name Nahuatl comes from the root *nahua* (*nawa*), which means "clear sound" or "command." See also Quetzalcoatl and Codex Vaticanus A.

BAAL-BERITH was a Canaanite deity of the covenant and of fertility and vegetation at Shechem, an ancient city between Mount Gerizim and Mount Ebal (the present-day West Bank). This deity might have been the guardian of a political treaty between the city-state of Shechem and other city-states or the local Israelite population; it might also have been party to a religious covenant. *Baal* is a Semitic word meaning "lord." In Canaan, Baal was the local manifestation of the deity of storms and was associated with annual revival. Baal's symbol was the bull. ABIMELECH (twelfth century BCE) was the son of Gideon (or Jerubbaal, one of the heroes or judges of the biblical Book of Judges); he became king of Shechem after his father's death and destroyed the city-state after being rejected by the people. Archaeological evidence points to the destruction of Shechem between 1150 and 1125 BCE.

To the east of the Mexica empire was the city-state of Tlaxcala, which, though closely related with the empire, was never actually conquered by it. The Mexica made an agreement with the Tlaxcala to have ritual battles. The goal of these battles was to capture prisoners to be sacrificed by a *teopixqui* (Nahua for priest; literally: "keeper of god"). The Tlaxcala later joined Cortés in his attack on the Mexica city-state of Tenochtitlan, which was destroyed in 1521.

HERODOTUS (c. 485-425 BCE), whom Cicero called "the father of history," is the author of the nine-volume *Histories*, a magisterial account of the historical context of ancient Greece, gathered on his extensive journeys through the world of the eastern Mediterranean. Herodotus's meticulous protohistorical study can be considered a polycentric and comparative narrative. It moves among different voices, stories, and points of view that together focus on the magnitude of the wars (492-479 BCE) between Greece and the expanding Persian empire, of whose genealogical and dynastic emergence Herodotus gives an extensive account. For centuries his historiographical achievements were regarded as the epitome of the empirical and critical gathering of historical knowledge.

The later Greek historian DIODORUS of Agyrium in Sicily (first century BCE) is best known for his forty-volume *Bibliotheca historica*, a compendium of universal history ranging from prehistorical mythologies to the reign of Gaius Julius Caesar (100-44 BCE). Aimed at a wider audience, Diodorus's monumental compilation has a tone and methodology very different from Hellenistic historiography in the tradition of Herodotus. Yet Diodorus Siculus provides the only surviving continuous narrative of events for long stretches of Greek history in the classical period. A complete set of Diodorus's *Bibliotheca* still existed in the imperial palace in Constantinople at the start of the Renaissance in the fifteenth century but perished when the city was sacked in 1453. Today, only fifteen books survive: books 1-5 and books 11-20.

25. In Mesoamerican studies, the word TOLTEC has been used to refer either to actual populations of pre-Columbian central Mexico or to mythical ancestors invented by the Aztecs. More recent scholars have come to believe that the Toltecs were not a distinct group. They cite as evidence the fact that among the Nahua peoples, the word *Tolteca* was synonymous with "artist," "artisan," or "wise man." "Toltecness" (*toltecayotl*) denoted art, culture, and civilization, as opposed to "Chichimecness" (*chimicayotl*), which referred to a savage, nomadic state.

CE (common era) and BCE (before the common era) are standard abbreviations for neutrally designating the Christian calendar system otherwise divided into AD (anno domini) and BC (before Christ). Both BCE/CE and BC/AD are based on a sixth-century estimate for the year in which Jesus was conceived or born, introduced in Rome by the Moldavian abbot Dionysius Exiguus (c. 470-556). The newer chronological standard would gradually be adopted throughout Europe and came to pre-dominate by the beginning of the second millennium. We have used it throughout these annotations, with CE as the default.

The ancient Greeks used the name PELASGIANS for populations that preceded the Hellenes in Greece. Since then, the term has come to signify more broadly all the autochthonous inhabitants of the Aegean lands before the advent of the Greek lan-guage. Archaeological excavations during the twentieth century found evidence sug-gesting that the Pelasgians were either a proto-Greek tribe or at least a tribe akin to the Greeks. The AUSONIANS were ancient inhabitants of middle or lower Italy.

Like Humboldt, the German historian August Ludwig von SCHLÖZER (1735–1809) was interested in politics and statistics. Schlözer is best known for laying the foundations for the study of Russian history. Humboldt refers here to *Vorstellung einer Universalgeschichte* (Idea of a universal history, 1772) and *Weltgeschichte nach ihren Haupttheilen im Auszug und Zusammenhänge* (Main elements of world history in excerpts and contexts, 1792–1801). See also Johann Christoph Gatterer.

One of the most important and complex deities of Mesoamerica, TEZCATLIPOCA ("Smoking Mirror" or "Obsidian Mirror") was the god of the nocturnal sky, of ances-tral memory, of time, and the deity of the North. Associated with Jupiter and destiny or fate, this omnipotent figure was the supernatural basis for political authority. As a polymorpous deity, Tezcatlipoca had several avatars and animal doubles. Illustra-tions that date from the tenth to the eleventh century depict him as a war god. At the time of the conquest, he was the patron of the Aztec Jaguar Warriors and the military schools/temples. His others avatars include Yohualli Ehecatl (Nocturnal Wind, god of sorcerers and metamorphoses); Necoc Yaotl (Enemy of the Two Sides, who pro-vokes men into conflicts on earth, associated with contamination and sin, and also generation and fertility); Tlatauhqui Tezcatlipoca ("Red Tezcatlipoca" or Xipe Totec, "Lord the Flayed One," associated with the Sun); Itztlacoliuhqui ("Curved Knife of Obsidian" or "Blue Tezcatlipoca," assimilated with Venus). In the cosmogonic myths, Tezcatlipoca collaborates with Quetzalcoatl in creating the sky and the earth but also argues with him on the ruling of the different epochs (suns). His calendrical names are 1 Death (Ce Miquiztli) and 2 Reed (Ome Acatl). Tezcatlipoca is usually depicted with an obsidian mirror or a snake in place of his right foot.

LAKE MOERIS was an ancient freshwater lake in the northwest of the Faiyum Oasis, fifty miles southwest of Cairo. It remains today as a smaller saltwater lake called Birket Qarun.

Lars PORSENA, or Porsenna, was an Etruscan king who ruled over the city of Clusium, the former Clevsin, in northern central Italy on the west side of the Apen-nines, presumably around 500 BCE. According to most accounts, Porsena was bur-ied in an elaborate tomb, described as having a fifteen-meter-high rectangular base with sides ninety meters long. It was adorned with pyramids. In 89 BCE the Roman general and politician Cornelius Sulla destroyed the tomb, together with the rest of the city.

MARCUS TERENTIUS VARRO (116–27 BCE), also known as Varro Reatinus, was one of the most influential Roman intellectuals, contributing to scholarship in almost all fields of study in his day. He completed about seventy-five different works, of which fifty-five are known. Varro wrote about history (*De vita populi Romani, De gente populi Romani*), geography, rhetoric, law (*De iure civili lib.* XV), philosophy (*De philosophia*), music, medicine, architecture, literary history (*De poetis, De comoediis Plautinis*), religion, agriculture, and language. Only two of his works survive almost completely: *De lingua latina* (Of the Latin language), in twenty-five books, and *De re rustica*, a treatise on farming written in dialogue form. According to **Pliny the Elder**, Varro wrote the first Roman illustrated book in 39 BCE; the *Hebdomades vel de imaginibus* (Sevens, or On portraits) has fifteen books with seven hundred portraits. Varro's *Disciplinae* (Disciplines, c. 34 BCE), an overview of nine areas of education (including medicine and architecture), influenced the development of the medieval educational trivium (grammar, dialectic, rhetoric) and quadrivium (arithmetic, geometry, music, and astronomy). Varro also established the accepted founding date of Rome, the standard list of the famous seven hills, and the definitive version of Rome's Trojan ancestry.

25n. Gaius Plinius Secundus (23–79 CE), also known as PLINY THE ELDER, was a Roman naturalist, politician, and naval commander. His monumental *Naturalis historia*, an encyclopedic work of thirty-seven books, is the most valuable compilation of ancient Greek and Roman science, covering the physical universe, geography, anthropology, biology, mineralogy, medicinal plants, and the fine arts. Humboldt, who first read Pliny during his studies at the University of Göttingen (1789–90), cites the Roman encyclopedist extensively and adopted Pliny's maxim for the motto of his own *Cosmos*: "Naturae vero rerum vis atque maiestas omnibus momentis fide caret si quis modo partes eius ac non totam conplectatur animo" (Indeed the power and majesty of the nature of the universe at every turn lacks credence if one's mind embraces parts of it only and not the whole; *Naturalis historia* 7.1).

26. The Mexican geographer and mathematician JUAN JOSÉ DE OTEYZA, or Oteiza (1777–1810), was one of Humboldt's key academic informants at Mexico City's Royal School for Mining. For both *Views of the Cordilleras* and the *Political Essay on the Kingdom of New Spain*, Humboldt consulted Oteyza's studies on Teotihuacan, which Humboldt himself never visited, and his data on Zacatecas, Durango, and the surroundings of Toluca. Oteyza had undertaken thorough research at the pyramid precinct in 1803, the same year in which both scientists had completed their calculations on the surface area of New Spain.

CHEPHREN (2572–2546 BCE), also known as Khafra or Khafre, was a pharaoh of the Fourth Dynasty (2639–2504 BCE) of Ancient Egypt. The three main rulers of this dynasty—Cheops, Chephren, and Mycerinus—each commissioned at least one pyramid to serve as a tomb, creating the colossal ensemble of the Pyramids of Giza. Chephren ordered the construction of the second largest of the Egyptian pyramids (originally 471 feet high and 706 feet long at the base) and is thought to have built the Great Sphinx.

A bishop in the Church of Ireland, RICHARD POCOCKE (1704–65) took a tour of Europe in the early 1730s, after which he embarked for the Middle East, where he traveled from 1736 until 1740. He visited Upper Egypt in 1737-38, journeying up the Nile to the Valley of Kings. He returned to Egypt in 1738/39 after a trip to Palestine. See also Frederik Ludvig Norden.

The Spanish Franciscan friar JUAN DE ZUMÁRRAGA (c. 1468-1548) is one of the few Spaniards referred to by name in several of the indigenous codices, notably the **Codex Telleriano-Remensis**, the Anales de Tlaxcala, and the Anales de Tlatelolco. After impressing **Charles V** (1500-1558) with his work as an inquisitor and witchcraft investigator in Pamplona, Zumárraga was appointed the first bishop of New Spain; he was also named Protector of the Indians. In 1534, Zumárraga founded the Colegio de Santa Cruz Tlatelolco, the first school of higher learning in America. It was to teach the indigenous nobility Latin, Spanish, rhetoric, logic, philosophy, music, and medicine. From 1535 to 1543, he conducted the first Inquisition in New Spain, with the goal of disciplining Indian idolaters and sorcerers. He is also credited with introducing the first printing press in New Spain (in 1536), and Zumárraga's *Breve y mas compendiosa Doctrina Cristiana en lengua Mexicana* (Brief compendium of the Christian doctrine in the Mexican language, 1539) was one of the first books printed in the Western Hemisphere. In 1546, Pope Paul III (1468-1549) named Zumárraga archbishop of New Spain, a position he held until his death.

26n. Humboldt repeatedly refers to the works by Louis Mathieu LANGLÈS (1763-1824), who, in addition to founding the École des Langues Orientales Vivantes (School of Living Oriental Languages) in Paris in 1795, published numerous French editions of travel accounts, to which he would add his own notes. Langlès's annotations in the first (and only) two volumes of *Asiatick Researches* to be translated into French were an appealing read for Humboldt. Another of Langlès's projects was a translation of *Voyage d'Égypte et de Nubie* (Voyage to Egypt and Nubia) by Frederik Ludvig NORDEN (1708-42), a Danish artist. In 1737, Norden had joined an expedition traveling to Egypt, where he drew maps and sketches of ancient and modern cities. Because of ill health, Norden stayed in Cairo rather than traveling on to Upper Egypt, which gave him an opportunity to make very precise measurements of the pyramid of Cheops. In the Nile valley Norden also met **Richard Pococke**. Upon his return to Denmark, Norden began preparing his two hundred drawings for publication, translating his notes into French. He did not see the book completed; it appeared posthumously between 1750 and 1755.

27. Built between 18 and 12 BCE in Rome, the PYRAMID OF CESTIUS was a tomb for Gaius (or Caius) Cestius Epulo (d. c. 18/12 BCE). One of the three inscriptions on the pyramid notes that it was built in 330 days, in accordance with Cestius's will. In the steeply pitched Nubian style, the pyramid was built of brick-faced concrete covered with white Italian marble. Two marble bases for the bronze statues that frame the burial chamber were rediscovered during excavations on the west side of the tomb in 1660-62. Plundered before then, the chamber contained only wall frescoes. In 1697, after the restorations ordered by Alexander VII (1655-67), Pietro Santi Bartoli (1635-

1700) made a record of the pyramid, its burial chamber's frescoes, and the statue bases with fragments of the bronze statue on top. This steep pyramid inspired Egyptian themes in the work of artists such as Andrea Mantegna (1431-1506), Raphael Sanzio of Urbino (1483-1520), Nicolas Poussin (1594-1665), and Antonio Canova (1757-1822).

When **Cortés** arrived in the area, CHOLULA was Mexico's second-largest city, with a population of about 100,000. Cortés and his troops slaughtered three thousand nobles of the city in the massacre of 1519, a strike that was supposed to preempt an attack on the Spaniards. Cholula had become a great socioreligious center with the demise of Teotihuacan (650-750), the largest city-state of the Valley of Mexico between 450 and 650, with a population of about 125,000 to 200,000 inhabitants.

The TEOCALLI OF CHOLULA, an architectural wonder of the Americas, is located on the Puebla-Tlaxcala valley right outside the city of Puebla, Mexico. This structure is the one of the largest monuments ever built, with a total volume significantly greater than that of the Great Pyramid of Giza. The pyramid at Giza is more than twice as high as the one in Cholula, but Cholula's footprint is four times larger. The base of the Cholula pyramid measures four hundred meters along its base; it is sixty-six meters high. Humboldt, who was the first to measure it, arrived at smaller measurements because accumulated layers of soil and vegetation led him to mistake one of the upper layers for the base, which was then underground. According to Eduardo Matos Moctezuma, the four angles of the pyramid correspond to the four cardinal points, with a slight deviation to the north (about 26°), which is linked to the position of the setting sun at the summer solstice. Excavations since the 1930s have revealed that the pyramid was expanded gradually in four major construction phases built over the course of about fifteen hundred years. The earliest stage of construction was probably toward the end of 1 BCE. Pyramids were constructed in stages of burial and reconstruction: the pyramid would then be entombed within earthen and stone fill, and a new facing would be built over these materials. Humboldt calls this pyramid *monte hecho a mano* (handmade mountain), referring to Tlamachihualtepetl in Nahuatl (*tla*, prefix for objects; *machihua* for handmade; and *tepetl* for mountain).

28. Born in Spain, Franciscan friar and historian Juan de TORQUEMADA (c. 1564-1624) moved as a child to Mexico City, where he learned Nahuatl from the indigenous scholar and poet Antonio Valeriano (1531-1605). After being named chronicler of the Franciscan order in 1609, Torquemada completed *Monarquía Indiana* (Indian monarchy) in 1612 (it was published in 1615). This work, which portrays pre-Hispanic Mexico as an advanced civilization, is an invaluable documentary source on preconquest and early colonial Mexico. For Torquemada, New Spain succeeded the Mexica empire (founded in 1325, not 1521) instead of beginning with the conquest. Since many of the book's chapters are translations of parts of the then-unpublished *Historia eclesiástica indiana* (Ecclesiastic Indian history, 1596) by Gerónimo de Mendieta (c. 1528-1604), Torquemada has been denounced as a plagiarist. His work served as a basis for other historians, notably **Clavijero.** Torquemada also wrote several other works, including *Vida y Milagros del Santo Confesor de Cristo Fray Sebastián de*

Aparicio (Life and miracles of the holy confessor of Christ, Friar Sebastian de Aparicio, 1602).

Canarian engineer AGUSTÍN DE BETANCOURT Y MOLINA (1758–1824), who became acquainted with Humboldt in Spain, served the Spanish crown until 1808 and the Russian empire from that time until his death. Betancourt is known for his inventions in early telegraphing and for his contributions to the *Essai sur la composition des machines* (1808), considered to be the first modern treatise on machines. With his coauthor, José María de Lanz y Zaldívar (1764–1839), from New Spain, Betancourt was the first to propose, in the *Essai*, a classification of mechanisms based on criteria for the transformation of motion. After leaving Spain to work for Czar Alexander I in 1808, he held various prestigious positions in Russia. In addition to authoring several manuscripts on engineering, Betancourt y Molina published the *Mémoire sur la force expansive de la vapeur de l'eau* (Report on the expansive power of steam, 1790).

In his *Historia verdadera de la conquista de la Nueva España* (*True History of the Conquest of New Spain*), BERNAL DÍAZ DEL CASTILLO (c. 1495–1584) narrates the Spaniards' activities before and after the taking of Tenochtitlan through the eyes of an ordinary soldier. Unlike **López de Gómara** and most other chroniclers of the time, Bernal had actually been in the New World and participated in the conquest. He arrived in America in 1514 on an expedition to Panama and was part of **Cortés**'s march to Tenochtitlan, where he witnessed the Noche Triste (the Spaniards' forced retreat to Tlaxcala) and the later destruction of the city. After the overthrow of the Mexica, Bernal became an *encomendado* (that is, he was entrusted by the crown with land) and lived most of his life in New Spain. He worked on his *Historia* for almost thirty years, from the 1550s until 1584. A copy of the manuscript, with alterations by Alonso Remón (1561–1632) was published in 1632; this corrupted version became the basis of subsequent editions and translations. The original manuscript remained in Guatemala until it was recovered by Género García (1867–1920), who transcribed and published the *Historia verdadera* as Bernal had intended it in 1904–5. See Pedro de Alvarado y Contreras.

29. The architectural feature created by PLACING STONE SLABS progressively to overlap and meet at the highest point to create the appearance of a dome is known as "corbelled vault." Extensively used in Mesoamerica, most famously in the lowland Maya cities, the technique was also employed in the construction of passage graves in Atlantic Europe (fifth to third centuries BCE) and Mycenaean tombs (third to second century BCE).

In fact, the construction of such GALLERIES or tunnels is precisely how the Great Pyramid at Cholula has been excavated in the twentieth century, beginning in the 1930s and continuing for the next fifty years. Tunneling through the base provides a means to pass through earlier "entombed" iterations of the pyramid, allowing for the investigation of the sequence of construction. About eight kilometers of tunnels have been cut through the pyramid; most of them are open to visitors.

While visiting the ancient site of Chanchan, the capital of the pre-Inca Chimu empire, Humboldt noted in his travel diary that the sixteenth-century Spanish adventurer GARCÍA GUTIÉRREZ DE TOLEDO had been able to discover the treasure of the Toledo chamber only because of the pity that his poor appearance generated in his friend Antonio Chayhuac, a native Indian of Trujillo and heir of the last cacique of Mansiche. Chayhuac, who showed Toledo the entrance to the grave's golden fortune, had asked the Spaniard to commit himself to a responsible way of life; only then would he also show him the way to the even larger treasure of the legendary chamber, *el peje grande*. Toledo, who, according to municipality records of Toledo, collected the gold in 1577 and 1578 and managed to spend his entire fortune in a very short time, later returned to Chayhuac, imploring him to reveal his secret; the Indian refused. Other versions claim that Chayhuac had meant to reveal the secret of the grave's location so that part of the revenue would benefit the local natives. The episode has become a Peruvian popular tale known as "El peje chico."

Scholars have long been intrigued by QUETZALCOATL's complex identity. At times Queztalcoatl (Plumed Serpent or Precious Serpent) is represented as a god, at others either as the priest of the Great Temple or a heroic leader. Recent scholarship distinguishes three distinct characters that have been conflated in Spanish accounts: (1) the deity of air or wind, Ehécatl Quetzalcoatl; (2) the high priest of the Great Temple of Tenochtitlan, named Quetzalcoatl Totec Tlamacazqui (priest of the plumed serpent); and (3) the heroic figure Ce Ácatl Topiltzin Quetzalcoatl. Ehécatl Quetzalcoatl is believed to be one of the creators of the universe and the founder of civilized life (associated with wisdom, arts, and philosophy). In the Codex Zouche-Nuttall and the *Historia de los Mexicanos por sus Pinturas* (History of the Mexica in their paintings, 1882), an anonymous sixteenth-century document, he is not a mortal but a deity associated with fertility, water, wind, life, and Venus, the morning star. When associated with Venus, he is worshiped as Tlahizcalpantecuhtli (Lord of the Dawn), a mighty, feared god believed capable of inflicting severe damage to people, maize, and water. Ehécatl Quetzalcoatl wears a beaklike mask, shell jewelry, and the *checailacaozcatl* (jewel of the wind), a cut conch pectoral. The temples built for Ehécatl Quetzalcoat were round to allow the free passage of the wind; among the most famous of these structures are the round temples in the Ceremonial Plaza of Tenochtitlan (located directly in front of the Great Temple) and Calixtlahuaca (Valley of Toluca, southwest of present-day Mexico City). Since Ehécatl Quetzalcoatl is the patron of priests and rulers, the Great Temple's priests are also named Quetzalcoatl. Equal in rank, they were called Quetzalcoatl Totec Tlamacazqui, the plumed serpent priest of Huitzilopochtl, the god of war and sun, and Quetzalcoatl Tlaloc Tlamacazqui, the plumed serpent priest of Tlaloc (god of rain and lightning). Each priest had different cult responsibilities that changed with the seasons, rainy or dry. Ce Ácatl Topiltzin Quetzalcoatl, the leader, was son of the earth/fertility goddess Chimalma(n)/Cihuacoatl/ Coatlicue/Coacueye and the quasi-divine conqueror and founder of the Toltec kingdom, Mixcoatl/Camaxtli/Totepeuh. Gerónimo de Mendieta (c. 1528-1604) and **Bernardino de Sahagún** portray Quetzalcoatl as a virtuous leader and a hater of war and human sacrifice, who practiced fasting and self-mortification to appease the wrath of the gods. Mendieta claims that Quetzalcoatl came from the Yucatán Peninsula, reigned in Cholula, and invented the calendar.

Sahagún depicts him as a leader of Tula, the first Nahua kingdom, who, after being tricked by his enemies into leaving his kingdom, set out on a voyage toward Tlillan Tlapallan ("place of the black and red colors" or "place of writing"). According to the Codex Vindobonensis Mexicanus I (housed in the Austrian National Library at Vienna), Quetzalcoatl dies and subsequently transforms into Venus. Building on and modifying Sahagún's narrative, **Clavijero** recounts Quetzalcoatl's leadership in Cholula after his exodus from Tula. According to Clavijero, Quetzalcoatl resumed his voyage to Tlapallan, but before leaving he promised to return and reign in peace. The *Historia de las Yndias de Nueva España y Yslas de la Tierra Firme* (History of the Indies, New Spain, and islands of the mainland) by Diego Durán (c. 1537–88) depicts Quetzalcoatl as a historical figure named Topiltzin, the last Toltec king. Torquemada portrays Quetzalcoatl as the leader of a group of bearded white-skinned men, a monotheist priest, miracle worker, and foreign missionary. Current scholarship has shown the story about Quetzalcoatl as a foreign white missionary to be largely an invention, much like his alleged opposition to human sacrifice. Both were products of the Spanish chroniclers and clearly served their purposes. In regions south of Central Mexico, such as Yucatán, Oaxaca, Puebla, Tlaxcala, and Guatemala, Ehécatl Quetzalcoatl would be equivalent to Kukulcán (Plumed or Feathered Serpent), Gucumatz, Nakxit. Quetzalcoatl's twin is Xolotl (the dog god), who is associated with the Evening Star.

30. SANNYASIN signifies a formal renunciation of worldly life. It is traditionally adopted only by men as the final stage in the Hindu social cycle. The *sannyasin*, who is expected to live an ascetic life while studying the scriptures, wears saffron garb and is forbidden the company of women. He may travel as a teacher or train disciples in an ashram (monastic hermitage).

Fray PEDRO DE LOS RÍOS (d. c. 1565), a missionary of the Dominican Order, was the compiler of the Codex Vaticanus A, also known as Codex Ríos or Codex Vaticanus 3738. He also annotated and purportedly supervised the assemblage of a related codex, the Codex Telleriano-Remensis, a term Humboldt coined in honor of its first known owner, **Charles-Maurice Le Tellier**. Combining indigenous with European elements, these codices focus on Aztec religion, ritual, mythology, history, and politics. Humboldt was the first to call attention to the similarities between the Codex Vaticanus A and the Codex Telleriano-Remesis, at the Vatican's Apostolic Library and the Bibliotéque Nationale de Paris, respectively.

30n. The German philosopher Karl Wilhelm Friedrich von SCHLEGEL (1772–1829), along with his brother August Wilhelm (1767–1845), was the originator of the core ideas that inspired early German romanticism. His conception of literary scholarship has had profound influence on the rise of what is still called *Geisteswissenschaften*, the sciences of the mind. After studying Sanskrit in Paris, Schlegel published *Über die Sprache und Weisheit der Indier* (About the language and the wisdom of the East Indians, 1808), an attempt at comparative Indo-Germanic linguistics, and the starting point of the study of Indian languages and comparative philology.

33n. WILLIAM VINCENT (1739–1815) established his reputation as a Latin scholar with his 1797 treatise on **Livy** and **Polybius**. But he was best known for his studies of ancient geography and commerce. Also in 1797, he published a commentary on **Arrian**'s *Voyage of Nearchus*—Nearchus had been an officer in the army of **Alexander the Great**. The Scottish geographer Alexander Dalrymple (1737–1808), the first hydrographer to the British admiralty, prepared charts for Vincent's *The Periplus of the Erythræan Sea* (1800 and 1805) and *The Commerce and Navigation of the Ancients in the Indian Ocean* (1807). Vincent was widely regarded as the most distinguished comparative geographer of his time, second only to James Rennell (1742–1830). As headmaster and later dean of Westminster School, Vincent became infamous for his brutal corporal punishment of the students. In 1792, he expelled the poet Robert Southey (1774–1843) for his satire on flogging (*The Flagellant*).

34. In his famous ten-volume *Periegesis of Greece*, the Greek geographer and historian PAUSANIAS (second century) describes a period when Greece had fallen peacefully to the Roman empire. While fragments from this period abound, Pausanias's *Periegesis* is the only fully preserved text of travel writing from this time to have survived. It has served scholars as a crucial link between classical literature and modern archaeology.

The work of Lucius Flavius ARRIANUS (c. 90–c. 160), also known as Arrian of Nicomedia, a Roman historian of Greek origin, is considered one of the most faithful historiographical sources on the age of **Alexander the Great**. His *Anabasis of Alexander* aimed at establishing the "true" historical account of Alexander's legendary conquest. Declaring most prior historical work useless, Arranius recongized only the great Xenophon (c. 430–353 BCE) and **Herodotus** as legitimate models for his own work. The only surving work of QUINTUS CURTIUS RUFUS (first century), another Roman historian, is the *Historiae Alexandri Magni Macedonis*, a biography of Alexander the Great (the first two of its ten books are lost, the remaining eight incomplete). Unlike Arrianus, whose expertise was admired during his lifetime and who was widely quoted, Curtius remained almost unrecognized until the Middle Ages and the Renaissance.

Humboldt refers frequently to the writings of STRABO (c. 64 BCE–c. 23), a Greek geographer and historian. Strabo's *Geography* is the only surviving work to discuss the peoples and countries known to the Greeks and Romans at his time. It was first published in a Latin translation in Rome around 1469.

The Mexican astronomer and archeologist ANTONIO LEÓN Y GAMA (1735–1802) was one of the founders of Aztec, or Mexica, archaeology. He was the first to examine and publish interpretive accounts of Sun Stone, the Coatlicue, and the Stone of Tizoc. León y Gama's *Descripción histórica y cronológica de las dos piedras* (Historical and chronological description of the two stones, 1792) was a key text for Humboldt's own work on these monuments. *Descripción* was only the first part of León y Gama's essay; the second part was published in 1832 by Carlos María de Bustamante. In explaining these sculptures, Léon y Gama sought to reconstruct the Mexica calendrical, mathematical, and writing systems. He also published the first exact observation of the longitude of Mexico, in his *Descripción orográfica del eclipse del sol, el 24 de Junio de 1778* (Orographic description of the solar eclipse of June 24, 1778), which the astronomer **Joseph-Jérôme Lefrançais de Lalande**, one of Humboldt close contacts, brought to wider attention. See also "Sacrificial Stone."

35. *Huaca* (Nahuatl: *wak'a*) does not mean "artificial hill," as Humboldt suggests; it means "sacred place or thing."

The Tumulus of ALYATTES is a Lydian funerary monument that **Herodotus** considered "by far the greatest work of human hands, outside the works of the Egyptians and the Babylonians." The archaeological site at Sardis in Smyrna (modern-day I'zmir) was fully excavated only in 1854 by Ludwig Peter Spiegelthal (1790–1858). CROESUS (c. 595–547 BCE) and Alyattes (609–560 BCE), also known as Hattusilis III, were the last two kings of the kingdom of Lydia (roughly the region of Turkish Anatolia), which fell with the defeat of Croesus by Cyrus II of Persia (c. 600–530 BCE) in 546 BCE.

The French diplomat Marie Gabriel Auguste Florent CHOISEUL-GOUFFIER (1752–1817) was ambassador to Constantinople (Istanbul) from 1784, when he also became a member of the French Academy, to 1792. He had previously resided in Russia, where he had fled after the outbreak of the French Revolution. After traveling in Greece and Asia Minor from 1776 to 1782, he published *Voyage pittoresque de la Grèce* (Picturesque voyage to Greece) in 1809 and 1822. His wife Sophie de Tisenhaus Choiseul-Gouffier (1752–1817) was also an author.

35n. In the mid-1730s, the French Jesuit Jean Baptiste DU HALDE (1674–1743) wrote a description of the Chinese empire which featured fifty plates of provinces, cities, and territories. The Paris-based Du Halde had never traveled to China himself; he collected and edited others' materials.

The Greek physician CTESIAS OF CNIDUS, who lived in the fifth century BCE, left historical accounts of Indian and Persian cultures; his chronologies are not, however, considered very reliable. Ctesias spent some time at the Persian court as a physician for rulers Darius II (d. 404 BCE) and Artaxerxes II (fl. fifth and early fourth centuries BCE).

John SHORE, the first Baron Teignmouth (1751–1834), was a governor-general of Bengal from 1793 to 1798 and subsequently founder and first president of the British and Foreign Bible Society. Both his father and grandfather had already been connected with the East India Company, and Shore followed in their footsteps, arriving in Bengal in 1769. It was in Bengal that Shore made the acquaintance of Sir **William Jones**, whom he would succeed as president of the **Asiatick Society** of Bengal upon the latter's death. Shore also published Jones's writings.

36. A figure of Greek mythology, CALLISTO is one of the chaste nymphs of Artemis, the Greek goddess of virginity, fertility, and the hunt. Seduced by Zeus in the Arcadian forests, Callisto bore a child by the name of Arcas, the future progenitor of the Arcadian people, and was banished by Artemis from her sacred circle of nymphs. In

Ovid's *Metamorphosis*, it is Zeus's wife Hera who, furious about her husband's betrayal, casts a spell on Callisto, turning her into a bear. One day, the she-bear Callisto meets Arcas, by then a young hunter. As she tries to embrace her beloved son, Callisto is on the verge of being kiilled by Arcas's spear when Zeus intervenes, turning both into stars: the Ursa Major and Ursa Minor constellations. As Humboldt points out, **Pausanias** mentions in his *Periegesis* that the Arcadians had erected a "heap of earth" as Callisto's tomb. On its summit he found a temple in honor of Artemis, which the Arcadians named Calliste, merging the two former adversaries into one figure of cultic worship.

The Inca TUPAC YUPANQUI, also known as Topa/Tupac/Tupa Inca Yupanqui or Thupa Inka Yupanki (c. 1448–c. 1493), was the tenth Sapa Inca (emperor) of Tawantinsuyu, the Inca Empire (1438–1527), from 1471 to 1493. His father, Pachacuti Inca Yupanqui or **Pachacutec** (d. 1471), ceded the empire's military command to him around 1463. A military genius, Tupac Yupanqui successfully extended the empire to the north and the south, conquering territories as far as central Ecuador, the Peruvian coast, the Bolivian *altiplano*, northern Chile, and northwest Argentina. Legend has it that Tupac Yupanqui was the Inca who embarked to the Pacific Islands with twenty thousand soldiers, a legend that propelled Spaniards to explore the Solomon Islands. Pedro Sarmiento de Gamboa (c. 1532–1608), author of *Historia Indica* (Indian history, 1572), reports that Tupac Yupanqui sailed from the Ecuadorian coast to Avachumbi (Fire Island) and Ninachumbi (outer island) with twenty thousand soldiers on balsa rafts. The name Tupac Inca Yupanqui means "Royal Honored Inca." He is an ancestor of **Garcilaso de la Vega, El Inca**.

37. At the close of the 1690s, the famous traveler Giovanni Francesco GEMELLI CARERI (1651–1725) published in Naples an account of his journeys, titled *Giro del mondo* (Voyage around the world). Careri began his travels in 1693, visting Egypt, Palestine, India, China, and the Philippines, from whence he sailed to Mexico. He arrived there in 1697 and stayed for over a year. While scholars have considered Careri's travelogue unreliable, Humboldt points out that his accounts of Mexico City in the sixth volume offer important historical information on pre-Hispanic codices and hieroglyphs (see Plate XXXII) and a careful study of Aztec pyramids and culture.

Educated at the University of Edinburgh, William ROBERTSON (1721–93) was a historian and minister of the Church of Scotland. Robertson's *History of America*, first published in 1777, was so popular that already by 1780 it had gone through nine editions. Robertson had started the *History* in 1769 upon completion of his biography of Charles V, a book that had sealed his reputation as one of the leading historians in Europe. But it had made him realize that the history of sixteenth-century Europe would be incomplete without an account of the colonization of the New World. The *History* was considered his masterpiece, although it does suffer from certain limitations, such as an unsurprisingly Eurocentric worldview. Robertson also published works on Scotland (1759) and India (1784). Humboldt points out that Robertson's and **Jean-Sylvain Bailly**'s scholarship on the Mexica calendar had already been rendered obsolete by the time *Views of the Cordilleras* was published.

XOCHICALCO is the pre-Columbian capital of an influential *altepetl* (city-state) located in the central Mexican highlands seventy kilometers southeast of present-day Mexico City. This fortified urban center, built atop a cluster of hills, had a population of ten to fifteen thousand at its peak during the Epiclassic in 650-900 BCE, a period of major cultural innovation and change. As the result of a power vacuum left after Teotihuacan's demise, Xochicalco became an influential religious and political center and the region's major market. Part of Xochicalco's industry was based on manufacturing obsidian tools (obsidian is volcanic glass). Xochicalco was characterized by a class-stratified state-level society that expanded its political domain through militarism, conquest, and the extraction of tribute. Although Humboldt did not have a chance to visit the site, he was the first to place Xochicalco correctly in historical time (between the Classic epoch and Toltec) by suggesting an early period of Maya influence. One of the most famous structures on Xochicalco's ceremonial plaza was the Pyramid of the Plumed Serpent, believed to have been built to commemorate the city-state's political power. On Plate IX, Humboldt depicts two serpents beautifully carved in relief on the lower level of the pyramid. The Feathered Serpent was the head of the state cult and the main deity of the warrior elites, in addition to being the god of fertility and abundance. Xochicalco was quickly abandoned around 900 and was never intensively reoccupied before the Spanish conquest. See also Quetzalcoatl.

37n. PAOLINO DA SAN BARTOLOMEO, born John Philip Wesdin and known as Paulinose Patiri (1748-1806), was an Austrian Carmelite missionary. Educated at the University of Prague and Rome in Oriental languages, he lived in India from 1776 to 1789. He was professor of Oriental languages at the Sacred Congregation for the Propagation of the Faith (founded in 1622 to train priests and missionaries and direct ecclesiastical matters overseas). As Paulinose Patiri, he wrote a large number of prose works in Malayalam, one of the four major languages of South India. After returning from India, he published several works on Sanskrit grammar and Hindu history, theology, and religion. In 1796 he published *Viaggio alle Indie Orientali* (Voyage to oriental India), the work that Humboldt mentions here.

38. José Antonio de ALZATE y Ramírez de Santillana (1737-99) played a pivotal role in disseminating and popularizing scientific knowledge in New Spain. As a clergyman in New Spain with an interest in philosophy, medicine, chemistry, geography, agriculture, mining, education, and archaeology, he was in contact with North American and European centers of research and was elected to the Académie Royale des Sciences of Paris. Alzate edited various influential scientific journals in New Spain, notably the *Diario Literario de México* (1768), the first literary journal (it published just eight numbers and was suppressed on the basis of vice-regal orders). Other journals included *Asuntos varios sobre ciencias y artes* (Various topics in the arts and sciences, 1772-73), *Observaciones sobre la física, historia natural y artes útiles* (Observations on physics, natural history, and the applied arts, 1787-88), and the *Gaceta de Literatura de México* (1788-95). In addition to translating Benjamin Franklin's (1706-90) works into Spanish, he delivered a eulogy in the year of Franklin's death. Humboldt refers to Alzate's description of Xochicalco, based on two visits to the site, published as *Descripción de las antigüedades de Xochicalco* (Description of the antiquities of Xochicalco) in 1791 as a supplement to volume 2 of the *Gazeta de Literatura de México*.

40n. Pedro José MÁRQUEZ (1741–1820) was forced to leave New Spain, the land of his birth, when Spain expelled the Jesuits from its colonies in 1767. In *Due antichi Monumenti* (Two ancient monuments), his short study of two pre-Columbian Mexica ruins (Tajín and Xochicalco), Márquez partly adapts Alzate's *Descripciones* and, in patriotic tones, expresses his deep concern for the indigenous cultures of Mexico. *Due antichi Monumenti* is among the many early historiographical and archaeological accounts of Mexico, the most famous of them being **Clavijero**'s *Storia antica del Messico*.

42. Swiss aristocrat and cartographer Jean-Frédéric D'OSTERWALD (1773–1850) initiated some of the first efforts to create reliable maps of Switzerland's many cantons. Johann Georg TRALLES (1763–1822) was a German physicist and mathematician whom Humboldt respected highly; he visited him in Berne in 1795, during his travels through Switzerland. Responsible for the first astronomical observatory in Berne, Tralles was a prominent representative of modern science with whom Humboldt discussed galvanic experiments and geomorphological measurements.

In Humboldt's day, establishing geographical measurements of a place and, by way of determining latitude and longitude, the ubication of that place was not a trivial exercise. One of Humboldt's methods of timekeeping drew on astronomical movement: lunar distances (measuring the angles between the moons of Jupiter and fixed stars) and observations of the Galilean moons of Jupiter. Determining lunar distances requires very accurate angular measurements, for which two different instruments had been perfected during the eighteenth century: the reflecting and repeating circle (also known as Borda circle) and the sextant. Humboldt had one of each: a sextant by Jesse RAMSDEN (1735–1800) and a Borda circle by Étienne Le Noir (1744–1832). The Ramsden sextant was probably Humboldt's most used instrument and one of his finest; he used to travel and work with the best scientific instruments available.

44. The French naturalist Horace Bénédict de SAUSSURE (1740–99) invented many of the scientific instruments of his time, among them the electrometer, the cyanometer, and the hair hygrometer. In the opening chapter of his *Relation historique*, Humboldt acknowledges the importance for his own writings of Saussure's account of his fourteen alpine expeditions, *Voyage dans les Alpes* (Travels in the Alps, 1769–96). Saussure's work—he was also one of the first to climb Mont Blanc—served Humboldt as a model for a scientific-personal narrative that combined the advances made through scientific exploration with the emotions and impressions of the traveler.

46. The Spanish conquistador Pedro de ALVARADO y Contreras (c. 1485–1541) commanded one of **Cortés**'s eleven ships; he helped defeat the Mexica and then invaded Maya territory. While in charge of Tenochtitlan, Alvarado caused the famous

Noche Triste (Night of Sorrows), the retreat of the Spaniards from the city in 1520. As both indigenous and Spanish accounts report, Alvarado had multitudes of unarmed Mexica noblemen massacred during a ceremony for **Toxcatl**, which resulted in a rebellion, Moctezuma's death, and the Spaniards' flight from Tenochtitlan. With indigenous allies, Alvarado was able to conquer the Quiche kingdom in present-day Guatemala. In 1524, after the surrender of the Quiche capital Utatlan, he founded the first Guatemala City (now Antigua). He also founded San Pedro Sula (Honduras) in 1536. **Charles V** appointed him governor of Guatemala and confirmed and extended his governorship in Honduras, despite the fact that his administration was characterized by revolts and illegal slave trading. In 1534 Alvarado led a poaching expedition to the Inca province of Quito, disembarking at Puerto Viejo in modern-day Ecuador. After being intercepted by Pizarro's then-lieutenant Diego de Almagro (1475–1538), he sold his ships and equipment and returned to Guatemala. He was crushed to death by a horse while searching for the chimera of the mythical Seven Cities of Cíbola in northern Mexico. His death is depicted in the **Codex Telleriano-Remensis**. Loathed by Mexica and Mayas for his notorious brutality, Alvarado, who was nicknamed Tonatiuh (sun in Nahuatl), is mentioned in surviving texts of the Quiche Maya.

Another Spanish conquistador, Pedro de Cieza de León (c. 1520–54), is regarded as the "prince of Peruvian chroniclers" for authoring the celebrated *Crónica del Perú*, the best firsthand account of the conquest of the Inca. It describes the land in detail and narrates the history of the Inca of Cusco. A year after watching Pizarro's ships and famed treasure unloading in Seville in 1535, Cieza set out for the New World; he was thirteen. He spent about fifteen years in South America as a soldier, keeping detailed notes about everything he saw: flora, fauna, landscapes, ruins, events, and so on. Being one of the first Europeans in the Andean region of South America to use native informants for his research, Cieza also amassed information about the indigenous peoples and their history. From 1548 to 1550, after being named official chronicler of the Indies, Cieza traveled in Peru, gathering information. He eventually collected thousands of manuscript pages that became the basis for his *Crónicas*. He would see only the first part of his *Crónicas* published, under the title *La crónica del Perú* (The Peru chronicle, 1553). The other parts were used by historians such as **Herrera y Tordesillas**. Because parts 2–4 were still unpublished when he died, Cieza included them in his will, requesting that they be either published by the executor or sent to Bartolomé de las Casas (1484–1566) for that purpose. This request was not carried out. Parts 2 and 4 were published in the late nineteenth century, part two as *El señorío de los Incas* (The dominion of the Inca, 1880) and part 4 as *Las guerras civiles del Perú* (Civil wars in Peru, 1877–81). Historians considered part 3 lost until Francesca Cantù, a modern history professor at the University of Studi Roma Tre, found the complete manuscript in the Apostolic Library of the Vatican in the 1970s. Cantù published part 3, which gives a detailed narrative of the Spaniards' presence in the Andes, as *Pedro de Cieza de León e il "descubrimiento" y conquista del Perú* (Pedro Cieza y León and the "discovery" and conquest of Peru, 1979), then as *Crónica del Perú: Tercera parte* (Peru chronicle: Third part) in 1989.

Garcilaso de la Vega El Inca (1539–1616) was related to the Inca Atahualpa

(d. 1533). Born Gómez Suárez de Figueroa, he was the son of the Spanish conquistador Sebastián de la Vega Vargas and the Inca princess Ñusta Chimpu Occlo. He adopted the epithet "El Inca" to signify pride in his indigenous ancestry. The first American to write in Spanish, Garcilaso lived in Spain from 1561 until his death. He authored an extensive history of the Inca, the *Comentarios reales de los Incas* (*Royal Commentaries of the Incas*), which was first published in Lisbon in 1609 and 1617. The *Comentarios* were banned in Peru in 1780 at the outset of a rebellion against colonial dominance led by José Gabriel Condorcanqui, better known as Tupac Amaru II (1742–81).

After the death of his father, **Inca Tupac Yupanqui**, HUAYNA CAPAC (Nahuatl: Wayna Qhapaq, Youthful Majesty; c. 1493–c. 1528) became the eleventh Inca emperor. Following in his father's footsteps, he extended the northern boundaries of the empire, conquering northeastern Peru and northern Ecuador, founding Quito as the second Inca capital, and pushing the Inca empire's frontier to the Ancasmayo River (Ecuador and Colombia's present-day boundary). Huayna Capac died during an epidemic of smallpox or measles, along with his heir Ninan Cuyuchi (d. c. 1528). Their death was followed by civil war and sibling rivalry over succession to the empire's throne. Huascar was eventually crowned in Cusco, and Atahualpa controlled Ecuador and parts of northern Peru. By 1532 Atahualpa had defeated his half-brother.

A young aristocrat from a wealthy hacendado family in Quito, Juan José Matheu Arias-Dávila y Herrera (1783–1850), COUNT OF PUÑONROSTRO and tenth MARQUIS OF MAENZA, accompanied Humboldt on his second of three ascents to Mount Pichincha (see Plate LXI) on May 26, 1802. During this expedition Humboldt and his Indian companion, Aldas, almost fell into the volcano's steaming sulfur crater while trying to cross a snow-covered passage. In his travel diaries Humboldt narrates the Pichincha episode in vivid detail, constantly referring to the accounts of **La Condamine**, who had seen and described the volcano's crater some fifty years earlier in his famous *Relation abrégée d'un Voyage fait dans l'Intérieur de l'Amérique Méridionale* (Abridged report of a voyage in the interior of South America, 1747). The second Pichincha ascent was subsequent to Humboldt's climbing Mount Cotopaxi on April 28, 1802. Later, in February 1803, Humboldt tried unsuccessfully to observe an eruption of Cotopaxi while canoeing back to Quito on the Río de Guayaquil (also Río Babahoyo).

47. Contrary to Humboldt's claims, Plate XI shows a stucco bas-relief from the northern part of the pre-Columbian Maya Palace at Palenque, a Maya city and ceremonial center located at the foot of the northernmost hills of the Chiapas highlands, overlooking the forest of the Gulf coast plain. The palace, with its impressive four-story tower, is the central and largest building of the site, which also includes a ball court and several temples. With several interior courts, galleries, and rooms, it served as the residence of most of Palenque's rulers. Its stucco sculpture, once painted, is one of the finest in the world. Palenque was occupied from about 1500 BCE until the eighth century. The drawing on which the plate is based was made by Ricardo or Ignacio Almendáriz (fl. c. 1787), the artist who accompanied Artillery Captain Antonio del Río (fl. 1786–89) on an excavation of the Maya site at Palenque in 1787. Almendáriz prepared thirty—often inaccurate—figures on twenty-six sheets; they were part of del Ríos's report published as *Description of the Ruin of an Ancient City Discovered near Palenque* (1822); the report's manuscript is housed at the Academía Real de Historia in Madrid, Spain. Humboldt corrects the mistake in his endnote to page 236.

Vicente CERVANTES (1758–1829), a pharmacist of Spanish origin and director of the capital's botanical garden, had originally been sent to New Spain for the purpose of inventorying American flora and transplanting its most useful species to Madrid's Jardín Botánico. Yet with the establishment of a chair of botany, his work went more in the direction of promoting knowledge, methods, and values to meet the needs of his new home. Among other achievements, Cervantes introduced a new chemical nomenclature for Mexico, fostering what recent historical scholarship has called "the vigor of scientific Creolization in New Spain." The Spanish physician and naturalist Martín de SESSÉ y Lacasta (1751–1808) was the head of a major botanical expedition across New Spain in 1787, and he founded the botanical garden in Mexico City in 1788. The expedition, which Sessé undertook together with Cervantes and the Creole botanist José Mariano Moziño (1757–1820), lasted until 1803 and resulted in several major publications, among them *Plantae Novae Hispaniae* (Plants of New Spain) and *Flora Mexicana*.

48. FRANCISCO HERNÁNDEZ DE CÓRDOBA (d. 1517) was the first Spanish conquistador to land (intentionally) on the east coast of Yucatán. Hernández de Córdoba took part in the conquest of Cuba under the command of Diego Velázquez de Cuéllar (c. 1461–1524). From Cuba he sailed with three ships in search of slaves, reaching Cozumel Island (Mexico), about 150 miles east of Cuba. He continued northwest to Cabo Catoche (present-day Quintana Roo, Mexico) and on February 23, 1517, landed in Campeche, where he was repeatedly wounded during a battle with the indigenous population. He succumbed to his wounds shortly after returning to Cuba with news about gold in the Yucatán. He is not to be confused with the other Francisco Hernández de Córdoba (d. 1526), also a Spanish conquistador, who sailed the Río Desagüadero (now Río San Juan) and founded the cities of Granada and León (Nicaragua).

The Spanish explorer Juan de GRIJALVA, or Grijalba (c. 1480–1527), Diego Velázquez's nephew, commanded a second expedition to the Yucatán. Departing Cuba in April 1518, Grijalva circumnavigated the peninsula and reached the Río Pánuco. On his return trip, he engaged in battle with the Campeche population to avenge Hernández de Córdoba's death and was himself wounded. He amassed considerable wealth during his trip, preparing the path for Cortés's invasion of Mexico. Grijalva was killed during an Indian rebellion in Nicaragua.

Antonio (or Anton) de ALAMINOS (b. c. 1485), who had sailed with Christopher Columbus on his fourth and final voyage to the Americas, was the pilot on both expeditions to the Yucatán. Alaminos believed that the Yucatán Peninsula was an island and named it Isla de Santa María de los Remedios—the first known map to show it as part of the mainland dates from 1527. In 1513, Alaminos led Juan Ponce de León's ill-fated discovery expedition to Florida. He is also reputed to have been the first to avail himself of the Gulf Stream, when delivering Cortés's letter to **Charles V** in 1519.

The Italian historian Lorenzo BOTURINI Benaducci (1702–55), a chevalier of the Holy Roman Empire, spent seven years in New Spain (1736–43). In the process of trying to gather evidence of the miracles of Our Lady of Guadalupe, he brought together the world's greatest collection of pictographs, linens, codices, maps, and other Mesoamerican artifacts. He copied, traded, and bought his collection of ancient indigenous documents from a variety of sources, including **Carlos de Sigüenza y Góngora**'s private library, the archives of the Chapter House of the Cathedral of Mexico, the Royal Tribunal, and the library of the University of Mexico, during his travels through central Mexico. Viceroy Pedro Cebrián y Agustín, count of Fuenclara (1687–1752), arrested Boturini, confiscated his collection, and expelled him from New Spain, alleging that he was requesting funds without proper authority from the crown. Once in Spain (after having been captured and released by pirates) and with the support of **Mariano Fernández de Echeverría y Veytia**, Boturini was able to argue his case successfully before the Council of the Indies. The council authorized him to publish a fifteen-volume treatise on Mesoamerican calendrics and chronology that would become *Idea de una nueva historia general de la America Septentorial* (Toward a new general history of northern America). But the council was about to withdraw its support, and he had to hasten the publication of his history, which included a catalog of his collection in New Spain, in 1746 (the council did withdraw support a few days after the book appeared in print). A year later, King Ferdinand VI (1713–59) confirmed Boturini as royal chronicler in the Indies and ordered the immediate return of his collection. But Boturini never saw his collection again and died before the first volume of his treatise *Ciclografía* was printed. He was among the first Europeans to propose a new interpretation of the indigenous documents and a reevaluation of pre-Columbian history. In addition to recognizing the value of the different ways in which Aztecs recorded history—putting pictorials and songs together with written accounts—he studied the rich metaphors of Nahuatl. *Ciclografía* was reissued as *Historia general de la América Septentrional* in 1949.

49n. In 1785 George DIXON (c. 1748–95) was appointed commander of the vessel *Queen Charlotte*, and accompanied by the *King George* under the command of Nathaniel Portlock (b. c. 1747–1817), he reached the west coast of North America a year later. Each captain published *A Voyage Round the World* in 1789. The bulk of Dixon's account was written by William Beresford (fl. 1788), the ship's supercargo. It was dedicated to Sir Joseph Banks (1743–1820) and is rather critical of the contradictory reports that John Meares (c. 1756–1809) published in *Voyages Made in the Years 1788 and 1789 from China to the North West Coast of America*. (Dixon had met and helped Meares in North America.) The former fur trader Dixon spent his remaining years at St George, Bermuda, running a jeweler's shop.

50. The work of French naturalist Georges Baron de CUVIER (1769–1832), especially his "theory on catastrophes," which Humboldt endorsed, was of great importance for the natural sciences. His bone comparisons also made Cuvier one of the founders of modern paleontology. Humboldt frequently consulted with Cuvier while working on his American travelogue.

The CARIB people once lived in the Lesser Antilles; the Spanish called them "Caribis," by which they meant "Canibales" (cannibals), a likely corruption of the Carib or Taíno words "Canibis" or "Caniba" (the brave ones), by which the Carib peoples referrred to themselves. Their reputation for ferocity stems from their countless wars with the Arawaks (Taínos). The Spanish did not fight the Caribs, but the British and the French did in the seventeenth century, when they became interested in the Lesser Antilles. In a peace treaty, the Caribs were given full possession of Dominica and St. Vincent. Today a few of them survive on Dominica.

50n. The German anthropologist Johann Friedrich BLUMENBACH (1752-1840) was considered an authority in the field of neuro- and cranioanatomy well into the nineteenth century. This discipline claimed a direct correspondence between the intelligence of an organism and the level of its structural organization. Studying this discipline was part of Blumenbach's interest in studying the natural variety of humankind, which he first discussed in his dissertation at the University of Göttingen in 1775: "De generis humani varietate nativa liber." Following the example of Carl von Linné (1707-8), Blumenbach established a system of four human types based on geological variables (Mongolian, American, Caucasian, and African), later adding the Malayan type. Blumenbach shared with Jean-Baptiste-Pierre-Antoine de Monet, chevalier of Lamarck (1744-1829), the view that the environment could modify an organism's morphology and that the resulting changes could be inherited, an idea known as Lamarckism. Humboldt acquired from Blumenbach an understanding of migration as a historical process that explains not just physical but also cultural variations of different human "races" (that is, peoples). This idea was taken up later by the anthropogeographer and naturalist Friedrich Ratzel (1844-1904), who, after extensive travels in the mid-1870s, authored a book about American geography. Ratzel is often discussed in comparison with Humboldt.

51. Claude ROBLET was a French surgeon aboard the vessel *La Solide* under the command of Captain **Marchand**. In 1791 the ship reached the northern parts of the Queen Charlotte Islands, or Haida Gwaii ("Islands of the People"), on the Pacific Northwest coast of Canada. Together with second captain Prosper Chanal, Roblet kept detailed accounts of several aspects of the customs of the Haida and Tlingit peoples. In describing the intricate designs that were painted on or carved in various artifacts, Roblet advanced the idea that these paintings or carvings were part of a system of writing and compared them to hieroglyphs.

Cox CANAL, or Cox's Channel, officially named Parry Passage, is located southwest of Victoria, between the northwestern end of Graham Island and Langara Island in British Columbia, Canada. In 1853 James Charles Prevost (1810-91), commander of the *Virago*, named the waterway after Sir William Edward Parry (1790-1855), the renowned Arctic explorer. Before then it had been known by several other names:

Cox's Canal, named by British commander William Douglas of the trading vessel *Iphigenia Nubiana*, who had remained in the passage for about a week in 1788; Cunneyah's Straits, on a map prepared by Joseph Ingraham (1762-1800) of the brig *Hope* in 1791-92; Puerto de Floridablanca in 1792, after the count of **Floridablanca**, by Jacinto Caamaño (1759-1825), Spanish commander of the *Aranzuzú*. The most extensive exploration of the shores of the passage was carried out by Captain **Marchand**.

The manuscript collection at the ROYAL LIBRARY OF BERLIN, known as "Humboldt Fragments," consists of sixteen manuscripts that are believed to have once belonged to **Boturini**'s collection. Humboldt bought and gave these fragments to the Royal Library of Berlin in 1806. He published fragements 3, 6, and 16 entirely and parts of fragments 1, 2, 8, and 10-14. Fragment 4, together with 3 (Plate XXXVIII), is part of the Códice de Huamantla, a sixteenth-century cartographic-historical painting on *amatl* paper, whose main topic is warfare. Fragment 5, which also dates from the sixteenth century, is a simple census from Central Mexico. Fragment 7 was part of the Boturini collection and belongs to the so-called Mizquiahuala sales receipts, six pictorial documents that are receipts of payment for goods and services provided to Spaniards. Fragements 9 and 15 record tribute. All the fragments were published in the heliogravure atlas *Historische Hieroglyphen der Azteken* (Historical Aztec hierogrlyps, 1892) by Eduard Seler (1849-1922). Seler published his commentaries on the Humboldt Fragments in *Die mexikanischen Bilderhandschriften Alexander von Humboldt's in der Königlichen Bibliothek zu Berlin* (Alexander von Humboldt's Mexica pictorial manuscripts in the Royal Library in Berlin, 1893). An illustrated English translation of Seler's commentary was published as *Mexican Picture-Writings of Alexander von Humboldt* by Charles P. Bowditch (1842-1921) in 1904. See also Codex Mendoza.

51n. The French Freemason ANTOINE COURT-DE-GÉBELIN (1725-84) was particularly interested in mythology, archaeology, and linguistics and believed that all modern languages could be traced back to one single origin. He left an unfinished study of ancient languages and mythology, titled *Monde primitif, analysé et comparé avec le monde moderne* (Primitive world analyzed and compared with the modern world). Court-de-Gébelin's scholarly reputation was adversely affected by his enthusiasm for animal magnetism and tarot cards, which he wrongly perceived to be an Egyptian book of wisdom; the popularity of tarot cards today may be traced back to him. See also Stephen Sewall.

53. The Spanish chronicler Francisco LÓPEZ DE GÓMARA (1511-60) was **Cortés**'s personal secretary. López de Gómara's *Historia general de las Indias: Crónica de la Nueva España* (General History of the Indies: Chronicle of New Spain), better known as *Historia de las Indias* (1553), is an unabashedly heroic treatment of the exploits of his patron. It was banned in Spain because of its uncritical glorification of Cortés's invasion of Mexico. **Bernal Díaz del Castillo** countered López de Gómara's version of the Spanish conquests in his *Historia verdadera de la conquista de la Nueva España* (*True History of the Conquest of New Spain*).

54. Carlos de SIGÜENZA y Góngora (1645-1700) was one of the most prominent scholars of baroque Mexico. His reputation as a historian, mathematician, and astronomer extended throughout the Americas and Europe. In his literary works, Sigüenza, a native of New Spain, promoted an outspoken *criollo* perspective on precolonial and colonial history. As an unrivaled expert on Mexico's pre-Columbian civilizations, Sigüenza wrote numerous historiographical studies, such as the *Historia del Imperio de los Chichimecas* (History of the Chichimec empire); almost none of them have survived. The famed Italian adventurer **Gemelli Careri** received copies of some of Sigüenza's manuscripts while visiting this collector of ancient Mexica codices in 1697 and published them in his *Giro del Mondo* (Tour around the world). On Plate XXXII, Humboldt depicts an inverted image of the "Map of Sigüenza," a Mesoamerican pictorial manuscript from the sixteenth century, which is now at the Museum of Anthropology in Mexico City. A Culhua-Mexica history in the form of a map, this rare document—also known as "Mapa de la peregrinación de los Aztecas" (Map of the Aztec migrations) and Códice Ramírez—spans the Mexica's migration from Aztlan-Culhuacan to the founding of Tenochtitlan.

56. The UNIVERSITY OF MEXICO was founded as the Real y Pontificia Universidad de México on January 25, 1553, by decree of **Charles V**, the third university in the New World after those of Santo Domingo (1538) and Lima (1551). It rapidly became the foremost institution of higher education in the Spanish American colonies, with its library having grown to over ten thousand volumes already in the sixteenth century. Modeled on the University of Salamanca, Spain's pride for scholastic excellence since 1218, the University of Mexico became the scholarly center of the colonies, and the capital became the largest city in Charles V's dominions. Teaching Amerindian languages, especially Nahuatl, had indeed, as Humboldt indicates, become an important aspect of colonial higher education. A knowledge of Nahuatl and Otomi was vital not only for legal issues but also for the missionary goal of evangelization, which could be achieved only by teaching the catechism in indigenous languages.

57. The judge Sir William JONES (1746-94) was one of the most noted British linguists and orientalist scholars. As a child Jones had been found to be gifted with languages. He studied at University College, Oxford, and quickly established a scholarly reputation in his field in the 1770s. Having been appointed judge at Fort William in Bengal, he arrived in Calcutta in 1783. In the following year Jones founded the **Asiatick Society**, which still exists today. He regularly contributed to its journal, the *Asiatick Researches*, which served as a platform for his at that time rather startling ideas, including his emphasis on roots and the grammatical structure of languages. By introducing a historical and comparative approach to linguistics, Jones effectively provided a basis for the idea of an Indo-Germanic language family, which would catch on in the nineteenth century. Jones's first biographer was **John Shore**.

HIMYARITIC language was a pre-Islamic language spoken by the Himyar (Himyarites or Homerites) peoples in present-day Yemen. Influenced by Sabaic literate culture from the fourth century, the Himyar spoke and wrote the now-extinct Sabaic language using a writing system known as the Ancient South Arabian monumental script. What Humboldt refers to as Himyarite characters is likely Sabaic, the most documented of the four Ancient South Arabian languages—the other three being Minaic, Qatabanic, and Hadramitic; it was the official language of the Himyar. Ancient South Arabian is distinct from the contiguous classical North Arabic or Ethiopic languages, and from Modern South Arabian languages. A segmental script with twenty-nine geometrically formed graphemes written horizontally from right to left, Sabaic has no punctuation marks. The letters stand independent of one another and in the early period are set in a fixed relationship of height and width; words are separated by a vertical line. Like the scripts of most Semitic languages, Sabaic lettering represents mainly consonants, rarely vowels. Some features of Himyaritic survive in modern Yemenite dialects.

SHRAMANAS, or *śrāmaṇas* (religious mendicants or homeless wanderers), were spiritual teachers in ancient India who renounced society and lived as itinerant ascetics in search of truth. *Shramanas* rejected the Veda as an authoritative text and the teachings of the religious superiority of the Brahmins, believing that anyone, regardless of race or social status, can close the cycle of reincarnation and suffering (*samsāra*). The most widely known *shramana* is the Buddha after whom Buddhism was named. According to myth, Buddha (Awakened One, also known as Shakyamuni) was Prince Siddhartha Gautama (c. 485-405 BCE), who became a *shramanas* to seek release from the world of perpetual rebirth and misery. Siddhartha later rejected extremes of self-indulgence and self-denial and became enlightened about the truth, or law (*dharma*), while meditating on the Tree of Awakening (Bodhi Tree). The shramanas are also precursors to Jainism, an unorthodox Indian movement devoted to the strict practice of nonviolence, and to the Ajivakas, an extinct heterodox Eastern religious movement that denied the existence of free will.

58n. Johann Christoph ADELUNG (1732-1806) was one of the most influential German-language scholars before Jacob Grimm (1785-1863). His works, including the first major dictionary, helped standardize the German language. In 1787, after almost a quarter-century of independent research, Adelung was appointed principal librarian in Dresden, Saxony. His *Mithridates, oder allgemeine Sprachenkunde* (Mithridates, or general linguistics), begun in 1806, was completed by **Johann Severin Vater** after Adelung's death. This study illustrates the connections between Sanskrit and major European languages.

59. The Swedish naturalist, agriculturalist, and topographer Pehr KALM (1716-79) was a disciple of Carl von Linné, who incorporated materials that Kalm had brought back from his travels in the Americas (1747-51) in his 1753 *Species plantarum*. Linné honored Kalm by naming the mountain laurel *Kalmia*. Kalm contributed to John Bartram's (1699-1777) account of Kalm's American travels and also published his own account, of which an English translation appeared in 1773. (The Kalm and Denon footnotes were switched in the French original. We corrected this mistake.)

As governor of New France (that is, Canada) from 1726 to 1747, the chevalier Charles de BEAUHARNOIS (also Beauharnais; c. 1670-1759) favored French expansion to the west. According to Kalm, Beauharnois ordered a Mr. de VERNADRIER to undertake an expedition across North America to the South Sea in 1746. Another supporter of scientific expeditions, the French statesman Jean Frédéric Phélypeaux, count of MAUREPAS (1701–81), held high offices in French politics from an early age, which allowed him to build up the French navy and exert influence over the Academy of Sciences.

60. The Dighton Rock has occupied the minds of New England intellectual elites for centuries. In the early twentieth century, this facination prompted a series of publications on the rock's history in the *Publications of the Colonial Society of Massachusetts*. The Reverend John DANFORTH of Dorchester (d. 1730) had made an initial drawing of the rock in 1680. Puritan minister Cotton MATHER (1663–1728) drew it in 1712. In 1730, Harvard professor Isaac GREENWOOD (1702–45) argued that the inscription, and by implication Native Americans, were of oriental origin. Ezra STILES (1727–95) was fascinated with the Dighton Rock and exchanged ideas about it with Stephen SEWALL (1734–1804), a Harvard professor of Hebrew and other Asian languages. Before becoming president of Yale College, Stiles had been a Congregational minister in Rhode Island and New Hampshire who helped found Rhode Island College in 1764 (it later became Brown University). Stiles, who prepared three separate drawings of the inscription on the rock, mentioned it in a 1783 sermon and in his diary, which was published as *Literary Diary* in 1901. Stiles's curiosity about Dighton Rock inspired Michael Lort (c. 1724–90) to make further inquiries in the 1780s, and Lort directed his readers to **Court-de-Gébelin**. Humboldt culled his information from Lort's 1787 "Account of an Ancient Inscription in North America."

60n. Peter Frederik SUHM (1728–98), who lived both in Norway and in Denmark, was the most significant historian in late-eighteenth-century Scandinavia. His approach to historical scholarship was typical for his day, combining documentary history with comparative philology. In 1775 Suhm opened his private Copenhagen library to the public, putting into practice his Enlightenment conviction that knowledge should be made broadly accessible. He is most remembered for his this extensive book collection.

61. From 1795 to 1804 Ramón BUENO (176?–18??) was a Franciscan friar in the missions of La Urbana and Tortugas, where Humboldt and Bonpland stayed for one night on April 6, 1800, during their trip on the Orinoco River. The rock carvings that the cleric showed to Humboldt were rare Amerindian petroglyphs, which led Humboldt, in his travel diaries, to believe he had come upon a certain use of alphabetical characters; later, however, he apparently discarded this speculation. It was on his visit to Bueno's mission that Humboldt also came across geophagy (soil eating) among the local Otomi population, a practice widely discussed at the time in Europe.

62. Pieter Jan TRUTER (John Trüter) was a member of the high court of justice of the Colony of the Cape of Good Hope. He led an expedition in South Africa together with military surgeon William Somerville (1771–1860), future husband of the

Scottish scientist and mathematician Mary Fairfax Somerville (1780–1872). Truter was father-in-law of the auditor-general of Cape Colony, John Barrow (1764–1848), who saw to it that English- and German-language editions of Truter's travel accounts were published. Barrow's close personal and professional connections with Sir Joseph Banks (1743–1820), the first president of the London Royal Society, ensured his role in that organization. Barrow is today best known for *Mutiny on the Bounty* (1831). During his lifetime, his travel accounts of eastern Asia and southern Africa were also widely read.

The famous Genovese naval captain and explorer Christopher COLUMBUS (1451–1506) initiated Europe's colonial era with his four voyages to the Americas between 1492 and 1504. On the first of his journeys to the New World, Columbus landed on the island of Cuba on October 27, 1492; believing he had set foot on the legendary island of Cipangu, he baptized it San Salvador. Humboldt was fascinated with Columbus and set his own voyage and scientific achievements as a parallel course to those of the first "discoverer" of the Americas. Named the "second discoverer" of the Americas by Simón Bolívar, Humboldt (whose mother's maiden name was Colomb) discusses Columbus throughout his writings. He read the Genovese captain's travel diaries while working on volume 3 of his *Relation historique*, and he emphasized in his own texts the importance of the *intellectual* dimensions of Columbus's "discovery," without denying the atrocities of the Spanish *conquista*. The parallels that Humboldt saw between Columbus and himself are particularly evident in his *Examen critique de l'histoire de la géographie du Nouveau Continent* (Critical examination of the history of the geography of the New Continent, 1834–38), a study of the New World that focuses on the fifteenth and sixteenth centuries. Here Humboldt describes Columbus as the prototype of the European discoverer, who—unlike a mere adventurer—followed an elaborate preconceived plan and was equipped with significant geographical knowledge and cartographical skills.

The DEVANĀGARIĪ, known today as Nāgarī, is syllabic rather than alphabetical: each symbol represents a syllable rather than a sound. The main script in which Sanskrit literature has been preserved since the nineteenth century, Nāgarī is considered the most important of the subcontinental Indian writing systems. In use since the eighth century, it is the sacred script of the Brahmans, the traditional spiritual leaders in Hinduism. Contrary to Humboldt's assumption on the history of Hindu literacy, the earliest Sanskrit texts date to the mid- to late second millennium BCE. They were ceremonial writings whose correct pronunciation was considered crucial to their religious efficacy.

MEGASTHENĒS (c. 350–290 BCE) was a Greek ambassador to India around 300 BCE. Seleucus I (c. 358–281 BCE), a general who had inherited the eastern part of **Alexander the Great**'s domains, appointed Megasthenēs envoy to the Maurya empire at Pātaliputra (near present-day Patna), after having been defeated by the founder of the Maurya empire, Chandragupta Maurya (c. 325–297 BCE), known to the Greeks as Sandrakottos. Chandragupta's empire was the first to extend from the Bay of Bengal to the Arabian Sea, dominating the entire Indo-Pakistani subcontinent (c. 325–185 BCE). Megasthenēs's *Indica* is a detailed firsthand report of his visit to

Chandragupta's court; it is considered the most comprehensive historical account of ancient India and survives as fragments in later works by classical authors such as **Diodorus Siculus**, **Strabo**, and **Arrian**. Megasthenēs's description of the imperial capital and his detailed explanation of the seven strata of Indian society have attracted particular attention.

The ESTRANGELO alphabet—alternate spelling Estrangelā, known as Old Syriac—is a form of Syriac-Aramaic writing. Estrangelo was used for writing exclusively in Syriac, the literary language of Christian writers east of the Euphrates. A Semitic script, Estrangelo has an alphabet of twenty-two letters that represent consonants; it is typically written from right to left (occasionally vertically downward). Early manuscripts in Estrangelo use diacritical points to indicate plurals and vowels. Estrangelo spawned two other types of script, the Nestorian in the east and the Jacobite (Serto or Sertā) in the west.

62n. Friedrich Justin BERTUCH (1747–1822) was a German publisher in Weimar, then a cultural center and home of some of the most famous German Enlightenment writers, scholars, and naturalists, including Johann Wolfgang von Goethe (1749–1832). See also Karl August Böttiger.

63. Contrary to what Humboldt claims, Diego VALADÉS (b. 1533) was born in Mexico as son to a conquistador father and an indigenous mother; he was the first recorded Mestizo friar of the Franciscan Order in New Spain. In 1570 the order sent him to the Vatican, where he wrote and illustrated his famous *Rhetorica Christiana* (Christian rhetoric), an extraordinary theological text that is the first printed account of the evangelization of Mexico. In addition to portraying the Franciscan experience in the New World, Valadés included descriptions of the indigenous deities and religious practices, with sumptuous illustrations depicting native life. This information is unavailable in other sources from this early stage of European colonization. For his catechism, Valadés developed a visual method for teaching alphabetic literacy to the indigenous peoples: the Roman letters were portrayed together with similarly shaped physical objects and pictorial representations of the complex system of numerical, calendrical, pictographic, ideographic, and phonetic glyphs from the symbolic script of the Aztec culture.

The Spanish Jesuit José de ACOSTA (1540–1600) was a missionary mainly in Peru, but he was also briefly stationed in Mexico. Acosta, who was trained in Greek philosophy, Latin rhetoric, and Christian theology, arrived in Peru in 1572, the year in which the Jesuit order was formally instituted in the Americas; he visited Cusco, Arequipa, La Paz, Charcas, Potosí, and Chuquisaca. His *Historia Natural y Moral de las Indias* (*Natural and moral history of the Indies*) was written in both Latin and Spanish in seven books. Its preliminary version was published under the title *De natura novi orbis* (Of the nature of the New World, 1588–89). The final edition was published in 1590 and translated into several European languages—Italian, French, Dutch, German, Latin, and English—between 1596 and 1604. Acosta's *Historia* was based on his direct interactions with the local peoples; he also benefited from conversations with and writings of Juan de Tovar (c. 1543–1626) and **Polo de Ondegardo**. In addition to publishing a catechism in Aymara and Quechua in 1583, the first book to be printed in Peru, Acosta wrote *De Christo revelato* (Of the revelation of Christ, 1590), *De temporibus novissimis* (Of modern times, 1590) and a book of sermons. In 1597, a decade after having returned to Spain, Acosta was appointed dean of the Jesuit College in Salamanca, a position he held until his death.

The German Jesuit Athanasius KIRCHER (1602–80) published widely on the Egyptian language and translated hieroglyphics. In 1633 he was appointed professor of mathematics and Oriental languages at the Roman College. After he was relieved of teaching duties in 1646, he devoted his time to researching, writing, and entertaining visitors.

William WARBURTON (1698–1779) was bishop of Gloucester and a religious controversialist. A self-congratulating, rancorous person, he was described as having "led the life of a terrier in a rat-pit, worrying all theological vermin." His major work, *Divine Legation of Moses Demonstrated* (1738–41), included explanations of Egyptian hieroglyphs, which became part of the French *Encyclopédie*.

The Greek philosopher PLUTARCH (Lucius Mestrius Plutarchus, c.45–120) was born in a small village of Chaeronea in Boeotia. He studied rhetoric and philosophy at the Platonic Academy in Athens and held one of the two priesthoods of Apollo in the sanctuary at Delphi for at least two decades. A Platonist and anti-Stoic, he traveled to Egypt and Italy and lectured in Rome. He became a Roman citizen, but although identified with Rome in his historical observations and approaches, he never referred to himself as a Roman. Plutarch's most famous work is *Parallel Lives*, a description of the virtues and vices of great Greek and Roman men. Equally well known are his *Moral Essays*, in which he discusses moral philosophy, religious beliefs, and literary criticism, among other subjects. "The Malice of Herodotus," one of his essays in the *Moral Essays*, criticizes **Herodotus** for sympathizing with the Persians, for Plutarch believed him to be biased against the Greeks. Of the 277 items in the *Catalogue of Lamprias*, a list of his works from around the fourth century, only 50 biographies and 78 miscellaneous works survive. Plutarch's popularity in Byzantine is responsible for the preservation of his works. Widely influential during the Renaissance, the English translation of *Lives* (1579) by Sir Thomas North (1523–1601) was Shakespeare's major source for *Julius Caesar*, *Antony and Cleopatra*, and *Coriolanus*.

The Greek theologian and early Christian philosopher ST. CLEMENT OF ALEXANDRIA (Titus Flavius Clement, c. 150–215) was one of the founding fathers of Christianity. A teacher and priest at Alexandria, from the death of his teacher Pantaenus (c. 190) until he fled persecution by Lucius Septimus Severus (145–211), Clement laid the foundations of the Alexandrian catechetical school of the second century. As the leader of that school, he significantly strengthened the influence of Christianity in Egypt, trying to reconcile the image of Alexandria, an important focus of the Hellenistic world at the time, with the teachings of early Christianity. Clement's principal achievements are *Logos protreptikos* (Hortatory discourse), an attack on paganism and exhortation to the Greeks to conversion; *Paidagogos* (Tutor), a guide offering moral counsel and etiquette for daily life; and *Stromateis* (Miscellanies), in which he addresses the role of philosophy and its relationship to Christian truth. Influenced in many respects by the first-century Stoic philosopher Gaius Musonius Rufus, Clement insisted that women were as capable as men of studying any subject. Likewise, he was opposed to the idea that marriage is incompatible with the spiritual life; he held up St. Paul as an example of a married spiritual leader. Moreover, he argued that the wealthy could also achieve salvation, since the proper use of wealth, rather than its mere possession, is what matters. Interestingly, however, in accordance with Middle Platonist antihedonistic thought, Clement equated asceticism with martyrdom, opposing opulence, pagan materialism, homosexuality, abortion, divorce, marital intercourse for pleasure, and eroticism.

64. Humboldt assumed that the HIEROGLYPHIC INSCRIPTIONS of the Egyptians might be readable in the same way as, for instance, Chinese characters, which shows that at the time when he wrote *Views of the Cordilleras*, Egyptian hieroglyphs still vexed scholars. It was only in 1822 that the Frenchman Jean-François Champollion (1790–1832), a student of **Silvestre de Sacy**, made a major breakthrough in deciphering them. The British scholar Thomas Young (1773–1829) had done important foundational work by correctly identifying the cartouche containing the name of Ptolemy. (*Cartouche* is the name that the French soldiers gave to an oval containing hieroglyphs within an inscription; the shape is reminiscent of the cartridges in guns.) Deciphering was made possible when the French scholars and scientists who accompanied Napoleon's Egyptian campaign (1798–1801) came into the possession of the so-called Rosetta Stone, which, it was soon discovered, contained the same inscription in three different scripts, including Egyptian hieroglyphs. Since the British, like the French, had more than just a scientific interest in Egypt, both national and personal sensitivities were stirred up when Europeans started eagerly to examine the inscriptions of the Rosetta Stone. An early enthusiast was the Swedish diplomat J. H. Åkerblad (1763–1819), who presented his ideas on the subject in *Lettre sur l'inscription égyptienne de Rosette, addressé au citoyen Silvestre de Sacy* (Letter about the Egyptian inscription of Rosetta, addressed to Citizen Silvestre de Sacy, 1802). As his correspondence with Young shows, Humboldt followed these developments with active interest. Following Young and Champollion, the next generation of Egyptologists included the German Carl Richard Lepsius (1810–84).

Responsible for inspiring some of the greatest achievements in Roman official art, Roman Emperor TRAJAN (53–117; r. 98–117) also commissioned the famous eponymous column. It was constructed between 107 and 108 from the spoils of the Dacian Wars and was dedicated in 113. Apollodoros of Damascus (c. 60/70–c.130), the column's architect, was Trajan's chief architect and military adviser, famous for building a bridge over the Danube during the wars. In style this column was the first of its kind. Its surface is decorated with a spiral band of low-relief carvings that wind around the shaft; they commemorate Trajan's two victorious campaigns against the Dacians (in 101–2; 105–6). With chronologically placed reliefs of the military campaign, the battles, and the defeated enemy, the column served as a sculpted historical narrative of Trajan's victory and a propagandistic tool. With Trajan's tomb as its base, the column has an interior passageway that leads via a spiral staircase to the top, where Trajan's bronze statue (grasping a warrior's spear and orb) was originally placed. That statute was replaced with one of St. Peter by Pope Sixtus V (1520–90), one of the column's restorers. Part of the Forum of Trajan and the best-preserved monument of that complex, the column was designed to be read: it stood between two libraries, one housing Greek works, the other Latin. This honorific column has been widely imitated, most notably in the Column of Marcus Aurelius (built 141–48).

Emperor Motecuhzoma Ilhuicamina, also known as MOCTEZUMA I or Moctezuma the Elder (c. 1397–1468), was the fifth Mexica ruler (1440/41–68/69). An experienced statesman when elected successor to his uncle, Itzcoatl (r. 1428–40), Moctezuma I strengthened Mexica military power and expanded territorial boundaries, consolidating the empire his uncle had began to establish. The **Codex Mendoza** reports his thirty-three conquests. Moctezuma I is credited with advancing the arts and commissioning sculptures and lavish botanical gardens. In addition to having a sacrificial stone (*temalacatl*) carved, he ordered the renovation of the great temple at Tenochtitlan and the building of the House of Eagles (Cuauhcalli). Deeply interested in history, he sent many expeditions in search of the seven caves of Chicomoztoc at Aztlan, the Mexica's mythical land of origin.

67. The Spanish Jesuit Lorenzo HERVÁS y Panduro (1735–1809) was a missionary in New Spain until his order was expelled in 1767. After returning to Rome, he compiled a six-volume catalog of the world's languages, originally published in Italian between 1784 and 1787 and in Spanish between 1800 and 1804/5. It included detailed information on numerous American languages, which he had solicited from his fellow missionaries. Certain inaccuracies notwithstanding, Hervás's work represents the culmination of Spanish colonial linguistics. Much like Humboldt himself, Hervás was interested in language studies as a key to human history.

The Venetian economist Giovanni RINALDO, count of Carli-Rubbi (1720–95), was a passionate collector of both Italian and Amerindian antiquities. In his famous *Lettere americane* (American letters), written between 1777 and 1779 and translated into French and German shortly after its publication, Carli gives a personal account of his admiration for the ancient Inca and Aztec cultures of Peru and Mexico. Carli regarded the well-structured Inca state as a projection of his own reformist spirit and an embodiment of his enlightened ideals; in his mind it was linked to the idea of Atlantis, the lost world from which the American and other populations supposedly had sprung. Although not primarily intended as a polemic, *Lettere* includes profound criticisms of **de Pauw** and **Robertson**, whose works denied the accomplishments of indigenous American societies with allegations of cultural inferiority and moral degeneration. The "Dispute on the New World" of which their work was part resurged in nineteenth-century Prussia as the so-called Berlin Debate, in which Humboldt participated vigorously. He opposed the writings of Georg Wilhelm Friedrich Hegel (1770–1831) and his concept of world history, which held that American cultures, especially those of Mexico and Peru, had to perish because of their supposed physical and spiritual inferiority.

69. Johann Phillipp Tabbert (1676–1747), named VON STRAHLENBERG after receiving his title in 1707, was a Swedish captain, cartograph, and geograph of German origin. He participated for two years in the Siberian expedition led by the German physician Daniel Gottlieb Messerschmidt (1685–1735) from 1721 to 1727. In 1730 von Strahlenberg published an important geographical, cultural, and linguistic study on northern Asia, titled *Das Nord und Östliche Theil von Europa und Asia* (The northern and eastern parts of Europe and Asia).

69n. The French Jesuit missionary Joseph-François LAFITAU (1681–1746) lived among the Iroquois in a Canadian mission at Sault Saint Louis, Québec, for nearly six years (1712–17). His two-volume *Moeurs des Sauvages amériquains* (Customs of the American savages) from 1724 is an ethnographic study with a large theoretical chapter on the likeness between New and Old World beliefs and symbols, stressing a "one world paradigm" that became increasingly popular in scholarly disputes of the time. Comparing the material of his own extensive fieldwork with results from other reports on American peoples, and on the ancient civilizations of Eurasia, Lafitau also advanced a theory about the universal religious instinct in human societies.

A missionary in China in the 1640s, the Austrian Jesuit MARTINO MARTINI (1614–61) had previously studied mathematics with **Athanasius Kircher** at the Roman College. Having been sent back to Rome in 1650, Kircher took advantage of the opportunity to prepare his historical and cartographical data on China for publication. He apparently based his maps on the *Mongol Atlas* (1311–12) by Chu Ssu-pen (1273–1337), which was later revised by Lo Hung-hsien (1504–64).

70. The mathematician and cartographer Antonio de ULLOA (1716–95) accompanied the 1735 French expedition to South America as a representative of the Spanish crown. After their return to Europe in 1745, Ulloa and **Jorge Juan** coauthored a confidential report to the crown, titled *Discurso y reflexiones políticas sobre el estado presente de la marina de los reynos del Perú* (Discourse and political reflections on the present state of the royal fleet of Peru, 1749), based on their observations during ther expedition. Denouncing colonial administrators' corrupt practices, the report was unofficially published in London as *Noticias secretas de América* (Secret news from America, 1826). Ulloa later held several high positions in the Spanish colonial administration, including governor of Louisiana (1766–68) and of Huancavelica (Peru, 1758–63), and lieutenant-general of the Spanish navy. In addition to several other joint publications with Juan, such as the *Relación histórica* (Historical report), Ulloa wrote a natural history titled *Noticias americanas* (American news, 1772). See also Bouguer and La Condamine.

The QIN DYNASTY (alternate spelling: Ch'in) ruled in China from 221 to 207 BCE. The Qin state, the predecessor of the dynasty, was a remnant state of the historical period known as the Warring States (656–221 BCE). Formed around 897 BCE, the Qin banned feudalism, introduced codified penal law, standardized the writing and measurement system, and built an impressive communication and road infrastructure, initiating the construction of the Great Wall. Developing increasingly sophisticated government institutions, the Qin reorganized and strengthened the military force until it dominated the neighboring states of Han, Chao, Wei, Ch'u, Yen, and Ch'i. The first emperor of Qin, Ying Cheng (c. 260–210 BCE), ruled as a king (*wang*) of the state before the unification of China in 221 BCE. He used the title emperor (*huang-ti*) until his death in 210 BCE. Although it lasted for only fifteen years, the Qin dynasty had an enduring impact, especially on the Han dynasty (206–220), which replaced it. Its system of centralized state bureaucracy, readily expanded and consolidated by the Han, burgeoned for another seventeen hundred years with only slight changes. Scholars believe it probable that the name China derives from Qin/Ch'in.

71. Étienne MARCHAND de la Ciotat (1755–1793) was the captain of *La Solide*, a three-hundred-ton French ship that sailed around the world on a fur trading expedition. *La Solide* departed Marseilles in 1790. In 1791, while anchored at Sitka Bay, Marchand traded with the Tlingit peoples and made a chart of Sitka Sound, titled *Plan de la Baie Tchinkitáné*, using for the first time the Tlingit-language ethnonym of the Tlingit nation: Lingít Aaní (Tlingit territory). Thereafter, Marchand traveled southward and stayed for a week in the lands bordering Parry Passage on the northern Queen Charlotte Islands (Haida Gwaii), British Columbia. The voyage's journals and maps, with their detailed descriptions of the Haida and Tlingit peoples and the Pacific islands visited, were edited by Charles Pierre Claret, count of Fleurieu (1738–1810), under the title *Voyage autour du Monde, pendant les années 1790, 1791, et 1792, par Étienne Marchand* (Voyage around the world during the years 1790, 1791, and 1792 by Étienne Marchand, 1798–1800). See also Claude Roblet and Cox Canal.

The German physician Johann LEDERER (1644–after 1673) emigrated to the New World as a young man. He became one of the first Europeans to explore the Piedmont and Blue Ridge Mountains of Virginia and Carolina and to leave a written record of his travels. Commissioned in 1669 by Virginia governor William Berkeley (1605–77), Lederer went on an exploration westward from the colony's outposts beyond the Tidewater settlements, trying to find a path through the Appalachians. Though highly disputed for their inadequacies, the narratives and maps of his three journeys shaped the perceptions of European geographers and ethnographers for many years. His friend Sir William Talbot, chief secretary of Maryland, later translated the accounts of his experiences, mostly with Native American cultures, languages, and trade, as *The Discoveries of John Lederer* (1672).

71n. While Louis de Lom D'Arce, baron of LAHONTAN (1666–1716), served as a military officer in the French colonial troops in Canada (1683–93), he established close relationships with leaders of the Hurons (Wendat), the Iroquois, and the Algonquin peoples; he even learned the Algonquin language. After deserting, Lahontan returned to Europe and published a three-volume work on his American experiences. The first volume consists of his travelogue, and the second volume—which Humboldt mentions—gives an ethnographic account of Canada's indigenous peoples. In his third volume, *Dialogues curieux* (Interesting dialogues), Lahontan narrates a fictional dispute between himself and his friend the Huron chief Adario-Kondiaronk, celebrating the Native American community as embodying a semiutopian ideal.

72. Niccolò (c. 1326–c.1402) and Antonio Zeno (d. c.1405), better known as the ZENI BROTHERS, purportedly explored North America at the end of the fourteenth century. In 1558 Niccolò Zeno (1515–65), a descendant of the brothers' family in Venice, published an account, together with a map, that was allegedly based on the adventurers' letters. The narrative claims that the two brothers were shipwrecked on Frislanda (Faroe or Iceland) around 1380. From there they explored farther west, reaching many North Atlantic islands, among them Engronelant (Greenland). The story and its accompanying map had profound implications for mapmakers of and voyages to the northwest Atlantic region. See also Drogeo.

While a missionary in Peru, the Franciscan monk Narciso GIRBAL y Barceló (1759–1827) explored the Amazonian rivers Ucayali and Marañón. In 1790, five years into his mission, Girbal began exploring the villages on the Amazonian rivers on his way to Sarayacu de Manoa, the ruins of a mission destroyed in 1767 during a revolt of the local population. After reestablishing the mission at Manoa, he took part in at least two other expeditions. Father Girbal's travel accounts were published in the *Mercurio Peruano* in 1791 and 1792. They were also included in Bernandino Izaguirre's famous *Historia de las misiones franciscanas y narraciones de los progresos de la geografía en el oriente del Perú, 1619–1921* (History of the Franciscan missions and accounts of the advances in geography in east Peru, 1924). Girbal met Humboldt in 1802, thirteen years before he returned to Spain.

72n. Known as the Venerable Bede, the theologian St. BEDE (also Beda or Bæda) (c. 673/74–735) was one of the most highly praised historians of the early Middle Ages. A voracious reader and prolific writer, he produced nearly forty written works on diverse subjects, from hagiography, biblical commentaries, translations of scriptural passages, and grammar to scientific works on chronology and natural history. Author of a treatise on music and a long poem, *On the Day of Judgment*, Bede wrote in medieval Latin. He is often referred to as the father or founder of English church history; his most famous work is the *Historia ecclesiastica gentis Anglorum* (*Ecclesiastical history of the English people*); he completed it in 731, and it was published much later, between 1474 and 1482 (the first English translation dates from 1565). The *Historia*, acclaimed for centuries for the quality of its scholarship and its aesthetics, focuses on the history of the Christian church in the Anglo-Saxon kingdom of Northumbria. Not only is the *Historia* the earliest significant history of England and the first major history of the English church; it is also regarded as the first important work of English literature. The author of many other writings, Bede also popularized the practice of dating years from the birth of Christ (*Anno Domini* or AD). Dante Alighieri (1265–1321) commemorates Bede in the *Divine Comedy* by placing him in canto X of *Paradiso*. In 1899, Pope Leo XIII (1810–1903) pronounced Bede *doctor ecclesiae* (doctor of the church); he was the only Englishman to receive this honor up to that time. Bede was canonized in 1935.

A Dominican monk from New Granada, Francisco NÚÑEZ DE LA VEGA (1632-1706) was appointed bishop of Chiapas and Soconusco in 1683, a position he held until his death. His *Constituciones diocesanas* (1702) discuss Maya religion and the predominance of a spirit cult called Nagualism in the native population of eighteenth-century southern Mexico. Núñez de la Vega was the first to mention Votan (Uotan: "heart"), a deity of highland Chiapas. In the seventeenth century, the peoples of Teopisca (who spoke Tzeltal) were regarded as the descendants of Votan. According to a myth related by Núñez de la Vega, Votan traveled through a serpent-made subterranean passage, bringing back several tapirs and a treasure, which he deposited in a dark house he constructed in Huehuetlán in Soconusco (a district in the state of Chiapas). In the Maya *tzolk'in* calendar, Votan was identified with the third day (*ak'b'al*, darkness).

73. A Jeronymite priest from the monastery at El Escorial who had arrived in Lima in 1773, Diego CISNEROS (c. 1740-1812) was a strong supporter of the Peruvian Enlightenment. He was a member of the Sociedad de los Amantes del País (Patriotic Society) and contributed to the *Mercurio Peruano* under the pseudonym Archidamo. He opened a library and circulated banned works by such authors as Voltaire, Rosseau, Diderot, d'Alembert, and Montesquieu, to which his political connections with the Spanish monarchy gave him access. Cisneros's clandestinely introduced works were often cited in the *Mercurio Peruano*, the twelfth and last volume of which he personally financed. In 1796 the Inquisition opened a fruitless trial against Cisneros for possessing prohibited books and upholding heretic ideas. A letter criticizing the Inquisition, allegedly by Cisneros, was published in the newspaper *El Investigador* (1813-14) after the abolition of this tribunal in 1812.

75. The Codex Borbonicus, which Humboldt calls the ESCORIAL COLLECTION, is either a preconquest or early sixteenth-century ritual-calendrical screenfold painted on one side on native paper. Currently at the Bibliothéque de l'Assemblée Nationale Francaise, Paris, it was initially at the Escorial library in Spain, where **Waddilove** found and described it, before pages 1-2 and 39-40 were lost. The remaining thirty-six pages, which are divided in four sections, show great detail, making it possibly the most brilliantly painted of surviving manuscripts from that period in Mexico. The date of this codex is controversial. Codex Borbonicus is also known as Codex du Corps Legislatif, Codex Legislatif, Codex Hamy, or Calendario de París.

The CODEX COSPI, referred to by Humboldt as the Bologna Collection or Codex Mexicanus of Bologna, is a deerskin screenfold currently housed at the Biblioteca Universitaria, Bologna, Italy (mss no. 4093). Part of the **Borgia** Group, it includes images of deities, offerings, and numerals in four segments folded into twenty square pages, of which thirteen recto pages and eleven verso pages are painted. The first three of four sections present a different aspect of the divinatory cycle; the interpretation of the fourth section, which is in a different style, is uncertain. This work is also known as Codex Cospianus, Códice Cospiano, Codici di Bologna, Códice de Bolonia, or Libro della China.

The CODEX BORGIA, also known as the Codex Borgianus or the Velletri Collection, is also a deerskin screenfold manuscript currently at the Biblioteca Apostolica Vaticana in Rome, Italy (Codex Borgia Messicano 1). Dating from before the conquest, it is one of the five pictorial manuscript parts of the Borgia Group. It is the most important and most intricate source for the study of Central Mexican gods, rituals, divination, calendar, religion, and iconography. Humboldt's details about this codex antedate all its other publications (see Plates XV, XXVII, and XXXVII). Lord Kingsborough (1795–1837) published the codex in *Antiquities of Mexico* (vol. 3, 1831–48), Eduard Seler between 1904 and 1909 as *Codex Borgia: Eine altmexikanische Bilderschrift der Bibliothek der Congregatio de propaganda fide* (An ancient Mexican pictograph from the library of the Congregation of Propaganda Fide, 3 vols).

Yet another deerskin screenfold that is also part of the Borgia Group, the CODEX VATICANUS B, is preserved at the Biblioteca Apostolica Vaticana in Rome. It is a complex presentation of the Mesoamerican divinatory calendar together with the 260-day almanac. Humboldt was the first to publish parts of this codex (see Plates XIII and LX). The Codex Vaticanus B is also known as Codex Vaticanus 3773, Codice Vaticano Rituale, or Códice Fábrega. Whenever Humboldt speaks of the Codex Vaticanus in *Views of the Cordilleras*, he means the Codex Vaticanus B.

Humboldt's "Vienna Collection" is better known as CODEX VIENNA, a ritual-calendrical and historical deerskin screenfold from western Oaxaca, painted on both sides of fifteen segments. The obverse side has fifty-two completely painted pages protected by wooden covers; an additional thirteen pages were later painted on the reverse side. Preserved at the National Library in Vienna since 1677, this codex is believed to have been sent to Europe by Cortés in 1519. A copy of the codex, furnished in Weimar (Germany) around 1655, is at the National Museum, Copenhagen. This manuscript contains mythological genealogies (time span corresponds to a period from the eighth to the fourteenth century), lists of place glyphs, dates, persons, gods, or priests. Humboldt reproduced four pages from it in Plates XLVI, XLVII, and XLVIII. This codex is also known as CODEX VINDOBONENSIS, Codex Vindobonensis Mexicanus 1, Codex Hieroglyphicorum Indiae Meridionalis, Codex Clementino, Codex Leopoldino, or Codex Kreichgauer.

As a young novice, Jesuit scholar José Lino FÁBREGA Bustamante (1746–97) from New Spain was forced into exile when the order was expelled from the Americas in 1767. In Italy, Fábrega eventually came under the protection of Cardinal **Stefano Borgia**, who allowed the young man to study the *tonalamatl*, a precolonial codex from the cardinal's private library; it has been known as the Codex Borgia ever since. Fábrega's studies, which later extended to the Roman and Vatican libraries, produced impressive findings: he found various other codices, notably the Codex Vaticanus B 3773, the Codex Ríos (which is also known as Codex Fábrega), and a rare copy of the Codex Cospi. All of them belong to the same family of precolonial ritual calendars, which later became known as the Codex Borgia Group, a term coined by the German Americanist Eduard Seler. Fábrega wrote a pioneering commentary on the Codex Borgia, which Humboldt discusses rather critically here (see Plate XV).

While visiting his brother Wilhelm in Rome in 1805, Humboldt was able to study the annotated Mexica manuscript of the Codex Borgia 1805, which Camillo Borgia (1773–1817), nephew of **Cardinal Stefano Borgia**, had brought to Rome from Velletri, home to the cardinal's private museum of classical, Egyptian, Asian, and precolonial art. Count Camillo, a military man in the armies of the Vatican, Austria, and France who was imprisoned twice for sympathizing with the French Revolution, eventually had to leave his family behind in Naples and flee to Tunisia, where he undertook several expeditions with the Dutch military engineer and archaeologist Jean Emile Humbert (1771–1839).

Robert Darley WADDILOVE (1736–1828) worked in Spain from 1771 to 1779, serving as embassy chaplain accompanying the British ambassador Thomas Robinson, second BARON GRANTHAM, aka Lord Grantham (1738–86). In Spain, Waddilove also collected scholarly materials for British friends, including **William Robertson**, who was then working on his *History of America*. In 1775 Waddilove was elected a fellow of the Society of Antiquaries, to which **Michael Lort** would report his findings about the Dighton Rock in Massachusetts a decade later. Lord Grantham was recalled when Spain declared war in 1779; he was subsequently involved in the peace negotiations with France, Spain, and the nascent United States of America.

Marquis Ferdinando COSPI (1606–85/86) was a Bolognese patrician and collector of curiosities and historical documents of non-European cultures, which he donated to the city of Bologna in 1657 as a public cabinet for the use of scholars. The book of ancient Mexica manuscripts that Valerio ZANI (d. 1696) gave Cospi as a Christmas gift in 1665 (instead of selling it, as Humboldt suggested), was first titled *Libro della China* (Book of China) and only later corrected to *Libro del Messico* (Book of Mexico), showing the confusion about foreign cultures that still endured among European scholars in the seventeenth century. Zani's last name is an anagram of the surname of Count Aurelio degli Anzi (d. 1696), a Milanese scholar and poet who edited *Il genio vagante* (The wandering genius, 1691–93), four volumes of accounts by Italian and foreign travelers from the seventeenth century, often accompanied by comments and brief reviews.

76. A promoter of colonial expansion, KING EMMANUEL OF PORTUGAL, or Manuel I of the House of Aviz (1469–1521), was the fourteenth monarch of Portugal (r. 1495–1521). Under his reign the Portuguese opened the sea route to India (1497–98) and landed in Brazil (1500). In 1496–97 King Manuel passed legislation expelling Jews and Muslims from Portugal, which resulted in many Jews' forced conversion to Christianity. From the profits of Portugal's monopoly over sea trade with Asia, the king bought exotic goods and objects from China, India, and other parts of the world. In addition, he commissioned the building of twenty-six monasteries and two cathedrals. Their architectural style, known as Manueline, used exuberant ornaments and symbolism inspired by the Portuguese maritime voyages of the fifteenth century.

Born out of wedlock as Giulio de' Medici (1478–1534), POPE CLEMENT VII (p. 1523–34) was a nephew of Pope Leo X. He was appointed vice chancellor of the Holy See in 1517. His indecisiveness about permitting Henry VIII (1491–1547) to divorce from from Catherine of Aragón (1485–1536) is thought to have accelerated the

Reformation, a movement against the Catholic Church that resulted in Protestantism and the appointment of Henry VIII as head of the Church of England. As a result of Clement's ambivalence and his failed attempt at reconciling Charles V and Francis I of France (1494–1547), imperial troops invaded Italy and sacked Rome in 1527. After securing Charles V's release from imprisonment, Clement crowned him Holy Roman Emperor in 1530. An art lover and patron of artists and intellectuals, including Niccolò Machiavelli (1469–1527), Francesco Guicciardini (1483–1540), Benvenuto Cellini (1500–71), Raphael (1483–1520), and Michelangelo (1475–1564), Clement commissioned the *Last Judgment* for the Sistine Chapel.

Cardinal Ippolito de' MEDICI (1511–35), Leo X's nephew and Clement VII's cousin, ruled Florence when Clement VII became pope. The last Florentine Republic was inaugurated with Ippolito's expulsion from Florence, along with his cousin Alessandro. Ippolito became archbishop of Avignon instead of duke of Florence. As a consolation, Clement VII appointed him cardinal in 1529, at the age of eighteen. In 1532 Ippolito traveled to Hungary to serve as a papal legate to Charles V. Titian (c. 1485/90–1576) drew Ippolito's portrait in 1533; it is titled *Ippolito de' Medici in Hungarian Costume* and is housed at the Galleria Palatina in Florence. Like other Medicis before him, Ippolito was a patron of the arts, sponsoring, among others, Alfonso Lombardi (1497–1543) and Giorgio Vasari (1511–74).

GIOVANNI DE' MEDICI (1475–1521), who became Pope LEO X (r. 1513–21), was made cardinal in 1488, at the age of thirteen, by Pope Innocent VIII (1432–92). In 1512 he secured the return of the Medicis to Florence with the aid of Pope Julius II (c. 1443–1513). A supporter of the arts and sciences, he set up Greek colleges in Rome and Florence and collected exotic goods and animals. By 1508 he had set up a private library in Rome that held rare manuscripts and books from Europe and the East. To raise funds for the construction of St. Peter's Basilica, he renewed and promoted the sale of indulgences (full or partial remission), which led Martin Luther (1483–1546) to oppose the church's practices in his ninety-five theses of 1517. After unsuccessfully trying to silence Luther, Pope Leo X published the bull *Exsurge Domine* (1520), condemning him on forty-one counts. Luther's excommunication in 1521 marked the beginning of the Reformation. Leo X was the patron of many artists, most notably Raphael and Michelangelo. Due to his extravagant lifestyle, Leo X left the papal treasury deeply in debt when he unexpectedly died of malaria.

The cardinal of CAPUANUS, Nikolaus von Schönberg (1472–1537), also known as Fra Niccolò, was a member of the Dominican Order. He was appointed archbishop in 1520, and Pope Paul III (1468–1549) made him cardinal in 1535. In 1510 Schönberg became professor of theology at Sapienza University in Rome, publishing *Orationes quinque de admiranda Christi pugna cum diabolo in deserto* in 1512. Leo X entrusted him with diplomatic missions to France, Hungary, and Poland. A trusted counselor of Clement VII, he was to broker peace between Charles V, Francis I, and Henry VIII in 1524. He became governor of Florence in 1530.

Peter LAMBECK, or Lambecius (1628–80), was a librarian at the Library of the Imperial Court in Vienna, in which capacity he served Emperor **Leopold I** for seventeen years. Leopold I (1640–1705), who concerned himself with antiquarian studies, was Holy Roman Emperor from 1658 to 1705.

Johann Georg I (1634–86) was DUKE OF SAXE-EISENACH from 1672 to 1686.

Samuel PURCHAS (bap. 1577–1626) was a Church of England clergyman who compiled much of *Hakluytus Posthumus, or Purchas his Pilgrims* (1624–25), the outgrowth of nearly twenty years of collecting oral and written travel accounts about Europe, Asia, Africa, and the Americas. See also Richard Hakluyt.

Successor to **Cortés**, Antonio de MENDOZA (c. 1490–1552), marquis of Mondéjar and count of Tendilla, was the first of sixty-four viceroys of New Spain (1535–51). The Codex Mendoza is named after him. A nobleman from Spain, Mendoza served **Charles V** as a diplomat before arriving in New Spain to replace the conquistadors and establish an administrative system that lasted until independence. Mendoza was successful in undermining Cortés's stronghold, and upon **Alvarado**'s death he took his men and equipment. Despite the 1542 New Laws for protecting the Indians, Mendoza supported both the *encomienda* system, under which the indigenous population provided tribute and labor for the elite, and the *repartimiento*, which required each village to provide a weekly quota of Indians for work on projects presumably for the public good. By successfully resisting the New Laws in 1545 with the support of the clergy, Mendoza averted violent protest from the holders of the *encomiendas*. Lured by accounts of mythic riches and spurred by his desire to expand his dominion, Mendoza backed several Rennaissance expeditions to the Mexican northwest and the Pacific. In 1536 he founded the mint in Mexico City to promote trade; he supported the creation of the Colegio de San Juan de Letrán for the indigenous nobility and the founding of the **University of Mexico**. Mendoza favored the collection of indigenous materials as a publicity tool for strengthening his political position in New Spain in relation to the Spanish courts. In the late 1500s he commissioned the anonymous *Relación de Michoacán* (Report of Michoacan)—delivered around 1541 but not published until 1869. The extant 277-page manuscript (Ms. C.IV.5, El Escorial, Spain) offers historical and ethnographic information about Michoacan for the years immediately before and after the conquest. The *Relación* depicts Mendoza as a wise and benevolent leader, beloved by Spaniards and Indians alike. Due to Mendoza's success as a New Spain administrator, Charles V appointed him viceroy of Peru in 1551. In that same year Mendoza commissioned from Juan Díez de Betanzos y Araos (c. 1510–76) a work on Inca history and customs. Betanzos's work, which was finished after Mendoza's death, is titled *Suma y narración de los Incas* (Overview and narrative of the Inca, 1880).

The Dominican friar André THÉVÊT (c. 1516–94) was royal cosmographer of France. The author of the popular and much-translated *Singularitez de la France antarctique* (The uniqueness of Antarctic France, 1557), he was one of the first French writers to describe North America. His work was published in English as a selection of chapters on North America titled *Thévêt's North America*. For a long time deemed an amalgam of plagiarized borrowings from the accounts of more experienced travelers, Thévêt's work has recently been rehabilitated by contemporary scholars arguing that much of his primary material is instead taken directly from oral sources. Thévêt's use of the **Codex Mendoza** allowed him to present material on Aztec creation myths and religious beliefs that were singular for his time.

77. A promoter of British settlement in North America, Richard HAKLUYT (c. 1552–1616) purchased a "Florida" manuscript (that is, the Codex Mendoza) and other materials from the heirs of **André Thévêt** in 1587. After showing the document to Sir Walter RALEIGH (1554–1618), Hakluyt hired a translator to prepare an English version of the Spanish text, which he included in his *Principal Navigations* (a book dedicated to Raleigh). Upon Hakluyt's death, both the translation and the codex passed into the hands of his assistant, the Anglican clergyman **Samuel Purchas**, who continued Hakluyt's work of collecting and publishing accounts of English voyages and discoveries around the world. Raleigh also requested that Hakluyt write *A Discourse of Western Planting* (1584) in support of his colonizing initiative in the Americas. Although Raleigh himself never traveled to Virginia, he was the mastermind behind several expeditions there, beginning in 1585. Thévêt claimed that Hakluyt had borrowed some of his materials; Hakluyt countered that Thévêt, a sympathizer with Spain, had neglected publication of the materials on purpose.

Alongside Raleigh, the British antiquarian Sir Henry SPELMAN (c. 1563–1641) was a member of an early British Society of Antiquaries in the late sixteenth century; he attempted in vain to revive the society early in the seventeenth century.

Having served as envoy to Genoa and Rome in the 1640s and 1650s, the polyglot French scientist and diplomat Melchisédec THÉVENOT (1620–92) became royal librarian to French King Louis XIV (1638–1715; crowned in 1643) in 1684. Purchas published Thévenot's *Relations de divers voyages curieux* (Report of various interesting voyages) in 1696; it was appreciated specifically for its strategic detail on the Low Countries.

The Mendoza Collection, better known as CODEX MENDOZA, was commissioned by Viceroy **Mendoza** and composed around 1541–42. A combination of pre-Columbian pictorial history with Spanish translations and annotations, the codex was drawn, colored, and interpreted by native scribes under the supervision of Spanish missionaries. Seventy-two annotated pictorial leaves and sixty-three pages of commentary relate the history and customs of the Mexica. As Humboldt points out, this codex is divided into three sections: sections 1 and 2 depict the victories of the Mexica emperors, the founding of Tenochtitlan in 1325, and tax demands from the thirty-eight provinces of the empire, while section 3 focuses on Mexica daily life from birth to adulthood. The first two sections were likely copied from preconquest sources; the third was created upon the viceroy's request. This codex, which is housed at the Bodleian Library at Oxford, was initially confused with Codex Telleriano-Remensis because its location was unknown. Humboldt was the first to point out that Purchas's illustrations did not belong to Codex Mendoza but to another codex altogether, the Codex Telleriano-Remensis, named after Charles-Maurice Le TELLIER (1642–1710), archbishop of Reims (r. 1671–1710). One of the most influential figures in French church affairs under Louis XIV (1638–1715), Le Tellier had the benefit of significant political connections: his father, Michel (1603–85), and his brother François-Michel of Louvois (1641–91) served as chancellor of France and war minister, respectively. In addition to having a brilliant ecclesiastic and political career, Le Tellier was an avid collector of rare manuscripts and printed books. He built his renowned private collection, known as the Le Tellier Library, by traveling to different parts of Europe in search of valuable manuscripts and having them copied or sent to him; he also sent connoisseurs on book-hunting expeditions. In 1701, Le Tellier donated around five hundred manuscripts, including the Codex Telleriano-Remensis, to the Bibliotheque du Roi (today's Bibliothèque Nationale of Paris). The Bibliothèque Sainte-Geneviève in Paris inherited sixteen thousand of his printed books. The catalog of his collection of printed works was published as *Bibliotheca Telleriana* (1693). See also Pedro de los Ríos.

77n. Jean Baptiste Michel PAPILLON (1698–1776) came from a family of French wood engravers and paper manufacturers. In 1766 he published *Traité historique et pratique de la gravure en bois* (Historical and practical treatise on wood engraving), in which he described contemporary methods of manufacturing wallpaper.

80. Under rule of the reformist King Charles III of Spain (1716–88), the Catholic Order of the Jesuits, known as the Society of Jesus, was expelled from Spain and the Spanish colonies in 1767. After major civil unrest in Madrid a year earlier, for which the Jesuits were declared to be the sole responsible agent, the order was accused of moral corruption, excessive self-enrichment, and insurrectionary political motives. The Jesuits' expulsion from Spain—they had been banned from Portugal and France several years earlier (in 1759 and 1764)—was the culmination of an ongoing struggle between civil and ecclesiastical powers over how to define and delimit their respective jurisdictions. During the period prior to their expulsion, the order had increasingly come to oppose the monarch's regency. After their removal, the majority of its members emigrated to Italy, and Spain and other countries lost an important group of intellectuals who were active in all branches of cultural and scholarly life.

Humboldt's reference to LORD ARCHER can be traced to **William Robertson**'s *History of America* (1803, 3:415–16). Robertson points out that Archer was a grandson of Edward Earl of Oxford, who had purchased **Moctezuma**'s cup in Cádiz while in harbor with the fleet under his command. Robertson notes that he derived his information from the judge, antiquary, and naturalist Daines Barrington (c. 1727–1800). Robertson's description fits Edward de Vere, the seventeenth earl of Oxford (1550–1604), who participated in the battle of Cádiz in 1596. Edward de Vere and the Lord Archers were all descendants of William I, king of England, that is, William the Conqueror (c. 1027–87). By 1867, Moctezuma's cup had passed to "Lord Amherst," or William Pitt, the second Earl Amherst of Arracan (1805–86).

81. Father José Antonio PICHARDO (1748–1812) was a linguist and antiquarian from New Spain. He met and collaborated with Humboldt on issues of Mesoamerican history. The Tira de Tepechpan, the Cozcatzin Codex, and the Codex en Cruz are among the numerous manuscripts and books Pichardo collected. Executor of the will of Antonio de **León y Gama**, Pichardo safeguarded those documents, which had been part of the **Boturini** collection. Father Pichardo made a copy of the Tira, the original and copy of which are now housed at the French National Library. To disprove the United States' claim that Texas had been part of the 1803 Louisiana Purchase, Pichardo wrote a treatise titled *Informe Pichardo sobre los Límites de Luisiana y Texas* (Pichardo's report on the borders of Louisiana and Texas). Part of this report, edited and published between 1931 and 1946 by Charles Wilson Hackett (1888–1951), can be found at the Archivo General de la Nación in Mexico City.

A self-taught astronomer, the New Spain Creole Joaquín Luciano Velázquez CARDENAS DE LEÓN (1732–86) was professor of mathematics at the **University of Mexico**. Among his notable astronomical observations was that of the transit of Venus in 1769 from Baja California. In 1773, Velázquez determined the longitude and latitude of Mexico and, for the first time, corrected New Spain's position on maps. A year later, he led topographic and geodesic surveys of the valley of Mexico. Velázquez requested that the crown create both the School of Mines (est. 1792) and the Mining Board; he directed the latter until his death. Among other reports, he authored *Conocimientos interesantes sobre la Historia Natural de las cercanías de Mexico* (Interesting knowledge about the natural history in and around Mexico). (He is not to be confused with another Joaquín Velázquez de León (1803–82), a later Mexican astronomer and mathematician.)

82. Michele MERCATI (1541–93), also known as Michael Mercatus and Mercator, was a physician and naturalist from Florence. He founded the Vatican's Botanical Garden and became famous for his collections of minerals, fossils, and archaeological artifacts.

85. NESTORIANISM is a theological doctrine that emphasized Christ's humanity, stressed problems of will and ethics to overcome sin, and rejected the epithet "Mother of God" for the Virgin Mary, preferring "Mother of Christ." Nestorios (c. 386–c.451), patriarch of the see in Constantinople in 428, developed the doctrine in the first half of the fifth century. After the Third Ecumenical Council at Ephesus (431) accused and condemned Nestorios for claiming for Jesus two distinct persons and natures (divine and human), his followers established the "Church of the East." The largest and most influential schismatic Christian community, the Church of the East spread throughout Persia, some regions of Syria, northern Arabia, India, Central Asia, and China. Small congregations still exist in Iran, Iraq, and the United States. In the late 1970s, the Nestorian patriarchate was in Tehran. See also Estrangelo.

DROGEO refers to a sixteen-century cartographic fiction for what was considered the discovery of New England (named "Estotiland") or Nova Scotia (named "Drogeo") by the Scottish nobleman Henry Sinclair (1350–1404). It was based on reports of an errant Nordish fisherman who, blown off course, found himself among the peoples of the fertile and populated island of Drogeo. This tale is based on a collection of travel reports illustrated by an influential woodcut map and published in 1558 by the Venetian writer, mathematician, and geographer **Niccolò Zeno**. The account of these discoveries purports to have been drawn from old letters and a sailing chart owned by the Zeno family. In his popular and influential narrative, Zeno recounts the story of his fourteenth-century ancestors Niccolò and Antonio Zeno, who had allegedly accompanied Sinclair on his expedition to the new mysterious islands overseas without ever reaching their destination. Hence what was recorded about Drogeo was based on the accounts of the fisherman, who had found the library of the island's king stocked with books in Latin—evidence of some earlier European contact—and had heard tales of more civilized peoples, cities, and great wealth lying to the south, possibly a reference to the early populations of North America. In 1898 the Zeni brothers' narrative was proved to be a compilation of earlier sixteenth-century works. Zeno's map, an amalgam derived from fourteenth- and fifteenth-century maps with imaginary islands and mainland added, contributed to the large pool of fictional geography that was reflected in and informed Europe's late medieval worldview and endured in later cartographical and historiographical perspectives on the Americas and the history of its discovery. Aside from these apocrypha of pre-Columbian discovery, archaeological studies in the 1960s proved that there had in fact been Norse exploration and scattered colonization of American soil in the late tenth and eleventh century. Evidence for this included Norse settlements in L'Anse aux Meadows in the Canadian provinces of Newfoundland and Labrador, which became the first site on UNESCO's World Heritage List in 1978. See also Zeni brothers.

Hernando de SOTO (c. 1500-1542) was a conquistador in Nicaragua and Peru (with Pizarro) who became governor of Cuba. After the failed expedition to Florida by Pánfilo de Narváez (1470-1528), the more experienced de Soto was sent on a second voyage to the tropical peninsula in 1539. On his famous trek through Florida, along the Alabama River, and across the Mississippi, de Soto searched for the mythical Seven Cities and a pathway to the Pacific—both without success. After his death in 1542, his expedition continued south across the Mississippi and along the coast to Tampico, where the remaining crew dispersed.

85n. The Jesuit priest Juan José de EGUIARA y Eguren (1696-1763) is best remembered as author of *Bibliotheca Mexicana* (Mexican library, volume A–C published in 1755). Eguiara started the *Bibliotheca* in 1735 while a professor at the Real y Pontificia Universidad de México, of which he became rector in 1749. An ambitious biobibliographical project, the *Bibliotheca* is a response to the *Epístolas* (Letters, 1756) by Manuel Martí (1663-1737), dean of the Cathedral of Alicante, in which he discourages a young disciple from visiting the "barbaric" Mexico and advised him instead to go to civilized Rome. In the twenty prologues of the *Bibliotheca*, written in Latin and titled *Anteloquia*, Eguiara y Eguren defends indigenous cultures against European criticisms, praising the Nahuatl language, with its sophisticated discourses and pictorial recordkeeping, and the intellectual achievements of Indians and New Spain Creoles after the conquest. This project was Eguiara's priority; in 1751 he declined a bishopric of Mérida to focus on this catalog of New Spain's literary and cultural production. Eguiara left in manuscript form volumes corresponding to letters D through J, now at the Library of the University of Texas, Austin. He is the author of 244 known works, of which only fifteen were published, including *Selectae Dissertationes Mexicanae* (Select Mexican dissertations, 1746) and *Praelectio Theologica* (Theological treaties, 1747).

87. The CODEX VATICANUS A, also known as Codex Ríos, Codex Vaticanus 3738, or Copia Vaticana, is a pictorial manuscript from the Valley of Mexico. Copied to European paper on 101 leaves, the text has seven major sections covering cosmogenic and mythological traditions, the divinatory almanac, calendrical tables for 1558–1619, an eighteen-month festival calendar, sacrificial and other customs, pictorial annals for 1195–1549, and year hieroglyphs for 1556–62. Reproducing twelve pages in Plates XIV, XXVI, Humboldt refers to this document as Codex Anonymous, Codex Vaticanus Anonymous number 3738, and Codex Mexicanus Anonymous. This codex's repository is the Vatican Library in Rome (Codex Vaticanus Latinus 3738).

The Benedictine monk Cardinal Francisco Antonio LORENZANA (1722–1804) was archbishop of New Spain from 1766 until 1772, in charge of the confiscation of Jesuit property after their **expulsion from Spanish colonies** in 1767. A reformist clergyman, he presided over the 1771 Fourth Mexican Church Council, which issued a resolution for the dissolution of the Society of Jesus. Particularly concerned with the demographic composition of parishes, Lorenzana ordered a census of parishioners between 1768 and 1769. He authored the widely cited *Cartas pastorals y edictos* (Pastoral letters and edicts, 1770), in which he emphasized the difficulties of teaching Castilian to the indigenous population and the inferiority of the many indigenous languages, especially Nahuatl. In addition to annotating Cortés's *Historia de Nueva España* (1770), he edited the proceedings of the First (1555), Second (1565), and Third (1585) Mexican Church Councils.

88. The term MAGNA GRAECIA is the Latin translation of the Greek "Megale Hellas," which initially referred to South Italy and over time came to include Sicily; it is believed to have developed in Greek intellectual circles at the beginning of the fifth century BCE. The Greek presence in southern Italy can be traced back to the eighth century BCE, when colonizers from a number of Hellenic city-states began to leave their country and sail west, progressively colonizing the islands and coasts of South Italy and Sicily. This complex cultural and economic phenomenon was vividly expressed in Homer's *Odyssey* and brought the great tradition of Greek vase-making to the newly colonized territories. Vases were produced for daily, religious, and funerary functions and often represented the collective experience of this dramatic migratory enterprise that had disseminated classical Greek culture across the Mediterranean Sea.

89. GIUSTINIANI was the name of a distinguished Italian family with Venetian and Genovese branches; it had among its members bishops, senators, poets, and historians. Vincenzo Giustiniani (1564–1637) was a famous art collector who built the family's palace in Rome. His collection of Italian baroque masterpieces was among the most important of the sixteenth and seventeenth centuries. It was sold to the Prussian king Friedrich Wilhelm III (1770–1840) in 1815 to become a cornerstone of the royal collection, today part of the *Gemäldegalerie* (gallery of paintings) in Berlin. The tale of **Cardinal Stefano Borgia**'s fortunate rescue of the precious Mexica manuscripts from the servants' children in the Giustiniani Palace was told for the first time in Humboldt's text and appears to be true. Ethnohistorical and archaeological studies have been discussing the manuscripts' origins for decades, suggesting most recently the Puebla-Tlaxcala region as the probable origins of the codices **Borgia**, **Cospi**, and **Vaticanus B**.

91. IATROMATHEMATICAL refers to a group of academicians, of whom René Descartes (1595–1650) was one of the foremost, who maintained that all physiologic processes were the result of physical laws. It is synonymous with the mechanistic school.

90. According to **Ixtlilxochitl**'s *Sumaria relación* (Summary report), HUEMATZIN (alternate spelling: Huemantzin), the sage astronomer of Tetzcoco, wrote the encyclopedic Teoamoxtli, the sacred book of the Toltecs, in the seventh century. Pre-Hispanic ritual codices of religious content based on oral traditions are called *teoamoxtli*, that is, "books of religious content." This divine book was said to contain the legends of the creation of the world up to that time; accounts of Toltec rulers, laws, governments, temples, idols, and rites; and a compendium of Toltec science and knowledge, their astronomy, philosophy, architecture, and other arts. Huematzin's role in crafting this teoamoxtli and his qualities as a venerable leader are also mentioned in **Boturini**'s *Idea*.

92. In the early 1770s, the explorer and fur trader Samuel HEARNE (1745–92) led expeditions in the part of North America that is now Canada; he was the first to see and cross the Great Slave Lake. A travel account was published posthumously in 1795 and translated into five languages. Another fur trader, Sir Alexander MACKENZIE (c. 1763–1820) from Scotland, whose family had emigrated to Canada in the 1770s, would by 1793 become the first person to cross the American continent north of Mexico. He was looking for a commercial route to the Pacific Ocean to compete with the Hudson Bay Company, in whose employ Hearne had explored North America earlier. The explorer Meriwether LEWIS (1774–1809) was born in Virginia, where his family moved in the same circles as **Thomas Jefferson**'s. When a soldier in the mid-1790s, Lewis met William Clark (1770–1838), who would accompany him on his expedition to the Pacific (1803–6). In 1801, newly elected President Jefferson asked Lewis to serve as his private secretary. Lewis thus became involved in planning the expedition across the continent that Jefferson had long had in mind, and that would become known as the Lewis and Clark expedition. An early publication based on the travel journals appeared in 1809.

92n. Friedrich Leopold Graf zu STOLBERG-Stolberg (1750–1819) was a German-Danish poet and translator. While working as a diplomat in Protestant Northern Germany until 1800, Stolberg published numerous poems with his brother Christian and also authored an Italian travelogue, both heavily influenced by their acquaintance with the literary circles around Johann Wolfgang von Goethe (1749–1832). Next to translating canonical works from Greek antiquity, such as Homer's *Iliad*, Stolberg's late works focus mainly on religious topics, the fifteen-volume *Geschichte der Religion*

Jesu Christi (History of the religion of Jesus Christ, 1806–18) being its centerpiece. With this historical compendium Stolberg aimed at a religious curriculum based on a profound historical analysis of the scriptures. Stolberg also argues that an original universal religion (whether conceived as Hindu or Christian) had spread from Asia to provide the religious foundations for the rest of the world.

93. The god of the infernal regions, STYGIAN JUPITER, or Stygian Jove, is Pluto in the *Aeneid* by Virgil (70–19 BCE). "Stygian" refers to the River Styx, one of the nine rivers of the underworld in Greek mythology.

94n. Born in Königsberg (Kaliningrad), David MILL (also Millius, 1692–1756) was a German Protestant theologian at Utrecht interested in, among other things, Hebrew antiquity.

97. Better known by the Nahuatl name MOTOLINIA (the one in misery), Toribio Paredes, or Toribio de Benavente (c. 1490–1569), is one of the legendary first twelve Franciscan missionaries in New Spain. He arrived in Tenochtitlan in 1524. Having been asked to write about the indigenous peoples' beliefs system before the conquest, Motolonia studied and documented indigenous religions and customs and the Franciscans' proselytizing efforts from 1524 to 1540. The extant part of this work is known as *Historia de los Indios de la Nueva España* (History of New Spain's Indians), and he completed it around 1541 (it was published in 1858). Motolonia's *Historia* includes an introductory letter (*epístola proemial*) about indigenous history and is divided into three treatises. For the *Historia* he drew on oral accounts and pictorial codices. In the introductory letter, Motolonia mentions as his sources five different kinds of books— *tonalamatls* (from *tonally*, day, and *amatl*, book)—that indigenous peoples used to record history and beliefs. These included the annals of history describing conquests, sucession of rulers, and other important events; the book of days and feasts; the book of dreams, illusions, superstitions, and omens; the book of baptism and the naming of infants; and the book of marriage rites and ceremonies. Other now lost writings by Motolinia, which go into more detail on topics found in the *Historia*, were published as *Memoriales* in 1903. Current scholars believe that he had actually completed a larger work about indigenous subjects, of which the *Historia* is merely an extract, and that the *Memoriales* are parts of earlier drafts. The title of this longer work was supposedly *De moribus indorum* (Of the customs of the Indians), *Libro de los ritos, costumbres y conversión de los indios* (Book of rites, customs, and the conversion of the Indians), or *Relación de las cosas, idolatrías, ritos y ceremonias de la Nueva España* (Account of the things, idolatries, rites, and ceremonies of New Spain). In addition to his work on the religious conquest, Motolonia authored a now lost account of the military conquest. Famous is his letter to **Charles V**, dated January 2, 1555, in which Motolinia attacks the Dominican friar Bartolomé de las Casas (1484–1566), bishop of Chiapas, and supports **Cortés**. Before his death, Motlonia had traveled as far south as Nicaragua and Guatemala.

98. The BATTLE OF CANNAE in 216 BCE was considered one of the most spectacular military defeats in the history of the Roman empire. In this historic confrontation between the Roman army and the Carthaginian army, led by Hannibal (246–183 BCE), the Roman Confederacy had assembled the largest number of soldiers ever, outnumbering Hannibal by far. Yet the general's multinational legions of mercenaries and allies managed to outflank and enclose their opponents, virtually annihilating at least fifty thousand Romans and Italians. Hannibal's army turned the battlefield of Cannae into the bloodiest loss that a European army has ever suffered in a single day. Cannae may well be the most studied battle in history.

99n. TERTULLIAN, or Quintus Septimius Florens Tertullianus (c. 160–240), was an early Christian apologist from Carthage (in present-day Tunisia), then the Roman capital in Africa. A pagan-born layman highly educated in rhetoric and history, Tertullian converted to Christianity, publishing several works in Latin and Greek (the dates of which are contested). Thirty-one of his Latin works still exist; all those he wrote in Greek are lost. *Apologeticum* (The apology, c.212) is a defense of Christians' human dignity and an appeal to the provincial governors of the Roman empire to tolerate Christianity. Responsible for much of the theological vocabulary of Western Christianity, Tertullian also argued about the threefold nature of God, possibily coining the terms *trinitas* (trinity), *substantia* (essence) and *persona* (face, expression, role, or mask). Condemning any reconciliation between Christianity and paganism, Tertullian rejected philosophy as the source of truth.

LACTANTIUS, or Lucius Caecilius (or Caelius) Firmianus (c. 50–c. 325), was a pagan-born African who studied rhetoric under Arnobius (c. 235–c. 330). The Roman emperor Diocletian (r. 284–305) appointed Lactantius chair of Latin rhetoric in Nicomedia, Bithynia, the eastern and most prominent capital of the empire at the time. By Diocletian's First Edict of February 23, 303, the beginning of the Great Persecution (303–13), Lactantius had converted to Christianity, which caused him to lose his position and spurred his major work, *Institutiones divinae* (*Divine Institutions*, c.303–13). The most comprehensive summary of Christian teaching written to convert an educated pagan audience, *Institutiones* is divided into seven books. In addition to providing extensive mythological genealogies, Lactantius refuted pagan religion and the philosophers, upheld monotheism, and insisted that God is the only source of real insight. Constantine I (c. 285–337) at length appointed him tutor of his oldest son Crispus (c. 305–26). Before *Institutiones*, Lactantius had authored several works that are now lost, including ten books of letters. Only four of his works survive, all of them Christian.

Gaius SUETONIUS Tranquillus (c. 70–c.140) was a biographer who practiced law and administered the imperial archives. He wrote in both Latin and Greek. His surviving works include *De viris illustribus* (On illustrious men), biographies of Roman grammarians, rhetoricians, poets, orators, historians, and philosophers; and *De vita Caesarum* (Of the life of Caesar), a set of twelve biographies from Caesar to Domitian.

100. Ennio Quirino VISCONTI (1751–1818) was an Italian art historian, archaeologist, and antiquarian. In 1782 he assisted his father, the papal antiquarian Giovanni Battista Visconti (1722–84), in compiling the first volume of the illustrated and annotated catalog of the Vatican's Museo Pio-Clementino and prepared volumes 2 to 6 in subsequent years. After having earned a reputation as an antiquarian, Visconti became conservator of the Capitoline Museum. An advocate of the French Revolution, he also served as the director of the Musée Napoléon in Paris (now the Louvre).

100n. "HYMNUS IN MERCURIUM" (Hymn to Hermes) is part of the collection known as *The Homeric Hymns*, thirty-three ancient Greek poems written in epic style addressed to various gods. Composed between the seventh and sixth centuries BCE by different poets, these poems were once attributed to Homer. The "Hymn to Hermes"—the date of its composition is not known but estimated at the end of the sixth century BCE—tells of the deeds and mischiefs of a young Hermes in 582 lines. The passage Humboldt has in mind is this: "He chose a stout laurel branch and trimmed it with the knife . . . held firmly in his hand: and the hot smoke rose up. For it was Hermes who first invented fire-sticks and fire. Next he took many dried sticks and piled them thick and plenty in a sunken trench: and flame began to glow, spreading afar the blast of fierce-burning fire" (lines 109–10, translation by Evelyn White). In addition to inventing fire sticks and the lyre, Hermes is god of thiefs and shepherds, of abundance, fertility, and prosperity. As the leader of dreams, Hermes was also thought to guide the souls of the dead to the underworld. Hermes's Roman counterpart is Mercury.

A Greek poet from Alexandria, APOLLONIUS RHODIUS, or Apollonius of Rhodes (c. 295–215 BCE), was the head of the legendary Royal Library at Alexandria under **Ptolemy II Philadelphus** (308–246 BCE). His only surviving work, *Argonautica*, is the only Greek hexameter epic poem to have been written between Homer and Virgil. Divided into four books, *Argonautica* tells the story of the voyage of Jason and the Argonauts in their quest for the Golden Fleece, and Jason's love for princess Medea. Apollonius's main influences were Homer, Pindar (c. 518–438 BCE), and Euripides (c. 408–406 BCE). *Argonautica* was a model for Virgil's *Aeneid* and is one of the source texts for Gaius Valerius Flaccu's *Argonautica* (first century). As a scholar, Apollonius wrote about Homer and **Hesiod**, among others.

Lucius Annaeus SENECA, better know as Seneca the Younger (4 BCE/1 CE–65 CE), was a Roman orator, writer, Stoic philosopher, and politician. After 49, Agrippina (15–59) appointed him tutor of the future emperor Nero (37–68), whom Seneca later served in the position of political adviser and minister. As a response to Nero's exihibionist and criminal behavior, Seneca retired from Roman public life in 64 CE and dedicated himself almost exlusively to writing and philosophy. During this period he wrote *Quaestiones naturales* (Natural questions), which deals with ethics and natural phenomena, and *Epistulae morales* (*Moral Letters*), twenty books of letters about philosophy. Seneca was forced to commit suicide for his alleged participation in the conspiracy to replace Nero with Gaius Calpurnius Piso (d. 65). His other surviving works include poems and tragedies that became important influences for Christopher Marlowe (1564-93), William Shakespeare (1564-1616), and Ben Jonson (1572-1637).

After having studied under Plato (c. 428-347 BCE) at the Academy at Athens, Tyrtamus of Eresus, better know as THEOPHRASTUS (c. 370-c. 287 BCE), replaced Aristotle (384-322 BCE) as the head of the Lyceum at Athens in 323 BCE. Diogenes Laërtius, Theophrastus's biographer, attributes to him 227 treatises on diverse subjects, ranging from biology and geognosy to ethics and metaphysics. His major botanical works, *De historia plantarum* (Of the history of plants) and *De causis plantarum* (Of the causes of plants), offered the first systematic descriptions and categorizations of the various parts of plants, earning Theophrastus renown as the father of scientific botany.

101. While being detained in France during the Napoleonic Wars from about 1802 to 1806, the British Orientalist Alexander HAMILTON (1762-1824) drew up a catalog of Sanskrit manuscripts at the Paris library. His release was procured thanks to the intervention of his colleague **Silvestre de Sacy**. Hamilton became the first professor of Hindu literature and the history of Asia at the future Haileybury College, founded in 1806 by the East India Company. **Friedrich Schlegel**'s *Über die Sprache und Weiheit der Indier* (About the language and wisdom of the [East] Indians, 1808) grew out of Schlegel's studies with Hamilton. Hamilton's breadth of knowledge and enthusiasm for Sanskrit scholarship earned him the nickname "Sanscrit Hamilton."

105. In Quito, Humboldt stayed in the mansion of Juan Pio Montúfar y Larrea, the **marquis of Selvalegre** (1756-1818). The latter's son, Carlos MONTÚFAR y Larrea (1780-1816), was to become one of Humboldt's most trusted and cherished travel companions in the Americas. Together with **Aimé Bonpland**, Montúfar accompanied Humboldt on his (partial) ascent of Chimborazo on June 23, 1802, a famous episode that Montúfar describes in his diary, which was recovered and published posthumously in 1889 (see *Boletín de la Sociedad Geográfica de Madrid* 25). In 1804 Montúfar sailed for Europe with Humboldt, staying first in Paris and then in Madrid. When Napoleon's army invaded Spain in 1808, the Creole Montúfar became involved in the military defense of his *patria*. Two years later he returned to Quito, where he became part of the fight for independence; Montúfar joined Bolívar's army and was captured and executed in 1816.

110. Charles VII, king of Naples, housed the excavated antiquities from the archaeological Roman sites of Herculaneum and POMPEII (east and southeast of Naples) in his Palazzo Reale at Portici near Naples. Both cities were destroyed in 79 CE by an eruption of Vesuvius, which preserved the antiquities under layers of ash and lava for posterity. Systematic excavations at Herculaneum began in 1738, in Pompeii ten years later. The collection of Portici became a private museum of ancient art, as the sale of the artifacts was prohibited and extensive popular knowledge about the collection unencouraged. The early Roman wall paintings (termed Pompeian or Campanian) that were housed at Portici spanned three centuries (second century BCE—late first century BCE). These wall paintings, which existed in most private houses and some public buildings and baths, are the best preserved and the most numerous of all the artifacts from Pompeii. Landscape paintings, intended to make a room or garden appear larger, were animated by mythological figures. The royal collection became the basis for the Museo Archeologico Nazionale at Naples.

111. The CONCHOCANDO OF LICÁN was a sovereign of the Puruaes (Puruays or Puruguayes) peoples. According to a letter Humboldt sent to his brother Wilhelm, dated November 25, 1802, Onaina Abomatha (alternate spellings: Ouanía, Ouainia) was the last *conchocando* (king) of the independent peoples of Licán (present-day Chimborazo province in Ecuador) before the Inca conquered them.

112. The Sanskrit word VAIVASVATA means "Vow of Truth" and was designated for Manu, son of the Sun God Vivasvat ("Shining Forth") and the seventh of the fourteen Manus, or Menous, who all belonged to previous universes. In Hindu mythology, Vaivasvata (like Noah) survived a great flood only because a fish told him to build an ark and fill it with the seeds of all creatures. The fish later incarnates as Brahma (or Vishnu), the Creator God, telling Manu to become the progenitor of the human race. The Hindu myth of Vaivasvata varies. At times it is SATYAVRATA who is chosen to lead the earth's creatures from the end of one age (to be deluged by a great flood) to another, and he later becomes the first human being and progenitor of the human race. This instance in Humboldt's text is a typical example of what contemporary scholarship has called a "bold comparison"; such comparisons are typical of the cosmopolitical and relational thinking at the core of Humboldtian science. The discussion of linguistic and representational similarities between Inca and Hindu cultures may well have been inspired by the works of Sir **William Jones**, the British Orientalist whom Humboldt cites in this paragraph and in many other parts of this book. In his "Essai sur le dieux de la Grèce, de l'Italie et de l'Inde" (On the gods of Greece, Italy, and India), presented at a conference at the **Asiatick Society** in Calcutta on March 24, 1785, Jones advanced a similar hypothesis.

116. BUGNATO, or rustication, is a texture and pattern applied to the surface of stones to decorate exterior walls, often with a deliberately rough surface effect. Dividing lines, channels, or joints are often exaggerated to emphasize the three-dimensional quality of the wall. Used in ancient Greek and Roman architecture, the technique became popular again during the Italian Renaissance.

117. Nicolas-Louis VAUQUELIN (1763–1829) was a French chemist who discovered chromium and beryllium in 1797 and 1798. He served as professor of chemistry at the Paris Faculty of Medicine from 1809, succeeding his teacher and friend François Fourcroy (1755–1809). Vauquelin himself supported Louis-Jacques Thenard (1777–1857), another soon-to-be famous chemist. During his time in Paris, Humboldt was in touch with all three men.

118. Known as the Stone of Tizoc, the SACRIFICIAL STONE (bas-relief around a cylindrical stone; see Plate XXI after León y Gama's drawings) is a masterpiece of intricate stone carving that depicts the victories of Tizoc (r. 1481–86), who is represented as **Tezcatlipoca**. The stone was discovered on December 17, 1791, below Mexico City's main plaza. **León y Gama** identified it as a sacrificial stone in the second part of *Descripción histórica y cronológica de las dos piedras* (Historical and chronological description of the two stones, 1792). Called *temalacatl*, the stone was likely used for gladiatorial sacrifices of important captured warriors. It might also have been used as a *cuauhxicalli*, a vessel in which to deposit the hearts of sacrificed victims. After its discovery, the stone was buried in the cathedral gardens, with only its top left visible, until it was disinterred and transferred to the university in 1824. It is on exhibit at the Museum of Anthropology in Mexico City's Chapultepec Park, along with the MEXICA CALENDAR STONE (Plate XXIII), also known as the Aztec Sun Stone (Piedra del Sol). The Sun Stone is a large monolithic sculpture that was also excavated in the main square of Mexico City on December 17, 1790. Carved in 1479 (the year 13 Acatl), it measures about 3.6 meters (12 ft) in diameter and 1.22 meters (4 ft) in thickness, and weighs 24 tons. **León y Gama**, the first to publish a detailed description and analysis, coined the name "Calendar Stone" when he argued that it was a combination sundial and calendar for marking the passage of the sun. Although his interpretation has been since revised, León y Gama correctly identified many of its features. Organized in concentric rings, the stone represents human sacrifice related to the cult of the sun god, Tonatiuh, identified as the stone's central image, which has also been interpreted as a representation of Tlaltecuhtli, the deity for the night sun of the underworld. The central deity is surrounded by the symbol of the fifth sun and the four previous creations. Like a cosmological map, it also depicts the four cardinal points of the universe and the twenty days of the month. The stone celebrates the creation of the present era of humankind (the fifth sun). A notable study of the Sun Stone is Eduardo Matos Moctezuma's *The Aztec Calendar and Other Solar Monuments* (2004). Elizabeth Hill Boone's *Cycles of Time and Meaning in the Mexican Books of Fate* (2007) discusses Mexica ritual calendars.

The colossal statue on Plate XXIX was found on August 13, 1790, in front of the National Palace; **León y Gama** associates it with TEOYAOMIQUI (alternate spelling: Teoyaomicqui). In Mesoamerican mythology, this goddess collects the souls of the warriors who die in sacred wars, along with the souls of those who are ceremoniously sacrificed. She is also known as Huahuantli, the deity of the sixth hour of the day. In the first volume of his study *México a través de los siglos* (Mexico across the centuries, 1888), Alfredo Chavero (1841–1906) identified this statute with Coatlicue, the earth goddess, symbol of fertility and mother of gods. The Coatlicue statue was reburied shortly after it was found, to prevent the common people from worshiping it. Humboldt was able to see it despite the religious administration's orders to keep it hidden. Like the two stone sculptures, this statue is at the Museum of Anthropology in Mexico City.

120. Juan Bautista RAMUSIO, or Giovanni Battista Ramusio (alternate spellings: Ranusio and Ramnusio) (1485–1557), was a historical geographer and collector of travel and exploration narratives from Venice, Italy. As a diplomat and administrator in the Venetian Republic, he was at the center of a network of correspondents who provided accounts of the most important geographic discoveries of his time. *Relazione d'un gentiluomo* (Account of a gentleman, 1565) is part of the third volume of Ramusio's collection of travel accounts, *Delle Navigazioni e Viaggi* (Of navigations and travels); it is exclusively dedicated to the Americas. Volume 1, published in 1550, is devoted to descriptions of Africa, Brazil, and the coasts of the Indian Ocean (as far as Japan). Volume 2, published posthumously in 1559, is concerned with Muscovy, Scandinavia, and Asia (from Persia to India). The *Relazione* was published as *Relación de las cosas de Nueva España y de la gran ciudad de Temestitán México* (Account of the state of affairs in New Spain and in the great city of Temestitán, Mexico) in the first volume of *Colección de documentos para la historia de México* (Collection of documents about the history of Mexico, 1858) by Joaquín García Icazbalceta. Ramusio was also the coeditor of Quintilian's works (1514).

121. Spanish historian Antonio de HERRERA Y TORDESILLAS (1559–1625) served as royal chronicler of the Indies under Philip II (1527–98) for almost thirty years, from 1596 to 1625. In his famous *Historia General* (General history, 1601–15), an apology for the conquest also known as *Décadas*, he rewrote the history of the Indies from published and unpublished sources. Herrera y Tordesillas never traveled to the Americas.

122. The French astronomer Louis GODIN (1704–60) was the first to present, and successfully argue to the Academy of Sciences at Paris, the idea of determining the earth's shape by traveling to South America to measure the length of a degree of meridian arc at the equator. The head of a 1735 expedition, he traveled with **La Condamine**, **Bouguer**, and seven other affiliates aboard the ship *Le Portefaix*; upon reaching Cartagena, the team joined their Spanish colleagues **Ulloa** and **Juan**. While in the viceroyalty of Peru, Godin shipped several specimens of new curiosities to the Natural History Cabinet at the Royal Botanical Garden in Paris. He stayed in the Audiencia of Quito as professor of mathematics at the University of San Marcos until he returned to France in 1751.

122. A member of the Académie de Peinture et Sculpture (in 1731) and the Académie des Inscriptions et Belles Lettres (in 1742), the French engraver Anne-Claude-Philippe de Tubières, COUNT OF CAYLUS (1692–1765), was the patron of artists such as Edmé Bouchardon (1698–1762) and Joseph-Marie Vien (1716–1809). The count's vast collection of coins and ancient artifacts were cataloged in seven volumes of engravings titled *Recueil d'antiquités égyptiennes, étrusques, grecques, romaines et gauloises* (Compendium of Egyptian, Etruscan, Greek, Roman, and Gallic antiquities, 1752–67), one of the most influential works on antiquarian research published in the eighteenth century. His biography of Jean-Antoine Watteau (1684–1721) was issued in 1748.

124. Pedro Romero de Terreros (1710–81), the first count of Regla and one of the wealthiest men in New Spain, owned the AMALGAMATION FACTORY at Regla. The

factory was a mine with sufficient water for refining silver ore. Privately owned for several generations, Reglas's mines became property of the Mexican government in 1948.

126. Bernardino de SAHAGÚN (c. 1499–1590) is known as the "the father of modern ethnography," having authored valuable ethnographic and linguistic texts about the indigenous cultures of the Valley of Mexico and adjacent territory at the time of the conquest. Sahagún arrived in New Spain in 1529, only eight years after the initial conquest. He learned Nahuatl in present-day Mexico City, where he taught the Indian nobility at the Imperial College of Santa Cruz de Tlatelolco. In 1540, while living in the monastery of Huexotzinco (today's Puebla), he finished his first book in Nahuatl, *Sermonario* (Homilies). Around 1558, Sahagún was commissioned to write about the indigenous cultures of his day. For this assignment, he spent several years interviewing informants from different social classes, gathering information about their religious practices, history, customs, and knowledge of their surroundings. Between 1563 and 1568, he crafted the *Historia universal (general) de las cosas de (la) Nueva España* (General universal history of the state of affairs in New Spain), a partial edition of which was published between 1829 and 1830. The work was written in three-column page format in Nahuatl, Spanish, and Latin. In 1569 Sahagún produced a now lost copy of the *Historia* in Nahuatl. There are only two surviving copies of the *Historia*. The first one is a Spanish-only version known as the *Manuscrito de Tolosa* or *Tolosano*, which was reportedly at the Franciscan monastery of Tolosa (Navarra, Spain) in 1732–33 and is now housed at the Real Academia de la Historia in Madrid. The second one is the Florentine Codex at the Biblioteca Medicea Laurenziana in Florence, Italy. The culmination of Sahagún's efforts, the Florentine Codex consists of 1,210 leaves, with about 1,846 color illustrations, brought together in 1564 under the title *Colloquios y doctrina cristiana con que los doze frayles de San Francisco embiados por el papa Adriano sesto y por el emperador Carlos quinto convirtieron a los Indios de la Nueva España, en lengua mexicana y española* (Conversations and the Christian doctrine with which the twelve friars of San Francisco dispatched by Pope Adrian VI and the emperor Charles V to convert the Indians of New Spain, in the Mexican and the Spanish languages); it is housed at the Vatican. The *Colloquios* narrates purported dialogues between surviving Mexica priests and the first twelve Franciscans in 1524. While Sahagún was preparing a bilingual Spanish/Nahuatl version of his *Historia*, Philip II ordered the Council of the Indies to inspect his work because of a 1577 ban on writing about matters relating to the superstitions of the *Indios* or their former way of life. As a result, at least one manuscript of the *Historia* was sent to Spain. Around 1585 Sahagún prepared and revised his *Kalendario mexicano, latino y castellano* (Mexican, Latin, and Castilian calendar) and the *Arte adivinatoria* (Art of divination), later copies of which can be found at the Biblioteca Nacional de México.

The Franciscan friar Andrés de OLMOS (c. 1491–c. 1571). In 1528 Olmos arrived in New Spain as **Zumárraga**'s aide. One of the primary scholars at the Colegio Imperial de la Santa Cruz de Tlatelolco from its inauguration in 1536, Olmos mastered several indigenous languages: Huastec, Totonac, and Nahuatl. His three works on Huastec and two on Totonac are lost. He authored the first Nahuatl grammar, titled *Arte de la Lengua Mexicana* (Art of the Mexican language, signed 1547). To illustrate the beauty and charm of Nahuatl, Olmos included in *Arte* a collection of ceremonial didactic speeches, or *huehuetlatolli* (meaning "ancient word" or "discourses of wisdom from the elder"). *Arte* was partially published in 1604 by Juan de Bautista (1555–1615); in 1875 it was printed in French as *Grammaire de la langue nahuatl ou mexicaine* by Rémi Siméon (1827–90). In 1533 Bishop Sebastián Ramírez de Fuenleal (c. 1490–1547) commissioned Olmos to write about the preconquest rites and antiquities of Mexico City, Tezcoco, and Tlaxcala. Based on interviews and conversations with indigenous informants—among them wise men, or *tlamatinime*—Olmos's 1539 *Tratado de antigüedades mexicanas* (Treatise on Mexican antiquities) was the first systematic study of the rituals, political practices, institutions, and literature of the indigenous peoples in New Spain. At least three copies and one original were sent to Spain but were lost. At the request of Fray Bartolomé de las Casas (1484–1566), Olmos wrote a *Summa*, an epilogue or summary, of the treatise, parts of which are included in *Histoyre du Mechique* (History of Mexico, 1905) and the *Codex Tudela* (1947). Olmos is the author of the earliest missionary play in Nahuatl, *El juicio final* (The final judgment), performed for fourteen years (1535–48) at the Church of San José de Naturales in Mexico City. Many anonymous works are attributed to Olmos. Among them are *Historia de los Mexicanos por sus pinturas* (History of the Mexicans in their paintings), which was written before 1536 as part of the anonymous sixteenth-century codex *Libro de Oro y Thesoro Indico* (1882) and *Costumbres, fiestas, enterramientos y diversas formas de proceder de los indios de Nueva España* (Customs, festivals, funerals, and various practices of the Indians of New Spain, 1945). Olmos's work served as an inspiration and source for other missionary ethnographers, notably **Sahagún**, Gerónimo de Mendieta (c. 1528–1604), **Motolinia**, and **Torquemada**.

A chronicler from the capital of Mexico, Cristóbal del CASTILLO (1526–1606) was fluent in Nahuatl and had a prodigious knowledge of astronomy and calendrics. Despite his Spanish name, Castillo was probably of Tezcocan descent and did not identify himself with the conquerors; yet his values were Catholic. To preserve fading indigenous records by using the Latin alphabet in writing Nahuatl, Castillo completed an account of the Mexica history from the migration from the legendary homeland of Aztlan to the conquest. His *Historia de la venida de los mexicanos y otros pueblos e Historia de la Conquista* (History of the arrival of the Mexica and other peoples and history of the conquest) shows the important role that Huitzilopochtli played as the guide of the Mexica during their pilgrimage. Written around 1600, the *Historia* remained in manuscript at the Bibliothéque Nationale in Paris until Francisco del Paso y Troncoso published it as *Fragmentos de la obra general sobre historia de los mexicanos* (Fragments of the general work about the history of the Mexica) in 1908.

A grandson of **Montezuma II**, the Nahua historian Fernando de Alvarado TEZOZOMOC (c. 1525–c. 1610) wrote his *Crónica Mexicana* (Mexica chronicles, 1598) in Spanish. His work covers the history of the Mexica from their beginnings up to the time of the conquest. While condemning Mexica beliefs and religious practices, the Christinanized Tezozomoc provided an indigenous point of view on the Mexica rise to power in the late fourteenth century. The second part of his now lost chronicle was to cover the years since the arrival of the Spaniards. Before he died, Tezozomoc wrote the *Crónica Mexicayotl* (1609) in Nahuatl. The later Mexica historian Domingo Francisco de San Antón Muñón CHIMALPAHIN Quauhtlehuanitzin (b. 1579) is the author of a manuscript edition of **Francisco López de Gómara**'s *La conquista de Mexico* (The conquest of Mexico), a heroic treatment of the exploits of his patron **Cortés**. Although the Spanish crown had banned Gómara's controversial book, copies were still smuggled into the colonies, which is how Chimalpahin gained access to it. The only known Spanish-language account of the conquest that includes extensive critical and corroborative commentary by an indigenous historian, Chimalpahin's manuscript contains more than forty chapters on Aztec culture that were not included in Leslie Byrd Simpson's abridged 1964 translation of this text, titled *Cortés: The Life of the Conqueror by his Secretary, Francisco López de Gómara*. A complete translation is forthcoming. Chimalpahin also wrote a detailed history of events in Nahuatl while employed as a fiscal administrator in Mexico City, a position held for thirty years. Chimalpahin's works were largely unknown, however, until the mid-eighteenth century, when his writings appeared in an inventory of the **Boturini** collection in Mexico City.

127. YUGAS (world ages) are Hindu units of time that make up the lifetime of a universe. There are four yugas; we are now in the fourth, called Kali Yuga after the goddess Kali, the great destroyer. The Kali Yuga is to last 432,000 years, some 5,000 of which have already passed. Once this yuga ends, a new cycle will start. The Sanskrit word KALPA denotes a time unit that refers to a "great length of time."

128n. An acquaintance of Humboldt's, the German mathematician Christian Ludwig IDELER (1766-1846) was a professor of astronomy, geography, and chronology in what is now Humboldt University in Berlin. His son Julius Ludwig Ideler (1809-42) would translate Humboldt's *Examen critique* (Critical examination) from French into German.

129. The French astronomer Guillaume-Joseph-Hyacinthe-Jean-Baptiste LE GENTIL de la Galaisière (1725-92) became the protagonist of one of the most peculiar accounts in historical astronomy. An urgent astronomical matter of his time was to ascertain the distance between the earth and the sun by measuring the transit of Venus, which was calculated to occur in 1761. For the transit to be measured accurately, hundreds of astronomers were to travel to different locations around the globe in order to observe the event from diverse angles. In 1760 Le Gentil, under royal orders, signed on with an astronomical expedition to the Indian Ocean, determined to reach the French colony of Pondicherry on the southeast coast of India to make his observations. Due to a British sea blockade, though, Le Gentil never reached Pondicherry. He saw the transit only from the sea and was forced to suspend his voyage, traveling instead to Île de France (Mauritius). In 1768 he finally managed to sail to Pondicherry, where he was granted permission to install an astronomical observatory designed to observe Venus's second transit one year later (its last one in over a century). Yet on the day of the transit and despite perfect weather conditions in the weeks before, clouds obstructed Le

Gentil's view of the planet's course, leaving him with no results for his efforts—for a second time. In his *Voyage dans le mer de l'Inde* (Voyage to the Indian Sea, 1782), he gives a vivid report of this and many other events during his decade-long expedition around the world, during which he undertook significant astronomical, cartographical, and cultural studies in the Phillipines, Madagascar, and India.

Pierre Simon Comte de L A P L A C E (1749–1827) was one of the most influential and controversial mathematicians and astronomers of his time. Humboldt had met and worked closely with Laplace during his years in Paris, and he highly respected him. Among Laplace's most important works are the *Exposition du système du monde* (System of the world, 1795) and his five-volume *Traité de Mécanique Céleste* (Treatise on celestial mechanics, 1798–1825). The first volume of the *Traité* had been sent right after its publication to Lima, where Humboldt received it in 1802 while working on the notes to his astronomical, magnetical, and barometrical observations, which were to be published in Paris between 1808 and 1811 under the coauthorship of Jabbo Oltmanns (1783–1833) as *Recueil d'observations astronomiques* (Compendium of astronomical observations). Contemporary scholarship has pointed out both Humboldt's detailed knowledge of Laplace's works and the resemblances in scientific approach and structural design between Laplace's *Exposition du système du monde* and Humboldt's *Kosmos*.

129. President of the French National Assembly in the early stages of the French Revolution, Jean Sylvain B A I L L Y (1736–93) was also an astronomer and a member of the French Academy of Sciences (1763) who published several studies on the history of science. Seminal were his successive publications on the history of astronomy: ancient astronomy (1775), modern astronomy (1779–82), and East Indian and Oriental astronomy (1787). Bailly was guillotined during the Reign of Terror. The lunar impact crater Bailly, estimated to be more than three billion years old, is named after him. See also William Robertson.

Another French astronomer, Joseph-Jérôme Lefrançais de L A L A N D E (1732–1807), accurately measured the distance to the moon in 1751, simultaneously with Nicolas Louis de Lacaille (1713–62). Having succeeded his professor Joseph-Nicolas Delisle (1688–1768) at the Collège de France in 1761, Lalande helped organize expeditions to observe the transit of Venus (in 1761 and 1769), using their results to calculate the solar parallax, that is, the angular width of the earth's equatorial radius as it would be seen from the center of the sun at the mean distance between the earth and the sun. His *Historie céleste française* (Celestial history of France) from 1801 cataloged over forty-seven thousand stars. One of them was the planet Neptune, which LaLande recorded in 1795, fifty-one years before the German astronomer Johann Gottfried (1812–1910) located it. Also part of the catalog is the fourth-closest star to the sun, named Lalande 21185. Among other Lalande works on astronomy and narratives about his travels to Italy and England, *Astronomie* (1764) and *Bibliographie astronomique* (1803) are notable. He established the Lalande Prize (1802–1970), which the French Academy of Science awarded for outstanding contributions to astronomy. See also Le Gentil.

The colonial administrator Polo de ONDEGARDO (c. 1510–75) arrived in the viceroyalty of Peru around 1546. In 1558 he was appointed *corregidor* (chief magistrate) of Cuzco, a position that he held for two terms (1558–60 and 1571–72). When the Spanish authorities requested information about the history and ritual practices of the Inca, Polo purportedly interviewed 475 surviving elders, officials, and priests of the former empire at Cuzco. In the process, he discovered several royal mummies that were still being worshiped. To destroy any remaining symbols of Inca imperial power, he had the mummies shipped to Lima along with his 1559 report, which was subsequently lost. The summary excerpt that remains of it was published as *De los errores y supersticiones de los indios* (Of the errors and superstitions of the Indians) in 1585, offering information about Inca political organization, economics, religion, and social customs. Recognized as the most important document produced in Cusco during the immediate post-Inca period, Polo's work served as a basis for policy design and reform during the sixteenth century and for other chroniclers, notably **Acosta** and the Jesuit Bernabé Cobo (1582–1657). Polo's other reports are the *Instrucción contra las ceremonias y ritos que usan los indios conforme al tiempo de su infidelidad* (Warning against the ceremonies and rites that the Indians use in accordance with the time of their faithlessness, 1567) and the *Relación de los fundamentos acerca del notable daño que resulta de no guardar a los Indios sus fueros* (Report on the causes of serious harm to the Indians which results from not keeping their traditional laws, 1571).

130n. The Swedish naturalist Carl Peter THUNBERG (1743–1828) studied botany under Carl von Linné at the University of Uppsala, where he developed a interest in Japan, a country that was then closed to most Western explorers and missionaries. Only the Dutch East India Company was allowed a trading station. In late 1771, Thunberg made arrangements with the latter to set out for the Cape of Good Hope, where he lived among Dutch settlers to learn their language and customs so that he could pass for a Dutchman. In 1775 Thunberg finally embarked for Batavia (Java, Indonesia), then went on to Japan, following in the footsteps of **Engelbert Kaempfer**. Like all foreigners, Thunberg was detained on tiny Deshima Island near Nagasaki. Having established cordial relations with the Japanese interpreters, with whom he exchanged medical and botanical information, Thunberg was eventually permitted to set foot in Nagasaki and subsequently traveled as far as Edo (Tokyo). In 1778 he returned to Sweden via Denmark and England, where he met Joseph Banks (1743–1820). In 1784 Thunberg succeeded to the chair that his teacher Linné had once occupied at Uppsala.

131. When he was crowned the ninth Sapa Inca, TITU-MANCO-CAPAC, or Inca Cusi Yupanqui (also Inca Yupanqui), took the name Pachacutec (or Pachacuti). The creator of the Tahuantinsuyu empire (known as the Inca empire), Pachacutec became ruler as a result of his victory over the Chanca people in 1438; he led the empire for about thirty-three years, until 1471. After conquering the neighboring states, Pachacutec turned over the command of the imperial armies to his son, Topa Inca (aka **Tupac Yupanqui**), and retired to Cusco, which he rebuilt and turned into the empire's political capital. In Cusco, which then housed only royalty, nobility, and religious monuments, Pachacutec ordered the construction of the Quri Kancha (or Coricancha, "enclosure of gold"), the Temple of the Sun that became the central sanctuary of the Inca empire.

133. According to Eduardo Matos Moctezuma, the order of the months did not vary, even if the first month of the year was not the same for all the peoples in the Valley of Mexico. The name of one of the eighteen months of the Aztec calendar, TOXCATL literally means "dry thing"; it is also known as Popochtli, Tepopochhuiliztli, or Tepopochtli and corresponds to days around May, when the Mexica held the major celebration to the omnipotent **Tezcatlipoca**: the king symbolically sacrificed himself in the form of a captive who was worshiped as the image (*ixiptla*) of the deity for one year. Selected from among the most handsome young prisoners, the avatar had to have an unblemished body, which the king personally adorned for the brutal ceremonial sacrifice. Humboldt is referring here to the Toxcatl Massacre, a turning point in the Spanish conquest of Mexico, which occurred when the Spaniards, treated as guests in Tenochtitlan, attacked and massacred the unprepared Mexica during the celebration of Toxcatl. The attack, led by **Alvarado**, was unprovoked. The outbreak of open hostilities that followed forced **Cortés** and his soldiers to flee the city during what became known as the Noche Triste (Night of Sorrows). The Spaniards barely escaped; many were killed, and all the survivors were wounded. See also Pedro de Alvarado y Contreras.

135. The Athenian poet ARISTOPHANES (c. 445-c. 385 BCE) was regarded as the finest playwright of his time. An example of the comedy of ideas, his play *Nubes* (*The Clouds*) was written in 419 BCE and originally produced four years later in the city of Dionysia. Humboldt's reference is to the chorus leader's speech about the need for counting the days according to the moon, that is, the lunar calendar (lines 610–25 in the Greek text).

136. The Athenian astronomer METON observed the summer solstice in 432 BCE to determine the length of the year more accurately. To align the lunar month with the solar year, in that same year he introduced the lunisolar calendaric cycle known as the Metonic cycle, a lunar cycle consisting of 6,940 days (19 solar years and 235 months). Still part of the Jewish calendar, the lunar cycle was adopted by the Greeks until the introduction of the Julian calendar in 46 BCE. Meton is one of the characters of **Aristophanes**'s *Birds* (414 BCE).

142n. Although he never traveled to China and Tibet, the Augustinian priest Agustin Antonio GIORGI, or Georgi (1711–97), of Rimini, Italy, a professor of the Holy Scripture in Rome, knew several Asian languages. In 1762 he published *Alphabetum Tibetanum* (Tibetan alphabet), a Latin-Tibetan dictionary based on the works of the Capuchin monk Francesco Orazio (1680–1747), who lived in Tibet from 1716 to 1732.

In this dictionary Giorgi also compared different idioms from ancient Egypt. The book has been criticized as inadequate and unreliable.

The French astronomer and mathematician Jean Baptiste Joseph DELAMBRE (1749-1822) was a student of the famous astronomer **Lalande**, whom he succeeded as professor of astronomy at the Collège de France in 1807. Humboldt had befriended Delambre early on and corresponded with him while traveling in America.

Chair of Persian at the Collège de France and professor of Arabic at the École des Langues Orientales, Antoine-Isaac Silvestre DE SACY (1758-1838) became interested in the study of Arabic early on in his life. He counted Humboldt among his disciples. Sacy achieved a first breakthrough in deciphering the Rosetta Stone that eventually led to the decoding of the **hieroglyphic inscriptions of the Egyptians**. See also Alexander Hamilton.

143n. Paris-born Huguenot Jean CHARDIN, aka Sir John Chardin (1643-1712), the son of a merchant-jeweler, traveled on business to Persia and India from 1665 to 1679. Encouraged by the success of his first voyage—which led to a 1671 publication—Chardin revisited both places in 1671, returning to Europe a minor celebrity in 1680. He settled in England. His travel account was published both in English and in French in 1686. Chardin's *Journal du voyage en Perse et aux Indes orientales* (Diary of travels to Persia and the East Indies) served as a source for the *Lettres persanes* (Persian letters, 1721) by Charles-Louis de Secondat, baron of La Brède and of Montesquieu (1689-1755). A fictional narrative by Persian noblemen about their views on French and European society, the *Persian Letters* became exemplary of the epistolary genre. It is one of the key texts of the European Enlightenment.

146. The French Republican (or Revolutionary) calendar started retrospectively with year I on September 22, 1792—the fall equinox but also the beginning of the Republican period during the French Revolution. The new calendar was not officially adopted until late in 1793, during the radical Republican phase of the French Revolution known as the Reign of Terror (September 1793 to July 1794). Reforming the society in its reference to time and space with the goal of counteracting superstition, fanaticism, and even Christian festivals, the calendar was composed by, among others, the mathematicians Charles Gilbert Romme (1750-95) and Gaspard Monge, count of Péluse (1746-1818), the poets André de Chénier (1762-94) and Philippe-François-Nazaire Fabre d'Eglantine (1755-94), and the painter Jacques-Louis David (1748-1825). **Lalande** was also involved in the creation of the calendar. Romme was sentenced to death, whereupon he committed suicide; Chénier and Fabre d'Eglantine were guillotined. Napoleon I abandoned the Republican calendar in 1806. Other reforms during the French Revolution included a committee on weights and measures.

149n. Engelbert KAEMPFER (1651-1716) was a German physician in the employ of the Dutch East India Company; he visited Edo (Tokyo) twice, in 1691 and 1692. One of the first to travel with scholarly ambitions, having studied available materials beforehand, Kaempfer had been to Iran in the 1680s and had also traveled to Siam (Thailand). Kaempfer's account of Siam is probably the most reliable one of his day. See also Carl Peter Thunberg.

151n. In 1729-32, the Jesuit Étienne SOUCIET (1671-1744) edited the *Observations mathématiques, astronomiques, géographiques, chronologiques et physiques: Tirées des anciens livres chinois ou faites nouvellement aux Indes et à la Chine, par les pères de la Compagnie de Jesus* (Mathematical, astronomical, geographical, chronological, and physical observations: Taken from the ancient Chinese books or made originally in the East Indies and in China, by the fathers of the Society of Jesus); many materials for the publication had been procured by Father Antoine GAUBIL (1689-1759). Gaubil had left for China as a missionary in 1721, arriving in 1723 in Beijing, where he remained until his death. He was a corresponding member of the Academies of Sciences in Paris, London, and St. Petersburg. Souciet—like **Nicholas Fréret**—was a leading member of the French Académie des Inscriptions et Belles-Lettres who dismissed Sir Isaac Newton's (1642-1727) astronomical and historical arguments. Souciet's own *Observations* contained mistakes and misprints, as even Gaubil acknowledged. See also Louis Jouard de la Nauze.

153. A dramatist and poet from India, KĀLIDĀSA is considered the most renowned poet and playwright of the classic Sanskrit tradition; his works are part of an ancient tradition of court poetry and drama. One of his long poems and two narratives, with three additional plays, survive, including the famous *Abhijñānaśākuntala* (The recognition of Śakuntalā or Śakuntalā). The other surviving works are *Meghadūta* (The cloud messenger), *Raghuvaṃśa* (The lineage of Raghu), *Kumārasambhava* (The birth of Śiva's son), *Mālavikāgnimitra* (Mālavikā and Agnimitra), and *Vikramorvaśīya* (Urvaśī won by valor). *Rtusaṃhāra* (Cycle of the seasons) is also attributed to Kālidāsa. Likely from the priestly class (Brahmin), he lived in northern India between the late fourth and mid-fifth century, probably under the patronage of the Gupta king Candragupta II (r. c. 375-415 BCE). His name translates as "servant of Kālī."

154. Libra as a separate constellation was incorporated into the zodiac as its twelfth sign during the reign of Gaius Julius Caesar (100-44 BCE), when the Romans reformed their astronomical and calendrical systems in consultation with the Greek astronomer SOSIGENES (first century BCE). Likely from Alexandria, Sosigenes wrote three calendrical treaties that have all been lost. The reform aimed at replacing the different temporal systems of the empire with a single standard that modified the 365-day Egyptian solar calendar by including an extra day every four years. See also Mercedonius.

154n. German philologist Philip Karl BUTTMANN (1764-1829) is best known for his *Griechische Grammatik* (Greek grammar, 1808), which was published in English by future Harvard president Edward Everett (1794-1865) in 1822. Buttmann's son Alexander (1813-93) followed in his father's footsteps. Like Humboldt, the elder Buttmann had studied at the University of Göttingen and later moved to Berlin, where he first worked as a librarian and eventually became a university professor.

155. Thanks to his father's connections, Henry Thomas COLEBROOKE (1765-c. 1836) was able to travel to India for the East India Company in 1782. Colebrooke was a prolific scholar with manifold interests and broad knowledge, ranging from Sanskrit to natural history. He may be seen as a successor to **William Jones**, especially as he took over the responsibility of preparing a digest of Hindu law after Jones's death. Colebrooke published twenty papers in *Asiatick Researches*, the journal of the **Asiatick Society** of Bengal.

156. The French draftsman and naturalist Pierre SONNERAT (1748-1814) published accounts of his travels to New Guinea in 1776 and to China and the East Indies in 1782. He brought back numerous specimens to Paris and likely exchanged some with Sir Joseph Banks (1743-1820) of London. Sonnerat had been trained under the king's naturalist, Philibert Commerson (1727-73).

The Vatican's antiquarian Francesco BIANCHINI (1662-1729) reconstructed a Greco-Egyptian planisphere in 1705. Found during excavations of the Palatine Hill (Mount Aventine) in Rome, the planisphere was initially at the Vatican Museum and is now at the Louvre in Paris.

Bianchini, who had traveled to Rome in 1684 under the protection of Cardinal Pietro Ottoboni the Elder (1610-91), the future Pope Alexander VII, directed excavations that discovered five statues in Egyptian style. With *La istoria universale provata con monumenti, e figurata con simboli de gli antichi* (Universal history demonstrated by monuments and depicted in symbols of the ancients, 1697), Bianchini made significant contributions to the study of the archaic history of Greece and Rome, and to the fields of cultural history and visual studies more broadly. Interested in a host of subjects, including mathematics, physics, and astronomy, he drew the first map of Venus in 1726, publishing his findings about the planet in *Hesperi et phosphori nova phaenomena sive observationes circa planetam Veneris* (New phenomena of Hesperus and Phosphorus, or rather observations concerning the planet Venus, 1728). He corresponded with Isaac Newton, whom he had met in London in 1713.

A close friend of Sir **William Jones,** with whom he also collaborated, Samuel DAVIS (1760-1819) was particularly interested in astronomy as represented in Sanskrit texts. An Orientalist who served in the East India Company, Davis also acquainted with Thomas and William **Daniell.**

157. ARATUS of Sycyon (fl. first half of third century BCE) is the author of the astronomical poem *Phainomena*, the most influential Greek poem after Homer's epics. The poem is based on the work of the Greek astronomer **Eudoxus of Cnidus.** Numerous subsequent authors have translated, rewritten, and commented on Aratus's work, which illustrates its overwhelming influence over the course of centuries. GERMANICUS translated Aratus's poem into Latin, adding some two hundred lines and correcting what he saw as errors. As Christian Ludwig Ideler points out in *Sternna-men* (The names of stars, p. xlii), there is some uncertainty about exactly what Germanicus was. Both Humboldt and Ideler describer him as a *scholiast*, thus indicating that he was a literary commentator. Germanicus is typically identified as Germanicus Julius Caesar (16/15 BCE–19 CE), whose original name was Nero Claudius Drusus Germanicus. The nephew and adopted son of Roman emperor Tiberius (42 BCE–37 CE; r. 14–37 CE), Germanicus was a successful general whose promising career was cut short by death. Tiberius presumably encouraged plots to murder Germanicus, seeing him as a threat. Among Germanicus's children were future emperor Gaius Caligula (Gaius Caesar Germanicus, 12–41 CE) and Julia Agrippina (also Agrippina the Younger), mother to Emperor Nero. GEMINUS OF RHODES (also Geminos; fl. c. 77 BCE) wrote an introduction to astronomy called *Elementorum astronomiae*, translated as *Introduction to the Phenomena*, an important work that includes a discussion of some of the works of the earlier astronomer **Hipparchus**.

158. The influential astronomer Claudius PTOLEMY (c. 83/90–c. 161) notably shaped European and Arabic cartography. He was of Alexandria and made his observations between 125 and 141 CE.

Persian physician, geographer, and historian Zakariyyā ibn Muhammad ibn Mahmud al-Qazwīnī (c. 1203–83) was the compiler of *Kitāb ajā'ib al-makhlūqāt wa-gharā'ib al-mawjūdāt*. Possibly translated as "Marvels of things created and miraculous aspects of things existing," the text was written in Arabic around 1263; an expanded second edition dates from about 1275. Widely copied and translated, *Kitāb* is considered one of the most popular and important natural history reference texts of its time. This cosmography includes mythological creatures, celestial phenomena, general geography, and the animal kingdom. al-Qazwīnī is also the author of an influential geography titled *Ajāib al-buldān* (The wonders of the land) from about 1262; another edition, *Āthār al-ibād wa-akhbār al-ibaād* (The striking features of the lands and the historical relations of humankind), was issued in 1275. Arranged alphabetically within the earth's seven climates, this geography includes descriptions of cities, churches, and statues, along with views of Byzantine society and Asia Minor. A 1537 Persian translation of *Marvels of things*, with more than 150 illustrations, is available online at the US National Library of Medicine. al-Qazwīnī is not to be confused with encyclopedist Hamdullah al-Mustafa al-Qazwini (b. 1281), author of *Nazhatu-l-Qulub* (Heart's Delight), written around 1340.

158n. A military engineer in the British East India Company, Sir John CALL (1732–1801) was appointed chief engineer and captain of the corps of engineers in 1757. Call, who served in India until 1770, kept journals of his work there; some of his drawings were included in *History of the Military Transactions of the British Nation in Indostan* (1763). Back in England, where he continued to pursue his interest in Indian affairs, he was elected to the Royal Society in 1775 and the Society of Antiquaries in 1785. He also designed a prison for Cornwall (built at Bodmin) and founded the banking firm of Pybus, Call, Grant & Co. in 1785.

159n. Although he never visited China himself, the French Jesuit Jean-Baptiste Gabriel Alexandre GROSIER (1743–1823) edited and published Chinese records that Joseph François Marie Anne de Moyriac de Mailla (1669–1748) had translated from Tartar-Manchu. Grosier collaborated with Michel-Ange-André Le Roux des Hauterayes (also Deshautesrayes; 1724–95) in the endeavor, which lasted from 1777 to 1785. Grosier also added one volume to this *Histoire générale de la Chine* (General history of China).

A professor of Syriac at the Collège Royale in 1757, Joseph DE GUIGNES (1721–1800) later served as royal censor and keeper of antiquities at the Louvre. He was a fellow of the London Royal Society and a member of the Academy of Inscriptions and Letters in Paris. Besides writing a history of the Huns—*Histoire générale des Huns* (1756)—de Guignes tried to prove that China had been a colony of Egypt.

162n. The Hungarian historian Johann Christian ENGEL (1770–1814) was a student of **Schlözer** and **Gatterer** at Humboldt's alma mater, the University of Göttingen. The German apothecary Johann Gottlieb (or Jean-Théophil) GEORGI (1738–1802) traveled as a natural scientist and geographer to Siberia in 1773 and 1774. Starting his journey in St. Petersburg, he joined the botanist Johann Peter Falk (1732–74), among others. A clerk in the royal courts of Hungary, János THURÓCZY (Johannes de Thurocz or János Turca, also Thwrocz; c. 1435–c. 1489) wrote his *Chronica Hungarorum* (Hungarian chronicle) in the late sixteenth century.

165. Originally an Arabic and Persian scholar, Antoine Léonard de CHÉZY (1773–1832) became the first Sanskrit scholar in France. In 1799 Chézy was appointed to the Cabinet of Oriental Manuscripts in the French National Library, where, having missed an opportunity to join Napoleon's Egyptian campaign on account of ill health, he taught himself Sanskrit. He met Humboldt during this time. He eventually held the first chair in Sanskrit at the Collège de France (created in 1814), the first among few of its kind in Europe. De Chézy published a number of translations of writing from Sanskrit and Persian, including a second edition of *Yajñadattavadha*, with an extensive grammatical analysis, the transcribed text as an appendix, and a literal Latin translation by Jean-Louis Burnouf (1775–1844). In 1827 De Chézy published *Théorie du Sloka*, as well as a translation of a puranic poem as "L'Ermitage de Kandou, poeme extrait et traduit du Brahma-pourana" (*Journal Asiatique* 1 [1822]: 1–16) and another lyric work, *Ghatakarpara* (*Journal Asiatique* 2 [1823]: 39–45). In 1830 his most famous work appeared, *La reconnaissance de Sacountala*. Using the pseudonym Apudy, he translated and published the lyric work *Amarus'ataka* as *Anthologie érotique d'Amarou* (1831).

168. Lucius CASSIUS DIO (also Dio Cassius, Dion Cassius, or Dio Cocceianus; c. 150–235 CE) was a Roman administrator and historian of Greek extraction. The various offices he held gave him great opportunities to investigate the past, allowing him to compose a history of Rome in eighty books. Books 36–55 are preserved almost intact, but the remaining books are merely available as fragments today.

170. The DECANS are stars, or groups of stars, that rise at twelve intervals during the course of the night and at ten-day intervals during the year. There are thirty-six of them. The decans appear to have formed the basis for the twenty-four-hour division of the day and may be traced back to ancient Egypt, above all to the tenth dynasty (c. 2100 BCE) and to the time of Seti I (1318-1304 BCE) and Ramses IV (d. 1150 BCE). The idea of decans traveled from Egypt to India to the Islamic world and, via Byzantium, returned to the West, where they can be found in Italy.

The French writer Bernard Le Bovier de FONTENELLE (1657-1757) was a nephew of poet Pierre Corneille (1606-84). A predecessor of the French philosophers of the eighteenth century, Fontenelle served as secretary to the French Academy of Sciences from 1699 until 1740, which allowed him to apply his scientific and literary talents in eulogies for academy members.

171. Attributed to the director of the Palatine Library, Gaius Julius HYGINUS (c. 64 BCE-c. 17 CE), the *Poeticon astronomicon* (1475) is a compilation of basic cosmographic information, astronomical and mythological details, and a catalog of over seven hundred stars. The *Poeticon* was the first printed atlas; the success of its 1482 edition came from its celestial mythology, not from any accuracy in illustrating the actual and described position of the stars. All of Hyginus's works, which included a commentary on Virgil and a treatise on agriculture, are now lost.

Interested in mathematics as an avenue to understanding astronomy, SRĪPETI, Srīpati, or Shripati (fl. 1039-1056 BCE) calculated planetary positions and transits and eclipses, and provided rules for mathematical problems. Among his astronomical works are *Dhikotidakarana* (Procedure giving intellectual climax, written c. 1039 BCE), on solar and lunar eclipses; *Ganitatilaka* (The ornament of mathematics), an arithmetical treatise; *Dhruvamanasa* (Permanent mind), on calculating planetary positions, transits, and eclipses; and *Siddhāntaśekhara* (The crest of established doctrines). Humboldt alludes to *Ratnamālā*, a commentary on astrology in twenty chapters, purportedly written around 1059 BCE.

171n. Guillaume du CHOUL (fl. mid-sixteenth century) from Lyon, who owned a collection of ancient medals and was interested in indigenous antiquities, was particularly intrigued by "pagan survivals" within Christianity. In his discussion of ancient religious history, he examined the recent discoveries of Gallo-Roman objects in light of his local knowledge. In this way he could supply interpretations of ancient artifacts despite the absence of textual evidence. Choul's work was much admired by his contemporaries, including **André Thévêt**. Choul published in his native French rather than only in Latin, a practice that did not become widespread until the Enlightenment, since in earlier days science and scholarship were deemed exclusive to the learned elites.

The Roman poet, astronomer, and astrologer Marcus MANILIUS (fl. first century) was the author of a didactic poem on astrology titled *Astronomica*. The poem is divided into five books and composed in occasionally twisted and strangely beautiful Latin. Manilius described the universe (the constellations, above all) and also the zodiacs and their influence on humans. His work's classical perspectives flourished in Europe during the Middle Ages and Renaissance.

172. Joseph HAGER (also Giuseppe; 1757–c. 1819) was born in Milan, Italy, to German parents. Having studied history and languages in Vienna and Rome, he spent two years in Constantinople (Istanbul) learning Arabic. After his return, he became interested in Chinese. His *Monument de Yu* (1802), from which Humboldt quotes, is a treatise on Chinese worship. Hager also analyzed elementary characters of Chinese and its ancient symbols.

Well known for his twelve years of studying the plants of North America, the French botanist André MICHAUX (1746–1802) traveled widely on orders of the French government. During his return trip to Paris in 1797, he lost most of his collected specimens in a shipwreck. Prior to his time in Canada, Nova Scotia, and the United States, Micheaux had traveled in Egypt and Persia to collect plants and grains (1782–85). Besides bringing mimosa, gingko, and camellia from Persia, he compiled a French-Persian dictionary. Some of Michaux's observations on North America were published by his son François André Michaux (1770–1855) in *Histoire des chênes de l'Amérique, ou Descriptions et figures de toutes les espèces et variétés de chênes de l'Amérique Septentrionale, considérées sous les rapports de la botanique, de leur culture et de leur usage* (History of the oak trees of America, or descriptions and drawings of all the species and varieties of oak trees in North America, considered under their botanical aspects and their cultivation and use, 1801) and *Flora Boreali-Americana* (1803). Between trips, Michaux was introduced to **Thomas Jefferson**, who asked him to organize an expetition to the American West in the 1790s. Micheaux got only as far as St. Louis, for at that point Jefferson recalled the expedition for political reasons.

Although not born in Italy, Ambrosius Theodosius MACROBIUS was a high-ranking official of the Roman empire between the fourth and fifth centuries. He has been identified with two possible imperial administrators: Macrobius (b. c. 360), prefect of Spain in 399 and proconsul of Africa in 410, and Theodosius (b. c. 390), the praetorian prefect for Italy in 430. Between 384 and 395 (or 430 and 440), Macrobius wrote the *Saturnalia*, seven books of dialogues that deal with a range of topics, including Roman religion and philosophy, etymologies, literary history and analysis (especially of Cicero and Virgil), medicine, physiology, and astronomy. In addition to writing a treatise on Greek and Latin words, Macrobius authored the *Commentarii in Somnium Scipionis* (Commentary on Scipio's dreams), an important source of Neoplatonic cosmology and philosophy. See also Nonnus.

173. Fontenelle's original passage reads as follows: "En général le Planisphère est plus Astrologique qu'Astronomique, & par là il n'est guère du ressort de l'Académie. Ce n'est pas que l'histoire des folies des Hommes ne soit une grande partie du savoir, & que malheureusement plusieurs de nos connoissances ne se réduisent là; mais l'Académie a quel-que chose de mieux à faire" (110). (In general, the Planisphere is more astrological than astronomical, and for this reason it is not within the Academy's purview. This is not to say that the history of the follies of men does not constitute a large portion of our learning, and that unfortunately much of our knowledge does not boil down to this, but the Academy has better things to do.)

The second ruler of the Ptolemaic dynasty, PTOLEMY II PHILADELPHUS (308–246 BCE), was king of Egypt from 285 to 246 BCE. He supported the sciences and made Alexandria a center for poets and scholars.

174n. In addition to being a theater director and journal editor, the German writer Johann Gottlieb RHODE (1762–1827) was a historian of religion in, among other places, Berlin, Breslau (Wrocław), and Riga. The book Humboldt mentions is *Versuch über das Alter des Thierkreises und den Ursprung der Sternbilder: mit erläuternden Kupfern* (Essay about the age of the zodiac and the origin of the zodiac signs, with illustrative copper engravings).

175. The Arab astronomer al-Battānī (c. 858–929) is also known by his Latin name ALBATEGINUS (or Albatenius and Albategni). His full name was Abu' abd Allah Muhammad Ibn Jabir Ibn Sinan Al-battani Al-harrani As-sabi'. Only one complete manuscript by him is preserved, at El Escorial (Spain). It shows that **Ptolemy**'s astronomy had been fully embraced in the Islamic East by then, replacing the Indian-Iranian tradition. Al-Battānī was born in Syria and died in Iraq. A crater on the moon is named Albategnius in his honor; it is near another mighty lunar crater named in honor of **Hipparchus**.

A descendant of the Acolhua kings of Tetzcoco through his maternal grandmother, the New Spain chronicler Fernando de Alva Cortés IXTLILXOCHITL (c. 1580–1650) held several positions in the viceroyalty's colonial administration, including that of judicial governor of Tetzcoco (1612–13) and of Chalco Tlalmanalco (1616–21); he was also a translator (*nahuatlato*) in the court system in Mexico City. The *Codex Ixtlilxochitl*, which consists of three unrelated parts of mixed Spanish and pictorial text, is named after him. Ixtlilxochitl's *Relaciones* (Reports) and his *Historia de la Nación Chichimeca* (History of the Chichimec people), his last and most extensive work on what is now known as the *Codex Xolotl*—Ixtlilxochitl calls it *La crónica de los reyes chichimecas* (The chronicle of the Chichimex kings)—was unpublished until the nineteenth century. Writing in Spanish, Ixtlilxochitl used indigenous pictorial manuscripts, hieroglyphic texts, and Nahua oral traditions as his main sources of information. In transcribing primary sources and providing detailed explanations and commentaries, which contemporary anthropologists have shown to be substantially accurate, he preserved for posterity materials that were believed lost or might have become unintelligible. *Historia* ends abruptly in midsentence, in a section that deals with the conquest through the first phase of the final siege of Tenochtitlan. Preserved in the Mexican Collection of the Bibliothèque Nationale in Paris, the *Codex Xolotl*, a historical and genealogical manuscript from the sixteenth century, details the Tetzcoco history in the Valley of Mexico from the arrival of the Chichimecs of Xolotl from around 1224 through the Tepanec War of 1427.

176n. A Jesuit missionary in China beginning in 1750, Jean Joseph-Marie AMIOT (1718–93) wrote and made accessible to Europeans the ideas and life of East Asians through his contribution to the Jesuit series *Mémoires concernant l'histoire, les sciences, les arts, les mœurs, les usages, &c. des chinois* (Accounts about the history, the sciences, the arts, the customs, the practices, etc. of the Chinese). Amiot became the confidant of the Qianlong emperor (r. 1736–95) of the Qing (Manchu) dynasty (1644–c. 1912), the last of the imperial dynasties of China. In 1772 he published the first translation into a European language (French) of *The Art of War* by the Chinese general, military theorist, and philosopher Sun-Tzu (fl. 400–320 BCE). In *Mémorie sur la Musique des Chinis tant anciens que moderns* (Account of Chinese music, ancient and modern, 1780), the earliest account of the theoretical basis of Chinese music available in a European language, Amiot advanced the view that acoustic knowledge had originated in the East (that is, China) in the year 2698 BCE; he argued that it had influenced the West, not the other way around. A prolific writer and translator, he died in Peking.

177. MERCEDONIUS was an intercalary month of varying length in the Roman republican calendar. Humboldt spells it Merkidinus, as did **Plutarch**. The Roman calendar was a system that evolved in the pre-Christian era. Legend has it that Romulus, mythical founder of the city of Rome, instituted it around 738 BCE. But it is probably closer to the truth that the system was derived from the ancient Greeks. The insertion of an occasional Mercedonius at the end of February ensured that the calendar remained in step with the seasons. It was Gaius Julius Caesar who, in 45 BCE, finally initiated a reform that led to the introduction of the Julian calendar, which, in 1582, was replaced by the current Gregorian calendar (also new-style calendar) under Pope Gregory XIII (Ugo Boncompagni, or Buoncompagni; 1502–85; r. 1572–85). Intercalation is now reduced to one leap day every four years. Despite its troublesome history of adoption until the seventeenth century, the Gregorian calendar remains the internationally most accepted civil calendar. See also Sosigenes.

179. HUIXACHTECATL mountain was an extinct volcanic crater whose name means "Hill of the Stars."

180. There appears to be some disagreement about the identity of ACHILLES TATIUS. One Achilles Tatius, who flourished in the second century CE, was a Greek rhetorician who authored *Leucippe and Cleitophon*, a romance that was widely translated during the Renaissance. Nineteenth-century sources attribute the *Isagoge ad Arati phaenomena* (Introduction to Aratus's *Phaenomena*), from which Humboldt quotes here, to this author. Lynn Sumida Joy points out, however, that there were two men by that name and that this commentary on **Aratus** should be attributed to the later Greek writer and astronomer Achilles Tatius (fl. third century). Little is known about either of them. **Denis Petau** reprinted the *Isagoge* in his *Uranologion* in 1703.

The son of classical scholar Jules-César (or Julius Caesar) Scaliger (1484–1558), Joseph Juste SCALIGER (1540–1609) had immense knowledge of both ancient and modern languages. In 1583 the French-born (and Italian-descended) philologist published a major work on earlier calendars, titled *De emendatione temporum* (On the improvement of time).

The Greek astronomer, mathematician, and geographer EUDOXUS OF CNIDUS (c. 395/90–342/37 BCE) was part of Plato's academy. None of his writings survive, but there are numerous references to and discussions of his contributions to scholarship throughout antiquity. For example, **Aratus**'s famous astronomical poem is based on his work. Humboldt mentions Eudoxus's "Octaëterides," a poem that **Censorinus** attributed to Eudoxus—Octaëterides refers to the eight-year cycle with which the ancient Greeks experimented in their search for a time unit that could reconcile solar with lunar cycles (see also **Mercedonius**). The Greek astronomer and poet **Eratosthenes of Cyrene** is also believed to have authored a commentary on the "Octaëterides," which is mentioned by **Geminus** in **Petau**'s *Uranologion*, a collection of Greek astronomical texts which Humboldt quotes repeatedly.

180n. A leading theologian in the sixteenth and seventeenth century, the French Jesuit Denis PETAU (Dionysius Petavius; 1583–1652) was best known for his edition of Greek astronomical texts, the *Uranologion* (1630). Alongside Scaliger's *De emendation temporum* (On the improvement of time), Petau's 1627 *Opus de doctrina temporum* (Work on the system of time) laid the foundations of modern studies in chronology. Petau was also a poet, orator, and dramaturgist.

181. The Greek physician and philosopher SEXTUS EMPIRICUS (c. second century CE) is best known as the foremost representative of empirical skepticism, or Pyrrhonism. This school of thought rejected dogmatism in scientific research and thus the possibility of any final state of knowledge. Forgotten for centuries, Sextus Empiricus's philosophical writings were rediscovered during the Renaissance, among them *Adversus mathematicos* (Against the professors). Written toward the end of the second century, its eleven books criticized specialized fields of science, including grammar, rhetoric, geometry, arithmetic, astrology, and music, along with philosophies such as epistemology, logic, philosophy of nature, and ethics. Sextus Empiricus is also the author of *Pyrrhōneioi hypotypōseis* (Theories of Pyrrhonism), which provides a general outline of Pyrrhonist skepticism.

181n. An early Christian convert, the Sicilian Iulius FIRMICUS Maternus (c. 300 CE) forcefully polemicized against paganism. He became best known for his long and ambitious handbook on astrology. Dedicated to his friend the Roman governor Lollianus Mavortius, this work was not only the largest but also the last theoretical treatise on astrology to appear in Western Europe before the arrival of the Arabs.

The theologian ORIGEN (Oregenes Adamantius; c. 185–c. 254 CE) was of Christian and (likely) Greek descent. A disciple of **Clement of Alexandria**, he is known as a synthesizer of early Christian theology.

182n. FATHER DUCROS (also du Croz; fl. 1720s) is a Jesuit missionary to Mauritius mentioned in volume 18 of *Lettres édifiantes et curieuses, écrites des missions etrangèes par quelques missionnaires de la Compagnie de Jesus* (Edifying and interesting letters written by some missionaries of the Society of Jesus on their foreign missions, 1728) edited, among others, by **du Halde.**

183. Mariano VEYTIA, whose full name is Mariano José Fernández de Echeverría Orcalaga Alonso Linaje de Veytia (1718-79), was a lawyer from Puebla, New Spain. A member of a wealthy family, he traveled to Spain in 1738 and met **Boturini,** who introduced him to Mesoamerican history and antiquities. In 1750 Veytia returned to his homeland, where he examined and copied documents from the Boturini Collection. Using these rare materials, Veytia wrote his *Historia Antigua de México* (Ancient history of Mexico) around 1769. The *Historia* was published partially in 1826 and completely in 1836, except for the prologue, which was published separately in 1927. Arguing that the Mesoamerican polities had devolved from monotheism to polytheism, Veytia began his history with the migration and settlement of seven Chichimec families in North America. He also claimed that the apostle St. Thomas had arrived in New Spain as a bearded white man dressed in a white robe with red crosses, barefoot, and carrying a staff. St. Thomas had thus infused the figure of **Quetzalcoatl** with the virtues of the doctrines of the Catholic Church and also prophesied the arrival of the Spaniards. Veytia is also the author of *Historia de la fundación de la Ciudad de Puebla de los Ángeles en la Nueva España* (History and founding of the City of Puebla de los Ángeles in New Spain), written around 1777 but published only in 1931.

184. Interested in mathematics and ancient languages, Nicolas Antoine BOULANGER (1722-59) was a contemporary of the French philosophers Denis Diderot (1713-84), Jean-Jacques Rousseau (1712-78), and François Marie Arouet, better known as Voltaire (1694-1778). Boulanger contributed to Diderot's *Encyclopédie,* and Diderot, in his turn, wrote a biography of Boulanger, which was included in the 1894 edition of Boulanger's works.

Nicolas FRÉRET (1688-1749) joined the Académie des Inscriptions et Belles-Lettres in 1714 and became its perpetual secretary in 1743. In a 1718 paper Fréret discussed Chinese ideograms, which he had learned about from the Chinese scholar Arcadio Huang, who came to Paris in 1711. Comparing Mexica pictograms to Chinese characters and to alphabetical writing, Fréret argued that the mind gains concepts through abstraction. He also wrote *Défense de la chronologie* (In defense of chronology, 1728), a commentary on Sir Isaac Newton's *Chronology of Ancient Kingdoms* (1728). See also Louis Jouard de la Nauze.

185. Humboldt is referring to the lunar and solar tables that the German astronomer and mapmaker Johann Tobias MAYER (1723-62) began calculating for the British government in 1753. Using lunar distances, these tables determined longitude at sea with an accuracy of half a degree. Mayer's method, together with a formula for correcting errors in longitude due to atmospheric refraction, was published posthumously in 1770 by Nevil Maskelyne under the title *New and correct tables of the motions of the sun and moon, by Tobias Mayer to which is added the method of finding the longitude improved*. Humboldt held up Mayer's method as exemplary for his own research, not only because it made gravitational equilibration a universal heuristic but also because it placed precise measurements at the center of any attempt to determine the laws of temperature or magnetism. No amount of theory could supply the basic terms of the equations that describe these lines. These laws, Humboldt emphasized, were fundamentally empirical and increased in accuracy with more precise, and more frequent, measurements. Humboldt viewed his own isothermal work as a continuation of Mayer's, which was published by Georg Lichtenberg (1742-99) as *Opera inedita* (Unpublished works) in 1775. In 1752, in his second year as professor of economics and mathematics at Göttingen, Mayer invented an improvement to the reflecting circle, which Charles Borda (1733-99) would later develop further.

189n. Knighted in 1833, Sir Charles WILKINS (bap. 1749-1836), a British merchant in the East India Company from 1770 to 1786, was the first to translate into English the Bhagavad Gītā, a small part of the epic Mahābhārata (which inspired **Kālidāsa**'s famous *Abhijñānaśākuntala*) and an influential Hindu text. A pioneering Sanskrit scholar, Wilkins established a printing press for Asian languages in 1778 and published *A Grammar of Sanskrit* in 1779. He was one of the founding members of the **Asiatick Society** of Bengal in Calcutta. Elected to the Royal Society in 1788, Wilkins became the librarian of the East India Company in 1800.

190. The private physician of King Philip II (1527-98), Francisco HERNÁNDEZ DE TOLEDO (c. 1515-87) was a renowned Spanish naturalist. In 1570 Hernández was appointed chief medical officer of the Indies; he arrived in New Spain a year later, charged with researching the indigenous flora and fauna of the New World. He used the opportunity to question Nahua intellectuals who knew Latin. In 1577 Hernández returned to Spain, where he prepared eleven volumes with illustrations of three thousand collected species. The work was written in Latin and struck Philip II as too philosophical. Nardo Antonio Recchi was asked to present a digest of it, listing only useful medicinal plants. Recchi's version, however, was not published until several decades later. Hernández began a Spanish translation and also ordered a Nahuatl version of his work. One of the two original manuscripts, Hernández's corrected draft, is divided between the Museum of Natural Sciences and the General Archive of the Ministerio de Hacienda in Madrid. His final report to the king was housed at the Escorial and was destroyed during a 1671 fire. The Spanish Jesuit Juan Eusebius NIEREMBERG (1595-1658), who taught physiology at the Royal Academy in Madrid, based his *Historia naturae* (Natural history, 1635) on Hernández original works, including the lost Escorial manuscript. Hernández also translated into Spanish and annotated Pliny the Elder's *Historia naturali* (*Natural History*).

193n. José Mariano MOZIÑO (1757-1820), also spelled Mociño and Muciño, was a botanist from New Spain and a member of the famous Royal Botanical Expedition of

New Spain, led by Martín de **Sessé** y Lacasta (1751–1808). Together with Juan Francisco de la Bodega y Quadra (1743–94), Moziño traveled to the northern borders of Spain's eighteenth-century territory in what is commonly called the Expedition of the Borders to the North of California. In 1792 the expeditioners arrived at Nootka (western Vancouver Island, Canada), where Moziño spent several months with the indigenous inhabitants. He narrated his observations in his *Noticias de Nutka* (News from Nootka), arguing that the Nootka—the Makah, Nitinat, Clayoquot, and Kyuquot—were better off without European goods. To this work he attached a brief dictionary of terms in Wakashan, the language of Nootka. In 1793 Moziño returned to New Spain and described the last eruption of the San Martín volcano at (San Andrés) Tuxtla, Veracruz. Under the penname José Velázquez de Vice Cotis, he was also a regular contributor to *Gazetas de Literatura*.

194. Dutch-born philosopher and historian Cornelius de PAUW (1739–99) was a disciple of Georges-Louis Leclerc, the famous count of Buffon (1707–88), author of the monumental anti-Americanist *Histoire naturelle* (Natural history, 1749–88). Humboldt's rejection of the "systematic ideas" of such authorities on European Enlightenment as de Pauw, Abbé Guillaume Thomas François RAYNAL (1713–96), and **Robertson** (1721–93) in his discussion of the mythological characteristics and cultural specifics of the Aztec Calendar Stone is no accident. In his very popular *Recherches philosophiques sur les Américains* (Philosophical researches on the Americans, 1768–69), de Pauw had advanced his degeneracy thesis regarding indigenous American cultures: in the "unfavorable" American environment it was inevitable that plants, animals, humans, and, by extension, human institutions, whether indigenous or transplanted from Europe, would eventually degenerate. The "natural people" of the Americas were therefore to be seen as naive children obeying only the impulses of their instincts; they were incurably lazy and incapable of any mental progress. Instead of just reading travel accounts as de Pauw did, Humboldt had actual travel experience and extensive studies to back up his counterarguments (see "The Art of Science," introducing this book). Raynal was a French Jesuit historian who contributed significantly to preparing the intellectual climate for the French Revolution. His most important work was the *Histoire philosophique et politique des deux Indes* (Philosophical and political history of the two Indies), which appeared in several editions between 1770 and 1789. The *Histoire* was a very popular work whose revolutionary tone became more noticeable in later editions. In 1774 it was placed on the Roman Catholic Church's Index of Forbidden Books, and Raynal was banished from Paris in 1781. His property having been confiscated, he died in poverty.

197. In the late eighteenth and early nineteenth centuries, Euro-Americans started exploring large caves in Tennessee and Kentucky, such as the Mammoth and Salt Caves, where they found traces of prehistoric peoples. Humboldt's Mr. Cutter was probably a local adventurer; we could not trace him. Humboldt adds to his remarks about Cutter a reference to the 1796 *American Pocket Atlas with a Concise Description of Each State* by Matthew Carey (1760–1839), a Philadelphia-based publisher of Irish extraction who had come to the United States of America in the 1780s. Carey was interested in mapmaking and had some twenty maps of the United States made for his atlas.

198. The Spanish conquistador Sebastián de BELALCÁZAR, or Benalcázar (c. 1480–1551), participated in the conquest of Nicaragua (1524–27) and in **Pizarro**'s expedition to Peru in 1531. Under Pizarro, he led the conquest of Quito in 1534 and subsequently founded Guayaquil. Pizarro's death sentence made Belalcázar flee Quito in search of the mythical El Dorado; in the process he also founded Cali and Popayán in what is now Colombia. In 1539 Belalcázar met **Jiménez de Quesada** in Bogotá. Because of territorial disputes with fellow conquistadors, he was finally indicted for executing Jorge Robledo (c. 1500–1546).

Juan LARREA y Villavicencio (1759–1824), who accompanied Humboldt on his climb of Cotopaxi, was a poet from the Royal Audiencia of Quito. A supporter of the 1809 Revolution of Quito, the first independence movement in Spanish America, Larrea was appointed treasurer of the Primera Junta de Gobierno Autónoma of 1809. A member of the Economic Society of Quito, he wrote a description of manufacturing in the northern and central highlands of the Audiencia of Quito, titled *Las Manufacturas de la Provincia de Quito* (1802).

200. Humboldt wrote about his rather risky June 23, 1802, EXCURSION to the top of Chimborazo, then believed to be the highest mountain in the world, in a little-known dramatic prose sketch from 1837, which he revised for his *Kleinere Schriften* (Shorter writings) in 1853.

201. Among other things, Humboldt's close friend Louis-Joseph GAY-LUSSAC (1778–1850) was one of several well-known French physicists and chemists who, following the 1783 ascent of the hot-air balloon *Montgolfière*, associated themselves with scientific ballooning. In 1804 Gay-Lussac collaborated with Jean-Baptiste Biot (1774–1862) to reach an altitude of twenty-three thousand feet in order to make magnetic and temperature observations. The following year Humboldt and Gay-Lussac, a student of the French chemist Claude Louis Berthollet (1748–1822), worked together in studying the composition of the atmosphere. Gay-Lussac's British peer and competitor in the analysis of the then-curious new substance iodine was the chemist and inventor Sir Humphry Davy (1778–1829).

During a British expedition to Nepal around the turn of the nineteenth century, Colonel Charles CRAWFORD (fl. 1801–3) was the first European to see the immense height of peaks of the Himalaya, or Soomoonang, the name used in Humboldt's day for the loftiest mountain passes in the north of Nepal and Bhutan at the border with Tibet. This range is now known as the Great or Higher Himalayas. Crawford briefly served as surveyor general of India during the 1810s. In 1818 the Sanskrit scholar **Colebrooke** published a paper about the height of Himalayan peaks, which was also based on Crawford's observations. The question of the height of the Himalaya peaks was settled by Sir Andrew Scott Waugh (1810–78) after his appointment as surveyor general in 1843: he determined the elevations of seventy-nine of the highest peaks in the Himalaya through triangulation. Up to then Chimborazo had been assumed to be the highest mountain in the world. Waugh named Peak XV of that range Mount Everest, after the geodesist and military engineer Sir George Everest (1790–1866). Mount Everest is 8,848 meters (29,029 ft) high.

202. Early on in the French Revolution, in year II of the Republican calendar, the architect Jean-Thomas THIBAULT (1757–1826), together with his now more famous colleague Jean-Nicolas-Louis Durand (1760–1834), designed a prize-winning temple to equality. It was a building without function but with high symbolic value; the design was discussed widely, not just in France.

202n. Professor of philosophy and later poetry and rhetoric at the University of Leipzig since 1798 and 1803/1809, respectively, Johann Gottfried Jakob HERMANN (1772–1848) was one of the most prominent scholars of Greek antiquity of his day. Future Yale president Theodore Dwight Woolsey (1801–89), who studied under Hermann in the 1820s, along with his cousin Henry Edwin Dwight (1797–1832), presented Hermann to a US audience as "the most distinguished Greek of Germany." Hermann stood for a philological tradition that focused on a thorough study of individual languages, especially of grammar and meter, which, he argued, was the only way to understand intellectual life in antiquity.

Edward MOOR (1771–1848) was a writer on Hindu mythology. He went to India at the age of eleven as a cadet in the East India Company. By the time he was seventeen, Moor was already noted for his proficiency in the language of the natives—though not Sanskrit. The 1810 *Hindu Pantheon* (which also included plates) was Moor's most significant scholarly study. For fifty years this work, by current standards prolix and in some respects inaccurate, remained an important first attempt at presenting the gods of the Hindus systematically to an English audience. A member of the **Asiatic Society** of Bengal, Moor had gathered many of the materials himself but also relied on the work of others, such as Sir **William Jones**, and on correspondence with the Sanskrit scholar Sir **Charles Wilkins**, among many others. Moor died in England.

203. HESIOD (fl. c. 700 BCE) is one of the most ancient Greek poets; he is the principal source of the earliest recorded phase of Greek ideas about the gods. Although he is often mentioned in the same breath with Homer, Hesiod lived in a very different world, both socially and spiritually. Among Hesiod's works, the *Theogony* gives the fullest account of myths about the origin of the gods.

204n. TITUS LIVIUS, or Livy (c. 59 BCE–c. 17 CE) was a historian from Padua, Italy. His famous *Ab urbe condita libri* (Books from the founding of the city), commonly known as *History of Rome*, covers the origins of Rome beginning in 9 BCE. Only 35 of the 142 books survive (books 1–10 and 21–45); other small excerpts, summaries (known as *periochae*), and passages can be found in the work of other writers. Livy aimed to chronicle the rise of Rome as the empire of Italy and the Mediterranean world. His writings inspired Shakespeare's *The Rape of Lucrece* (1594), and Dante alludes to his accounts of the war of Hannibal in the *Inferno*.

207. The Zend Avesta is the principal collection of sacred texts of Zoroastrianism, written in Avestan, a language from eastern Iran. The French scholar Abraham-Hyacinthe ANQUETIL-DUPERRON (1731–1805) brought back a collection of Zoroastrian manuscripts from a trip to India, publishing them in 1771. Zoroastrianism is the ancient pre-Islamic religion of Iran that also survives in India among Persian immigrants (which is why it is also called Parsiism); it was named after the Iranian prophet and reformer Zoroaster (fl. sixth century BCE). When Anquetil-Duperron published his book, it created a sensation in France, despite the fact that it was overburdened with irrelevant detail (mainly about himself). It also provoked **William Jones**'s anger for its boastful tone and its insulting remarks about Oxford's low standards. Although Jones's counterattack in an anonymous open letter was not exactly a paragon of objective criticism, it won him the admiration of his British peers. Anquetil-Duperron had left for the French possessions in India in the mid-1750s, desiring to study ancient texts. Together with his compatriots, he was expelled by the British in 1761, which might explain some of his resentments.

208. Born in Egypt, NONNUS (fl. fifth century) was the most notable Greek poet from the Roman period. His most famous poem, *Dionysiaca*, tells the story of the god Dionysus, who, according to legend, set out from Greece to conquer India. The poem abounds in astrological references. See also Macrobius.

210n. *Timaeus* is a theological work by Plato (c. 429–c. 347 BCE), written in the form of a dialogue and first published in 360 BCE. Humboldt quotes it from a 1578 edition. Continuously influential since antiquity, *Timaeus* is an important document in the history of European thought, giving an account of how the cosmos and everything therein came into being.

211. LUIS DE VELASCO, the marquis of Salinas (1535–1617), served as viceroy in Peru from 1590 to 1595, and in Mexico City from 1607 to 1611. In 1611 he was recalled to Madrid to become president of the Royal Council of the Indies.

211n. The German classical scholar and bibliographer Johann Albert FABRICIUS (1668–1736) was most famous for his *Bibliotheca Graeca* (Greek library, 1705–28), which was revised and continued by the German literary scholar and philologist Gottlieb Christoph Harless (or Harles; 1738–1815) between 1790 and 1812. This comprehensive multivolume study covers Greek literature from pre-Homeric times until the fall of Constantinople (Istanbul) in 1453.

214. Andrés Manuel DEL RÍO (1764–1849) was a mineralogist educated in Spain, France, England, and Germany (at the Freiberg Mining Academy). In 1794 he became professor of mineralogy at the School of Mines in Mexico City. In 1801 he discovered the element vanadium, a silvery metal used mainly for alloys (such as steels), which he named erythronium.

Based on its characteristics and the inscriptions, Eduardo Matos Moctezuma believes that the ax Humboldt depicts on Plate XXVIII, known as the Humboldt celt, comes from the Zapoteca peoples of the Valley of Oaxaca. Investigating the symbolic meaning of its inscriptions, scholars also commonly attribute it to the Olmec and date it circa 900 BCE, arguing that its four-part motif is an Olmec representation of the four cardinal directions.

215n. The avid traveler POLYBIUS (c. 200–c. 118 BCE) was a Greek historian and statesman who wrote about the rise of the Roman empire. His declared interest was not simply to write Greek history but to provide an account of how Rome took over the "inhabited" world.

216n. Born in Paris, Antoine-Chrysostome QUATREMÈRE DE QUINCY (1755-1849) originally studied law but eventually switched to sculpture. In 1785 he won a prize from the French Académie d'Inscriptions et Belles-Lettres for an essay on Egyptian architecture, a revised version of which was published in 1803. As Sylvia Lavin has pointed out, the widely held assumption that the 1803 publication is identical with the earlier essay is wrong—substantial changes had been completed by 1801.

218. A clergyman from Spain, Primo Feliciano MARÍN de Porras (d. 1815) was appointed bishop of Monterrey in October 1800 and arrived in New Spain in 1803. Two years later he would visit present-day southern Texas. Bishop Marín de Porras argued in Humboldt's favor to have the statue of **Coatlicue** disinterred for anaylisis.

224. The Berlin-born naturalist Peter Simon PALLAS (1741-1811) studied at the universities of Berlin, Halle, Göttingen, and Leiden and explored natural history collections in London, Amsterdam, and The Hague. In 1768, the year when Russian empress Catherine II (1729-96) invited him to accept a professorship in St. Petersburg, Pallas undertook a scientific expedition through Russia and Siberia. He went on a second voyage in 1793-94, publishing his travelogue between 1799 and 1801.

Henry SALT (1780-1827) had initially hoped to be a portraitist, but his talent was not sufficient to make a success of it. In 1802 he shifted gears and accompanied George Annesley Mountnorris (Lord Valentia; c. 1769-1844) as secretary and draftsman on a journey to India, Ceylon, and the Red Sea. In 1805 Salt first traveled to Abyssinia; he was sent back there in 1809. His travels also brought him to Egypt, where he served as British consul general beginning in 1815. Besides publishing his *Twenty-Four Views in St. Helena, India, and Egypt* (1809), Salt coauthored Lord Valentia's *Travels in India* that same year; he also provided the book's illustrations. *A Voyage to Abyssinia* was not published until 1814.

226. John BENTLEY (d. before 1825), **William Jones**, and **Jean-Sylvain Bailly** all studied the Hindu **yugas**. Bentley was a member of the **Asiatick Society** and gave a first paper in 1799. Humboldt had access to Bentley's publication on the Hindu astronomical cycles in the *Asiatick Researches*; Bentley's book on the subject was published posthumously in 1825. Bentley was involved in scholarly quarrels, as is evident from a forceful 1826 public letter by **Colebrooke** in reply to an attack on him in Bentley's 1825 book.

227. BEROSUS (also Berossus, Berossos, or Berosos; b. c. 330–323 BCE, fl. c. 290 BCE) was a Chaldean priest who wrote a history of Babylonia in three books. An important resource for later historians on account of its information on Babylon's origins, Berosus's work is preserved only in fragmentary citations. Modern scholars have devoted much attention to his concepts of chronology and history.

228. The translation of the Nahuatl proper noun AZTLAN, or Aztatlan, as "the resting place of herons" comes from the *Crónica mexicáyotl* by Fernando Alvarado **Tezozomoc**. The exact meaning of the name remains uncertain; another possible translation is "place of whiteness." As Aztlan was surrounded by water and mist, the Aztecs migrated to and settled in Tenochtitlan to re-create their legendary homeland in the Valley of Mexico. Scholars have debated for centuries over the actual location of Aztlan, all to no avail.

231. A diplomat and army officer for the East India Company, Samuel TURNER (c. 1749–1802) first traveled to India in 1780, then spent almost his entire military career there, except for the years 1783 and 1784, when he led a mission to Tibet and Bhutan. An account of this mission, the first of these regions in English, was published in 1800. The mission was underwritten by Turner's cousin Warren Hastings (1732–1818), governor general of Bengal, who was eager to promote British-Indian trade across the Himalayas. Turner's papers are preserved at the Bodleian Library in Oxford, England.

234. On September 19, 1648, on top of the Puy-de-Dôme observatory in Clermont, some two hundred miles south of Paris, Florin PÉRIER (1605–72) verified Pascal's law of pressure, discovered by his brother-in-law the French mathematician Blaise PASCAL (1623–62). Pascal constructed mercury barometers that allowed him to measure air pressure. As Humboldt implies here, it also allowed for determining elevation.

235. See *Codex Borgia* above.

236. The first of fourteen Capetian kings, HUGUES CAPET (also Hugh; c. 938/41–96) was king of France from 987 to 996. He succeeded the last of the French Carolingian kings; the Carolingian dynasty, whose most famous representative was Charlemagne (747–814), had ruled western Europe, including modern France and Germany, since 750. The name Capet was derived from Hugues's nickname, Cape (Lat. Capa). The Capetians carried the hand of justice in their ceremony of coronation and anointing, which they depicted on their seal.

236n. Bernard de MONTFAUCON (1655-1741) was one of the first scholars to base historical inquiry on the study not only of texts but also of buildings and other monuments.

Claude-François MÉNESTRIER (1631-1705) was a Jesuit antiquarian whose extensive work includes, among many other things, a study of the history of dance, for which he is best remembered today.

The French Jesuit priest Marc Gilbert de VARENNES (1591-1660) wrote about heraldry and was best known for the work from which Humboldt quotes here.

The historian Antonio Agustín (also AUGUSTINUS and Agostino; 1517-86) was archbishop of Tarragona. His work about medals and other antiques was published posthumously in 1587.

A moral and political theorist, the Flemish humanist and classics scholar Justus LIPSIUS (Joest Lips; 1547-1606) held the chair of history and philosophy at the University of Jena. He subsequently served as professor of history and law at the University of Leiden and finally represented history and Latin at Leuven (Louvain). A leading editor of Latin prose texts, Lipsius was also noted for his antiquarian and historical studies.

238. The identity of Lord HILLSBOROUGH cannot be established with absolute certainty. He might have been either Wills Hill (1718-93), viscount of Kirwalin and earl of Hillsborough, or his son and successor, Arthur Hill (1753-1801), who was also known as Lord Kilwarlin and after 1789 as the earl of Hillsborough. Wills Hill, who is typically referred to as "Lord Hillsborough," served as the British secretary of state in charge of American affairs from 1768 until 1772, in which capacity he aggravated Benjamin Franklin (1706-90). The Hill family's Kilwarlin estate and castle, which dates back to the seventeenth century, is likely to have housed the collection to which Humboldt is referring here.

Gustavus BRANDER (1719/20-87) was a British merchant of Swedish descent who was interested in antiquities and in science, as is evident from his publication about fossils. A collector of historical objects, he eventually served as the curator of the British Museum.

238. The British politician, governor of Massachusetts (1757-60), and amateur antiquarian Thomas POWNALL (1722-1805) also contributed to *Archaeologia: or Miscellaneous Tracts Relating to Antiquity*, the journal of the London Society of Antiquaries. Pownall compared North American Indian cultures with British antiquity, arguing that humans as natural beings will always be the same under the same circumstances.

241. The French physician Louis de RIEUX was named commissioner for quinine production and resided in New Granada. Together with his son, he accompanied Humboldt and Bonpland for some time; they had first met in Havana, where Humboldt had noticed that the son could sketch very well and had sought to interest him in joining the expedition as a painter. The Rieuxes went with Humboldt and Bonpland up the Magdalena River, but when the elder Rieux fell ill, he happily remained behind, guarding his mistress, as Humboldt noted in his diary. Humboldt had mixed feelings about the older Rieux. He described him as an enemy of **José Celestino Mutis** who rejoiced at the news of Mutis's death.

The Inquisition unsuccessfully tried to convict Mariano Luis de URQUIJO y Muga (1768–1817) for his 1791 translation of Voltaire's *La mort de César* (Caesar's death), which had been banned in 1762. As Spanish secretary of state (1798–1800), Urquijo limited the Inquisition and the papacy's power and passed a decree allowing bishops to grant matrimonial dispensations (known as Cisma de Urquijo). He also served as minister of state (1808–13) under Joseph I, king of Spain. Urquijo supported Humboldt's request to travel to Spanish America. In 1799 Goya painted a portrait of Urquijo; it now resides at the Real Academia de la Historia, Madrid.

242. During an expedition to Lapland in 1736, the French mathematician Pierre-Louis Moreau de MAUPERTUIS (1698–1759) took measurements verifying the Newtonian notion that the earth is an oblate spheroid. He was subsequently dubbed "the man who flattened the earth." As **Lalande** reported, the Swedish Academy in Stockholm sent a Mr. SWANBERG to Lapland in 1799 to find the measuring points used in 1736 and, if possible, to correct Maupertuis's results. A Segar SWANBERG, master of mines, was mentioned in an 1811 English edition of Carl von Linné's account of his journey to Lapland in 1732.

244. José Domingo DUQUESNE de la Madrid (c. 1748–1822), from New Granada, studied at the Colegio de San Bartolomé in Bogotá in present-day Colombia. For more than two decades, this ethnographer served as priest in the villages of Lenguazaque and Gachancipá. Based on his experience working with the Muisca, he wrote "Disertación sobre el Calendario de los muyscas indios naturales de este Nuevo Reino de Granada" in 1795. The "Disertación" mentions a Muisca calendar stone and sets forward Duquesne's understanding of the Muisca numbering and time-counting system, arguing that the Muisca had a calendar just as complex as the Aztecs'. In addition to authoring a now-lost manuscriped titled *Comento del Apocalipsis* (Comment on the Apocalypse), Duquesne wrote *Historia de un Congreso Filosófico tenido en Parnaso por lo tocante al imperio de Aristóteles* (History of a philosophical conference held in Parnassos about the imperium of Aristotle) in 1791.

245. The botanist José Celestino MUTIS y Bosio (1732–1808) headed the Royal Botanical Expedition in the New Kingdom of Granada (that is, the modern countries of Colombia, Panama, Venezuela, and Ecuador) in 1782. Mutis cataloged new plants, built an astronomical observatory, trained painters and young natural philosophers, and supported the development of agriculture, commerce, and culture in the viceroyalty. Humboldt exchanged botanical information with Mutis when he and Bonpland stayed at Mutis's house in Bogotá from July to September 1801. See also Francisco Antonio Zea.

245n. The Spanish poet Juan de CASTELLANOS (1522-1607) traveled to America as a soldier around 1535. Around 1554 he settled in as the priest of Santiago de Tunja in present-day Colombia. He wrote *Elegías de Varones Ilustres de Indias* (Elegies for illustrious men of the Indies) over a period of two decades, narrating the conquest in verse. One of the longest chronicles in verse, *Elegías* is made up of fifty-five cantos—150,000 hendecasyllabic lines—and was published in four parts in the nineteenth century: parts 1-3 in 1847 and part 4, as *Historia del Nuevo Reino de Granada* (History of the New Kingdom of Granada), in 1886. The *Discurso del capitán Francisco Draque* (Discourse of Captain Francis Drake), an integral part of *Elegías*, was suppressed by the Council of the Indies; it was not published until 1921.

The Spanish Franciscan missionary and historian Fray Antonio MEDRANO (d. c. 1569) arrived in America before 1560 and settled in the Province of Santa Fé de Bogotá, where he worked on a chronicle of New Granada, later titled *Recopilación Historial* (Historical collection). Upon Medrano's death during **Quesada**'s failed attempt at locating El Dorado, the manuscripts were transferred to a fellow Franciscan, Pedro de AGUADO (b. 1513/30-after 1589), who had arrived in South America around the same time as Medrano and became the provincial minister of Santa Fé de Bogotá in 1573. Aguado expanded and concluded the work that Medrano had started, producing the *Recopilación historial resolutoria de Santa Marta y Nuevo Reino de Granada* (Historical compendium of the establishment of Santa Marta and the New Kingdom of Granada). In 1575 he made a brief trip to Spain, where he gained authorization to publish *Recopilación* in 1582.

247. The Dominican missionary and philologist Fray Bernardo de LUGO taught Muisca at the Rosario Convent in Santa Fé de Bogotá, writing the first Chibcha grammar between 1617 and 1618. *Gramática de la Lengua general del Nuevo Reyno, llamada Mosca* (Grammar of the common language of the New Kingdom, known as Mosca, 1619) describes the grammatical aspects of Chibcha in detail. Concurring with **Duquesne**, Lugo mentions series of ordinal numbers and units of time. The first Chibcha-Spanish dictionary had been compiled by an anonymous clergyman, possibly in the first half of the seventeenth century; the manuscript remains at the Biblioteca Nacional de Colombia (MS 158). It was published by María Stella González de Pérez as *Diccionario y gramática chibcha* in 1987 (Chibcha dictionary and grammar) in 1987. Ezequiel Uricoechea (1834-80) had included the language's first dictionary in his *Gramática, vocabulario, catecismo y confesionario de la lengua chibcha* (Grammar, vocabulary, cathechism, and confessional of the Chibcha language, 1871).

249. Abbé Lorenzo HERVÁS y Panduro (1735-1809) was a Spanish Jesuit who, in his six-volume *Catálogo de las lenguas de las naciones conocidas* (Catalog of the languages of the known peoples, 1800-1805), discussed the major linguistic theories of his day and provided detailed descriptions of the languages in the world. He had gathered much of the information from his missionary colleagues.

251n. Having studied medicine at Edinburgh University in 1788, the British explorer Mungo PARK (1771–1806) traveled to Sumatra in 1793 to collect plant specimens; he was a surgeon's mate on an East India ship. Upon his return to England a year later, the African Association (founded 1788) dispatched Park to explore the Niger River; he left for Gambia in 1795 and returned to England in 1797. His travel accounts were published in 1799. Park died during a second voyage to explore the course of the Niger River in 1805–6.

254n. The Swedish diplomat Nils Gustaf PALIN (1765–1842) served as ambassador to Constantinople (Istanbul) in 1828. He produced several books on the Rosetta Stone and went so far as to claim that he had deciphered the **hieroglyphic inscriptions of the Egyptians.**

255. The Alexandrian scholar Horus, also known as Niliacus HORAPOLLON or Horapollon the Younger (c. fifth century), attempted to decipher hieroglyphic writings. Following a Greek tradition, he based his work on the wrong premise by assuming that the hieroglyphs were symbols and allegories rather than phonetic signs.

Edme François JOMARD (1777–1862) accompanied Napoleon's expedition to Egypt in 1798 as a topographical engineer. Jomard contributed several volumes to the monumental *Description to l'Égype* (Description of Egypt), a collaborative work that took decades to complete. A member of the Académie royale des inscriptions et belles-lettres (French Academy of Inscriptions and Letters) beginning in 1818, Jomard was one of the founders of the Geographical Society in 1821. In 1828 he was appointed to the position of curator of the maps and charts preserved in the Royal Library of Paris, of which he later became the main librarian. Jomard was a friend of Humboldt's for more than half a century, and his connections provided Humboldt with access to many a private collection of antiquities.

258n. The Roman grammarian CENSORINUS (fl. c. 238) wrote *De die natali* (translated as *The Birthday Book*) as a present for his best friend, Quintus Caerellius, who was then between forty-nine and fifty-six years of age. *De die natali* is a very concise text celebrating harmony and order in the universe; it touches on almost every subject of cosmic importance, such as mathematics, music, history, astronomy, and embryology, making it a treasure trove for scientists, poets, and scholars alike. The book also provides a solid chronology of dates referring to ancient history.

260. Flavius Valerius Constantinus, or CONSTANTINE I (also Constantine the Great; 274–337) was the first Roman emperor to profess Christianity. Constantine was brought up in the court of Emperor Diocletian (245–316; r. 284–305) at Nicomedia (now Izmit, Turkey). He maintained the indictions, or fiscal years, that may be traced back to Diocletian's reforms and his introduction of an *indictio* (annual levy or tribute) in 314, starting retroactively in 312. During a fifteen-year cycle, indictions were to be paid for one year starting on September 23 (later changed to September 1 and ending on August 31). In the course of the fifth century, indictions came into use as a means to ascertain dates in the Byzantine empire, a late incarnation of the Roman empire; they are mentioned in church documents through the Middle Ages.

266. The Codex Dresden — Humboldt's Dresden manuscript, or the Codex Mexicanus of Dresden — is a ritual-calendrical screenfold on *amatl* paper housed at the Sächsische Landesbibliothek (State Library of Saxony) in Dresden. One of the most important surviving pictorial Maya manuscripts, the Codex Dresden contains divinatory almanacs, representations of many ceremonies and deities, multiplication tables for synodical revolutions of Venus and various numbers, tables of eclipse of Venus, and addresses other matters, such as disease and agriculture. This codex is probably a copy with Mexica influences, dating from the twelfth century; it suffered deterioration during World War II. It is one of only four Maya hieroglyphic manuscripts in existence, the others being the Codices Madrid, Paris, and Grolier. Humboldt published five of its thirty-nine pages on Plate XLV.

Karl August BÖTTIGER (1760–1835) was a headmaster in Weimar, a hub for the German literary elite of that time. From 1795 to 1803 he edited the Weimar *Journal des Luxus und der Moden* (Journal of luxury and fashions). Böttiger is often quoted in connection with his intimate knowledge of the German theater scene but less so for in relation to his work as an archaeologist and a naturalist. Böttiger delivered a lecture on Pliny at a Berlin conference for German naturalists and physicians that Humboldt organized in 1828. At that time, long before Charles Darwin (1809–82) had written his influential *On the Origin of Species* (1859), Böttiger spoke up against the widely held belief, shared by Humboldt, that species do not disappear. Böttiger backed up his argument by pointing to Carl von Linné's research, which suggested that some species known to antiquity were no longer extant. See also Friedrich Justin Bertuch.

The Italian nobleman Camillo MARCOLINI (1739–1814) enjoyed a brilliant political career under Frederick Augustus I (1750–1827), king of Saxony. Marcolini became the director of fine arts of the Electorate of Saxony and its successor state, the Kingdom of Saxony. In 1774 he was appointed director of the famous Porcelain Factory at Meissen, Germany, the first European ceramics factory to produce hard-paste porcelain.

Beginning in 1734, Johann Christian GÖTZE (1692–1749) was royal librarian in Dresden, the capital of the German kingdom of Saxony. Having been educated in Vienna and Rome, Götze discovered in 1739 what became known as the **Dresden Codex** in Vienna, in the hands of a private collector who did not understand its significance. Götze was instantly aware of the document's value and included it in his library's collection. This Codex Mexicanus Vindobonensis (that is, the Codex Vienna) is the Mexica manuscript to which Humboldt refers here. It had been sent to Europe during the early colonial period and is now preserved at the Austrian National Library in Vienna. This is the only surviving precontact Mixteca codex. (The Mixteca lived in Central America in what is now northern and western Oaxaca, and also in southern Mexico in the states of Guerrero and Puebla).

267. *Quipu* (or *khipu*) are record-keeping and counting devices that the preconquest Inca made from knotted textile strings. The largest collection of *quipu* is reputedly housed at the Museum für Völkerkunde (Museum of anthropology) in Berlin—it was renamed the Ethnological Museum in 1999. The Mexica did not use *quipu*, although they did have a counting device similar to the Chinese abacus that was in use circa 500 BCE (its earlier origins are Japanese). The Aztec abacus, which archaeologists date around 900–1000, consists of a wooden frame on which were mounted strings threaded with kernels of corn.

Editor of the Weimar-based *Asiatisches Magazin* (Asiatic journal), Julius Heinrich von KLAPROTH (1783–1835) knew several Asian languages and engaged in comparative linguistics and philology. He was also interested in geography, ethnography, and Asian history. Klaproth traveled to China in 1805–6 and 1806–7; he also taught in St. Petersburg, Russia, for a time. In 1821 he became a foreign associate of the new Société Asiatique (Asiatic Society) in Paris, where he had resided since 1815. Humboldt was in fruitful contact with Klaproth—an association that dated back to Humboldt's frustrated attempts to visit Asia in the 1810s.

268. Fluent in ten languages, Joseph von HAMMER-PURGSTALL (1774–1856) served as an interpreter for languages of the Middle East. Educated in Graz and Vienna, he joined the Austrian diplomatic service in 1796, and, in 1799 was sent to Constantinople (Istanbul) as an interpreter. He moved on to Egypt, where he participated in the English and Turkish campaigns against the French. His notable contribution to literary history was the journal *Fundgruben des Orients* (Treasure troves of the Orient); it was published in six volumes in Vienna between 1809 and 1819.

268n. Daniel von NESSEL (1644–99) was librarian to the emperor of Austria. In 1690 Nessel compiled the catalog of Greek and Asian manuscripts to which Humbold refers here, on the basis of the work of **Peter Lambeck**. Nessel was a friend of the German philosopher Gottfried Wilhelm Leibniz (1646–1716).

273. An officer in the Russian Navy, Adam Johann von KRUSENSTERN (1770–1846) received permission for a circumnavigation of the world which took place from 1803 to 1806. Krusenstern thus came to head the first Russian expedition that fully explored the Pacific Ocean. Aboard one of his ships was Otto von Kotzebue (1787–1846), who was himself to head two expeditions around the world between 1815 and 1826. Humboldt made use of Kotzebue's travelog when preparing the freestanding edition of his *Political Essay on the Island of Cuba* (1826).

274. During his American journey Humboldt not only ascended Mount Teide in Tenerife while on his way to the New World and Chimborazo near Quito but also climbed many other volcanic mountains in the Andes. Indian guides were indispensable for most of these expeditions. Among the many helpers, most anonymous, Felipe ALDAS stands out as the only one to follow Humboldt up to the volcanic crater of Rucupichincha, the famous peak of Mount Pichincha (see Plate LXI), on his second of three ascents on May 26, 1802. The expedition turned out to be so danger-ous that Humboldt and his Indian companion almost fell into the volcano's steaming sulfur crater while trying to cross a snow-covered passage. Humboldt narrates the Pichincha episode in vivid detail in his travel diaries, constantly referring back to the accounts of **La Condamine**, who had seen the volcano's crater some fifty years ear-lier and described it in his famous *Relation abrégée d'un Voyage fait dans l'Interieur de l'Amerique Méridionale* (Abridged account of a voyage in the interior of South America, 1747).

275. During the time when Humboldt was writing *Views of the Cordilleras*, the QUEEN OF PRUSSIA was Louise Augusta Wilhelmina Amalia von Mecklenburg-Strelitz (1776–1810), who posthumously came to embody the virtues of the good Prussian woman. Reportedly she had had a positive influence on King Friedrich Wilhelm III (1770–1840), whom she married in 1793. They were enthroned in 1797. Queen Lou-ise (or Luise), as she is commonly called, knew the Humboldts, particularly the older brother Wilhelm (1767–1835), who was instrumental in establishing a new type of university in Berlin around the time of Louise's death. The Prussian state reorganized itself, especially its system of education, after military defeat at the hands of Napoleon (1769–1821) in 1806, focusing on the production of knowledge as a means to regain the prestige that had been lost on the battlefield.

276. The career of Pierre Louis Antoine CORDIER (1777–1861), a French geologist and inspector of mines, closely mirrored Humboldt's own, at least to the extent that Cordier was a central figure in Parisian scientific circles and also a traveler who had seen much of Europe, Egypt, and, as Humboldt notes here, Tenerife. Cordier and Humboldt formed a close lifelong friendship. Cordier had been part of the Napole-onic campaign in Egypt and had also traveled to Gibraltar and Tenerife in 1803. In the Canary Islands he climbed the Teide volcano, which had been the starting point for Humboldt's American journey four years earlier.

280. CODEX TELLERIANO-REMENSIS, also known as Codex (Mexicanus) Telle-rianus and Códice Le Tellier, is a pictorial manuscript on European paper in fifty leaves, or one hundred pages—Humboldt miscounted, failing to include three end pages and one blank page. This codex has three major pictorial sections in various in-digenous styles and is annotated in Spanish in several handwritings. The manuscript includes an eighteen-month calendar, a 260-day divinatory calendar, pictorial annals for 1198–1562 in two major styles, and historical notices in Spanish, without drawings, for 1519–57. Humboldt reproduces this codex for the first time in Plates LV and LVI. He was also the first to provide informed comments on its history and contents and to point out the similarities between this codex and Codex Vaticanus A; some leaves that are missing from Codex Telleriano-Remensis are preserved in Codex Vaticanus A.

281. Núño Beltrán de Guzmán, or NÚÑO DE GUZMÁN (c. 1490–1544), was presi-dent of the First Audiencia (created in 1528) at Veracruz, the judicial and administrative body that replaced **Cortés**'s rule in New Spain. In this role he repeatedly clashed with Cortés and the Franciscans, including Bishop **Zumárraga**, who was responsible for his removal from power in 1530. In 1525 Beltrán de Guzmán had been appointed governor of the northern province of Panuco, where he was notorious for selling the indigenous population as slaves to traders in Saint-Domingue. Beltrán de Guzmán began the expansion of Spanish control into northern and western Mexico through murder and pillage of indigenous settlements. His actions prompted the famous rebellion against the Spanish known as the Mixton Wars (1532–42).

282. Fascinated by comets, French Catholic priest and astronomer Alexandre Guy PINGRÉ (1711–96) observed the transits of Venus in 1761 from the Island of Rodriguez in the Indian Ocean, between Mauritius and Madagascar, and in 1769 from Saint-Domingue. Pingré included the results from his Venus observations in his 1783 book on cometography.

286n. STOBAEUS, also known as John of Stobi (fl. fifth century), compiled an anthology of excerpts from Greek (pagan) poets and prose writers in the early fifth century. Intended for the instruction of his son Septimius, the excerpts—on subjects such as philosophy, physics, rhetoric, poetry, ethics, and politics—were arranged in four books grouped later under the titles *Eklogai* (Selections) and *Anthologion* (Anthology). The latter still exists, preserving quotations from works that would have otherwise have been unknown.

291. Juan Pio Montúfar y Larrea (1758–1818), MARQUIS OF SELVALEGRE, was the head of the 1809 Junta of Quito. His father, Juan Pio Montúfar y Frasso (d. 1761), had been president of the Quito Audiencia from 1753 to 1761. His son Carlos Montúfar (1780–1816) would accompany Humboldt and Bonpland to climb Pichincha and Chimborazo, and also during their journey to New Spain and back to Europe. In 1810 Carlos returned from Spain to Quito, where, as a representative of the Spanish Council of Regents, he tried to reach an agreement with members of the 1809 junta. Carlos managed to create a government of reconciliation, naming his father vice president.

296. One of the representatives of the Spanish crown for the famous scientific expedition led by **Godin**, the Spanish naval officer and mathematician George Juan, or JORGE JUAN y Santacilia (1713–73), reached Cartagena in 1735. Like his colleague **Ulloa,** who was part of the same expedition, Juan wrote an account of his travels, the five-volume *Relación histórica del viaje a la américa meridional* (Historical narrative of the voyage to South America, 1748). The *Relación* included a series of original maps and plans of cities in South America. In addition to several works published together with Ulloa, Juan wrote *Compendio de navegación* (Navigational compendium, 1757), *Examen marítimo theórico práctico* (Theoretical and practical examination of nautical subjects, 1771), and *Estado de la Astronomía en Europa* (The state of astronomy in Europe, 1774). Juan eventually became Spanish ambassador to Morocco.

298. Owner of a coffee plantation near Caracas, Andrés de IBARRA met Humboldt in 1800, when the Prussian scientist was exploring the New Granada region and was preparing to climb Mount Silla. While Humboldt's account of the Silla expedition (in chapter 13 of his *Relation historique*) only mentions Ibarra briefly, calling him a "gentle host," the notes of his travel diary tell a different story. Disgusted by the way that most of the Creole landowners in Venezuela were treating their African slaves, Humboldt harshly criticized Ibarra—along with many other *latifundistas*—for his open racism, commenting on his attempts to prohibit the exercise of local crafts by free people of color.

300. Humboldt's reference here is to a phrase—sometimes translated as "I squatted a-knee"—in the short dialogue "Lexiphanes" (Word-monger) by the Assyrian orator LUCIAN of Somosata (c. 115–c. 200), who wrote about eighty works in Greek, most of them satirical in tone. They include essays, speeches, letters, dialogues, and stories. His literary dialogues fuse old comedy and popular philosophy. Lucian's works were important in the development of three literary genres: satirical dialogue, the imaginary voyage (including very early science fiction), and the dialogue of the dead. He influenced several Renaissance artists, among them Sandro Botticelli (1445–1510) and Sodoma (1477–1549).

EROTIANUS, or Erotian, was a grammarian, possibly a physician, from the first century CE. Author of *Glossaria in Hippocratem* (Hippocratic glossary), the most famous Hippocratic lexicon of antiquity. Eortianus most likely lived in Rome during the reign of Nero (54–68).

The philologist Tiberius HEMSTERHUIS (1685–1766) was a professor of Greek affiliated with various Dutch universities, including the one in Leiden. Among his writings were notes to "Alberti's Hesychius." Hesychius of Alexandria (fl. fifth century) compiled a lexicon, basing it on older works but also adding original terminology. Joannes Alberti (1698–1762), professor of theology in Leiden, published a new edition of the lexicon in 1746. Tiberius's son François Hemsterhuis (also Franciscus or Frans; 1721–90) wrote on aesthetics and moral philosophy; he influenced important German thinkers such as Johann Gottfried von Herder (1744–1803). See also David Rhunken.

300n. Today an art gallery, VILLA BORGHESE was a state museum in Rome. A villa with surrounding gardens, it displays Italian baroque paintings and ancient sculpture. The building was begun by Flaminio Ponzio (c. 1559–1613) and, after Ponzio's death, continued by the Dutch architect Jan van Santen (Giovanni Vasanzio; 1550–1621). The paintings were mainly collected by Pope Paul V (Camillo Borghese; 1552–1621) and his nephew Cardinal Scipione Cafarelli (1576–1633), whom Camillo had adopted into the Borghese family. Scipione Borghese, as Cafarelli is typically called, had the

Villa Borghese built. The antiquities, in turn, were primarily obtained by Marcantonio Borghese IV (1730–1800), who had the Villa Borghese renovated, and by Francesco Borghese (1776–1839). The Italian government acquired the collection and the property from the Borghese family in 1902.

303. David RHUNKEN (1723–1798) was born in Pomerania and studied Greek under **Tiberius Hemsterhuis** at Leiden. He had previously studied at Göttingen, one of Humboldt's alma maters. By 1757 Rhunken had become Hemsterhuis's assistant, and in 1761 he was called to a chair of Latin and history at Leiden.

304. DAEDALUS is a mythical Greek architect to whom buildings and statues were attributed but whose origins have been lost. Daedalic sculpture, a phase in early Greek art, is named in his honor; the Greek *daedalus* means "skillfully wrought." Daedalus cursed his skill after he made waxen wings for his son Icarus, who died flying too closely to the sun with them.

THESEUS is a hero of Attic legend who lived through many adventures, such as triumphing over a minotaur, a creature half man and half bull. Theseus and his second wife Phaedra live on in Greek tragedies.

Born is what is now Spain, Lucius Iunius Moderatus COLUMELLA (fl. 50 CE) was a Roman soldier turned farmer in Italy. The owner of several estates near Rome, Columella is the author of the most systematic agricultural manual that survives. Written around 60–65 CE in twelve books, *De re rustica* (On agriculture) advocated a particular type of agricultural management: slave-staffed estates characterized by capital investment, close supervision by the owner, and integration of crop cultivation and animal husbandry. Another of his extant works is a book on trees, *Liber de arboribus*, which was probably part of a shorter manual on agriculture. His works criticizing astrologers and on religion in the context of agriculture did not survive.

307. Swiss-born Johann Leonhard HUG (1765–1846) was an important Catholic New Testament scholar and professor of Asian languages at the German University of Freiburg im Breisgau. He is particularly noteworthy for having established bridges between Catholic scholarship and Protestant biblical research.

308. The French mathematician Jean Baptiste Joseph FOURIER (1768–1830) was part of the 1798 French scientific expedition to Egypt which led to the discovery of the Rosetta Stone, key to deciphering the **hieroglyphic inscriptions of the Egyptians,** which Jean François Champollion (1790–1832), one of Fourier's disciples, accomplished in 1822. Previous work by Thomas Young (1773–1829) needs to be acknowledged here as well: Young had, for example, identified the name Ptolemy on a cartouche in the Rosetta Stone by 1816. Fourier was one of the authors of the multivolume *Description de l'Égypte* (Description of Egypt), a revolutionary study of Egypt that appeared between 1809 and 1828.

NABONASSAR was a Babylonian king who ascended to power in 747 BCE. He appears to have initiated the tradition of keeping systematic astronomical records in

Babylon, the ancient cultural center in southeast Mesopotamia between the rivers Euphrates and Tigris (today southern Iraq).

309. A Greek mathematician best known for his discovery of the precession of the equinoxes, HIPPARCHUS (c. 170/190–c. 120 BCE) is credited with inventing trigonometry while transforming astronomy from a theoretical to a practical science, uniting Greek and Babylonian astronomical traditions. Hipparchus's recorded astronomical observations ranged from 141 to 127 BCE; they resulted in a catalog of about 850 stars based on their brightness. Hipparchus accurately estimated the lunar distance and the length of the year and the month; he predicted celestial positioning and calculated lunar and solar eclipses. He was purportedly the first to use latitude and longitude systematically. The inventor of several instruments and the founder of observatories in Rhodes and Alexandria, Hipparchus wrote several now-lost treatises that survive only through references by **Ptolemy**. His only surviving work is his commentary on **Aratus**'s astronomical poem *Phaenomena*, based on a treatise by Eudoxus by the same name.

VITRUVIUS (fl. 70–15 BCE) was a Roman architect and engineer and the author of the ten-book *De architectura*—commonly known as *Ten Books on Architecture*. The only major treatise on classical antiquity to survive, *De architectura* contains a plethora of information on Greek and Roman art and architecture. A continuing authoritative source on classicism, Vitruvius's work was possibly written between 33 and 14 BCE. *De architectura* has been enormously influential since the Renaissance. The famous drawing *Vitruvian Man* by Leonardo da Vinci (1452–1519), in which the naked male body is circumscribed in a circle and a square, is based on Vitruvian's image of a man as the measure of proportion.

311. There were two men by the name of Aldus MANUTIUS—the Elder (1449–1515) and the Younger (1547–97). Humboldt refers to the Elder, whose Italian name was Aldo Manuzio il Vecchio, originally Teobaldo Manucci. The founder of a printing and publishing dynasty (Aldine Press), the Italian Manutius was a leading figure in printing, publishing, and typography. He was responsible for many early print editions of Greek and Latin classics, notable for their high quality at low prices. Aldine Press produced some one thousand editions during the course of the sixteenth century.

Postumius Rufius Festus AVIENUS (fl. c. 400) was a Roman writer of about forty-two fables and the proconsul of Africa in 366 CE. His three didactic poems are extant in whole or in part. His first work was an expanded translation, from the Greek, of **Aratus**'s poem *Phaemonema*, along with more than seven hundred lines of ancient commentaries. Avienus's geographical poem *Descriptio orbis terrae* (Description of the world) is a translation of Dionysius's *Periegetes: Ora maritima*, which partially survives in 703 iambic lines. *Descriptio* is a description of the coasts of the Atlantic, Mediterranean, and Black Seas.

One of the characters in **Macrobius**'s *Saturnalia*, SERVIUS, or Maurus or Marius Servius Honoratus, was a Latin grammarian and literary commentator of the fourth century. Servius's commentary on Virgil survived in two forms: a shorter and a longer version. Servius used the shorter version for didactic purposes; the augmented version is known as *Servius Auctus*, *Servius Danielis*, or *Scholia Danielis*.

Born in Judea, ST. EPIPHANIUS (c. 310-403) was appointed bishop of Salamis (Constantia, Cyprus) in 367 CE. Intolerant of even the slightest hint of heresy, he defended orthodox belief in his monumental *Panarion* (Medicine chest), commonly known as *Against All Heresies*. In this treatise he mentions that the Pharisees translated the zodiac sign Libra as "Moznaim." One of the chief editions of Epiphanius's works is the one by **Petau** from 1622.

ERATOSTHENES of Cyrene (c. 285-194 BCE) succeeded **Apollonius of Rhodes** as the head of the Alexandrian Library. Most of his work survived through Strabo and in quotations and references from later writers. Eratosthenes is renowned for his work on chronology, mathematics, and descriptive geography. His *On the Measurement of the Earth*, part of his *Lectures on Astronomy*, used mathematical geography to closely calculate the actual circumference of the earth. It has survived only in the writings of Cleomedes (fl. first century BCE), who used Eratosthenes's measurements for his *Meteora graece et latine*.

312. The ARAUCANIANS (now known as Mapuche) are a group of indigenous peoples in present-day central-southern Chile and adjacent parts of Argentina. The Araucanians have struggled for their independence since the arrival of the Spanish. As early as 1553, they inflicted a crushing defeat on the Spaniards at Tucapel, and their resistance to the conquest checked the movement of the Spanish south of Peru. The Mapuche's struggle for independence, which has lasted for more than three centuries, officially ended around 1881-85 during what is known in Chile as the "Pacification of Araucania" and in Argentina as the "Desert Campaign." In Chile, the Mapuche were defeated and dispersed into small rural communities in a lengthy process (1884-1929). In Argentina, the Desert Campaign resulted in the death of more than thirteen hundred Mapuche; about twelve thousand were taken captive and one thousand resettled. Around 1910, the Argentine state launched another such campaign against indigenous peoples in the north. Today the Mapuche continue to struggle for the recognition of their land and religious rights.

313. Savilian Professor of Astronomy at Oxford beginning in 1619, John BAIN-BRIDGE (1582-1643) was an astronomer and physician whose major work, *Canicularia*, was published posthumously in 1648. During his time at Oxford, he became interested in Arabic astronomical works and studied the language so as to gain access to that knowledge and apply it in tracing inaccuracies in ancient Greek astronomical observations.

Louis Jouard de la NAUZE (1696-1773) published several articles in the *Mémoires* of the French Royal Academy in the mid-eighteenth century. He joined the Jesuit order for a time and, in 1729, became a member of the French Academy. Many of his publications deal with ancient chronology, a topic about which he liked to argue with

Nicolas Fréret, though rarely with any success. Nauze also exchanged letters with the Jesuit Father **Souciet**.

The Greek mathematician and astronomer THEON of Alexandria (fl. c. 360-80) was a fourth-century commentator on **Ptolemy**. Both his *Petit commentaire* (Little commentary) and his *Grand commentaire* (Long commentary) on Ptolemy's *Handy Tables* survive, the latter in fragments; they were translated into French in 1978 and 1999, but not into English. We have only parts of Theon's commentary on Ptolemy's *Syntaxis* (Arabic *Almagest*). Theon's writings were translated into Arabic in the early ninth century, making Ptolemy's work available to Islamic scientists and preserving it for medieval Europe.

A contemporary of **Ptolemy**, VETTIUS VALENS (fl. c. 140-170) worked as a professional astrologer in Alexandria, Egypt, and had an astrological school. He authored *Anthologiae* (possibly as a textbook), in which he drew on Egyptian mythology and presented complete horoscopes.

Johann Christoph GATTERER (1727-99) held the chair in history at the University of Göttingen for forty years. Humboldt was apparently influenced by Gatterer, whose ideas on geography and chronology he would first have encountered at Göttingen in 1789-90. Humboldt applied Gatterer's ideas of drainage basins to the rivers of Mexico. Gatterer was also a teacher of **August Ludwig von Schlözer**.

A student at the seminary in Rimini and the University of Bologna, Luigi Gaetano MARINI (1740/42-1815) was a Vatican archivist and librarian. The 1795 publication to which Humboldt is referring here was a commentary on marble sculpture fragments of the so-called Acts of the Arval Brothers, dating from the time of Roman emperor Elagabalus (also Heliogabolus, byname of Caesar Marcus Aurelius, original name Varius Avitus Bassianus; 203-22). The Arval Brothers, or Fratres Arvales, were a college or priesthood in ancient Rome, chiefly occupied with annual public sacrifice for the fertility of the fields. The fragments had been discovered in 1776 when Pope Pius VI (1717-99) set out to have foundations laid for a new sacristy for St. Peter's in Rome. Marini had worked in the Vatican Museum and Library since 1772; in 1800 he became the head of the library and the archives. When Napoleon Bonaparte (1769-1821) took the Vatican archives to Paris in 1810, Marini was ordered to come along. He died before the archives were returned after Napoleon's fall. Marini sold his own library and abandoned his studies for meditation.

314. Lucius Cornelius SULLA, "Felix" (the Fortunate, 138-78 BCE), is one of the most controversial figures of the late Roman Republic. Both a skillful diplomat and an unbeaten military leader, he has often been portrayed as a political despot whose march on Rome in 83 BCE exposed the weakness and limited power of the Senate, eventually initiating the political populism that was to culminate in the dictatorship of Julius Caesar. This critical perspective on Sulla's legacy can be traced back to his first biographer, **Plutarch**, who—while assigning him a place among the great Roman leaders in his series *Parallel Lives*—depicts him as a ruthless warlord. Recent scholarship has given closer attention to Sulla's republican beliefs, the political reforms during his short restorative dictatorship (82–80 BCE), and his skillful "politicization of the Roman army" (Arthur Keaveney).

315. According to **Ptolemy**'s mathematical treatise *Almagest* (The great book), TIMO-CHARIS of Alexandria made astronomical observations between 295 and 272 BCE, recording the positions of stars, lunar occultations, and the passage of Venus. The earliest examples of systematic Greek astronomical observations, his work served as a basis for subsequent astronomical work, notably that of Ptolemy and **Hipparchus**. A crater of the moon is named after him.

Abbé Noël Antoine PLUCHE (1688–1761) was a professor of humanities and subsequently of rhetoric in Rouen. His *Histoire du ciel* (History of the heavens) was, among other things, intended to refute the theories of modern thinkers such as Sir Isaac Newton (c. 1642–1727) and René Descartes (1596–1650). Pluche's writings on natural history have a theological foundation and were widely reprinted and translated.

On July 5, 1802, the abbot Domenico TESTA delivered a paper on the Egyptian zodiac to a meeting of the Academy of the Roman Catholic Religion; it was published in Italy that same year and came to Humboldt's attention. From 1793 to 1795 Testa was involved in a debate with two other abbots, Giovanni Battista Alberto Fortis (1741–1803) and Giovanni Serafino Volta (late eighteenth century). A "Testa D." was active in Italian geology in the Venice region in 1793.

The Egyptian priest MANETHO (fl. c. 300 BCE) was the author of *Aegyptiaca*, a history of Egypt written in Greek, which might have been commissioned by **Ptolemy II Philadelphia**. Only fragments survive, showing that Manetho used oral and local written sources to supplement **Herodotus**'s account of Egypt. Manetho's work is not of use so much for chronology as for dividing the Egyptian rulers into thirty-one dynasties.

A native of North Africa, Martianus Minneus Felix CAPELLA (fl. fifth century) wrote prose and poetry that had a notable cultural influence until the late Middle Ages. His *De nuptiis Philologiae et Mercurii* (*On the Marriage of Philology and Mercury*) is an allegorical treatise on different types of classical learning.

Between the second and fourth centuries, the Roman Lucius AMPELIUS authored the *Liber memorialis* (Book of memory), from which Humboldt quotes. It is mostly a geographical and mythological work, but it also includes a historical summary up to the time of **Trajan**.

315n. The Swiss antiquarian and diplomat Friedrich Samuel VON SCHMIDT (aka Baron Smith; 1737–96) was director of the library of the grand-duke of Baden in Germany. Little is known about the Swedish astronomer and philologer Carl Gottlieb SCHWARTZ (1757–1824), whose name some, including Humboldt, spelled Swarz. He studied at the University of Uppsala. His *Recherches sur l'origine et la signification des constellations de la sphère grecque* (Research on the origin and the meaning of the constellations in the Greek sphere) was originally written in Swedish and appeared in two French editions in 1807 and 1809. The French historian of mathematics Jean Etienne MONTUCLA (1725–99) wrote the first comprehensive history of mathematics,

Histoire des mathématiques dans laquelle on rend compte de leurs progrès depuis leur origine jusqu'à nos jours (History of mathematics with an account of its development from the origin to our days). This study, which also discusses astronomical topics, saw print between 1799 and 1802.

Johann Konrad SCHAUBACH (1764–1849) was a historian of ancient astronomy and a professor at Meiningen, Germany. He authored *Geschichte der Griechischen Astronomie bis auf Eratosthenes* (History of Greek astronomy up to Eratosthenes, 1802).

316. Governor of Grenada (West Indies) and Madras (Chennai), Lord George MACARTNEY (1737–1806), the first Earl Macartney, headed the first formal British diplomatic mission to China between 1792 and 1794. On his China mission, which was far less successful than his earlier activities as an envoy to Russia had been, Macartney was accompanied by Sir George L. Staunton (1737–1801), a diplomat with a French medical education. The two had first met in 1779, when Staunton had negotiated Macartney's release (he had been taken prisoner by the French when governor of Grenada). Both men were of Irish extraction. In 1797 Staunton published the book about their diplomatic mission, *Authentic Account of an Embassy from the King of Great Britain to the Emperor of China*, to which Humboldt is referring here.

A disciple of Johann Gottfried von Herder (1744–1803) and the leading orientalist of his day besides **Silvestre de Sacy**, Friedrich MAJER (1772–1818) was an important Indic encyclopedist and catalyst for romanticism in the German university town of Jena. Humboldt refers here to Majer's *Allgemeines Mythologisches Lexicon* (General mythological encyclopedia) from 1803–4. Majer introduced the philosopher Arthur Schopenhauer (1788–1860), whom he had met in 1813–14, to Indian antiquity. It is also likely that Majer inspired **Friedrich Schlegel** in his pioneering study of Sanskrit.

317. The Spanish Dominican friar Juan de MARIETA was the author of the *Historia eclesiástica de España* (Ecclesiatic history of Spain, 1596). Carlos de Tapia ZENTENO (1698–c. 1770), a clergyman from New Spain, wrote Nahuatl and Aztec grammars, notable among them *Arte novissima de la lengua mexicana* (Latest grammar of the Mexican language, 1753) and *Noticia de la lengua Huasteca* (Grammar of the Huastec language, 1767). A priest at the Tampamolon parish, Tapia Zenteno became the professor of Nahuatl at the **University of Mexico**.

319. The Dominican friar Gregorio GARCÍA (fl. 1607–27, d. 1627) was a missionary in New Spain and Peru; he lived in New Spain for about twelve years and in Peru for about eight. Interested in the earliest history of Peru, he studied extensively the origin of the indigenous American peoples. In *Origen de los Indios del Nuevo Mundo e Indias Occidentales* (Origin of the Indians of the New World and West Indies, 1607), he argued that the early inhabitants of the Americas could only have come from Europe, Africa, or Asia. On the basis of a supposed similarity between the Jews' and the Amerindians' character, nature, and customs, García proposed that the latter were the descendants of the ten lost tribes of Israel who had reached America by sea and land. He was also the author of *Historia eclesiástica seglar de la India Oriental y*

Occidental (Ecclesiastic and lay history of the East and West Indies, 1625), in which he discussed evidence of pre-Hispanic evangelization in America.

320. PAPPUS of Alexandria (fl. 320 CE) was the most important mathematician writing in Greek in the late Roman empire. His voluminous account of the most important mathematical works in ancient Greek was called *Synagoge* (Collection). After *Synagoge* had been printed in a late-sixteenth-century Latin edition, Pappus became widely known among European mathematicians. His influence may be traced in the works of René Descartes (1596-1650) and Isaac Newton (1642-1727), among many others.

French clergyman and historian François DU CREUX (1596-1666) wrote a history of Canada in Latin from archival sources. His *Historiae Canadensis* (1664) includes a map titled "Tabula Novae Franciae," which illustrates the location of Jesuit missions in Canada. This map is available online at the Wisconsin Historical Society's Historical Maps Collection. Du Creux was admitted to the Society of Jesus in 1614. After teaching for twelve years, he worked on Greek and Latin grammars and on publishing *Historiae*, an English edition of which appeared in 1951.

321. William WILSON is the compiler of *A Missionary Voyage to the Southern Pacific Ocean* (1799), which Humboldt mentions here. James Wilson was the captain of that journey, which took place in 1796, 1797, and 1798 on the ship *Duff*. The actual driving force behind the book's publication was John Love (1757-1825), a Presbyterian minister from London and secretary of the Glasgow Missionary Society.

The historian Barthold Georg NIEBUHR (1776-1831) was a son of the well-known Danish explorer and geographer Carsten Niebuhr (1733-1815). In his *Römische Geschichte* (Roman history), one of the seminal works of nineteenth-century German historiography, Niebuhr introduced critical methods into historical research. Niebuhr knew the Humboldts, particularly Alexander's older brother Wilhelm, who had been the Prussian ambassador to Rome.

James TASSIE (1735-99) was a Scottish modeler and portrait medallionist who, in good family tradition, had started out training as a mason. He soon switched to sculpture. In the early 1760s he moved from Glasgow to Dublin, where he met the physician Henry Quin (1717/8-91), an enthusiastic amateur of classical cameos and intaglios. Tassie joined up with Quin and relocated to London, where he applied the results of his work with Quin and was an almost immediate success in the jewelry business. In 1781 Catherine the Great of Russia (1729-96) placed an order with him. Tassie visited Parisian collections in search of new items for reproduction; he also turned to making portraits of contemporaries. In 1791 the German mineralogist Rudolf Erich RASPE (1737-94), better known as the author of *Baron Münchausen's Marvellous Travels*, published a catalog of Tassie's stock. Tassie had engaged Raspe's services in 1785, the year that the first edition of the Münchhausen stories was published. Raspe, who had been a librarian at several German-speaking universities, had fled to England in 1775 after being accused of stealing from the gem collection of the *landgraf* of Kassel.

PLATES
FROM THE ORIGINAL EDITION

I. Page 18. *Buste d'une Prêtresse Aztèque*. Drawing from the Academy of Painting in Mexico City, based on the original sculpture from the collection of Guillaume Dupaix. Engraved in Paris by the French printmaker Raphael Urbain Massard (1775–1843).

II. Page 19. *Buste d'une Prêtresse Aztèque. Vue par derrière*. Drawing from the Academy of Painting in Mexico City, based on the original sculpture from the collection of Dupaix. Engraved in Paris by Massard.

III. Page 23. *Vue de la grande place de Mexico*. Drawn by Raphael Ximeno y Planes (1759–1825), who taught painting at the Academy of San Carlos in Mexico City. Etched in Paris by the engraver Louis Bouquet (1765–1814).

IV. Page 26. *Ponts naturels d' Icononzo*. Drawing based on a sketch by Humboldt and engraved by Wilhelm Friedrich Gmelin in Rome.

V. Page 32. *Passage du Quindiu, dans la Cordillère des Andes*. Drawn by Josef Anton Koch (1768–1839) in Rome, on the basis of a sketch by Humboldt. Engraved in Stuttgart by Christian Friedrich Traugott Duttenhofer (1778–1846).

VI. Page 37. *Chute du Tequendama*. Drawing based on a sketch by Humboldt and engraved by Wilhelm Friedrich Gmelin in Rome.

VII. Page 44. *Pyramide de Cholula*. Drawing by Wilhelm Friedrich Gmelin, based on a sketch by Humboldt. Engraved by the Berlin artist Anton Wachsmann (1765–1836) and Friedrich Arnold (1786–1854) in Berlin.

VIII. Page 56. *Masse détachée de la Pyramide de Cholula*. Drawing by Pierre Jean François Turpin (1775–1840) in Paris, based on a sketch by Humboldt. Engraved by Pietro Parboni (1783–1841) in Rome.

IX. Page 58. *Monument de Xochicalco*. Drawing by Francisco Aguera Bustamante (1779–1820) in Mexico City in 1791. Engraved by Pinelli in Berne.

X. Page 63. *Volcan de Cotopaxi*. Drawing by Wilhelm Friedrich Gmelin in Rome, based on a sketch by Humboldt. Engraved by Friedrich Arnold in Berlin.

XI. Page 69. *Relief Mexicain trouvé à Oaxaca*. Drawing provided by Vicente Cervantes. Engraved by F. Pinelli in Rome.

XII. Page 74. *Pièce de procès en Écriture hiéroglyphique—Généalogie des Princes d'Azcapozalco*. Drawing provided by Vicente Cervantes. Engraved by Pinelli in Rome. Colorized.

XIII. Page 87. *Manuscrit hiéroglyphique Aztèque conservé a la Bibliothèque de Vatican*. Colorized.

XIV. Page 108. *Costumes dessinés par des Peintres Mexicains du Temps de Montezuma*. Colorized. These images correspond to Codex Vaticanus A 3738; they represent those of the codex but are not exactly the same. Humboldt amalgamated the images of folios 8 through 15 in one page. He did not incorporate the comments on the pages.

XV. Page 111. *Hiéroglyphes Aztèques du Manuscrit de Veletri.* Drawn and engraved by Fr. Pinelli in Rome.

XVI. Page 123. *Vue du Chimborazo et du Carguairazo.* Drawn by Wilhelm Friedrich Gmelin in Rome, based on Humboldt's sketch, and engraved by Friedrich Arnold in Berlin.

XVII. Page 129. *Monument Péruvien du Cañar.* Drawn by Wilhelm Friedrich Gmelin in Rome, based on Humboldt's sketch, and engraved by Louis Bouquet in Paris.

XVIII. Page 134. *Rocher d'Inti-Guaicu.* Drawn by Josef Anton Koch in Rome, based on Humboldt's sketch, and engraved by Christian Friedrich Traugott Duttenhofer in Stuttgart.

XIX. Page 137. *Ynca-Chungana du Jardin de l'Inca près de Cañar.* Drawn by Wilhelm Friedrich Gmelin in Rome, based on Humboldt's sketch, and engraved by the painter François Morel (Francesco Morelli) in Rome.

XX. Page 140. *Intérieur de la Maison de l'Inca au Cañar.* Drawn by Wilhelm Friedrich Gmelin in Rome, based on Humboldt's sketch.

XXI. Page 144. *Bas-relief Aztèque de la Pierre des Sacrifices, trouvée sous le pavé de la grande Place de Mexico.* Drawn by Guillaume Dupaix in Mexico City, based on Humboldt's sketch, and engraved by Raphael Urbain Massard in Paris.

XXII. Page 149. *Rochers basaltiques et Cascade de Relga.* Drawn by Wilhelm Friedrich Gmelin in Rome, based on Humboldt's sketch, and engraved by Louis Bouquet in Paris.

XXIII. Page 155. *Relief en basalte représentant de Calendrier Mexicain.* Engraved by Jean-Baptiste-Antoine Cloquet (d. 1828) in Paris.

XXIV. Page 222. *Maison de l'Inca, à Callo, dans le Royaume de Quito.* Drawn by Wilhelm Friedrich Gmelin in Rome, based on Humboldt's sketch, and engraved by Louis Bouquet in Paris.

XXV. Page 225. *Le Chimborazo vu depuis le Plateau de Tapia.* Colorized. Drawn by Jean-Thomas Thibault (1757–1826) on the basis of Humboldt's sketch and engraved by Louis Bouquet.

XXVI. Page 234. *Époques de la Nature, d'après la Mythologie Aztèque de la Bibliothèque du Vatican.* Drawn by Pinelli in Rome.

XXVII. Page 239. *Signes hiéroglyphiques des jours de L'Almanach Mexicain—Peinture Hiéroglyphique tirée du Manuscrit Borgien de Veletri.* Colorized. Drawn by Pinelli de Roncalli in Rome.

XXVIII. Page 242. *Hache Aztèque.* Drawn and engraved by Friedrich Arnold in Berlin.

XXIX. Page 245. *Idole Aztèque de porphyre basaltique, trouvée sous le pavé de la grande Place de Mexico.* Drawn in Mexico City by Francisco Aguera Bustamante and engraved by Jean-Baptiste-Antoine Cloquet in Paris.

XXX. Page 251. *Cascade du Rio Vinagre, près de Volcan de Puracé.* Drawn by Josef

HUMBOLDT'S LIBRARY

Humboldt's library consisted of nearly seventeen thousand volumes, most of which were destroyed in a warehouse fire at Sotheby's in London in 1865. We have no complete record, then, of the books Humboldt actually owned. Although Henry Stevens's sales inventory from 1863— *The Humboldt Library: A Catalogue of the Library of Alexander von Humboldt*—predates this fire, it is also incomplete because Stevens, who had bought Humboldt's library in 1860, had already sold off many items prior to the scheduled auction in 1865.

The following list of the sources Humboldt referenced directly or indirectly in *Views of the Cordilleras and Monuments of the Indigenous Peoples of the Americas* is a humble attempt at a partial reconstruction of his library. We have checked his references to the extent possible and have indicated variations and other noteworthy information in the bracketed comments that follow some of the entries. Titles are typically given in the original spelling (translations of many titles have been included in the annotations). Humboldt's often abbreviated references in the text of *Views* have been completed to help readers locate items in this bibliography.

Académie de Trévoux, Jesuits at the.1771. *Dictionnaire universel françois et latin: vulgairement appelé dictionnaire de Trévoux, contenant la signification & la définition des mots de l'une & de l'autre langue, avec leurs différens usages; les termes propres de chaque état & de chaque profession: la description de toutes les choses naturelles & artificielles; leurs figures, leurs espèces, leurs propriétés: L'explication de tout ce que renferment les sciences & les arts, soit libéraux, soit méchaniques, &c. Avec des remarques d'érudition et de critique; Le tout tiré des plus excellens auteurs, des meilleurs lexicographes, étymologistes & glossaires, qui ont paru jusqu'ici en différentes langues.* 8 vols. Paris: Compagnie des Libraires Associés. [The dictionary was published in four editions: 1704, 1721, 1752, and 1771. It was a revised edition of Antoine Furetière's 1690 *Dictionaire universel, contenant generalement tous les mots françois tant vieux que modernes, & les termes de toutes les sciences et des arts.* 3 vols. The Hague: Arnout & Renier Leers, 1690.]

Achilles Tatius. 1703. "Ex Achille Tatio, isagoge ad Arati phaenomena." In *Uranologion*, edited by Dionysius Petavius [Denis Petau]. Antwerp: G. Gallet. *See* Petavius, Dionysius.

Acosta, José de. 1584. *Doctrina christiana, y catecismo para instrvccion de los Indios, y de las de más personas, que han de enseñadas en nuestra sancta fé. Con vn confessionario, y otras cosas necessarias para los que doctrinan, que se contienen en la paginga siguiente. Compvesto por avctoridad del Concilio prouincial, que se cele-*

bro en la Ciudad de los Reyes, el año de 1583. Y por la misma traduzido en las dos lenguas generales, de este reyno, quichua, y aymara. Ciudad de los Reyes [Lima]: Antonio Ricardo. [Facsimile edition: Acosta, José de. 1985. *Doctrina christiana y catecismo para instrucción de los indios.* Madrid: Consejo Superior de Investigaciones Científicas.]

———. 1589. *De natura novi orbis libri duo, et De promulgatione Evangelii, apud barbaros, sive De procuranda Indorum salute libri sex.* Salmanticae: apud Guillelmum Foquel.

———. 1590. *Josephi Acostae e Societate Jesu De Christo revelato: libri novem.* Romae: apud J. Tornerium.

———. 1591. *Historia natural y moral de las Indias: en que se tratan las cosas notables del cielo, y elementos, metales, plantas, y animales dellas: y los ritos, y ceremonias, leyes, y gouierno, y guerras de los Indios.* Barcelona: Iayme Cendrat. [English edition: 2002. *Natural and Moral History of the Indies.* Edited by Jane E. Mangan, translated by Frances López-Morillas. Durham, NC: Duke University Press.]

———. 1592. *Josephi Acostae ex Societate Jesu De Temporibus novissimis libri quatuor.* Romae: ex typographia J. Tornerii.

Adelung, Johann Christoph, Johann Severin Vater, Friedrich von Adelung, and Wilhelm von Humboldt. 1806–17. *Mithridates, oder allgemeine Sprachenkunde.* 4 vols. Berlin: Vossische Buchandlung.

Aguado, Pedro de. 1916–7. *Historia de Santa Marta y Nuevo Reino de Granada.* Edited by Jerónimo Becker. 2 vols. Madrid: Jaime Ratés.

———. 1918–19. *Historia de Venezuela.* Edited by Jerónimo Becker. 2 vols. Madrid: Jaime Ratés. [Written around 1575–79. Manuscript is in two parts: "Primera parte de la recopilación historial resolutoria de Sancta Marta y Nuevo Reino de Granada de las Indias del Mar Oceano" and "Segunda parte de la istoria que conpuso fray Pedro de Aguado . . . descubrimiento y fundación de la gouernacion y provincia de Uenecuela." Not published until the twentieth century; first part (partial): *Recopilación historial* (Bogotá: Imprenta Nacional, 1906); second part: *Historia de Venezuela,* 2 vols. (Caracas: Imprenta Nacional, 1913). Humboldt mentions Aguado as one of Piedrahita's sources. *See* Piedrahita, Lucas Fernández de; Medrano, Antonio.]

Agustín, Antonio (Augustinus). 1654. *Antiquitatum Romanarum Hispanarumque in nummis veterum dialogi XI.* Edited by Andreas Schott. Antwerp: Henricum Aertssens. [First published in Tarragona (Spain) as *Dialogos de medallas, inscriciones y otras antiquedades* (1587).]

Albategnius. *See* Battānī, Muh̩ammad ibn-Jābir al-.

Alva Ixtlilxochitl, Fernando de. 1975–77. *Obras históricas.* Edited by Edmundo O'Gorman. 2 vols. Mexico City: Universidad Nacional Autónoma de México. [Written around 1600–40 in two parts: *Relaciones* and *Historia de la nación Chichimeca.* Lorenzo Boturini Benaducci found and copied the manuscripts in the 1730s; they are the basis for subsequent complete or partial editions. Humboldt must have consulted a copied manuscript bound together with fragmentary texts.]

Alvarado Tezozomoc, Fernando [Hernando] de. 1975. *Crónica mexicana.* Edited by Manual Orozco y Berra. Mexico City: Editorial Porrúa.

————. 1975. *Crónica Mexicáyotl: De F. Alvarado Tezozomoc*. Translated by Adrián León. Mexico City: Universidad Nacional Autónoma de Mexico. [*See also* Alvarado Tezozómoc, Fernando, and Adrián León. 1998. *Crónica mexicáyotl*. Primera serie prehispánica 3. Mexico City: Universidad Nacional Autónoma de México, Instituto de Investigaciones Históricas. *Crónica mexicana* was written in Spanish in 1598 and fully published in Mexico in 1878; *Crónica mexicáyotl* in Nahuatl in 1609. Humboldt quotes León y Gama as source for Alvarado Tezozomoc on his page 186. Partially published in Kingsborough's *Antiquities of Mexico*. *See also* Chimalpahin, León y Gama, Antonio de; Veytia, Mariano.]

Alzate y Ramírez, José Antonio. 1791. *Descripción de las antigüedades de Xochicalco. Dedicada a los señores de la acutal expedicion marítima al rededor del Orbe. Suplemento a la Gazeta de Literatura de México*. 2 vols. Mexico City: Felipe de Zuñiga y Ontiveros.

Amiot, Joseph Marie, et al. 1776–1814. *Mémoires concernant l'histoire, les sciences, les arts, les mœurs, les usages, &c. des chinois: Par les Missionnaires de Pékin*. 17 vols. Paris: Nyon. [Humboldt cites volume 2 (1777). Other contributors are François Bourgeois, Pierre Martial Cibot, Aloys Kao, Aloys de Poirot, Charles Batteux, and Louis-Georges-Oudart Feudrix de Bréquigny.]

————. 1789–90. *Dictionnaire tartare-mantchou françois: composé d'après un dictionnaire mantchou-chinoise*. Edited by Louis-Mathieu Langlès. 3 vols. Paris: F. A. Didot.

Ampelius, Lucius, and Lucius Annaeus Florus. 1783. *L. Annaei Flori Epitome rerum Romanarum. L. Ampelii Liber memorialis. Praemittitur notitia literaria studiis Societatis Bipontinae*. Biponti: ex typographia Societatis.

Anders, Ferdinand, Maarten Evert Reinoud Gerard Nicolaas Jansen, and Luis Reyes García. 1993. *Códice Borgia (Museo Borgia P. F. Messicano 1); Biblioteca Apostólica Vaticana*. 2 vols. Madrid: Sociedad Estatal Quinto Centenario; Graz, Austria: Akademische Druck-und Verlagsanstalt; Mexico City: Fondo de Cultura Económica. [Vol. 1: facsimile of the original manuscript (Codex Borgianus) folded accordion-style; volume 2: commentary volume with special title *Los templos del cielo y de la oscuridad: oráculos y liturgia: libro explicativo del llamado Códice Borgia (Museo Borgia P.F. Messicano 1) Biblioteca Apostólica Vaticana. See also* Nowotny, Karl Anton. 1976. *Codex Borgia*. Graz: Akadem. Druck-u. Verlagsanst.]

Anquetil-Duperron, Abraham-Hyacinthe, trans. 1771. *Zend-Avesta, ouvrage de Zoroastre: contenant les idées théologiques, physiques & morales de ce législateur, les cérémonies du culte religieux qu'il a établi, & plusieurs traits importans relatifs à l'ancienne histoire des Perses*. Paris: N. M. Tilliard. [English edition: 1880–87. *The Zend-Avesta*. Translated by James Darmesteter. Oxford: Clarendon.]

Apollonius of Rhodes. 1810–13. *Apollonii Rhodii Argonautica: Ex recensione et cum notis Richard Fr. Phil. Brunckii. Acc. scholia graeca ex codice biblioth. imperial. Paris nunc primum evulgata*. Edited by Richard Franz P. Brunck. 2 vols. Leipzig: Fleischer. [Includes *Scholia graeca*. First edition published in 1496. English edition: Apollonios Rhodios. 2007. *The Argonautika*. Edited by Peter Green. Berkeley: University of California Press.]

Aratus. 1793–1801. *Aratou Soleōs phainomena Kai diosēmeia. Arati Solensis Phaenom-ena et diosemea graece et latine ad codd. mss. et optimarvm edd. fidem recensita. Accedunt Theonis Scholia vvlgata et emendatiora e Cod. Mosqvensi, Leontii de Sphaera aratea libellvs, et versionvm Arati poeticarvm Ciceronis, Germanici, et R.F. Avieni qvae svpersvnt.* Edited by Johann Gottlieb Gerhard Buhle. Leipzig: Weidmann. [English edition: Aratus Solensis. 1997. *Phaenomena.* Edited by Douglas Kidd. Cambridge: Cambridge University Press.]

Aristophanes. 1788. *Aristophanis nubes graece et latine una cum scholiis graecis edidit et animadversionibus illustravit Theophilus Christophorus Harles.* Edited by Gottlieb Christoph Harles. Leipzig: Weidmann.

Aristotle. 1639. *Aristotelis opera omnia Græce et Latine: In tomos quatuor noua et ex-peditiori partitione distributa, & selectiorum interpretum, . . . studio, censurá & notis longè emendatissima. Adjecta est synopsis analytica.* Edited by Guillaume Du Val. Paris: Apud Ægidium Morellum. [English translation: 1995. *The com-plete works of Aristotle: The revised Oxford translation.* Edited and translated by Jonathan Barnes. Bollingen Series 71, no. 2. Princeton, NJ: Princeton University Press.]

Arrate y Acosta, José Martín Félix de. 1830. *Llave del Nuevo Mundo, antemural de las Indias Occidentales: La Habana descripta; Noticias de su fundación, aumentos y estado.* Havana: Sociedad Económica de Amigos del País. [Written in 1761. Hum-boldt had access to Arrate's 1750 manuscript.]

Arrian, Flavius. 1802. *Histoire des expéditions d'Alexandre: Reédigée sur les mémoires de Ptolémée et d'Aristobule, ses lieutenans.* Translated by Pierre Jean-Baptiste Chaussard. Paris: Genets. [English edition: 1983. *Arrian: Anabasis Alexandri.* Ed-ited by Peter Astbury Brunt. Cambridge, MA: Harvard University Press. *See also* Vincent, William.]

Asiatick Researches: or, Transactions of the Society, Instituted in Bengal, for Inquiring into the History and Antiquities, the Arts, Sciences, and Literature of Asia. 1788–1839. Published by the Asiatick Society of Bengal (Calcutta). 20 vols. Calcutta: Bengal Military Orphans Press, etc. [*Asiatick Researche*s was very popular jour-nal published originally in Calcutta from 1788 to 1839 (a total of 20 volumes; vols. 15 to 20 used the modernized title *Asiatic Researches*). That the Calucutta edition was reprinted in London and pirated numerous times in different places accounts for the discrepancies in dates in relation to volume numbers. We refer the reader to the index published in the *Journal of the Asiatic Society of Bengal*, 73 vols., Cal-cutta: Baptist Mission Press, 1839–46. *See also* Jones, William.]

Asiatisches Magazin. 1802. Edited by Julius Heinrich von Klaproth. Weimar: Verlag des Landes.

Bailly, Jean Sylvain. 1775. *Histoire de l'astronomie ancienne depuis son origine jusqu'à l'établissement de l'école d'Alexandrie.* Paris: Frères Debure.

———. 1779–82. *Histoire de l'astronomie moderne depuis la fondation de l'école d'Alexandrie, jusqu'à l'époque de M.D.CC.XXX.* Paris: Chez les frères de Bure.

———. 1787. *Traité de l'astronomie indienne et orientale: ouvrage qui peut servir de suite á l'Histoire de l'astronomie ancienne.* Paris: Debure l'aîné.

Bainbridge, John. 1648. *Cl. v. Iohannis Bainbridgii, astronomiae in celeberrima Academia Oxonensi Professoris Saviliani, Canicularia una cum demonstratione ortus Sirii heliaci, pro parallelo inferioris Aegypti.* Edited by John Greaves. Oxford: Excudebat Henricus Hall, impensis Thomæ Robinson.

Barrow, John. 1806. *A Voyage to Cochinchina in the Years 1792 and 1793 Containing a General View of the Valuable Productions and the Political Importance of This Flourishing Kingdom and Also of Such European Settlements as Were Visited on the Voyage: With Sketches of the Manners, Character, and Condition of Their Several Inhabitants; To Which Is Annexed an Account of a Journey, Made in the Years 1801 and 1802, to the Residence of the Chief of the Booshuana Nation, Being the Remotest Point in the Interior of Southern Africa to Which Europeans Have Hitherto Penetrated—The Facts and Description Taken from a Manuscript Journal with a Chart of the Route.* London: Printed for T. Cadell and W. Davies. [The description of the journey to the Bechuana was based on the original manuscript of the Durch traveler Pieter Jan Truter. *See also* the excerpted German translation by Friedrich Justin Bertuch.]

Barton, Benjamin Smith. 1797. *New Views of the Origin of the Tribes and Nations of America.* Philadelphia: John Bioren.

Barton, Benjamin Smith, and Thomas Beddoes. 1803. *Hints on the Etymology of Certain English Words and on Their Affinity to Words in the Languages of Different European, Asiatic, and American (Indian) Nations, in a Letter from Dr. Barton to Dr. Thomas Beddoes.* Philadelphia. [Reprinted in *Transactions of the American Philosophical Society* 6 (1809): 145–58.]

Battānī, Muḥammad ibn-Jābir al- (Albategnius). 1645. *Mahometis Albatenii de Scientia stellarvm liber. Cum aliquot additionibus Ioannis Regiomontani. Ex bibliotheca Vaticana transcriptus.* Edited by Joannes Regiomontanus. Bologna: Typis Haeredis V. Benatij. [This work is also known as *De motu stellarum* and *De numeris stellarum et motibus*.]

Bede, the Venerable (St.). 1722. *Historiæ ecclesiasticæ gentis Anglorum una cum reliquis ejus operibus historicis in unum volumen collectis.* Edited by John Smith. Cambridge. [First published as *Historia ecclesiastica gentis Anglorum* (1474). English edition: 1991. *Bede's Ecclesiastical History of the English People.* Edited and translated by Bertram Colgrave and R. A. B. Mynors. Oxford: Clarendon; New York: Oxford University Press.]

Benavente, Toribio de (Motolinía). 1858. *Historia de los indios de Nueva España.* In *Colección de documentos para la historia de México,* edited by Joaquín García Icazbalceta. Mexico City: Andrade. [Written around 1541. Humboldt had access to Motolinía's work via Torquemada. English edition: 1951. *Motolinía's History of the Indians of New Spain.* Edited and translated by Francis Borgia Steck. Washington, DC: Academy of American Franciscan History.]

Bentley, John. 1798. "XXI. Remarks on the Principal Æras and Dates of the Ancient Hindus." *Asiatick Researches* 5:315–44.

———. 1799. "XIII. On the Antiquity of the Surya Siddhanta, and the Formulation of the Astronomical Cycles Therein Contained." *Asiatick Researches* 6:537–58.

———. 1805. "VI. On the Hindu Systems of Astronomy, and Their Connection with History in Ancient and Modern Times." *Asiatick Reseaches* 8:193–244.

Berdan, Frances F., and Patricia R. Anawalt. 1992. *The Codex Mendoza*. 4 vols. Berkeley: University of California Press. [Vol. 1, interpretation; vol. 2, description; vols. 3 and 4 facsimile and pictorial parallel image replicas. Also known as *Códice Mendocino*, *Colección Mendoza*, or *Peintures hiéroglyphiques du Recueil de Mendoza*. *See also* Purchas, Samuel.]

Berosus the Chaldean. 1510. *Berosus Babilonicus de his quae praecesserunt inundationem terrarum: Item. Myrsilus de origine Turrenorum. Cato in fragmentis. Archilocus in epitheto de temporibus. Metasthenes de judicio temporum. Philo in breviario temporum. Xenophon de equivocis temporum. Sempronius de divisione Italiae. Q. Fab. Pictor de aureo seculo & origine urbis Romae. Fragmentum itinerarii Antonini Pii. Altercatio Adriani Augusti & Epictici.* Edited by Giovanni da Viterbo Nanni. Paris: Collegium Plessiacum. [English edition: Burstein, Stanley Mayer, ed. 1980. *The Babyloniaca of Berossus*. Edited by Stanley Mayer Burstein. Malibu, CA: Undena. *See also* Verbrugghe, Gerald P., and John M. Wickersham. 1996. *Berosso and Manetho, Introduced and Translated: Native Traditions in Ancient Mesopotamia and Egypt.* Ann Arbor: University of Michigan Press.]

Bertuch, Friedrich Justin. 1798–1816. *Allgemeine Geographische Ephemeriden*. 50 vols. Weimar: Verlage des Landes-Industrie Comptoirs.

———, trans. 1807. "John Barrows Auszug aus handschriftliches Tagebuche einer in den Jahren 1801 und 1802 unternommenen Reise nach Litäku, der Residenz des Hauptes der Buschuana, als dem äussersten Punkte des inneren Süd-Afrika, wohin bis jetzt Europäer gedrungen sind." *Allgemeine Geographische Ephemeriden* 22 (February): 140–79, (March): 257–306. [Humboldt cites vol. 12, p. 67. Truter and the Bechuanas (Buschwanen or Booshuana) are also mentioned in vol. 21 (1806) in a critique to John Barrows's 1806 *A Voyage to Cochinchina*. Vol. 22 depicts Humboldt as frontispiece.]

Bianchini, Francesco. 1697. *La istoria universale provata con monumenti, e figurata con simboli de gli antichi.* Rome: Antonio de Rossi.

———. 1728. *Hesperi et Phosphori nova phaenomena, sive, Observationes circa planetam Veneris vnde colligitur: I. Descriptio illius macularum, seu celidographia. II. Vertigo circa axem proprium, vel perieilesis spatio dierum 24. cum triente. III. Parallelismus axis in orbita octimestri circa solem. IV. Et quantitas parallaxeos methodo Cassinianâ explorata.* Rome: apvd Joannem Mariam Salvioni. [English edition: 1996. *Observations concerning the Planet Venus.* Translated by Sally Beaumont. Berlin: Springer.]

———. 1737. *Francisci Blanchini Veronensis Astronomicae, ac geographicae observationes selectae Romae, atque alibi per Italiam habitae, ex eius autographis excerptae una cum geographica meridiani romani tabula a mari supero ad inferum, ex iisdem observationibus collecta et concinnata cura et studio Eustachii Manfredi in Bononiensi Scientiarum Instituto Astronomi.* Edited by Eustachio Manfredi. Verona: Typis Dyonisii Ramanzini, Bibliopolae, apud S. Thoman. *See* Fontenelle, Bernard Le Bouyer de.

Blumenbach, Johann Friedrich. 1808. *Decas quinta collectionis sua craniorum diversarum gentium illustrata.* Göttingen: H. Dieterich. [Part 5 of 7. Complete series published 1790–1828.]

Böttiger, Karl August. 1811. *Ideen zur Archäologie der Malerei: Erster Theil, Nach Masgabe der Wintervorlesungen im Jahre 1811.* Dresden: Verlage der Waltherschen Hofbuchhandlung.

Boturini Benaducci, Lorenzo. 1746. *Idea de una nueva historia general de la América Septentrional fundada sobre material copioso de figuras, symbolos, caratères, y geroglificos, cantares, y manuscritos de autores indios, ultimamente descubiertos.* Madrid: Juan de Zuñigá. [Includes *Catálogo del museo histórico indiano.* Facsimile edition: 1974. *Idea de una nueva historia general de la América Septentrional.* Edited by Miguel León Portilla. Mexico City: Editorial Porrúa. *See also* 1990. *Historia general de la América Septentrional.* Edited by Manuel Ballesteros Gaibrois. Mexico City: Instituto de Investigaciones Históricas, UNAM.]

Bouguer, Pierre. 1729. *De la méthode d'observer exactement sur mer la hauteur des astres: piece qui a remporte' le prix proposé par l'Academie Royale des Sciences pour l'année 1729.* Paris: Claude Jombert.

———. 1746. *Traité du navire, de sa construction et de ses mouvemens.* Paris: Jombert. [Expanded edition, *see* 1792. *Nouveau traité de navigation, contenant la théorie et la pratique du pilotage.* Edited by Nicolas Louis de LaCaille and Joseph Jérôme Le Français de Lalande. Paris: Desaint. English translation: 1765. *A treatise on shipbuilding and navigation.* Edited by Mungo Murray. London: Millar.]

———. 1748. *Entretiens sur la cause de l'inclinaison des orbites des planetes: où, L'on répond à la question proposée l'Academie Royale des Sciences, pour le sujet du prix des années 1732 & 1734.* Paris: C[harles] A[ntoine] Jombert.

———. 1748. "Relation abrégée du voyage fait au Pérou par Messieurs de l'Académie Royale des Siences, pour mesurer les degrés du méridien aux environs de l'Equateur, & en conclure la figure de la terre." In *Histoire de l'Académie royale des sciences avec les mémoires de mathématique et de physique pour la même année tirés des registres de cette académie, Année 1744,* 249–97. Paris: L'Imprimerie Royale.

———. 1749. *La figure de la terre, déterminée par les observations de Messieurs Bouguer & de La Condamine, de l'Académie royale des sçiences, envoyés par ordre du roy au Pérou, pour observer aux environs de l'equateur: avec une relation abregée de ce voyage, qui contient la description du pays dans lequel les opérations ont été faites.* Paris: Charles-Antoine Jombert. *See also* La Condamine, Charles Marie de.

———. 1752. *Justification des mémoires de l'Académie Royale des Sciences de 1744: et du livre de La figure de la terre: déterminée par les observations faites au Pérou, sur plusieurs faits qui concernent les opérations des académiciens.* Paris: Charles-Antoine Jombert.

———. 1754. *Lettre à Monsieur *** dans laquelle on discute divers points d'astronomie pratique, et où l'on fait quelques remarques sur le Supplement au Journal historique du voyage à l'Equateur de M. de la C.* Paris, H. L. Guerin & L. F. Delatour.

———. 1757. *De la manœuvre des vaisseaux, ou, Traité de méchanique et de dynamique dans lequel on réduit a des solutions très-simples les problêmes de marine les plus difficiles, qui ont pour objet le mouvement du navire.* Paris: H. L. Guerin & L. F. Delatour.

———. 1760. *Traité d'optique sur la gradation de la lumiere.* Paris: H. L. Guerin & L. F. Delatour. [Revised and expanded version of his *Essai d'optique, sur la gradation de la lumiere* (1729). English edition: 1961. *Optical Treatise on the Gradation of Light.* Translated by W. E. Knowles Middleton. Toronto: University of Toronto Press.]

Boulanger, Nicolas Antoine. 1761. *Recherches sur l'origine du despotisme oriental.* Paris. [English edition: 1764. *The Origin and Progress of Despotism. In the Oriental, and Other Empires of Africa, Europe, and America.* Translated by John Wilkes. Amsterdam.]

———. 1792. *Œuvres de Boullanger.* Paris: J. Servieres. [Vol. 2 of 8: *L'Antiquité dévoilée par ses usages.*]

Buttmann, Philipp. 1806. In *Historische Untersuchungen über die astronomischen Beobachtungen der Alten* by Ludwig Ideler, 373–78. Berlin: C. Quien.

———. 1808. *Griechische Grammatik.* Berlin: Mylius. [English edition: 1822. *Greek Grammar.* Translated by Edward Everett. Boston: O. Everett.]

Call, John. 1772. "A Letter from John Call, Esq; To Nevil Maskelyne, F. R. S. Astronomer Royal, Containing a Sketch of the Signs of the Zodiac, Found in a Pagoda, Near Cape Comorin in India." *Philosophical Transactions (1683–1775)* 62:353–56.

Carey, Mathew. 1796. *Carey's American Pocket Atlas with a Concise Description of Each State.* Philadelphia: Printed for Mathew Carey by Lang and Ustick.

Carli, Gian Rinaldo. 1788. *Lettres américaines, dans lesquelles on examine l'origine, l'etat civil, politique, militaire & religieux, les arts, l'industrie, les sciences, les mœurs, les usages des anciens habitans de l'Amérique; les grandes epoques de la nature, l'ancienne communication des deux hémisphères, & la dernière révolution qui a fait disparoître l'Atlantide.* Translated by J. B. Lefebvre de Villebrune. 2 vols. Paris: chez Buisson. [Originally in Italian; published in 1780.]

Castellanos, Juan de. 1589. *Elegías de varones ilustres de Indias. Primera parte.* Madrid: Casa de la viuda de Alonso Gomez. [Part 1 of 4. Parts 1–3 published as *Elegías de varones ilustres de Indias* (1847); part 4 in 1914. *See* Castellanos, Juan de. 1921. *Discurso del capitán Francisco Draque.* Madrid: Instituto de Valencia de D. Juan; 1997. *Elegías de varones ilustres de Indias.* Edited by Gerardo Rivas Moreno. Bogotá: G. Rivas Moreno; and 2004. *Antología crítica de Juan de Castellanos: Elegías de varones ilustres de Indias.* Edited by Luis Fernando Restrepo. Bogotá: Editorial Pontificia Universidad Javeriana. *See also* Piedrahita, Lucas Fernández de.]

Castillo, Cristóbal del. 1991. *Historia de la venida de los mexicanos y de otros pueblos: E Historia de la conquista.* Edited by Federico Navarrete Linares. Mexico City: Instituto Nacional de Antropología e Historia. [Humboldt had access to Castillo's 1599 manuscript. Partial English edition: 1908. *Fragmentos de la obra general sobre historia de los mexicanos escrita en lengua náuatl por Cristóbal del Castillo á fines del siglo XVI.* Translated by Francisco Paso y Troncoso. Florencia: S. Landi.]

Censorinus. 1805. *De die natali*. Edited by Johann Sigmund Gruber. Nuremberg. [First printed in 1497. English edition: 2007. *The Birthday Book*. Translated by Holt Neumon Parker. Chicago: University of Chicago Press.]

Cervantes, Vicente de, et al. 1788. *Exercicios públicos de botánica, que tendrán en esta real y pontificia Universidad el B.D. Joseph Vicente de la Peña, Don Francisco Giles de Arellano, y Don Joseph Timoteo Arsinas; dirigiéndolos Don Vicente Cervantes el jueves 11de diciembre a las quatro de la tarde*. Mexico City: [Felipe de] Zúñiga y Ontiveros.

———. 1794. *Discurso pronunciado en el Real Jardin Botánico el 2 de junio. Suplemento a la Gazeta de Literatura de México 2 de julio de 1794*. Mexico City: Felipe de Zúñiga y Ontiveros.

———. 1805. "Account of the Ule-Tree, and of Other Trees Producing the Elastic Gum." In *Tracts Relative to Botany, Translated from Different Languages*, edited and translated by Charles Konig, 229–39. London: Phillips and Fardon.

———. 1817. *Flore du Mexique collection des plantes rares ou peu connues observes au Mexique et dans la nouvelle Espagne*. 13 vols. Geneva. [Humboldt had knowledge of this publication. Other contributors include Martín Sessé, José Mariano Moziño, Atanasio Echeverría, and Juan Cerda.]

Chardin, Jean. 1711. *Voyages de Monsieur le Chevalier Chardin en Perse et autres lieux de l'Orient*. 3 vols. Amsterdam: J. L. de Lorme. [Vol. 1 was partially published in London as *Journal du voyage du Chevalier en Perse et aux Indes Orientales, por la Mer Noire et par la Colchide. Premiere Partie qui Contient le voyage de Paris a Isfaham* (1686), with a concurrent English translation. A 1735 edition in 4 vols. includes Chardin's *Le Couronnement De Soleïman Troisiéme roy de Perse* (1671). The most complete edition was published by Langlès (10 vols, 1811). *See* 1971. *Sir John Chardin's Travels in Persia*. Edited by Percy Sykes. Amsterdam: N. Israel. *See also* Langlès, Louis-Mathieu.]

Chézy, Antoine Léonard de, trans. 1805. *Medjnoun et Leila, poëme traduit du persan de Djami* by Abd al-Rahmān ibn Ahmad Nūr al-Dīn Ǧāmī. 2 vols. Paris: A L'imprimerie de Valade. See also Silvestre de Sacy, Antoine Isaac.

———. 1810. "Notice de l'ouvrage institulé *A Grammar of the Sanskrite Language* by Charles Wilkins, c'est-à-dire, *Grammaire Sanskrite*." *Moniteur Universel* 146 (May 26): 578–80.

Chimalpahin Quauhtlehuanitzin, Domingo Francisco de San Antón Muñón. 1998. *Las ocho relaciones y el Memorial de Colhuacan*. Edited and translated by Rafael Tena. Mexico City: Consejo Nacional para la Cultura y las Artes.

———. 2001. *Diario*. Edited and translated by Rafael Tena. Mexico City: Consejo Nacional para la Cultura y las Artes. [Chimalpahin wrote around 1605–31. For a recent English edition: 2006. *Annals of His Time: Don Domingo de San Antón Muñón Chimalpahin Quauhtlehuanitzin*. Edited and translated by James Lockhart, Susan Schroeder, and Doris Namala. Stanford, CA: Stanford University Press. English edition of manuscripts housed in England: 1997. *Codex Chimalpahin: Society and Politics in Mexico, Tenochtitlan, Tlatelolco, Texcoco, Culhuacan, and Other Nahua Altepetl in Central Mexico: The Nahuatl and Spanish Annals and Accounts Collected and Recorded by don Domingo de San Antón Muñón Chimalpahin Quauhtlehuanitzin.* 2 vols. Edited and translated by Arthur J. O. Anderson, Susan Schroeder, and Wayne Ruwet. Norman: University of Oklahoma Press. *See also* 2010. *Chimalpahin's Conquest: A Nahua Historian's Rewriting of Francisco López de Gómara's "La conquista de México."* Edited by Susan Schroeder et at. Stanford, CA: Stanford University Press.]

Choiseul-Gouffier, Marie-Gabriel-Auguste-Florent. 1782–1824. *Voyage pittoresque de la Grèce.* 2 vols. Paris: J. J. Blaise. [Vol. 1 was published in 1782; vol. 2 part 1 in 1809, part 2 in 1822.]

Cicero, Marcus Tullius. 1740–42. *M. Tullii Ciceronis opera. Cum delectu commentariorum.* Edited by Pierre Joseph Thoullier d'Olivet. Vol. 3 of 9. Paris: Coignard. [*De divinatione* bk. 2 in vol. 3 (1740), 49–98, notes on 469–80.]

Cieza de León, Pedro de. 1554. *La chronica del Peru.* Antwerp: Martin Nucio. [First published in 1553 in Seville; part 1 of 4. *See* 1998. *The Discovery and Conquest of Peru: Chronicles of the New World Encounter.* Edited and translated by Alexandra Parma Cook and Noble David Cook. Durham, NC: Duke University Press.]

Cisneros, Diego. *See* Sociedad Académica de Amantes de Lima.

Clavijero, Francisco Javier [Saverio]. 1780–81. *Storia antica del Messico: cavata da' migliori storici spagnuoli, e da' manoscritti, e dalle pitture antiche degl' indiani; divisa in dieci libri, e corredata di carte geografiche, e di varie figure: e dissertazioni sulla Terra, fugli Animali, e fugli abitatori del Messico.* 4 vols. Cesena, Italy: Gregorio Biasini. [English edition: 1807. *The History of Mexico.* Translated by Charles Cullen. 2 vols. London: J. Johnson.]

Clement of Alexandria. 1715. *Klementos Alexandreos ta eyriskomena = Clementis Alexandrini opera quæ extant.* Edited by John Potter. 2 vols. Oxford: e theatro Sheldoniano. [English edition: 1919. *Clement of Alexandria.* Edited by George William Butterworth. Cambridge, MA: Harvard University Press.]

Clément, François. 1783–87. *L'Art de vérifier les dates des faits historiques, des chartes, des chroniques, et autres anciens monumens, depuis la naissance de Notre-Seigneur, par le moyen d'une table chronologique.* 3 vols. Paris: A. Jombert jeune. [Begun in 1743 by Maur François Dantine. After his death in 1746, it was completed and published in 1750 by Charles Clémencet and Ursin Durand. Appended to vol. 3 is "Supplément fourni par M. Ernst."]

Cleomedes. 1605. *Cleomedis Meteora Graece et Latine.* Edited by Robert Balfour. Bordeaux: Milangius. Includes Eratosthenes's *Measurement of the Earth.* [English edition: 2004. *Cleomedes' Lectures on Astronomy. A Translation of "The Heavens."* Edited by Alan C. Bowen and Robert B. Todd. Berkeley: University of California Press.] *See also* Eratosthenes.

Codex Vaticanus anonymous. *See* Ríos, Pedro de los.

Colebrooke, Henry Thomas. 1809. "VI. On the Indian and Arabian Divisions of the Zodiack." *Asiatick Researches* 9:323–76. [First published in 1807 in Calcutta.]

———. 1818. "VI. On the Notions of Hindu Astronomers, concerning the precession of the Equinoxes and Motions of the Planets." *Asiatick Researches* 12:210–52. [First printed in Calcutta.]

Columella, Lucius Iunius Moderatus. 1794–97. *Scriptorvm Rei Rvsticae Vetervm Latinorvm Tomvs primvs-[qvartvs]*. Edited by Johann Gottlob Schneider. Leipzig: Fritsch. [Vol. 2 of 4. Vol. 2 in 2 parts; part 1: *L. Ivnii Moderati Colvmellae De re rvstica libri XII et Liber de arboribvs;* part 2: *Commentarivs*. English edition: 1941–55. *On agriculture, with a recension of the text and an English translation.* 3 vols. Edited and translated by Harrison Boyd Ash (vol. 1), E. S. Forster, and Edward H. Heffner (vols. 2–3). Cambridge, MA: Harvard University Press; London: W. Heinemann.]

Cordier, Louis. 1801. "Traité de minéralogie d'après les principes de Werner, par le cit. Brochart. Extrait." *Journal de Physique, de Chemie, d'Historie Naturelle et des Arts* 52:228–44.

———. 1803. "Lettre de L. Cordier, ingénieur des mines de France, au cit., Devilliers fils., extrait : Aux îles Canaries de Santa-Cruz de Ténérife, le 1 mai 1803." *Journal de Physique, de Chemie, d'Historie Naturelle et des Arts* 57:55–63.

———. 1807. "Recherches sur différens produits volcaniques." *Journal des Mines* 21, no. 124: 249–60.

———. 1808. "Suite des recherches sur différens produits des volcans : Second mémoire." *Journal des Mines* 23, no. 133:55–74.

———. 1808. "Sur le Dusodile, nouvelle espèce minérale." *Journal des Mines* 23, no. 136:271–74.

———. 1809. "Description du Dichroïte nouvelle espéce minérale." *Journal de Physique, de Chemie, d'Historie Naturelle et des Arts* 68:298–304. [Also published in *Journal des Mines* 23, no. 136: 217–24 and 25, no. 146: 129–38. For English abstracts of Cordier's papers in the *Journal de Physique* and the *Journal des Mines*, see *Retrospect of Philosophical, Mechanical, Chemical, and Agricultural Discoveries: Being an Abridgment of the Periodical, and Other Publications, English and Foreign, Relative to Arts, Chemistry, Manufactures, Agriculture, and Natural Philosophy; Accompanied, Occasionally, with Remarks on the Merits and Defects of the Respective Papers; and, in Some Cases, Shewing to What Other Useful Purposes Inventions May Be Directed, and Discoveries Extended, Beyond the Original Views of Their Authors* 5 (1810): 244–46, 354, 507–8. London: W. H. Wyatt at the Repertory of Arts and Patent-Office.]

Cortés, Hernán (Fernando). 1525. *La quarta relacion q[ue] Ferna[n]do Cortes gouernador y capitan general porsu majestad enla Nueua España d[e]1 mar oceano embio al muy alto [y] muy potentissimo inuictissimo señor don Carlos emperador semper angusto [sic] y rey de España nuestro señor: enla qual estan otras cartas [y] relaciones que los capitanes Pedro de Aluarado [y] Diego Godoy embiaron al dicho capitan Fernardo [sic] Cortes.* Toledo: Gaspar de Auila.

———. 1770. *Historia de Nueva-España escrita por su esclarecido conquistador Hernan Cortes: Aumentada con otros documentos, y notas.* Edited by Francisco Antonio Lorenzana. Mexico City: Imprenta del Superior Gobierno, J. A. de Hogal. [English edition: 2001. *Letters from Mexico.* Edited and translated by Anthony R. Pagden. New Haven, CT: Yale Nota Bene. Humboldt quotes from the Second Letter on p. [31].]

Court-de-Gébelin, Antoine. 1773–82. *Monde primitif, analysé et comparé avec le monde moderne.* 9 vols. Paris: Chez l'auteur. [New edition of nine volumes published in 1777–96.]

Creuxius, Franciscus. *See* Du Creux, François.

Cuvier, Georges. 1798. *Tableau élémentaire de l'histoire naturelle des animaux.* Paris: Baudouin.

———. 1799. "Mémoire sur les espèces d'éléphans vivantes et fossiles." *Mémoires de l'Institut National des Sciences et Arts: Sciences mathématiques et physiques* 2:1–22.

———. 1800–1805. *Leçons d'anatomie comparée.* Edited by C. Duméril and G. L. Duvernoy. 5 vols. Paris: Crochard, Fantin, Baudouin, imprimeur de l'Institut. [Expanded edition published in 8 volumes (1835–46). English edition: 1802. *Lectures on Comparative Anatomy.* Edited by James Macartney, translated by William C. Ross. London: Wilson for T. N. Longman and O. Rees.]

———. 1806. "Sur le grand mastodonte, animal très-voisin de l'éléphant." *Annales du Muséum d'Histoire Naturelle* 8:270–312.

———. 1806. "Sur les elephans vivants et fossiles." *Annales du Muséum d'Histoire Naturelle* 8:1–58, 93–155, 249–69.

———. 1806. "Sur différentes dents du genre des mastodontes, mails d'espèces moindres que celles de l'Ohio, trouvés en plusieurs lieux des deux continents." *Annales du Muséum d'Histoire Naturelle* 8:401–24.

———. 1807. *Recherches anatomiques sur les reptiles regardés encore comme douteux par les naturalistes : Faites à l'occasion de l'axolotl, rapporté par M. de Humbolt du Mexique.* Paris: L. Hausmann.

———. 1812. *Recherches sur les ossemens fossiles de quadrupèdes, ou l'on rétablit les caractères de plusieurs espèces d'animaux que les révolutions du globe paroissent avoir détruites.* 4 vols. Paris: Deterville. [English edition: 1997. *Georges Cuvier, Fossil Bones, and Geological Catastrophes: New Translations & Interpretations of the Primary Texts.* Translated by Martin J. S. Rudwick. Chicago: University of Chicago Press.]

Daniell, Thomas, William Daniell, and James Wales. 1795–1807 [1808]. *Oriental Scenery: Twenty-Four Views in Hindoostan.* London: Bowyer. [Contains 144 color plates and 8 plans. Plates on parts 1–5 engraved by the Daniell brothers and part 6 by Wales.]

Dante Alighieri. 1785. *L'enfer: poëme du Dante: traduction nouvelle.* 2 vols. Edited and translated by Antoine Rivarol. Paris: Chez P. Fr. Didot le jeune, Mérigot le jeune.

———. 1796. *La Divine Comédie de Dante Alighieri: contenant la description de l'Enfer, du Purgatoire et du Paradis.* Edited by Marie-François Sallior. Paris: Gueffier. [English edition: 1965. *The Divine Comedy.* Edited and translated by Geoffrey L. Bickersteth. Cambridge, MA: Harvard University Press.]

Davis, Samuel. 1790. "XV. On the Astronomical Computations of the Hindus." *Asiatick Researches* 2:225–87.

———. 1792. "IX. On the Indian Cycle of Sixty Years." *Asiatick Researches* 3:209–27. Calcutta: Printed and Sold by T. Watley; Sold at London by P. Elmsly.

La Décade Egyptienne, Journal Littéraire et d'Économie Politique. 1798–1801. Institut d'Egypte. Cairo: Imprimerie nationale.

Delambre, Jean Baptiste Joseph. 1792. "Tables astronomiques calculées sur les observations les plus nouvelles pour servir à la troisième édition de l'Astronomie." In *Astronomie* by Joseph Jérôme Le Français de Lalande, vol. 1 of 3. Paris: Chez la veuve Desaint, De l'imprimerie de P. Didot l'ainé.

———. 1807. "De l'arithmétique des Grecs." In *Œuvres d'Archimède*, edited and translated by François Peyrard, 571–601. Paris: Chez François Buisson.

Delambre, Jean Baptiste Joseph, and Johann Tobias Bürg. 1806. *Tables astronomiques. Première partie/publiées par le Bureau des Longitudes de France.* Paris: Chez Courcier.

Denon, Dominique Vivant. 1802. *Voyage dans la Basse et la Haute Égypte pendant les campagnes du Général Bonaparte.* 2 vols. Paris: De l'Imprimerie de P. Didot l'Aîné. [English edition: 1986. *Travels in Upper and Lower Egypt during the Campaigns of General Bonaparte.* Edited by Edward Augustus Kendal. London: Darf.]

Díaz del Castillo, Bernal. 1795. *Historia verdadera de la conquista de la Nueva España.* Madrid: B. Cano. [Originally published in 1632; contains passages by Alonso Remón. Guatemala manuscript: 1904–5. *Historia verdadera de la conquista de la Nueva España.* 2 vols. Edited by Genaro García. Mexico City: Oficina tipográfica de la Secretaría de Fomento. Critical edition: 2005. *Historia verdadera de la conquista de la Nueva España: Manuscrito "Guatemala."* Edited by José Antonio Barbón Rodríguez. Mexico City: Colegio de México UNAM. Unabridged English edition of Guatemala manuscript: 1908–16. *The True Story of the Conquest of New Spain.* Edited and translated by Alfred Percival Maudslay. 5 vols. London: Hakluyt Society. *See also* 2008. *The History of the Conquest of New Spain.* Edited by Davíd Carrasco. Albuquerque: University of New Mexico Press.]

Dio Cassius. 1750–52. *Tōn Diōnos tu Kassiu tu Kokkeianu Rōmaikōn historiōn ta sōzomena = Cassii Dionis Cocceiani Historiae Romanae quae supersunt/Graeca ex codd. mss. et fragmentis supplevit, emendavit, Latinam versionem Xylandro-Leunclavianam limavit, varias lectiones notas doctorum et suas cum appartu et indicibus adiecit Hermannus Samuel Reimarus.* Edited by Johann Albert Fabricius, Hermann Samuel Reimarus, Henri de Valois, and Wilhelm Xylander. 2 vols. Hamburg: Sumtibus Christiani Heroldi. [English edition: 1970–87. *Dio's Roman history.* Translated by Earnest Cary. 9 vols. Cambridge, MA: Harvard University Press; London: W. Heinemann.]

Diodorus (Diodore de Sicile). 1745. *Diodōru tu Sikeliōtu bibliothēkēs historikēs ta sōzomena = Diodori Siculi Bibliothecae historicae libri qui supersunt.* Edited by Petrus Wesseling. 2 vols. Amsterdam: Sumptibus Jacobi Wetstenii. [A revised Wesseling edition was published in 11 volumes in 1793–1807.]

Dixon, George, ed. 1789. *Voyage autour du monde et principalement à la côte Nord-Ouest de l'Amérique, fait en 1785, 1786, 1787 et 1788 à bord du King-George et de la Queen-Charlotte, par les Capitaines Portlock et Dixon: Dédié, par permission, à Sir Joseph Banks, Baronet.* Translated by P. L. Lebas. 2 vols. Paris: Maradan. [*See* Dixon, George, ed. 1789. *A Voyage Round the World; But More Particularly to the North-West Coast of America Performed in 1785, 1786, 1787, and 1788, in the "King George" and "Queen Charlotte," Captains Portlock and Dixon.* London: Printed for J. Stockdale and G. Goulding. *See also* Portlock, Nathaniel, ed. 1789. *A Voyage Round the World: But More Particularly to the North-West Coast of America Performed in 1785, 1786, 1787, and 1788, in the "King George" and "Queen Charlotte," Captains Portlock and Dixon.* London: Printed for J. Stockdale and G. Goulding.]

Du Choul, Guillaume. 1556. *Discours de la réligion des anciens Romains: Illustré d'un grand nombre de médailles, et de plusieurs belles figures retirées des marbres antiques, qui se treuvent à Rome, et par nostre Gaule.* Lyon: Guillaume Rouille.

Du Creux, François. 1664. *Historiae Canadensis, seu Novae-Franciae Libri Decem.* Paris: S. Cramoisy et S. Mabre-Cramoisy. [English edition: James B. Conacher, ed. *The History of Canada or New France.* Translated by Percy J. Robinson. 2 vols. Toronto: Champlain Society, 1951. *See also* Diodorus and Herodotus.]

Du Croz. *See* Souciet, Étienne, and Antoine Gaubil.

Du Halde, Jean Baptiste. 1736. *Description géographique, historique, chronologique, politique, et physique de l'empire de la Chine et de la Tartarie chinoise, enrichie des cartes générales et particulieres de ces pays, de la carte générale et des cartes particulieres du Thibet, & de la Corée; & ornée d'un grand nombre de figures & de vignettes gravées en tailledouce.* 4 vols. The Hague: Scheurleer. [English edition: 1741. *The general history of China. Containing a geographical, historical, chronological, political and physical descritpion of the empire of China, Chinese-Tartary, Corea, and Thibet. Including an exact and particular account of their customs, manners, ceremonies, religion, arts and sciences.* Translated by Richard Brookes. London: J. Watts.]

Dupuis, Charles-François. 1795. *Origine de tous les cultes, ou Religión universelle: (suivie d'un tableau historique, explicatif et nominatif des signes du zodiaque).* Paris: Chez H. Agasse. [An abridged French edition was published in 1798, a revised edition of 7 volumes in 1822. English edition: 1872. *The Origin of All Religious Worship: Translated from the French of Dupuis. Containing Also a Description of the Zodiac of Denderah.* New Orleans.]

———. 1806. *Mémoire explicatif du zodiaque chronologique et mythologique: Ouvrage contenant le tableau comparatif des maisons de la lune chez les différens peuples de l'Orient, et celui des plus anciennes observations qui s'y lient, d'après les egyptiens, les chinois, les perses, les arabes, les chaldéens et les calendriers grecs.* Paris: Chez Courcier.

———. 1806. "Observations sur le zodiaque de Dendra. " *La Revue Philosophique, Littéraire et Politique* 15 (May): 321–38.

Duquesne de Madrid, José Domingo. 1848. "Disertación sobre el calendario de los Muyscas, indios naturales de este Nuevo Reino de Granada, Dedicada al S. D. D. José Celestino de Mutis, director general de la expedición botánica por S. M." In *Compendio histórico del descubrimiento y colonización de la Nueva Granada en el siglo décimo sexto,* by Joaquín Acosta, 405–17. Paris: Imprenta de Beau. [Written in 1795; Mutis showed Humboldt the manuscript in Bogotá. English translation of the compendium: 1912. *The Conquest of New Granada*. Translated by Clements Robert Markham. London: Smith, Elder.]

Eguiara y Eguren, Juan José de. 1755. *Bibliotheca mexicana, sive, Eruditorum historia virorum qui in America Boreali nati, vel alibi geniti, in ipsam domicilis aut studijs asciti, quavis lingual scripto aliquid tradiderunt eorum praesertim qui pro fide catholicâ & pietate ampliandâ fovendâque, egregiè factis & quibusvis scriptis floruere editis aut ineditis*. Mexico City: Ex novâ typographiâ in aedibus authoris editioni ejusdem Bibliothecae destinatâ. [Only first volume was published. Spanish edition of complete works: 1986–89. *Biblioteca mexicana*. Edited and translated by Benjamín Fernández Valenzuela, Ernesto de la Torre Villar, and Ramiro Navarro de Anda. Mexico City: Universidad Nacional Autónoma de México.]

Engel, Johann Christian von. 1797. *Geschichte des alten Panoniens und der Bulgarei, nebst einer allgemeinen Einleitung in die ungarische und illustrische Geschichte. Geschichte des ungrischen Reichs und seiner Nebenländer*. Vol. 1. Halle, Germany: Gebauer. [Vol. 2: *Staatskunde und Geschichte von Dalmatien, Croatien und Slavonien* (1798); vol. 3: *Geschichte von Serbien und Bosnien* (1801); vol. 4: *Geschichte der Moldau und Walachei* (1804).]

———. 1811. *Geschichte des Königreichs Ungarns: Neu übersehen und verbessert*. Vol. 1 of 5. Vienna, Austria: In der Camesinaschen Buchhandlung. [Vols. 2–5 published as *Geschichte des ungrischen Reichs* (1813–14).]

Epiphanios. 1544. *Tou hagiou Epiphaniou episkopou Kōnstanteias tēs Kyprou Kata haireseōn ogdoēkonta to epiklēthen Panarion, eitoun kibōtion, eis biblous men g', tomous de hepta diērēmenon. Tou autou hagion Epiphaniou Logos agkyrōtos, pasan tēn peri tēs theias pisteōs didaskalian en heautō dialambanōn. Tou autou Tōn tou Panariou hapantōn anakephalaiōsis. Tou auto Peri metrōn kai stathmōn. D. Epiphanii episcopi Constantiae Cypri, Contra octoginta haereses opus eximium, Panarium siue capsula medica appellatum, & in libros quidem tres, tomos uerò septem diuisum. Eivsdem D. Epiphanii Liber ancoratvs, omnem de fide christiana doctrinam complectens. Eivsdem Contra octoginta haereses operis a se conscripti summa. Eivsdem libellus De ponderibus & mensuris*. [See also Epiphanius, Denis Petau, et al. 1622. *Sancti patris nostri Epiphanii Constantiae, siue Salaminis in Cypro, Episcopi, Opera omnia: in duos tomos distributa*. Paris: Sumptibus Michaelis Sonnii, Claudii Morelli, et Sebastiani Cramoisy. English edition: 1990. *The Panarion of St. Epiphanius, Bishop of Salamis*. Translated by Philip R. Amidon. New York: Oxford University Press.)

Eratosthenes. 1795. *Eratosthenis Catasterismi: cum interpretatione latina et commentario curavit Io. Conrad Schaubach; epistola C. G. Heyne cum animadversionibus in Eratosthenem; et cum tabulis aere incifis*. Edited by Johann Conrad Schaubach and Christian Gotlobb Heyne. Göttingen: Apud Vandenhoeck et Ruprecht. [Critical edition: 1897. *Pseudo-Eratosthenis Catasterismi*. Edited by Alexander Olivieri. Vol. 3, *Mythographi graeci: Bibliotheca scriptorum Graecorum et Romanorum Teubneriana*. Leipzig: G. B. Teubner.]

Erotianus. 1780. *Erotiani Galeni et Herodoti Glossaria in Hippocratem ex recensione Henrici Stephani graece et latine Accesserunt emendationes Henrici Stephani Bar-*

tholomaei Eustachii Adriani Heringae etc. Recensuit varietatem lectionis ex manuscriptis codd. Dorvillii et Mosquensi addidit suasque animadversiones. Edited by Johann Georg Friedrich Franz. Leipzig: Sumt. Iohannis Friderici Iunii. [*See also* Erotianus. 1918. *Erotiani vocvm Hippocraticarvm collectio cvm fragmentis.* Edited by Ernst Nachmanson. Gothenburg: Eranos' förlag.]

Eudoxus. *See* Achilles Tatius.

Fábrega, José Lino. 1900. *Interpretación del Códice Borgiano.* Edited by Alfredo Chavero and Francisco del Paso y Troncoso. Mexico City: Museo Nacional de México. [Humboldt had access to the manuscript. *See also* Seler, Eduard. 1904–9. *Codex Borgia, eine altmexikanische Bilderschrift der Bibliothek der Congregatio de Propaganda Fide.* 3 vols. Berlin: Druck von Gebr. Unger.]

Fabricius, Johann Albert. 1795. *Bibliotheca Graeca.* 12 vols. Hamburg: Carolum Ernestum Bohn. [Humboldt quotes from vols. 1 and 4.]

Firmicus Maternus, Iulius, et al. 1503. *Julii Firmici Astronomicorum libri octo integri & emendati, ex Scythicis oris ad nos muper allati. Marci Manilii astronomicorum libri quinque. Arati. Phænomena Germanico. Cæsare interprete cum commentariis & imaginibus. Arati phænomenon fragmentum. Marco T. C. interprete, Arati Phænomena Ruffo festo Avienio paraphraste.* Colophon, Turkey: Rhegii Lingobardiæ, expensis F. Mazalis. [Reprint of the 1499 Aldus Manutius (Manuzio) edition published as *Scriptores astronomici veteres* (Venice). Critical edition: Firmicus Maternus, Julius. 1897–1913. *Firmici Materni matheseos, II, VIII.* Edited by W. Kroll, F. Skutsch, and K. Ziegler. Leipzig: Teubner.]

Fontenelle, Bernard Le Bovier de. 1709. "Planisphère céleste Egyptien et Grec 1705." In *Histoire de l'Académie Royale des Sciences. Année MDCCVIII. Avec les mémoires de mathématique & physique, pour la même année. Tirés des registres de cette académie.* Paris: Jean Boudot. [Humboldt mentions that Fontenelle and others misidentified Bianchini's planisphere as Egyptian; he attributes this brief note to Fontenelle.]

Fourier, Jean Baptiste Joseph. 1809. "Recherches sur les sciences et le gouvernement de l'Égypte." In *Descriptions de l'Égypte ou recueil des observations et des recherches qui ont été faites en Égypte pendant l'expédition de l'armée française*, edited by Edme François Jomard, Antiquités, Mémoires 1, 803–24. Paris: Imprimerie Impériale.

———. 1818. "Premier mémoire sur les monumens astronomiques de l'Égypte." In *Descriptions de l'Égypte ou Recueil des observations et des recherches qui ont été faites en Égypte pendant l'Expédition de l'Armée Française*, edited by Edme François Jomard, Antiquités, Mémoires 2, 71–86. Paris: Imprimerie Impériale. [Humboldt had access to the manuscript, noting that Fourier's work was yet to be published. *See also* Jomard, Edme François.]

Fréret, Nicolas. 1758. *Défense de la chronologie: fondée sur les monumens de l'histoire ancienne, contre le système chronologique de M. Newton.* Edited by J. P. de Bougainville. Paris: Chez Durand. [English edition: 1728. *Some Observations on the Chronology of Sir Isaac Newton To Which Is Prefixed, His Chronology; Abridged by Himself. Done from the French, by a Gentleman.* London: printed for T. Warner.]

———. 1796. *Œuvres complètes. Édition augmentée de plusieurs ouvrages inédits, et ré-digée.* 20 vols. Edited by Leclerc de Sept-Chênes. Paris: Chez Dandré. [Humboldt quotes from vols. 10 and 12.]

García, Gregorio. 1607. *Origen de los Indios del Nuevo Mundo, e Indias Occidentales.* Valencia: Pedro Patticio Mey. [Critical edition: 2005. *Origen de los indios del Nuevo Mundo e Indias Occidentales.* Edited by Carlos Baciero et al. Madrid: Consejo Superior de Investigaciones Científicas.]

Garcilaso de la Vega. *See* Vega, Garcilaso de la, El Inca.

Gatterer, Johann Christoph. 1777. *Abriss der Chronologie.* Göttingen: J. C. Dieterich.

———. 1785–87. *Weltgeschichte in ihrem ganzen Umfange.* 2 vols. Göttingen: Vandenhoeck. [Vol. 1 is subtitled *Von Adam bis Cyrus: ein Zeitraum von 3652 Jahren*; volume 2 *Von Cyrus bis zu und mit der Völkerwanderung: ein Zeitraum von mehr als 1000 Jahren.*]

Gaubil, Le Père Antoine. *See* Souciet, Étienne, and Antoine Gaubil.

Gay-Lussac, Louis-Joseph. 1804. "Extrait de la relation d'un voyage aérostatique, fait par MM. Guy-Lussac et Biot, lue à la Classe des Sciences mathématiques et physiques de l'Institut national, le 9 fructidor an 12." *Journal de physique, de chimie et d'histoire naturelle* 59:314–20. [English translation: *Philosophical Magazine* 19 (1804): 371–9; 21 (1805): 220–27.]

Gazette Littéraire de Göttingen. 1811. "Voyage de Humboldt et Bonpland: Première partie, relation historique, atlas pittoresque; Auch unter dem speziellen Titel, *Vues de cordilières* [*sic*] *et monumens des peoples de l'Amérique, par Alexandre de Humboldt.*" *Göttingische Gelehrte Anzeigen unter der Aufsicht der königlichen Gesellschaft der Wissenschaften*, September 30, 1553–66. Göttingen: Heinrich Dieterich. [This review appeared in the year following the publication of *Vues.* The item of interest is in the *Anzeigen* on p. 1563.]

Gemelli Careri, Giovanni Francesco. 1699–1700. *Giro del mondo.* 6 vols. Naples: Guiseppe Roselli. [Vol. 1, *Turchia*; vol. 2, *Persia*; vol. 3 *Indostan*; vol. 4, *Cina*; vol. 5, *Isole Filippine*; vol. 6, *Nova Spagna.* A 1711 edition was the basis for subsequent editions and translations. English translation: 1745. "A voyage round the world: in six parts, viz. I Of Turky. II. Of Persia. III. Of India. IV. Of China. V. Of the Philippine Islands. VI. Of New Spain." In *A Collection of Voyages and Travels*, vol. 4, edited by A. and J. Churchill. London: Henry Lintot and John Osborn.]

Geminus of Rhodes. 1703. "Elementorum astronomiae." In *Uranologion*, edited by Dionysius Petavius [Denis Petau]. Antwerp: G. Gallet. [English edition: 2006. *Geminos's "Introduction to the Phenomena": A Translation and Study of a Hellenistic Survey of Astronomy.* Edited and translated by James Evans and J. Lennart Berggren. Princeton, NJ: Princeton University Press. *See* Petavius, Dionysius]

Georgi, Agustin Antonio. 1762. *Alphabetum tibetanum missionum apostolicarum.* Rome: Typis sacræ congregationis de propaganda fide. [Facsimile edition: 1987. *Alphabetum tibetanum missionum apostolicarum commodo editum.* Cologne: Editiones Una Voce.]

Georgi, Johann Gottlieb, trans. 1765. *Reise nach Ostindien und China.* By Pehr Osbeck et al. Rostock: Johann Christian Koppe. [Georgi translated the book from

Swedish into German. Originally published in 1757 as *Dagbok öfwer en ostindisk resa åren 1750. 1751. 1752. Med anmärkningar uti naturkunnigheten, främmande folkslags språk, seder, hushållning, m.m.* English edition: Osbeck, Peter. 1771. *A voyage to China and the East Indies.* Translated by John Reinhold Forster. London: Benjamin White.]

———. 1775. *Bermerkungen einer Reise im Russischen Reich im Jahre 1772.* 2 vols. St. Petersburg: Gedruckt bey der Kayserl Academie der Wissenschaften. [Vol. 2 has the title *Bemerkungen einer Reise im Russischen Reich in den Jahren 1773 und 1774.* See also Osbeck, Pehr, et al.]

Germanicus Caesar, trans. 1728. *Germanici Caesaris opera omnia; Aratea phaenomena, prognostica epigrammata et fragmenta.* Lüneburg: Typis Ortmannianis. [English edition: 1976. *The Aratus Ascribed to Germanicus Caesar.* Edited and translated by David Bruce Gain. London: Athlone. See also Ideler, Ludwig.]

Geroglyficos de que usavan los Mexicanos (Codex Telleriano-Remensis). See Quiñones Keber, Eloise.

Girbal y Barceló, Narciso. 1791. "Peregrinación por los Ríos Marañon y Ucatali á los Pueblos de *Manoa,* hecha por el Padre Predicador Apostólico Fray Narciso Girbal y Barceló en el año pasado de 1790." *Mercurio Peruano* 3, nos. 75–77: 49–72.

———. 1792. "Segunda Peregrinación del Padre Predicador Apostólico Fray Narciso Girbal y Barceló, á los Pueblos de Manoa." *El Mercurio peruano* 5, nos. 150–52: 89–115. Lima: Impr. de los Niños Huerfanos.

———. 1792. "Noticias Interesantes á la Religion y al Estado." *El Mercurio peruano* V (No. 153): 116–23. Lima: Impr. de los Niños Huerfanos.

Girbal y Barceló, Narciso, et al. 1809. *Voyages au Pérou, faits dans les années 1791 à 1794, par les pp. Manuel Sobreviela, et Narcisso y Barcelo.* 2 vols. Paris: J. G. Dentu.

Götze, Johann Christian. 1743–44. *Die Merckwürdigkeiten der Königlichen Bibliotheck zu Dreßden.* 6 vols. Dresden: Georg Conrad Walther. [Humboldt references the title as *Denkwürdigkeiten der Dresdner Bibliothek, erste Sammlung.* The first book is the Dresden Codex.]

Grosier, Jean-Baptiste, ed. 1777–85. *Histoire générale de la Chine; ou, annales de cet empire, traduites du tongkien-kang-mou.* 13 vols. Translated from the Chinese by Joseph Anne Marie de Moyria de Maillac. Paris: Pierres. [English edition: 1788. *A General Description of China: Containing the Topography of the Fifteen Provinces Which Compose This Vast Empire; That of Tartary, the Isles, and Other Tributary Countries; the Number and Situation of Its Cities, the State of Its Population, the Natural History of Its Animals, Vegetables and Minerals. Together with the Latest Accounts That Have Reached Europe, of the Government, Religion, Manners, Customs, Arts and Sciences of the Chinese.* London: Robinson.]

Guignes, Joseph de. 1756–58. *Histoire générale des Huns: des Turcs, des Mogols, et des autres Tartares occidentaux, &c. avant et depuis Jésus-Christ jusqu'a présent; précédée d'une introduction contenant des tables chronol. & historiques des princes qui ont régné dans l'Asie. Ouvrage tiré des livres chinois, & des manuscrits orientaux de la Bibliotheque du Roi.* 4 vols. Paris: Chez Desaint & Saillant.

Hager, Joseph. 1802. *Monument de Yu, ou La plus ancienne inscription de la Chine :*

Suivie de trente-deux formes d'anciens caractères chinois, avec quelques remarques sur cette inscription et sur ces caractères. Paris: Chez Treuttel et Wurtz, libraires, de l'Imprimerie de Pierre Didot l'aîné. [English edition: 1801. *Explanation of the Elementary Characters of the Chinese: With an Analysis of Their Ancient Symbols and Hieroglyphics.* London: Phillips.]

————. 1811. *Illustrazione d'uno zodiaco orientale del cabinetto delle medaglie di Sua Maestaea a Parigi: Scoperto recentemente presso le sponde del Tigri in vicinanza dell'antica Babilonia; Monumento, che serve ad illustrare la storia dell'astronomia, ed altri punti interessanti dell'antichità.* Milano: Gio. Giuseppe Destrefanis.

————. 1811. "Memoria sulle cifre arabiche attribuite fin'ai giorni nostri agli Indiani: Ma inventate in un paese più remoto dell' India." In *Fundgruben des Orients. Bearbeitet durch eine Gesellschaft von Liebhabern = Mines de l'Orient, exploitées par une société d'amateurs,* edited by Joseph von Hammer-Purgstall, 2:65–81. Vienna: Anton Schmid, K. K. Privil. [Humboldt refers to this publication as "Memoria sulle cifre de la Cina."]

Hakluyt, Richard. 1598–1600. *The principal navigations, voiages, traffiques and discoueries of the English nation: made by sea or ouer-land, to the remote and farthest distant quarters of the earth, at any time within the compasse of these 1500 yeeres: deuided into three seuerall volumes, according to the positions of the regions, whereunto they were directed: this first volume containing the woorthy discoueries, &c. of the English toward the north and norteast by sea . . . together with many notable monuments and testimonies of the ancient forren trades, and of the warrelike and other shipping of this realme of England in former ages, whereunto is annexed also a briefe commentarie of the true state of Island, and of the Northern seas and lands situate that way, and lastly, the memorable defeate of the Spanish huge Armada, anno 1588, and the famous victorie atchieued at the citie of Cadiz, 1596.* 3 vols. London: George Bishop, Ralph Newberie, and Robert Barker.

Hamilton, Alexander, and Louis-Mathieu Langlès. 1807. *Catalogue des Manuscrits Samskrits [sic] de la Bibliothêque Impériale, Avec des notices du contenu de la plupart des ouvrages, etc.* Paris: Imprimérie bibliothgraphique.

Hammer-Purgstall, Joseph, Freiherr von, ed. and trans. 1804. *Encyklopädische Uebersicht der Wissenschaften des Orients, aus sieben arabischen, persischen und türkischen Werken übersetzt: Den Freunden und Kennern der orientalischen Literatur, gewidmet von einem derselben Beflissenen.* 2 vols. Leipzig: Breitkopf und Härtel.

————, trans. 1806. *Die Posaune des heiligen Kriegs aus dem Munde Mohammed Sohns Abdallah des Propheten.* Leipzig: Gleditsch.

————. 1806. *Ancient Alphabets and Hieroglyphic Characters Explained: With an Account of the Egyptian Priests, Their Classes, Initiation, and Sacrifices, in the Arabic Language by Ahmad bin Abubekr bin Wahshih.* London: W. Bulmer and Co. and sold by G. and W. Nicol.

————, trans. 1809. *Schirin: Ein persisches romantisches Gedicht nach morgenländischen Quellen* by Niżāmī Ganjavī. Leipzig: G. Fleischer der Jüngere.

————, ed. 1809–18. *Fundgruben des Orients: Bearbeitet durch eine Gesellschaft von*

Liebhabern (*Mines de l'Orient, exploitées par une société d'amateurs*). 6 vols. Vienna: A. Schmid.

Hemsterhuis. *See* Hesychius.

Hermann, Martin Gottfried. 1787–95. *Handbuch der Mythologie aus Homer und Hesiod, als Grundlage zu einer richtigen Fabellehre des Alterthums, mit erläuternden Anmerkungen begleitet.* 3 vols. Berlin: Stettin.

———. 1801–2. *Mythologie der Griechen.* 2 vols. Berlin: Voß.

Hernández, Francisco. 1615. *Quatro libros. De la naturaleza, y virtudes de las plantas, y animales que estan recevidos en el uso de medicina en la Nueva España, y la methodo, y correccion, y preparacion, que para administrallas se requiere con lo que el doctor Francisco Hernandez escrivio en lengua latina.* Edited and translated by Francisco Jiménez. Mexico City: En casa de la viuda de Diego Lopez Davalos. [This book, a medicinal handbook, contains 478 chapters translated into Spanish from Latin, with annotations and additions by Jiménez. English edition: 2000. *The Mexican Treasury: The Writings of Dr. Francisco Hernández.* Edited by Simon Varey, translated by Rafael Chabrán, Cynthia L. Chamberlin, and Simon Varey. Stanford, CA: Stanford University Press.]

Herodotus, et al. 1763. *Hērodotu Halikarnēssēos Historiōn Logoi 9 Epigraphomenoi Musai = Herodoti Halicarnassei Historiarvm Libri IX Musarum Nominibus Inscripti. Accedvnt Praeter Vitam Homeri varia ex priscis scriptoribus de Persis, Aegyptiis, Nilo, Indisque Excerpta et praesertim ex Ctesia.* Edited by Petrus Wesseling. Amsterdam: Sumptibus Petri Schovtenii. [Other contributors include Ctesias, Lorenzo Valla, Thomas Gale, Jacobus Gronovius, and Lodewijk Caspar Valckenaer.]

Herodotus and Ctesias. 1802. *Histoire d'Hérodote: Traduite du grec, avec des remarques historiques et critiques, un essai sur la chronologie d'Hérodote, et une table géographique.* Edited and translated by Pierre-Henri Larcher. 9 vols. Paris: G. Debure l'aîné. [2nd edition; first published in 1786.]

Hernández, Francisco, Nardo Antonio Recchi, et al. 1651. *Rerum medicarum Novae Hispaniae thesaurus; seu, Plantarum animalium mineralium Mexicanorum historia.* Rome: Ex typographeio Vitalis Mascardi.

Herrera y Tordesillas, Antonio de. 1601–15. *Historia general de los hechos de los castellanos en las Islas i Tierra Firme del Mar Oceano.* 9 vols. Madrid: En la Emplenta Real. [French edition: 1600–1671. *Histoire générale des voyages et conquestes des Castillans, dans les isles & terre-ferme des Indes Occidentales.* 3 vols. Paris. English edition: Herrera, Antonio de. 1740. *The general history of the vast continent and islands of America, commonly call'd, the West-Indies, from the first discovery thereof: with the best accounts the people could give of their antiquities. Collectd from the original relations sent to the kings of Spain.* Translated by John Stevens. 6 vols. London: Wood and Woodward.]

Hervás y Panduro, Lorenzo. 1786. *Aritmetica delle nazioni e divisione del tempo fra l'orientali. Opera dell abbate Don Lorenzo Hervás.* Cesena, Italy: Per Gregorio Biasini all'Insegna di Pallade. [Humboldt references one volume of the multi-volume *Idea del Universo che contiene la storia della vita dell'uomo, elementi cos-*

mografici, viaggio estatico al mondo planetario e storia della terra e delle lingue (1779–87).]

———. 1807. "Letter from Abbè Don Lorenzo Hervas, to the Author, upon the Mexican Calendar." In *The History of Mexico. Collected from Spanish and Mexican Historians, from Manuscripts and Ancient Paintings of the Indians: Ilustrated by Charts, and Other Copper Plates, To which are Added, Critical Dissertations on the Land, the Animals, and Inhabitants of Mexico*, by Abbé D. Francisco Saverio Clavigero, 1:465–76. Translated by Charles Cullen. 2 vols. London: Printed for J. Johnson by Joyce Gold.

Hesiod. 1701. *Hesiodi Ascraei quae exstant: ex recensione Joannis Georgii Graevii cum ejusdem animadversionibus & notis auctoribus. Accedit commentarius nunc primùm editus Joannis Clerici, et notae variorum, scilicet Josephi Scaligeri, Danielis Heinsii, Francisci Guieti, & Stephani Clerici, ac Danielis Heinsii Introductio in doctrinam operum et dierum nec non index Georgii Pasoris.* Edited by Jean Le Clerc (Clerici). Amsterdam: Apud G. Gallet, praefectum typographiae Huguetanorum. [*See* 2006–7. *Hesiod.* Edited and translated by Glenn W. Most. 2 vols. Cambridge, MA: Harvard University Press.]

Hesychius. 1746–56. *Hesychii Lexicon, cum notis doctorum virorum integris, vel editis antehac, nunc auctis & emendatis, Hadr. Junii, Henr. Stephani, Jos. Scaligeri, Claud. Salmasii, Jac. Palmerii, Franc. Guyeti, Godefr. Sopingii, Jo. Fungeri, Jo. Cocceji, Jo. Fred. Gronovii, Jo. Casp. Suiceri, Tanaq. Fabri, Corn. Schrevelii, Ed. Bernardi, etc. Vel ineditis Henr. Valesii, Dan. Heinsii, Phil. Jac. Maussaci, Thom. Brunonis, Isaaci Vossii, Jo. Viti Pergeri, Thom. Munkeri, Marc. Meibomii, Jo. Verweji, etc. . . . Ex autographis partim recensuit, partim nunc primum edidit, suasque animadversiones perpetuas adjecit Joannes Alberti . . . Cum ejusdem prolegomenis, et adparatu Hesychiano.* Edited by Joannes Alberti. 2 vols. [Vol. 2 edited by David Ruhnkenius.] Leiden: Apud Samuelem Luchtmans, et filium, academiae taypographos.

Hipparchus. 1703. "Commentarii in Aratum." In *Uranologion.* Edited by Dionysius Petavius [Denis Petau]. Antwerp: G. Gallet. *See* Petavius, Dionysius.

Homer. 1805. "XVIII. Hymnus in Mercurium." In *Homeri Hymni et Batrachomyomachia,* edited by Augustus Matthiae. Leipzig: Weidmann. [In English edition: 1864. *Poems and Translations,* edited by Edward Vaughan Kenealy, 295–326. London: Reeves and Turner. *See also* Ruhnken, David.]

Horapollon. 1779. *Hiéroglyphes dits d'Horapolle. Ouvrage traduit du grec par M. Requier.* Edited by Charles-Hippolyte de Paravey. Amsterdam: J. F. Bastien.

Hug, Johann Leonhard. 1801. *Die Erfindung der Buchstabenschrift, ihr Zustand und frühester Gebrauch im Alterthum: Mit Hinsicht auf die neuesten Untersuchungen über den Homer.* Ulm, Germany: Wohler.

Humboldt, Alexander von. 1806. "Ueber die Urvölker von Amerika, und die Denkmähler, welche von ihnen übrig geblieben sind. Vorgelesen in der Philomathischen Gesellschaft." *Neue Berlinische Monatsschrift* [ed. Johann Erich Biester] 15 (January-June [March]): 177–208. Berlin: Nicolai. [New edition: 2009. *Ueber die Urvölker von Amerika und die Denkmähler welche von ihnen übrig geblieben sind anthropologische und ethnographische Schriften.* Edited by Oliver Lubrich. Fund-stücke 21. Hannover, Germany: Wehrhahn.]

———. 1808. *Ansichten der Natur mit wissenschaftlichen Erläuterungen Bd. 1 (1808).* [One volume appeared in 1808. The second German edition from 1826 came in 2 volumes. A French edition in 2 volumes appeared also in 1808 with the same pub-lisher: *Tableaux de la nature ou considérations sur les déserts, sur la physionomie des végétaux et sur les cataractes de l'Orénoque,* trans. Jean Baptiste Benoît Eyriès.]

———. 1808–11. *Essai politique sur le royaume de la Nouvelle-Espagne avec un atlas physique et géographique, fondé sur des observations astronomiques, de mesures trig-onométriques et des nivellemens barométriques.* Paris: F. Schoell.

———. 1811. *Atlas géographique et physique du Royaume de la Nouvelle-Espagne: Fondé sur des observations astronomiques, des mesures trigonométriques et des nivellemens barométriques.* Paris: Schoell. *See also* Velásquez, Joaquin.

Humboldt, Alexander von, and Aimé Bonpland. 1805–9. *Recueil d'observations de zoologie et d'anatomie comparée: Faites dans l'océan Atlantique, dans l'interieur du nouveau continent et dans la mer du sud, pendant les années 1799, 1800, 1801, 1802 et 1803.* Paris: Levrault, Schoell. [Includes contributions by Baron Cuvier and P. A. Latreille.]

———. 1807. *Essai sur la géographie des plantes, accompagné d'un tableau physique des régions équinoxiales, fondé sur des mesures exécutés, depuis le dixième degré de lati-tude boréale jusqu'au dixième degré de latitude australe, pendant les années 1799, 1800, 1801, 1802 et 1803 par Al. Humboldt et A. Bonpland.* Paris: Fr. Schoell. [En-glish edition: 2010. *Essay on the Geography of Plants.* Edited by Stephen T. Jack-son, translated by Sylvie Romanowski. Chicago: University of Chicago Press.]

Humboldt, Alexander von, and Joseph Louis Gay-Lussac. 1805. "Expériences sur les moyens eudiométriques et sur la proportion des principes constituants de l'atmosphère." *Annales de Chimie* 53:239–59. [Also published in the *Journal de Physique* 60 (1805): 129–68.]

Humboldt, Alexander von, and Jabbo Oltmanns. 1809. *Nivellement barométrique fait dans les régions équinoxiales du nouveau continent, en 1799, 1800, 1801, 1802, 1803, 1804.* Paris: F. Schoell. [Part of the *Receuil d'observations astronomiques* (1808–11).]

———. 1810. *Recueil d'observations astronomiques, d'opérations trigonométriques et de mesures barométriques, faites pendant le cours d'un voyage aux régions équinoxiales du nouveau continent, depuis 1799 jusqu'en 1803, par Alexandre de Humboldt, ré-digée et calculées, d'après les Tables les plus exactes, par Jabbo Oltmanns. Ouvrage auquel on a joint des recherches historiques sur la position de plusieurs points im-portans pour les navigateurs et pour les géographes.* Vol. 1. Paris: F. Schoell.

Hyginus, Gaius Julius. 1535. *C. Ivili Hygini Avgvsti Liberti Fabvlarvm Liber, Ad Om-nivm poëtarum lectionem mire necessarius & antehac nunquam excusus. Eivsdem Poeticon Astronomicon, libri quatuor. Quibus accesserunt similis argumenti. Pa-laephati de fabulosis narrationibus, liber I.F. Fvlgenti Placiadis Episcopi Carthag-inensis Mythologiarum, libri III. Eivsdem de uocum antiquarum interpretatione. liber. I. Arati "Phainomenōn" fragmentum, Germanico Caesare interprete. Eivs-*

dem Phaenomena Graece, cum interpretatione latina. Procli de sphaera libellus, Graece & Latine. Index rerum & Fabularum in his omnibus scitu dignarum copiosissimus. Edited by Jacobus Micyllus. Basel: Apud Ioan. Hervagivm.

———. 1742. *Auctores mythographi latini. Cajus Julius Hyginus, Fab. Planciad. Fulgentius, Lactantius Placidus, Albricus philosophus, cum integris commentariis Jacobi Micylli, Joannis Schefferi, et Thomae Munckeri, quibus adcedunt Thomae Wopkensii emendationes ac conjecturae. Curante Augustino van Staveren, qui & suas animadversiones adjecit.* Edited by Augustinus van Staveren. Leiden: S. Luchtmans; Amsterdam: J. Wetstenium et G. Smith.

Ideler, Ludwig. 1806. *Historische Untersuchungen über die astronomischen Beobachtungen der Alten.* Berlin: C. Quien.

———. 1809. *Untersuchungen über den Ursprung und die Bedeutung der Sternnamen: Ein Beytrag zur Geschichte des gestirnten Himmels.* Berlin: Bey Johann Friedrich Weils.

Jameson, Robert. 1804–8. *System of Mineralogy.* 3 vols. Edinburgh.

Jefferson, Thomas. 1784–85. *Notes on the state of Virginia: written in the year 1781, somewhat corrected and enlarged in the winter of 1782, for the use of a foreigner of distinction, in answer to certain queries proposed by him respecting; 1782.* Paris. [Observations by Charles Thompson. French edition: 1786. *Observations sur la Virginie.* Translated by André Morellet. Paris: Barrois, l'aîné, libraire, rue du Hurepoix, pres le pont Saint-Michel.]

Jiménez de Quesada, Gonzalo. *See* Piedrahita, Lucas Fernández de. [Humboldt states that Piedrahita's work was "composed following Quesada's manuscripts." Quesada also wrote the 1568 *Tres Ratos de Suesca* and the 1539 *Epítome del Nuevo Reino de Granada.* The *Epítome* has since been reproduced in Ramos Pérez, Demetrio. 1972. *Ximénez de Quesada en su relación con los cronistas y el "Epítome de la conquista del Nuevo Reino de Granada."* English edition: 1922. *The conquest of New Granada, being the life of Gonzalo Jimenez de Quesada.* Edited by Graham Cunninghame and Robert Bontine. London: W. Heinemann. Variations in spelling: Giménez, Ximénez, Jiménez.]

Jomard, Edme François, ed. 1809–28. *Description de l'Égypte, ou Recueil des observations et recherches qui ont été faites en Égypte pendant l'expedition de l'armée Française.* 21 vols. Paris: De l'Imprimerie impériale. [*See also* Fourier, Joseph, who oversaw the publication and wrote the "Préface historique." Jomard discusses passages from Achilles Tatius in *Descriptions de l'Égypte,* vol. 7, 2nd ed. (1822).]

Jones, William. 1790. "XXVII. A Supplement to the Essay on Indian Chronology." *Asiatick Researches* 2:389–403. [This particular article comes from the London reprint of 1801–18, of which only 12 volumes appeared.]

———. 1792. "The Eighth Anniversary Discourse, Delivered 24 February 1791. By the President." *Asiatick Researches* 3:1–16.

———. 1792. "XVI. Discourse the Ninth. On the Origin and Families of Nations. Delivered 23 February, by the President." *Asiatick Researches* 3:479–92.

———. 1801. *Supplemental Volumes Containing the Whole of the Asiatick Researches Hitherto Published Excepting Those Papers Already in His Works.* London: G. G.

and J. Robinson. [*See also* Teignmouth, John Shore, and Samuel Charles Wilks. 1835. *Memoirs of the Life, Writings, and Correspondence of Sir William Jones.* London: J. W. Parker.]

———. 1805. "VIII. Sur les dieux de la Grèce, de l'Italie et de l'Inde; dissertation composée en 1784, et revue depuis, par le Président." *Recherches Asiatiques* 1:162–213. [The French edition includes "Notes de M. Langlès sur le Mémoire precedent" (1798). English: "IX. On the Gods of Greece, Italy, and India, Written in 1784, and since revised by the President." *Asiatick Researches* 1221–75.]

———. 1805. "XXIV. III.ᵉ Discourse anniversaire, Prononcé, le 2 Février 1786, par le Président." *Recherches Asiatiques* 1:497–519. [English: 1801. "XXV. The Third Anniversary Discourse, Delivered 2 February, 1786, by the President." *Asiatick Researches* 1:415–31.]

———. 1805. "V.ᵉ Discours anniversaire, prononcé le 21 février 1788, par le Président. Sur les tartars." *Recherches Asiatiques* 2:35–69. [English: 1801. "II. The Fifth Anniversary Discourse, Delivered 21 February, 1788, by the President." *Asiatick Researches* 2:19–42.]

———. 1805. "VII. Sur la chronologie des hindous; Lu, au mois de Janvier 1788, par le Président." *Recherches Asiatiques* 2:164–97. [English: 1801. "VII. On the Chronology of the Hindus. Written in January, 1788, by the President." *Asiatick Researches* 2:111–47.]

———. 1805. "XVI. Sur l'antiquité du zodiaque indien, par le Président." *Recherches Asiatiques* 2:332–347. [English edition: 1801. "XVI. On the Antiquity of the Indian Zodiac." *Asiatick Researches* 2:289–306.]

Juan, Jorge, and Antonio de Ulloa. 1748. *Observaciones astronomicas, y phisicas: hechas de orden de S. Mag. en los reynos del Perù.* Madrid: Impresso de orden del rey nuestro señor, por J. de Zuñiga. *See also* Ulloa, Antonio de.

———. 1749. *Dissertacion historica, y geographica sobre el meridiano de demarcacion entre los dominios de España, y Portugal, y los parages por donde passa en la America Meridional, conforme à los tratados, y derechos de cada estado, y las mas seguras, y modernas observaciones.* Madrid: Impr. de A. Marin.

———. 1752. *Voyage historique de l'Amérique méridionale: fait par ordre du roi d'Espagne.* Translated by E. de Mauvillon. Amsterdam: Arkste'e & Merkus. [Published in Spanish as *Relación histórica del viaje a la América Meridional: hecho de orden de S. Mag. para medir algunos grados de meridiano terrestre, y venir por ellos en conocimiento de la verdadera figura, y magnitud de la tierra, con otras varias observaciones astronómicas y físicas* (1748). Includes Garcilaso de la Vega's "Resumen historico del origen, y succession de los Incas, y demas soberanos del Perú." English edition: 1964. *A Voyage to South America.* Translated by John Adams. New York: Alfred A. Knopf.]

Julián, Antonio. 1787. *La perla de la America, provincia de Santa Marta, reconocida, observada y expuesta en discursos historicos, por el sacerdote don Antonio Julian, á mayor bien de la Católica monarquia, fomento del comercio de España, y de todo el Nuevo Reyno de Granada, é incremento de la Christiana religion entre les naciones barbaras, que subsisten todavia rebeldes en la provincia.* Madrid: A. de Sancha.

[First volume of a trilogy: volume 2, *El paraíso terrestre en la América meridional y Nuevo Reino de Granada*; volume 3, *Historia del río Grande: por otro nombre Magdalena, y río de Santa Marta*. Facsimile edtion: 1980. *La perla de la América: Provincia de Santa Marta*. Edited by Luis Duque Gómez. Bogotá: Academia Colombiana de Historia. *See also* Julián, Antonio. 1994. *Monarquía del diablo en la gentilidad del nuevo mundo americano*. Edited and translated by Mario Germán Romero. Santafé de Bogotá: Instituto Caro y Cuervo.]

Kaempfer, Engelbert. 1729. *Histoire naturelle, civile, et ecclesiastique de l'empire du Japon*. Translated by John Gaspar Scheuchzer et al. The Hague: Chez P. Gosse & J. Neaulme. [English edition: Kaempfer, Engelbert, and John Gaspar Scheuchzer. 1727. *The History of Japan, Giving an Account of the Ancient and Present State and Government of that Empire; of Its Temples, Palaces, Castles and Other Buildings; of Its Metals, Minerals, Trees, Plants, Animals, Birds and Fishes; of the Chronology and Succession of the Emperors, Ecclesiastical and Secular; of the Original Descent, Religions, Customs, and Manufactures of the Natives, and of Their Trade and Commerce with the Dutch and Chinese. Together with a Description of the Kingdom of Siam*. London: The translator. *See also* 1999. *Kaempfer's Japan: Tokugawa culture observed*. Edited and translated by Beatrice M. Bodart-Bailey. Honolulu: University of Hawaii Press.]

Kalm, Pehr. 1754-64. *Beschreibung der Reise, die er nach dem noerdlichen Amerika unternommen hat: eine Uebersetzung*. Edited by Johann Andreas Murray. 3 vols. Göttingen: Vandenhoeck. [Originally published in Swedish as *En resa till Norra America* (1753-61). English version: Forster, John Reinhold, trans. 1972. *Travels into North America by Peter Kalm*. Translated by John Reinhold Forster. Barre, MA: Barre, Massachusetts Imprint Society.]

Kircher, Athanasius. 1650. *Athanasii Kircheri e Soc. Iesu. Obeliscus pamphilius, hoc est, interpretatio noua & hucusque intentata obelisci hieroglyphici quem non ita pridem ex Veteri Hippodromo Antonini Caracallae Caesaris, in agonale forum transtulit, integritati restituit & in urbis Aeternae ornamentum erexit Innocentius X. Pont. Max. In quo post varia Aegyptiacae, Chaldaicae, Hebraicae, Graecanicae antiquitatis, doctrinae que quà sacrae, quà profanae monumenta, veterum tandem theologia, hieroglyphicis inuoluta symbolis, detecta é tenebris in lucem asseritur*. Rome: Typis Ludouici Grignani.

———. 1652-54. *Athanasii Kircheri e Soc. Iesv, Oedipus aegyptiacus hoc est Vniuersalis hierolglyphicae veterum doctrinae temporum iniuria abolitae instauratio: opus ex omni orientalium doctrina & sapientia conditum, nec non viginti diuersarium linguarum, authoritate stabilitum, felicibus auspicijs Ferdinandi III, Austriaci sapientissimi & inuictissimi Romanorum imperatoris semper Augusti è tenebris erutum, atque bono reipublicae literariae consecratum*. 3 vols. Rome: Ex typographia Vitalis Mascardi. [Vol. 2 has subtitle: *Gymnasium sive phrontisterion hieroglyphicum in duodecim classes distributum, in quibus encyclopaedia Aegyptiorum, id est, veterum hebraeorum, Chaldaeorum, Aegyptiorum. Graecorum, coeterorumque orientalium recondita sapienta, hucusque temporum iniuria pedita, per artificiosum sacrarum sculpturarum contextum demonstrata, instauratur*. Volume 3 has subtitle: *Theatrum hieroglyphicum, hoc est, noua & hucusque intentata obelisco-rum coeterorumque hieroglyphicorum monumentorum, quae tùm Romae, tùm in Aegypto ac celebrioribus Europae musaeis adhuc supersunt interpretatio iuxta sen-sum physicum, tropologicam, mysticum, historicum, politicum, magicum, medi-cum, mathematicum, cabalisticum, hermeticum, sophicum, theosophicum; ex omni Orientalium doctrina & sapientia demonstrata. See also* Zoëga, Johann Georg.]

Klaproth, Julius Heinrich von. 1811. "Ueber die Sprache und Schrift der Uiguren." In *Fundgruben des Orients: Bearbeitet durch eine Gesellschaft von Liebhabern (Mines de l'Orient, exploitées par une société d'amateurs).* Edited by Joseph von Hammer-Purgstall, 2:167–95. Vienna: Anton Schmid, K. K. Privil.

Krusenstern, Adam Johann von. 1810–12. *Reise um die Welt in den Jahren 1803, 1804, 1805 und 1806 auf Befehl Seiner Kaiserl. Majestät Alexanders des Ersten auf den Schiffen Nadeshda und Newa unter dem Commando des Capitäns von der Kaiserl. Marine A. J. Krusenstern.* Berlin: Haude und Spener. [Originally published in Russian, 1809–13. Humboldt quotes from the 2nd edition published in German in 1810–12. *See also* Krusenstern, Adam Johann von. 1814. *Atlas zur Reise um die Welt: Unternommen auf Befehl seiner Kaiserlichen Majestät Alexander des Ersten auf den Schiffen Nadeshda und Neva; unter dem Commando des Capitains von Krusenstern.* St. Petersburg: Verf.]

La Condamine, Charles Marie de. 1745. *Relation Abrégée d'un Voyage Fait dans l'intérieur de l'Amérique Méridionale: Depuis la Côte de la Mer du Sud, jusqu'aux Côtes du Brésil & da la Guiane, en descendant La Rivière des Amazones; Lûe à l'Assemblée publique de l'Académie des Sciences, le 28. Avril 1745; Avec une Carte du Maragnon, ou de la Rivière des Amazones, levée par le même.* Paris: Pissot.

———. 1748. "Mémoire sur quelques anciens monumens du Pérou, du tems des Incas." *Histoire de l'Académie Royale des Sciences et des Belles Lettres [Berlin] année 1746*[1748]: 435–56. Berlin: Ambroise Haude.

———. 1751. *Mesure des trois premiers degrés du méridien dan l'hémisphere austral tirée des observations de M.rs de l'/Académie Royale des Sciences, envoyés par le roi sous l'équateur.* Paris: Imprimerie Royale.

———. 1751. *Histoire des Pyramides de Quito, elevées par les académiciens envoyés sous l'Equateur par ordre du Roi.* Paris.

———. 1752–54. *Supplément au Journal historique du voyage à l'équateur, et au livre de la Mesure des trois premiers degrès du méridien: servant de reponse à quelques objections.* Paris: Durand [et] Pissot.

La Condamine, Charles Marie de, and Pierre Bouguer. 1751–54. *Journal de voyage fait par ordre du Roi à l'équateur servant d'introduction historique à la mesure des trois premiers degrés du méridien.* 2 vols. Paris: Impr. Royale. [Vol. 1, *Introduction historique: Ou journal des travaux des Académiciens envoyés par ordre du Roi sous l'Équateur.* Vol. 2, *Mésure des trois premiers degrés du méridien dans l'hémisphere austral.*]

La Nauze, Louis Jouard de. 1743. "L'année vague des Égyptiens." *Histoire de l'Académie royale des inscriptions et belles-lettres, avec les Mémoires de littérature tirés des registres de cette Académie, depuis l'année M. DCCXXXVIII jusques & compris l'année M. DCCXL* 14:334–75.

——. 1751. "De l'année solaire des Égyptiens, dite l'année Alexandrine." *Histoire de l'Académie royale des inscriptions et belles-lettres, avec les Mémoires de littérature tirés des registres de cette Académie, depuis l'année M. DCCXLI jusques & compris l'année M. DCCXLIII* 16:170–92.

——. 1751. "L'année lunaire des Égyptiens." *Histoire de l'Académie royale des inscriptions et belles-lettres, avec les Mémoires de littérature tirés des registres de cette Académie, depuis l'année M. DCCXLI jusques & compris l'année M. DCCXLIII* 16:193–204.

Lactantius. 1777. *L. Coelii Firmiani Lactantii Divinarum institutionum liber quintus; sive, De justitia.* Edited by David Dalrymple. Edinburgh: Ex officina A. Murray et J. Cochran. [English edition: 2003. *Divine Institutes.* Translated by Anthony Bowen and Peter Garnsey. Liverpool: Liverpool University Press.]

Lafitau, Joseph F. 1724. *Moeurs des sauvages américains comparées aux moeurs des premiers temps.* 2 vols. Paris: Saugrain l'aîné et Charles Étienne Hochereau. [English edition: 1974–77. *Customs of the American Indians Compared with the Customs of Primitive Times.* Edited and translated by William N. Fenton and Elizabeth L. Moore. 2 vols. Toronto: Champlain Society.]

Lahontan, Louis Armand de Lom d'Arce. 1703. *Nouveaux voyages de Mr. le baron de Lahontan dans l'Amérique Septentrionale, qui contiennent une relation des différens peuples qui y habitent; la nature de leur gouvernement; leur commerce, leurs coutumes, leur religion, & leur manière de faire la guerre. L'intérêt des François & des Anglois dans le commerce qu'ils font avec ces nations; l'avantage que l'Angleterre peut retirer dans ce pais, étant en guerre avec la France. Le tout enrichi de cartes & de figures.* 2 vols. The Hague: Chez les frères l'Honoré. [Vol. 2 is titled *Mémoires de l'Amérique Septentrionale, ou La suite des voyages de Mr. le baron de Lahontan: Qui contiennent la description d'une grande étendue de pais de ce continent, l'intérêt de François & des Anglois, leurs commerces, leurs navigations, les mœurs & les coutumes des sauvages &c.; avec un petit dictionnaire de langue du pais; le tout enrichi de cartes & de figures.* English edition: 1703. *New voyages to North-America.* 2 vols. London: Printed for H. Bonwicke, T. Goodwin, M. Wotton, B. Tooke, and S. Manship.]

——. 1704. *Suite du voyage, de l'Amerique, ou Dialogues de Monsieur le baron de Lahontan et d'un sauvage, dans l'Amerique. Contenant une desciption exacte des moeurs & des coutumes de ces peuples sauvages. Avec les voyages du meme en Portugal & en Danemarc, dans lesquels on trouve des particularitez trés curieuses, & qu'on n'avoit point encore remarquées. Le tout enrichi de cartes & de figures.* Edited by Nicolas Gueudeville. Amsterdam : Chez la veuve de Boetman, et se vend a Londres, chez D. Mortier. [Third or supplementary volume of Lahontan's voyages; divided in two sections: *Dialogues de Monsieur le Baron de Laontan et d'un Sauvage de l'Amerique* and *Voyages de Portugal et de Danemarc.*]

Lalande, Joseph Jérôme Le Français de. 1792. *Astronomie.* 3 vols. Paris: Desaint & Saillant. [3rd edition; 1st published in 1764, 2nd in 1771–81.]

———. 1803. *Bibliographie astronomique: Avec l'histoire de l'astronomie depuis 1781 jusqu'à 1802*. Paris: l'Imprimerie de la République.

Lambeck, Peter. 1776. *Petri Lambecii Hamburgensis Commentariorum de Augustissima Bibliotheca Caesarea Vindobonensi*. Vol. 3 of 8. Vindobonae: J. T. nob. de Trattnern. [First volume published in 1665, 8th in 1679.]

Langlès, Louis-Mathieu. 1789–90. *Dictionnaire tartare-mantchou françois composé d'après un dictionnaire mantchou-chinois*. See Amiot, Joseph.

———. 1795–98. Notes to *Voyage d'Égypte et de Nubie*. See Norden, Frederik Ludvig.

———. 1796. Notes to *Voyages de C. P. Thunberg, au Japon*. See Thunberg, Carl Peter.

———. 1804. *Rituel des tatars-mantchoux, rédigé par l'ordre de l'empereur Kien-Long, et précédé d'un discours préliminaire composé par ce souverain, avec les dessins des principaux ustensiles et instruments du culte chamanique: Ouvrage traduit par extraits du tatar-mantchou, et accompagné des textes en caractères originaux*. Paris: L'imprimerie de la république.

———. 1805. Notes to *Recherches Asiatiques*. See Asiatick Society of Bengal (Calcutta) and Jones, William.

———. 1807. *Alphabet mantchou: Rédigé d'après le syllabaire et le dictionnaire universel de cette langue*. Paris: Impr. Impériale.

———. 1811. "Notice chronologique de la Perse, depuis les temps les plus reculé jusqu'à ce jour." In *Voyages du chevalier Chardin en Perse, et autres lieux de l'Orient: enrichis d'un grand nombre de belles figures en taille-douce, représentant les antiquiés et les choses remarquables du pays* by John Chardin, edited by Louis-Mathieu Langlès, 151–239. Vol. 10 of 10. Paris: Le Normant, Imprimeur-Libraire.

Laplace, Pierre Simon de. 1799–1825. *Traité de mécanique céleste*. 5 vols. Paris: Impr. de Crapelet; A Paris, Chez J. B. M. Duprat. [Vols. 1–4 published 1799–1805; vol. 5 in 1825. English edition: 1829–39. *Celestial Mechanics by the marquis de La Place*. Translated by Nathaniel Bowditch. 4 vols. Boston: Hillard, Gray, Little, and Wilkins.]

———. 1808. *Exposition du système du monde*. 2 vols. 3rd ed. Paris: Chez Courcier. [English edition: 1830. *The System of the World by M. le marquis de Laplace*. Translated by Henry H. Harte. 2 vols. Dublin: Printed at the University Press for Longman, Rees, Orme, Brown, and Green.]

Larrea y Villancico, Juan de. 1996. "Las manufacturas de la Provincia de Quito." Edited by Christian Büschges. *Procesos: Revista Ecuatoriana de Historia* 9, no. 2: 139–43. [Written in 1802.]

Le Gentil de la Galaisière, Guillaume Joseph Hyacinthe Jean Baptiste. 1775. "Mémoire dans lequel on fait voir que de France à Canton, par le Nord-est, les voyages feroient presqu'aussi longs qu'ils le sont par le Cap de Bonne-espérance." In *Histoire de l'Académie royal des sciences. Année 1772. Première Partie*, 452–55. Paris: Imp. Royale.

———, 1780–81. *Voyage dans les mers de l'Inde, fait par ordre du roi, à l'occasion du passage de Vénus, sur le disque de soleil, le 6 juin 1761, & le 3 du même mois 1769*. 5 vols. Switzerland: Chez Les Libraires Associés. [English edition: 1964. *A Voyage to the Indian Seas.* Translated by Frederick C. Fischer, introduction by William Alain Burke Miailhe. Manila: Filipiniana Book Guild.]

Lederer, John. 1681. "Extrait d'une lettre escrite de Lyon d'l'Auteur du Journal par M. Spon fils D.M. contenant quelques choses particulaires des Americains de la Virginie, tirées des Memoires de Jean Lederer de Hambourg qui revenoü de ce Pays là apres dix ans de sejour qu'il y avoit fait." *Journal des Sçavans* 6:71–2. [Periodical published since 1665; title varies slightly: 1665–1790, *Journal des Sçavans*; 1791–1832, *Journal des Savans*; 1833–, *Journal des Savants*. English edition: 1958. *The Discoveries of John Lederer, with Unpublished Letters by and about Lederer to Governor John Winthrop, Jr., and an Essay on the Indians of Lederer's Discoveries by Douglas L. Rights and William P. Cumming.* Edited by William P. Cumming. Charlottesville: University of Virginia Press.]

León y Gama, Antonio de. 1792. *Descripción histórica y cronológica de las dos piedras: que con ocasión del nuevo empedrado que se está formando en la Plaza Principal de México, se hallaron en ella el año de 1790. Explícase el sistema de los calendarios de los indios, el método que tenian de dividir el tiempo, y la correcion que hacian de él para igualar el año civil, de que usaban, con el año solar trópico. Noticia muy necesaria para la perfecta inteligencia de la segunda piedra: a que se añaden otras curiosas e instructivas sobre la mitología de los mexicanos, sobre su astronomía, y sobre los ritos y ceremonias que acostumbraban en tiempo de su gentilidad.* México: Imprenta de don Felipe de Zúñiga y Ontiveros. [Also in an expanded 1832 edition edited by Carlos María de Bustamante. Facsimile edition: 2009. *Descripción histórica y cronológica de las dos piedras: Que con ocasión del nuevo empedrado que se está formando en la plaza principal de México, se hallaron en ella el año de 1790.* Facs. of 2nd (1832) ed. Edited by Eduardo Matos Moctezuma. Mexico City: Instituto Nacional de Antropología e Historia.]

Linné [Linnaeus], Carl von. 1764. *Species plantarum exhibentes plantas rite cognitas, ad genera relatas : cum differentiis specificis, nominibus trivialibus, synonymis selectis, locis natalibus, secundum systema sexuale digestas.* 2 vols. Vienna: Typis Joannis Thomæ de Trattner. [1st edition in 1753. *See also* Richter, Hermann E., 2003. *Codex botanicus Linnaeanus.* 2 vols. Edited by John Edmondson, translated by Sten Hedberg. Ruggell: A. R. G. Gantner Verlag.]

———. 1789–91. *Genera plantarum: eorumque characteres naturales secundum numerum, figuram, situm, et proportionem omnium fructificationis partium.* 2 vols. Edited by Johann Christian Daniel Schreber. Frankfurt-am-Main: Sumtu Varrentrappii et Wenneri.

———. 1805. *Systema vegetabilium: secundum classes, ordines et genera a clar. Willdenowio partim, partique a summa plantarum desumpta adjecta Appendice plantarum officinaliam cum characteristics et differntiis specificis &c.* Edited by Josue Scannagatta. Bonn: Jacous Marsigli.

Lipsius, Justus. 1595–96. *Iusti Lipsi de militia romana libri quinque. Commentarius ad Polybium. E parte prima historicae facis.* Antwerp: Ex Officina Plantiniana, Apud Viduam, & Ioannem Moretum.

Livius [Livy], Titus. 1735. *T. Livii Patvini historiarvm: libri qvi svpersunt ex editione et cvm notis Ioannis Clerici; adiecta est divesitas lectionis Gronovianae; cum prae-fatione Io. Matthiae Gesneri in academia Jottingensi.* 3 vols. Leipzig: Weidmann. [English edition: 2008. *The Rise of Rome: Books One to Five.* Edited by T. J. Luce. Oxford: Oxford University Press.]

López de Gómara, Francisco. 1553. *Hispania Victrix: primera y seguna parte de la historia general de las indias con todo el descubrimiento, y cosas notables que han acaesido don de que se ganaron hasta el año de 1551: con la conquista de Mexico y de la nueva España.* Medina del Campo, Spain: Guillermo de Millis. [Another 1553 edition was published published in Zaragoza, where the first edition had appeared in 1552. For a facsimile of the 1552 edition see 1978. *Historia de las Indias y con-quista de México: Zaragoza, 1552.* Mexico City: Centro de Estudios de Historia de México Condumex. Abridged English edition: 1965. *Cortés: The Life of the Con-queror by His Secretary Francisco Lòpez de Gòmara.* Lesley Byrd Simpson. Berkeley: University of California Press.]

Lorenzana, Francisco Antonio. *See* Cortés, Hernán (Fernando).

Lort, Michael. 1787. "XXV. Account of an Ancient Inscription in North America. Read November 23, 1786." *Archaeologia: or, Miscellaneous Tracts Relating to Antiquity. Published by the Society of Antiquaries of London* 8:290–301. London: J. Nichols, Printer to the Society. [Lort mentions this as his source for the inscription Court-de-Gébelin's *Monde Primitif* (p. 292). *See also* Court-de-Gébelin, Antoine.]

Lucian of Samosata. 1789. *Œuvres de Lucien: traduites du grec, avec des remarques historiques & critiques sur le texte de cet auteur, et la collation de six manuscrits de la Bibliotèque du roi.* Translated by Jacques Nicolas Belin de Ballu. 6 vols. Paris: Jean-François Bastien. [English edition: 1968–88. *Lucian in eight volumes.* Edited and translated by Austin Morris Harmon et al. Cambridge, MA: Harvard University Press; London: Heinemann.]

Lugo, Bernardo de. 1619. *Gramática de la lengua general del Nuevo Reyna, llamada Mosca.* Madrid: por Barnardino de Guzmā.

Macartney, George, et al. 1804. *Voyage en Chine et en Tartarie.* Translated by Jean-Baptiste Joseph Breton de la Martinière. 7 vols. Paris: chez la veuve Lepetit. [Vol. 7 is titled *Atlas du voyage en Chine et en Tartarie.* Originally published in English as *An Authentic Account of an Embassy from the King of Great Britain to the Emperor of China: Including Cursory Observations Made, and Information Obtained, in Travelling through that Ancient Empire, and a Small Part of Chinese Tartary, together with a Relation of the Voyage Undertaken on the Occasion by His Majesty's Ship the Lion, and the Ship Hindostan, in the East India Company's Service, to the Yellow Sea, and Gulf of Pekin, as well as of Their Return to Europe, with Notices of the Several Places where They Stopped in Their Way out and Home, Being the Islands of Madeira, Teneriffe, and St. Jago, the Port of Rio de Janeiro in South America, the Islands of St. Helena, Tristan d'Acunha, and Amsterdam, the coasts of Java, and Sumatra, the Nanka Isles, Pulo Condore, and Cochin China* (1797).] *See also* Staunton, George L.

Macrobius, Ambrosius Theodosius. 1556. *Macrobii Ambrosii Aurelii Theodosii, viri consularis, & illustris, In Somnium Scipionis, lib. II., Saturnaliorum, lib. VII., ex varijs, ac vetustissimis codicibus recogniti, & aucti.* Leiden: Seb. Gryphium. [English edition: 1990. *Commentary on the Dream of Scipio.* Translated by William Harris Stahl. New York: Columbia University Press. *See also* 1969. *The Saturnalia.* Translated by Percival Vaughan Davies. New York: Columbia University Press.]

Majer, Friedrich. 1803–4. *Allgemeines Mythologisches Lexicon.* Weimar: Im Verlage des Landes-Industrie-Comptoirs.

———. 1811–13. *Mythologisches Taschenbuch, oder Darstellung und Schilderung der Mythen, religiösen Ideen und Gebräuche aller Völker: Nach den besten Quellen für jede Klasse von Lesern entworfen.* Weimar: Im Verlage des Landes-Industrie-Comptoirs.

Manetho. 1698. *Manethonis Apotelesmaticorum libri sex nunc primum ex bibliotheca Medicea editi cura Jacobi Gronovii qui etiam Latine vertit ac notas adiecit.* Edited by Jakob Gronovius. Leiden: Frederic Haaring.

Manetho and J. V. D. Galateajus. 1815. *Manethonis Sebennytae series regum Aegypti: In XXX dynastias distributa . . . ; adjiciuntur nonnullis regibus nomina alia, quibus Herodotus, Diodorus, Eratosthenes et orientales historici eosdem reges distinxerunt, in hoc opusculo continentur; clavis et introductio ad chronologiam et historiam Aegypti.* (English edition: Manetho. 1971. *Manetho's History of Egypt.* Translated by W. G. Waddell. London: W. Heinmann.]

Manilius, Marcus. 1679. *Astronomicon: Interpretatione et notis ac figuris illustravit Michael Fayus, Jussu Christianissimi nissimi Regis, in usum serenissimi Delphini.* Edited by Michael Fay. Paris: Fredericum Leonard.

Marchand, Étienne. 1797–99. *Voyage autour du monde: pendant les anées 1790, 1791, et 1792.* Edited by Charles Pierre Claret de Fleurieu. 6 vols. Paris: L'Imprimerie de la République. [English edition: 1801. *A Voyage round the World, Performed during the Years 1790, 1791, and 1792.* London: Printed for T. N. Longman and O. Rees.]

Marieta, Juan de. 1596. *Historia eclesiastica de todos los santos, de España: Primera, Segunda, Tercera y Quarta parte; Con dos tablas.* Cuenca, Ecuador: Pedro del Valle.

Marini, Luigi Gaetano. 1795. *Gli atti e monumenti de' Fratelli Arvali scolpiti gia' in tavole di marmo ed ora raccolti, diciferati e comentati.* Rome: Presso Antonio Fulgoni.

Márquez, Pietro. 1804. *Due antichi Monumenti di architettura messicana illustrati.* Rome: Presso il Salomini.

Martianus Capella. 1499 [1498]. *De nuptiis Philologiae et Mercurii.* Edited by Franciscus Vitalis Bodianus. Vicenza: Rigo di Ca'Zeno. [English edition: 1977. *The marriage of Philology and Mercury.* Translated by William Harris Stahl, Richard Johnson, and E. L. Burge. New York: Columbia University Press. *See also* 1994–98. *The Berlin commentary on Martianus Capella's De nuptiis Philologiae et Mercurii.* Edited by Haijo Jan Westra. 2 vols. Leiden: E. J. Brill.]

Martini, Martino. 1654. *Histoire de la gverre des Tartares, contre la Chine. Contenant les reuolutions estranges qui sont arriuées dans ce grand royaume, depuis quarante ans*. Paris: I. Henavlt. [Originally in Latin. English edition: 1654. *Bellum Tartaricum; or, The Conquest of the Great and Most Renowned Empire of China, by the Invasion of the Tartars, who in these Last Seven Years, Have Wholy Subdued that Vast Empire. Together with a Map of the Provinces, and Chief Cities of the Countries, for the Better Understanding of the Story.* London: John Crook. Martini was a disciple of Athanasius Kircher.]

Mayer, Tobias, and Nevil Maskelyne. 1770. *Tabulae motuum solis et lunae novae et correctae; auctore Tobia Mayer quibus accedit methodus longitudinum promota, eodem auctore. Editae jussu praefectorum rei longitudinariae* = *New and correct tables of the motions of the sun and moon, by Tobias Mayer to which is added the method of finding the longitude improved ; by the same author.* Published by order of the Commissioners of Longitude. London: typis Gulielmi et Johannis Richardson; prostat venalis apud Johannem Nourse, Johannem Mount et Thomam Page.

Mayer, Tobias, and Georg Christop Lichtenberg. 1775. *Opera Inedita.* Göttingen: J. Ch. Dieterich. [English edition: 1972. *Unpublished Writings of Tobias Mayer.* Edited by E. G. Forbes. Göttingen: Vandenhoeck & Ruprecht.]

Meares, John. 1790. *Voyages Made in the Years 1788 and 1789 from China to the North West Coast of America to Which Are Prefixed an Introductory Narrative of a Voyage Performed in 1786 from Bengal in the Ship Nootka, Observations on the Probable Existence of a North West Passage, and Some Account of the Trade between the North West Coast of America and China, and the Latter Country and Great Britain.* London: Printed at the Logographic Press and sold by J. Walter.

Medrano, Antonio. 1598. *Historia del Nuevo Reyno de Granada.* Unpublished manuscript. [Humboldt notes that Medrano is one of Piedrahita's sources. *See* 2003. "Descripción del Nuevo Reino de Granada." Edited by Michael Francis. *Anuario Colombiano de Historia Social y de la Cultura* 30:341–60. The manuscript is at the Bancroft Library, Berkeley, California. *See also* Piedrahita, Lucas Fernández de, and Aguado, Pedro de.]

Ménestrier, Claude-François. 1750. *La nouvelle méthode raisonnée du blason; reduite en leçons, par demandes et par résponses. Enrichie de figures en taille-douce planches.* Lyon: Frères Bruyset.

Mercati, Michele. 1589. *Considerationi sopra gli Avvertimenti del Sig. Latino Latini intorno ad alcune cose scritte nel libro degli Obelischi di Roma.* Rome.

El Mercurio peruano. 1791–95. 12 vols. Edited by Jacinto Calero y Moreira. Sociedad Académica de Amantes de Lima. Lima: Impr. de los Niños Huerfanos. [Diego Cisneros contributed to this periodical].

Michaux, André. 1801. *Histoire des chênes de l'Amérique: ou, Descriptions et figures de toutes les espèces et variétés de chênes de l'Amérique Septentrionale, considérées sous les rapports de la botanique, de leur culture et de leur usage.* Paris: De l'imprimerie de Crapelet. [English edition: 1904. *Travels west of the Alleghanies: Made in 1793–96 by André Michaux, in 1802 by F. A. Michaux, and in 1803 by Thaddeus Mason Harris.* Translated and edited by Reuben Gold Thwaites. Cleveland: A. H. Clark.]

———. 1803. *Flora boreali-americana: Sistens caracteres planatarum quas in America septentrionali collegit et detexit: tabulis aeneis 51 ornata.* 2 vols. Paris: Apud fratres Levrault.

Mill, David. 1743. *Dissertationes selectae, varia s. litterarum et antiquitatis orientalis capita, exponentes et illustrantes. Curis 2., novisque dissertationibus, orationibus, et miscellaneis orientalibus auctae.* Leiden: Wishoff. [1st edition 1724.]

Montfaucon, Bernard de. 1729–33. *Les monumens de la monarchie françoise: qui comprennent l'histoire de France, avec les figures de chaque regne que l'injure des tems a epargnées.* 5 vols. Paris: Gandouin et Giffart. [English edition: 1750. *A collection of regal and ecclesiastical antiquities of France, in upwards of three hundred large folio copper plates.* 2 vols. London: W. Innys, J. and P. Knapton, R. Manby, and H. S. Cox.]

Monthly Magazine. See Phillips, Richard.

Montucla, Jean Étienne. 1799–1802. *Histoire des mathématiques dans laquelle on rend compte de leurs progrès depuis leur origine jusqu'à nos jours, où L'on expose le tableau et le développement des principales découvertes dans toutes les parties des mathématiques, les contestations qui se sont élevées entre les mathématiciens, et les principaux traits de la vie des plus célèbres.* 4 vols. Paris: H. Agasse. [1st edition published in 2 volumes in 1758.]

Moor, Edward. 1810. *The Hindu Pantheon.* London: J. Johnson.

Moziño Suárez de Figueroa, José Mariano. 1970. *Noticias de Nutka: An Account of Nootka Sound in 1792.* Edited and translated by Iris Higbie Wilson. Seattle: University of Washington Press. [Humboldt had access to Moziño's manuscript. *See also* Mociño, José Mariano. 1998. *Las "Noticias de Nutka."* Edited by Fernando Monge and Margarita del Olmo. Aranjuez, Mexico: Ediciones Doce Calles; Madrid: Consejo Superior de Investigaciones Científicas.]

Nessel, Daniel von. 1690. *Catalogus, sive recensio specialis omnium codicum manuscriptorum Graecorum, nec non linguarum orientalium augustissimae Bibliothecae caesareae vindobonensis.* 7 parts. Vienna: L. Voigt & J. B. Endteri.

Niebuhr, Barthold Georg. 1811–12. *Römische Geschichte.* 2 vols. Berlin: Verlag von G. Reimer. [English edition: 1835. *The History of Rome.* Translated by Julius Charles Hare and Connop Thirlwall. 2 vols. Philadelphia: Thomas Wardle.]

Nieremberg, Juan Eusebio, and Balthasar Moretus. 1635. *Ioannis Eusebii Nierembergii Madritensis ex Societate Iesu in Academia Regia Madritensi physiologiæ professoris Historia naturae, maxime peregrinae, libris XVI. distincta: In quibus rarissima naturæ arcana, etiam astronomica, & ignota Indiarum animalia, quadrupedes, aues, pisces, reptilia, insecta, zoophyta, plantæ, metalla, lapides, & alia mineralia, fluuiorumque & elementorum conditiones, etiam cum proprietatibus medicinalibus, describuntur; nouae & curiosissimae quaestiones disputuntur, ac plura sacrae scripturae loca eruditè enodantur. Accedunt de miris & miraculosis naturis in Europâ libri duo: item de iisdem in terrâ Hebræis promissâ liber vnus.* Antwerp: Ex officina Plantiniana Balthasaris Moreti.

Nonnus of Panopolis. 1809. *Dionysiacorum libri sex ab octavo ad decimum tertium, Res Bacchicas ante expeditionem Indicam complectentes.* Edited with notes by G. H. Moser, preface by Fridericus Creuzer. Heidelberg: Mohr et Zimmer. [English edition: 1940–42. *Dionysiaca*. Translated by W. H. D. Rouse. 3 vols. Cambridge, MA: Harvard University Press; London: W. Heinemann.]

Norden, Frederik Ludvig. 1795–98. *Voyage d'Égypte et de Nubie*. Edited by Louis-Mathieu Langlès. 3 vols. Paris: Imprimerie de Pierre Didot l'ainé. [First published in 1752; another edition in 1755, translated from the Danish by J. B. des Roches de Parthenay. English edition: Norden, Frederick Lewis.1757. *Travels in Egypt and Nubia*. Translated by Peter Templeman. London: L. Davis and C. Reymers.]

Núñez de la Vega, Francisco. 1702. *Constituciones diocesanas del Obispado de Chiappa*. Rome: Caietano Zonobi. [*See also* 1988. *Constituciones diocesanas del Obispado de Chiapas*. Edited by León Cázares, María del Carmen, and Mario Humberto Ruz. Meexico City: Universidad Nacional Autónoma de México, Centro de Estudios Mayas.]

Olmos, Andrés de. 2002. *Arte de la lengua mexicana: Concluído en el convento de San Andrés de Ueytlalpan en la provincia de la Totonacapan que es en la Nueva España, el 1. de enero de 1547*. Edited by Ascensión Hernández de León-Portilla y Miguel León-Portilla. Mexico City: Universidad Nacional Autónoma de México. [First published in 1885 as *Arte para aprender la lengua mexicana*. Humboldt had access to the 1547 manuscript, which is currently housed at the Latin American Library at Tulane University. *See also* Maxwell, Judith Marie. 1992. *Of the Manners of Speaking That the Old Ones Had: The Metaphors of Andrés de Olmos in the TULAL Manuscript "Arte para aprender la lengua mexicana," 1547; With Nahuatl/English, English/Nahuatl concordances*. Salt Lake City: University of Utah Press.]

Origen. 1733–59. *Opera omnia quae graece vel latine tantum exstant et ejus nomine circumferuntur, ex variis editionibus, & codicibus manu exaratis, Gallicanis, Italicis, Germanicis & Anglicis collecta, recensita, latine versa, atque annotationibus illustrata, cum copiosis indicibus, vita auctoris, & multis dissertationibus*. Edited by Charles Delarue. 4 vols. Paris: Jacobi Vincent. [Includes *Contra Celsum*. Vol. 2 published in 1738; vol. 3 in 1740; and vol. 4 in 1759.]

Palin, Nils Gustaf. 1802. *Lettre sur les hiéroglyphes. Davum me, non oedipum*. [Paris.]

———. 1804. *Analyse de l'inscription en hiéroglyphes du monument trouvé à Rosette, contenant un décret des pretres de l'Egypte en l'honneur de Ptolémée Epiphane*. Dresden: Walther.

———. 1804. *Essai sur les hiéroglyphes, ou Nouvelles lettres sur ce sujet*. Weimar: Bureau d'industrie.

———. 1812. *De l'étude des hiéroglyphes: Fragmens*. Paris: Delaunay.

Pallas, Peter Simon. 1788–93. *Voyages de M. P. S. Pallas en différentes provinces de l'empire de Russie, et dans l'Asie septentrionale*. Translated by M. Gauthier de la Peyronie. 5 vols. Paris: Lagrange. [Originally published in German as *Bemerkungen auf einer Reise in die südlichen Statthalterschaften des Russischen Reichs in den Jahren 1793 und 1794* (1771). English edition: 1812. *Travels through the Southern Provinces of the Russian Empire: In the Years 1793 and 1794*. Translated by Francis William Blagdon. 2 vols. London: Printed for John Stockdale.]

Paolini da Sancto Bartholomaeo. 1791. *Systema Brahmanicum liturgicum, mytho-*

logicum, civile, ex Monumentis Indicis Musei Borgiani Velitris Dissertationibus historico-criticis illustravit Paullinus a S. B. Rome: apud Antonium Fulgonium. [Alternate spelling: Paulinus a S. Bartolomaeo. Name is John Philip Wesdin; known as Paulinose Patiri.]

———. 1793. *Musei Borgiani Velitris Codices Manuscripti Avenses, Peguani, Siamici, Malabarici, Indostani, animadeversionibus historico-criticis castigati et illustrati Accedunt Monumenta inedita, et Cosmogonia Indico-Tibetana.* Rome: Antonium Fulgonium.

———. 1796. *Viaggio alle Indie Orientali umiliato alla santita di N.S. Papa Pio Sesto P.M.* Rome: Presso Antonio Fulgoni. [English edition: Paolino da San Bartolomeo. 1800. *A Voyage to the East Indies: Containing an Account of the Manners, Customs, &c. of the Natives, with a Geographical Description of the Country; Collected from Observations Made during a Residence of Thirteen Years, between 1776 and 1789, in Districts Little Frequented by the Europeans.* Edited by John Reinhold Forster, translated by William Johnston. London: Printed by J. Davis.]

Papillon, Jean-Michel. 1766. *Traité historique et pratique de la gravure en bois.* 2 vols. Paris: Pierre Guillaume Simon. [Vol. 1, *La partie historique*; vol. 2, *Tous les principes de cet art.*]

Pappus of Alexandria. 1660. *Pappi Alexandrini Mathematicae collectiones a Federico Commandino Vrbinate in latinum conuersae & commentarijs illustratae: In hac nostra editione ab innumeris, quibus scatebant mendis, & praecipue in graeco contextu diligenter vindicatae et serenissimo principi Leopoldo Gvlielmo, archidvci Avstriae, &c. dicatae.* Bologna: Ex typographia HH. de Duccijs. [English edition: 1986. *Book 7 of the Collection by Pappus Alexandrinus.* Translated by Alexander Jones. New York: Springer-Verlag.]

Park, Mungo. 1799–1800. *Voyage dans l'intérieur de l'Afrique, fait en 1795, 1796 et 1797, avec des éclaircissemens sur la géographie de l'intérieur de l'Afrique.* Translated by Jean-Henri Castéra. 2 vols. Paris: chez Dentu, Carteret. [Originally in English. *See also* 2000. *Travels in the Interior Districts of Africa.* Translated by Kate Ferguson Marsters. Durham, NC: Duke University Press.]

Pausanias. 1696. *Pausaniae Graeciae descriptio accurata, qua ceu manu per eam regionem circumducitur cum latina Romuli Amasaei interpretatione. Accesserunt Gvl. Xylandri & Frid. Sylburgii annotationes, ac novae notae Ioachimi Kuhnii.* Edited by Wilhelm Xylander, translated by Romulo Amaseo. Leipzig: Thoma Fritsch. [*See also* 1931–55. *Description of Greece.* Translated by W. H. S. Jones and H. A. Omerod. 5 vols. Cambridge, MA: Harvard University Press; London: Heinemann.]

Pauw, Cornelius de. 1768–69. *Recherches philosophiques sur les américains, ou mémoires intéressants pour servir à l'histoire de l'espèce humaine.* 2 vols. Berlin: George Jacques Decker, imp. du Roi. [English edition: 1806. *A General History of the Americans, of Their Customs, Manners, and Colours; An History of the Patagonians, of the Blafards, and White Negroes; History of Peru; An History of the Manners, Customs, &c. of the Chinese and Egyptians.* Edited by Daniel Webb. Rochdale, MA: T. Wood.]

Petavius, Dionysius [Denis Petau]. 1627. *Opus De Doctrina temporum*. Paris: Sebastian Cramoisy.

———. 1703. *Uranologion Sive Systema Variorum Auctorum, Qui de Sphæra Ac Sideribus eorumqve motibus Græce commentati sunt, quorum nomina post Præfationem leguntur. Omnia vel Græce ac Latine nunc primum edita, vel ante non edita; Item Variarum Dissertationum Libri VIII. Accesserunt in hac Nova Editione Ejusdem Petavii Et Jc. Sirmondi Dissertationes de Annmo Synodi Sirmiensis [et] Fidei formulis in ea editis, Petavii Lib. II de Lege et Gratia, Elenchus Theriacæ Vincentii Lenis, Dissertatio de Adjutorio sine quo non, et Adjutorio quo, quæ in Dogmatibus Theologicis omissa fuerunt; Ac Denique Epistolarum Libri Tres.* Antwerp: G. Gallet. [A compilation of different ancient authors, including a Tatius fragment in a Latin translation first published in 1567. Humboldt quotes various authors from this edition.]

Piedrahita, Lucas Fernández de. 1688. *Historia general de las conquistas del Nuevo Reyno de Granada: a la S.C.R.M. de D. Carlos Segundo, Rey de las Españas, y de las Indias.* Edited by Juan Baptista Verdussen. 2 vols. Amsterdam.

Phillips, Richard, ed. 1801. "Varieties, Library and Philosophical. Including Notices of Works in Hand, Domestic and Foreign." *The Monthly Magazine, or British Register* 11 (January-June): 336-45.

Pingré, Alexandre Guy. 1783-4. *Cométographie, ou, Traité historique et théorique des comètes.* 2 vols. Paris: Imprimerie royale.

Plato. 1578. *Platōnos hapanta ta sōzomena = Platonis opera quæ extant omnia: Ex nova Ioannis Serrani interpretatione, perpetuis eiusde[m] notis illustrata: quibus & methodus & doctrinæ summa breuiter & perspicue indicatur. Eivsdem Annotationes in quosdam suæ illius interpretationis locos. Henr. Stephani de quorundam locorum interpretatione iudicium, & multorum contextus Græci emendatio.* Edited by Henri Estienne, translated by Jean de Serres. 3 vols. [Geneva]: Henr. Stephanus.

Pliny the Elder. 1778-91. *Caii Plinii Secundi Historia Naturali. Cum interpretatione et notis integris Iohannis Harduini itemque cum commentariis et adnotationibus Hermolai Barbari, Pintiani, Rhenani, Gelenii, Dalechampii, Scaligeri, Salmasii, Is. Vossii, I.F. Gronovii, et variorum. Recensuit varietatemque lectionis adiecit Ioh. Georg. Frid. Franzius.* 10 vols. Leipzig: Impensis Guilielmi Gottlob Sommeri. [English edition: 1967-75. *Natural History.* Translated by H. Rackham. 10 vols. Cambridge, MA: Harvard University Press.]

Pluche, Noël Antoine. 1740. *Histoire du ciel, où l'on recherche l'origine de l'idolatrie, et les méprises de la philosophie, sur la formation des corps célestes, & de toute la nature.* 2 vols. Paris: chez la veuve Estienne & Fils. [English edition: 1752. *The History of the Heavens: Considered According to the Notions of the Poets and Philosophers, Compared with the Doctrines of Moses: being an inquiry into the origine of idolatry, and the mistakes of philosophers, upon the formation and influence of the celestial bodies.* Translated by John Baptist De Freval. 2 vols. London: J. Wren.]

Plutarch. 1624. *Plvtarchi chæronensis omnivm qvae exstant opervm. Tomvs primvs, continens vitas parallelas. Cum Latina interpretatione Crvserii, & Xylandri: Et doctorum virorum notis: Et libellis variantium lectionum ex mss. codd. diligenter collectarum: Et indicibus accuratiss. Eiusdem Plutarchi Liber de fluuiorum montiúmque nominibus, antehac non editus: cum versione & notis Mavssaci. Accedit nvnc primùm Plutarchi vita, ex ipso, & aliis vtriusque linguae scriptoribus, à Ioan Rvaldo collecta digestáque. Eiusdem Rvaldi animaduersiones ad insignia Plutarchi sphalmata, siue lapsiones II. & LXX.* Edited and translated by Hermann Cruser, Wilhelm Xylander, Philippus Jacobus Maussacus, and Jean Ruault. 2 vols. Lutetiae Parisiorum: Typis Regiis, apud Societatem graecarum editionum. [English edition: Babbitt, Frank Cole, et al., trans. 1927–2004. *Plutarch's Moralia.* 17 vols. Cambridge, MA: Harvard University Press; London: W. Heinemann.]

———. 1716–74. *Plutarchi Chaeronensis quae supersunt, omnia, graece et latine: principibus ex editionibus castigavit, virorumque doctorum suisque annotationibus instruxit.* Edited by Johann Jacob Reiske. 12 vols. Leipzig: In libraria Weidmannia. [Vols. 1–5, *Vitarum parallelarum*; vols. 6–10, *Operum moralium et philosophicorum*; vols. 11–12, indexes.]

Pococke, Richard. 1772–73. *Voyages de Richard Pococke, membre de la Société Royale, & de celle des Antiquités de Londres, &c. en orient, dans l'Égypte, l'Arabie, la Palestine, la Syrie, la Grece, la Thrace, &c. &c.* 2nd ed. Translated by Eydous and de La Flotte. 6 vols. Neufchâtel: Aux depens de la Société typographique. [Originally published in 1745 in English (2 vols., London).]

Polybius. 1609. *Polybii Historiarum libri qui supersunt Isaacus Casaubonus ex antiquis libris emendavit, latine vertit & commentariis illustravit; Æneæ vetustissimi Tactici commentarius De toleranda obsidione Is. Casaubonus primus vulgavit, Latinam interpretationem ac notas adiecit.* Edited and translated by Isaac Casaubon. [Frankfurt-am-Main]: Typis Wechelianis apud Claudium Marnium & haeredes Iohannis Aubrii. [English edition: 1954. *The Histories.* Translated by W. R. Paton. 6 vols. Cambridge, MA: Harvard University Press.]

Pownall, Thomas. 1779. "XXXII. Observations Arising from an Enquiry into the Nature of the Vases found on the Mosquito Shore in South America." *Archaeologia: or Miscellaneous Tracts Relating to Antiquity. Published by the Society of Antiquaries of London* 5:318–24.

Prévost, A. F., ed. 1747–80. *Histoire générale des voyages; ou, Nouvelle collection de toutes les relations de voyages par mer et par terre, qui ont été publiées jusqu'à présent dans les différentes langues de toutes les nations connues . . . pour former un système complet d'histoire et de geographie moderne, qui repr'sentera l'état actuel de toutes les nations: enrichie de cartes géographiques.* 25 vols. La Haye : P. de Hondt. [Continued by A. G. Meusnier de Querlon, A. Deleyre, and J. P. Roussellot de Surgy, with subsequent additions by J. P. J. Du Bois and others.]

Ptolemy. *See* Theon.

Purchas, Samuel, et al. 1625. *Haklvytvs Posthumus or Purchas His Pilgrimes: Contayning a History of the World, in Sea Voyages, & Lande-Trauells, by Englishmen and Others. Some Left Written by Mr. Hakluyt at His Death, More since Added, His also Perused, & Perfected. All Examined, Abreuiated, Illustrated w[i]th Notes, Enlarged w[i]th Discourses. Adorned w[i]th Pictures, and Expressed in Mapps.*

In Fower Parts, Each Containing Five Books. 5 vols. London: Printed by William Stansby for Henry Fetherstone.

Quatremère de Quincy, Antoine-Chrysostome. 1803. *De l'architecture égyptienne, considérée dans son origine, ses principes et son goût et comparée sous les mêmes rapports à l'architecture grecque, dissertation qui a remporté en 1785, le prix proposé par l'Académie des inscriptions et belles lettres.* Paris: Barrois.

———. 1805. "Sur l'idéal dans les arts du dessin." *Archives Littéraires de l'Europe, ou Mélanges de Literature, d'Histoire et de Philosophie : Par une Société de Gens de Lettres ; Suivi d'une Gazette Littéraire Universelle* 7:289–337. Paris: Henrichs; Tübingen: Cotta.

Qazwīnī, Zakarīyī ibn Muhammad. 1848–49. *Zakarija ben Muhammed ben Mahmud el-Cazwini's Kosmographie.* Edited by Ferdinand Wüstenfeld. Göttingen: Verlag der Dieterichschen Buchhandlung. [Reprint: 1967. Wiesbaden: Martin Sändig oHG.]

Quintus Curtius Rufus. 1801. *Q. Curtii Rufi De rebus gestis Alexandri Magni libri: Cum supplementis Io. Freinshemii praemittitur notitia literaria accedit index studiis Societatis Bipontinae.* 2 vols. Argentorati: ex typographia Societatis. [English edition: 1946. *Quintus Curtius [History of Alexander].* Translated by John C. Rolfe. 2 vols. Cambridge, MA: Harvard University Press.]

Quiñones Keber, Eloise. 1995. *Codex Telleriano-Remensis: Ritual, Divination, and History in a Pictorial Aztec Manuscript.* Austin: University of Texas Press. [This codex surfaced in the seventeenth century in France and was at first wrongly believed to be the source of Purchas's illustrations in the *Codex Mendoza* (1625); Humboldt was the first to challenge this view. The codex is preserved in the Bibliothèque Nationale de France in Paris. *See also* Purchas, Samuel.]

Ramond, Louis-François Elisabeth. 1801. *Voyages au mont-perdu et dans la partie adjacente des Hautes-Pyrénées.* Paris: Belin.

Ramusio, Giovanni Battista. 1556. "Relatione d'alcune cose della Nuova Spagna, & della gran citta di Temistitan Messico, fatta per un gentil'huomo del Signo Fernando Cortese." In *Terzo volume delle navigationi et viaggi.* Venice: Stamperia de Givnti.

———. 1613. *Delle navigationi et viaggi, in tre volumi divise.* Venice: Giunti.

Recherches Asiatiques, ou memoires de la société établie au Bengale : Pour faire des recherches sur l'histoire et les antiquités, les arts, les sciences et la literature de l'Asie. 1805. Translated by Antoine-Gabriel Griffet de La Baume. 2 vols. Paris: L'Imprimerie Impériale. [Translation of the *Asiatick Researches*; only two volumes were published.]

Revillagigedo, Juan Vicente Güémez Pacheco de Padilla Horcasitas y Aguayo, conde de. 1979. "Officio del Officio del 5 septiembre 1790." In *Trabajos arqueológicos en el centro de la Ciudad de México: Antología* by Leopoldo Batres and Eduardo Matos Moctezuma, 30. Mexico City: Instituto Nacional de Antropologia e Historia.

Rhode, Johann Gottlieb. 1809. *Versuch über das Alter des Thierkreises und den Ursprung der Sternbilder: mit erläuternden Kupfern.* Breslau [Wrocław]: Korn.

Ríos, Pedro de los. 1979. *Codex Vaticanus 3738 ("Cod. Vat. A," "Cod. Ríos") der Biblioteca apostolica Vaticana: Farbreproduktion des Codex in verkleinertem Format.* Graz, Austria: Akademische Druck- und Verlagsanstalt. [*See also* Anders, Ferdinand, Maarten Jansen, Luis Reyes García, ed. 1996. *Religión, costumbres e historia de los antiguos mexicanos: Libro explicativo del llamado Códice Vaticano A, Codex Vatic. Lat. 3738 de la Biblioteca Apostólica Vaticana.* Edited by Ferdinand Anders, Maarten Jansen, and Luis Reyes García. Graz, Austria: Akademische Druck- und Verlagsanstalt; Mexico City: Fondo de Cultura Económica.]

Robertson, William. 1803. *The History of America.* 4 vols. London: Strahan. [1st edition 1777.]

Ruhnken, David. 1749. *Davidis Ruhnkenii Epistola critica 1. In Homeridarum hymnos et Hesiodum, ad virum clarissimum, Ludov. Casp. Valckenarium.* Leiden: Cornelium de Pekker.

Sahagún, Bernardino de. 2000. *Historia general de las cosas de Nueva España.* Edited by Alfredo López Austin and Josefina García Quintana. 3 vols. Mexico City: Consejo Nacional para la Cultura y las Artes. [Also known as Florentine Codex. Humboldt had access to Sahagún via Torquemada. *See* 1970–82. *General History of the Things of New Spain.* Edited by Arthur J. O. Anderson and Charles E. Dibble. 12 vols. Santa Fe, NM: School of American Research; [Salt Lake City]: University of Utah. *See also* Torquemada, Juan de.]

Salt, Henry. 1809. *Twenty-Four Views in St. Helena, the Cape, India, Ceylon, the Red Sea, Abyssinia and Egypt, from Drawings.* London: W. Miller.

———. 1814. *A Voyage to Abyssinia and Travels into the Interior of that Country, Executed under the Orders of the British Government, in the Years 1809 and 1810: In Which Are Included, an Account of the Portuguese Settlements on the East Coast of Africa, Visited in the Course of the Voyage; A Concise Narrative of Late Events in Arabia Felix; and Some Particulars Respecting the Aboriginal African Tribes, Extending from Mosambique to the Borders of Egypt; Together with Vocabularies of Their Respective Languages.* London: F. C. and J. Rivington.

Saussure, Horace Bénédict de. 1780–96. *Voyages dans les Alpes, précédés d'un essai sur l'histoire naturelle des environs de Genève.* 8 vols. Neuchâtel: Fauche-Borel. [Abridged English translation: 1793 [1790]. *A Description of the Two Albinos of Europe, (One Twenty-One, the Other Twenty-Four Years of Age.) Extracted from M. Sassure's Journey to the Alps, in the Year 1785.* Liverpool: printed by Thomas Johnson. *See also* Carozzi, Albert V. 1995. *Horace-Bénédict de Saussure, Forerunner in Glaciology: New Manuscript Evidence on the Earliest Explorations of the Glaciers of Chamonix and the Fundamental Contribution of Horace-Bénédict de Saussure to the Study of Glaciers between 1760 and 1792.* Geneva: Editions Passé Présent; and Saussure, Horace Bénédict de. 2000. *Manuscripts and Publications of Horace-Bénédict de Saussure on the Origin of Basalt, 1772–1797 = Manuscrits et publications de Horace-Bénédict de Saussure sur l'origine du basalte, 1772–1797: Italy, 1772–73, Auvergne and Vivarais, 1776, Alps, 1779–96, Provence, 1780–87, Brisgau, 1791–94, Des Basaltes, 1794, Auvergne, 1795, Agenda, 1796, last note, 1797.* Geneva: Zoé.]

Scaliger, Joseph Juste. 1598. *Iosephi Scaligeri Iuli Cæsaris f. opus de emendatione temporum: castigatius & multis partibus auctius, ut nouum videri possit. Item veterum Græcorum fragmenta selecta, quibus loci aliquot obscurissimi chronologiæ sacræ & Bibliorum illustrantur, cum notis eiusdem Scaligeri.* Leiden: ex officina Plantiniana, Francisci Raphelengij.

Schaubach, Johann Konrad. 1802. *Geschichte der griechischen Astronomie bis auf Eratosthenes.* Göttingen: J. F. Röwer.

Schlegel, Friedrich von. 1808. *Über die Sprache und Weisheit der Indier: Ein Beitrag zur Begründung der Altertumskunde.* Heidelberg: Mohr und Zimmer.

Schlözer, August Ludwig von. 1772–73. *Vorstellung seiner Universal-Historie.* 2 vols. Göttingen: Bey Johann Christian Dieterich.

———. 1792–1801. *Weltgeschichte nach ihren Haupttheilen im Auszug und Zusammenhange.* 2 vols. Göttingen: Vandenhoeck und Ruprecht.

Schmidt, Friedrich Samuel von. 1759. *Dissertatio de zodiaci nostri origine aegyptia.* Bern: Societe litteraire.

Schwartz, Carl Gottlieb. 1807. *Recherches sur l'origine et la signification des constellations de la sphère grecque.* Paris: Migneret.

Seetzen, Ulrich Jasper. 1810. *A Brief Account of the Countries Adjoining the Lake of Tiberias, the Jordan, and the Dead Sea.* Bath: Printed and sold by Meyler and Son; London: Hatchard, Piccadilly.

Seetzen, Ulrich Jasper, and Franz Xaver von Zach. 1803. *Reise-Nachrichten des Russisch-Kaiserlichen Kammer-Assessors U. J. Seetzen.*

Seneca, Lucius Annaeus. 1658–59. *L. Annaei Senecae philosophi Opera omnia ex ult,* vol. 1, *Lipsii & I. F. Gronovii emendat. et M. Annaei Senecae rhetoris quae exstant ex. And. Schotti recens.* 3 vols. Amsterdam: Apud Elzevirios. [Vol. 2 (1658) has the title *L. Annaei Senecae philosophi Tomus secundus, in quo Epistolae, & Quaestiones naturales*; vol. 3 (1658), *M. Annaei Senecae rhetoris, Suasoriae, Controversiae, cum Declamationum excerptis, tomus tertius.* English edition: 1971–72. *Naturales quaestiones.* Translated by Thomas H. Corcoran. 2 vols. Cambridge, MA: Harvard University Press.]

Servius. *See* Virgil.

Sextus Empiricus. 1718. *Opera Græce et Latine Pyrrhoniarum Institutionum Libri III; Cum Henr. Stephani Versione et Notis; Contra Mathematicos, Sive Disciplinarum Professores, Libri VI; Contra Philosophos Libri V; Cum Versione Gentiani Herveti. Græca Ex Mss. Codicibus Castigavit, Versiones Emendavit Supplevitqve, et Toti Operi Notas Addidit Jo. Albertus Fabricius.* Edited by Henri Estienne. Leipzig: Gleditsch. [English edition: 1949–57. *Sextus Empiricus.* Translated by R. G. Bury. 4 vols. Cambridge, MA: Harvard University Press.]

Shore, John. 1801. "XVII. Account of the Kingdom of Nepal. By Father Giuseppe, Prefect of the Roman Mission. Communicated by John Shore, Esq." *Asiatick Researches* 2:307–22. London: G. Auld. *See* Asiatick Society of Bengal (Calcutta).

Sigüenza y Góngora, Carlos de. 1959. *Libra astronómica y filosófica.* Edited by Bernabé Navarro. Mexico City: Centro de Estudios Filosóficos, Universidad Nacional Autónoma de México.

Sigüenza y Góngora, Carlos de, and María Casteñada de la Paz. 2006. *Pintura de la peregrinación de los culhuaque-mexitin, Mapa de Sigüenza: Análisis de un documento de origen tenochca.* Zinacantepec, Mexico: Colegio Mexiquense. [*See* Gemelli Careri, Giovanni Francesco. *See also* 1963. *Documentos inéditos de don Carlos de Sigüenza y Góngora: La Real Universidad de México y don Carlos de Sigüenza y Góngora [y] El reconocimiento de la Bahía de Santa María de Galve.* Edited by Irving A. Leonard. Mexico City: Centro Bibliográfico Juan José de Eguiara y Eguren; and, Keen, Benjamin, and Juan José Utrilla. 1984. *La imagen azteca: en el pensamiento occidental.* Mexico City: Fondo de Cultura Económica.]

Silvestre de Sacy, Antoine Isaac. 1806. *Chrestomathie Arabe, ou extraits de divers écrivains Arabes tant en prose qu'en vers, à l'usage des élèves de l'école spéciale des langues Orientales des langues Orientales vivantes.* 3 vols. Paris: Imprimirie Impériale. [First edition. In French and Arabic, the first volume has the Arabic text, the second and third volume the French translation by Antoine Léonard de Chézy.]

———. 1810. *Grammaire arabe à l'usage des élèves de l'École Spéciale des Langues Orientales Vivantes.* 2 vols. Paris: Imprimerie Impériale. [*See also* 1834. *Principles of General Grammar.* Translated by David Fosdick. Andover, MA: Flagg, Gould and Newman.]

Solís, Antonio de. 1783-84. *Historia de la conquista de México, poblacion y progesos de la America Septentrional, conocida por el nombre de Nueva España.* 2 vols. Madrid: Antonio de Sancha. [English edition: 1753. *The History of the Conquest of Mexico by the Spaniards.* Edited by Nathaniel Hooke, translated by Thomas Townsend. 2 vols. London: H. Lintot. *See also* Arocena, Luis A. 1963. *Antonio de Solís, cronista indiano: Estudio sobre las formas historiográficas del barroco.* Buenos Aires: Editorial Universitaria de Buenos Aires.]

Sonnerat, Pierre. 1806. *Voyage aux Indes Orientales et a la Chine, fait par ordre de Louis XVI, depuis 1774 jusqu'en 1781: Dans lequel on traite des mœurs, de la religion, des sciences et des arts des Indiens, des Chinois, des Pégouins et des Madégasses; suivi d'observations sur le Cap de Bonne-Espérance, les îles de France et de Bourbon, les Maldives, Ceylan, Malacca, les Philippines et les Moluques, et de recherches sur l'histoire naturelle de ces pays, etc., etc.* 4 vols. Paris: Dentu. [English edition: Magnus, Francis, trans. 1788-89. *A Voyage to the East-Indies and China performed by order of Lewis XV between the years 1774 and 1781. Containing a description of the manners, religion, arts.* Translated by Francis Magnus. 3 vols. Calcutta: Stuart and Cooper.]

Sosigenes. *See* Buttmann, Philipp, and Ideler, Ludwig.

Souciet, Étienne, and Antoine Gaubil. 1729-32. *Observations mathématiques, astronomiques, géographiques, chronologiques et physiques: tirées des anciens livres chinois ou faites nouvellement aux Indes et à la Chine, par les pères de la Compagnie de Jesus.* 3 vols. Paris: Chez Rollin libraire. [Vol. 1 includes an excerpt from a letter dated August 20, 1728, from Father du Croz (also Ducros) to Souciet titled "Le zodiaque des indiens" (243-46). Vol. 2 carries the special title *Histoire de l'astronomie chinoise avec dissèrtations* and vol. 3 *Traité de l'astronomie chinoise.*

Humboldt refers to this book as *Traité de l'astronomie chinoise, Observations astronomiques,* or *Observations mathématiques.*]

Srīpeti. 1957. *Jyotisa-ratna-mālā of Srīpati Bhatta : A Marathi ṭīkā on His Own Sanskrit Work.* Edited by Murlidhar Gajanan Panse. Poona, India: Deccan College Postgraduate and Research Institute. [Written c. 1059 BCE. Humboldt references this work as *Ratnamālā,* citing as his source the *Recherches Asiatiques. See also* Jones, William, and Asiatick Society of Bengal (Calcutta).]

Staunton, George Leonard. 1797. *Authentic Account of an Embassy from the King of Great Britain to the Emperor of China; including Cursory Observations made, and Information obtained, in travelling through that Ancient Empire and a small part of Chinese Tartary &. Together with a Relation of the Voyage Undertaken on the occasion by his Majesty's Ship The Lion, and the Ship Hindostan, in the East India Company's Service, to the Yellow Sea, and the Gulf of Peking: as well as of their return to Europe. With Notices of the several places where they stopped in their way out and home; being the Islands of Madeira, Teneriffe, and St. Jago; the Port of Rio de Janeiro in South America; the Islands of St. Helena, Tristan d'Aclinba, and Amsterdam; the Coast of Java, and Sumatra, the Nanka Isles, Pulo-Condore, and Cochin-china. Taken chiefly from the papers of His Excellency the Earl of Macartney.* 3 vols. London: Printed by J. Nicol.

Stobaeus. 1797. *Ioannis Stobaei Sermones.* Edited by Nicol Schow. Leipzig: In libraria Weidmannia. [Alternate spellings: Stobée or Stobaei.]

Stolberg, Friedrich-Leopold Graf von. 1806. *Geschichte der Religion Jesu Christi.* Hamburg: Perthes.

Strabo. 1805–19. *Géographie de Strabon, traduite du Grec en Français.* Translated by François Jean Gabriel de La Porte du Theil, Adamantios Koraēs, Antoine-Jean Letronne, and Pascal François Joseph Gossellin. 5 vols. Paris: L'Imprimerie Imperiale. [English edition: 1960–70. *The Geography of Strabo.* Translated by Horace Leonard Jones. 8 vols. London: W. Heinemann; Cambridge, MA: Harvard University Press.]

Suetonius Tranquillus. 1802. *Opera: Textu ad codd. mss. recognito cum Io. Aug. Ernestii animadversionibus nova cura auctis emendatisque et Isaaci Casauboni commentario edidit Frid. Aug. Wolfius; Insunt reliquiae Monumenti ancyrani et Fastorum praenestinorum.* Edited by Friedrich August Wolf. 4 vols. Leipzig: Casp. Fritsch. [Humboldt references *De vita Caesarum.* English edition: 2000. *Lives of the Caesars.* Translated by Catharine Edwards. Oxford: Oxford University Press.]

Suhm, Peter Frederik, et al., ed. 1779–84. *Samlinger til den Danske Historie.* 2 vols. Kiøbenhavn, Denmark: A. H. Godiche.

Tapia Zenteno, Carlos de. 1753. *Arte novissima de lingua Mexicana.* Mexico City: Viuda de D. Joseph Bernardo de Hogal. [*See also* Tapia Zenteno, Carlos de. 1985. *Paradigma apologético y noticia de la lengua huasteca: Con vocabulario, catecismo y administración de sacramentos* by Carlos de Tapia Zenteno. Edited by Rafael Montejano y Aguiñaga and René Acuña. Mexico City: Universidad Nacional Autónoma de México, Instituto de Investigaciones Filológicas.]

Tassie, James, and Rudolf Erich Raspe. 1791. *A Descriptive Catalogue of a General*

Collection of Ancient and Modern Engraved Gems, Cameos as well as Intaglios,
Taken from the Most Celebrated Cabinets in Europe; and Cast in Coloured Pastes,
White Enamel, and Sulphur, / Catalogue raisonné d'une collection generale, de
pierres gravées antiques et modernes, tant en creux que camées, tirées des cabinets
les plus celébres du l'Europe. Moulées en pâtes de couleurs à l'imitation des pierres,
emaux blancs, et soufres, par Jacques Tassie, sculpteur. Mis en ordre et le texte rédigé
par R. E. Raspe. Orné de planches gravées. Au quel on ajouté un discours prelimi-
naire sur les differents usages de cette collection, sur l'origine de l'art de graver les
pierres dures, et le progrès de l'invention des pâtes. 2 vols. London: Printed for and
Sold by James Tassie and J. Murray, C. Buckton, printer.

Tertullian. 1684. *Apologétique de Tertullien, ou, Défense des Chrétiens contre les accu-*
sations des gentils. Paris: Chez Jean Jombert. [English edition: 1917. *Q. Septimi*
Florentis Tertvlliani Apologeticvs: The text of Oehler. Edited by John E. G. Mayor,
translated by Alex Souter. Cambridge: Cambridge University Press. *See also* Terul-
liani, Quinti Septimii Florentis. 1889. *Apologeticus adversus gentes pro christianis.*
Edited by T. Herbert Bindley. Oxford: Clarendon.]

Testa, Domenico. 1802. *Dissertazione dell'abate Domenico Testa sopra due zodiaci no-*
vellamente scoperti nell'Egitto letta in una adunanza straordinaria dell'Accademia
di religione cattolica, il dì 5. luglio 1802. Rome: Dalla Stamperia dell'Accademia.

Themistius. *See* Stobaeus.

Theon of Alexandria. 1538. *Kl. Ptolemaiou Megalēs syntaxeōs bibl. 13 = Claudii Ptol-*
emaei Magnae constructionis, id est Perfectae coelestium motuum pertractationis,
lib. XIII. Theōnos Alexandreōs Eis ta hauta [i.e. auta] hypomnēmatōn bibl. iā =
Theonis Alexandrini in eosdem commentariorum lib. XI. Basel: apud Ioannem
Valderum. [Includes Theon's *Commentariorum lib. XI* and Ptolemy's *Almagest,*
known in Greek as *Syntaxis.* English edition: Ptolemy. 1998. *Almagest.* Edited and
translated by G. J. Toomer. Princeton, NJ: Princeton University Press.]

Theon of Alexandria et al. 1822. *Theōnos Alexandreōs Hypomnēma eis tous Ptolemaiou*
Procheirois kanonas: Commentaire de Théon d'Alexandrie, sur les tables manuelles
astronomiques de Ptolemée, jusqu'a présent inedites. Paris: chez Merlin.

Theophrastus. 1613. *Theophrasti Evesii graece & latine opera omnia. Daniel Heinsius*
textum graecum locis infinitis partim ex ingenio partim e libris emendavitihiulca
supplevit, male concepta recensuit: interpretationem passim interpolant. Cum in-
dice locupletissimo. Edited by Daniel Heissius. Leiden: Ex typographia Henrici ab
Haestens, impensis Iohannis Orlers, And. Cloucq, Ich. Maire.

———. 1644. *De historia plantarum libri decem, græcè & latinè. In quibus textum græ-*
cum variis lectionibus, emendationibus, hiulcorum supplementis, latinam Gazæ
versionem nova interpretatione ad margines, totum opus absolutissimis cum notis,
tum commentariis, item rariorum plantarum iconibus illustravit J. B. à Stapel.
Accesserunt J. C. Scaligeri in eosdem libros animadversiones et Roberti Constantini
annotationes, cum indice. Amstelodami, apud Henricum Laurentium. [English
edition: 2003. *Enquiry into Plants.* Translated by A. F. Hort. Heinemann; Cambridge, MA: Harvard University Press.]

Thévenot, Melchisédec. 1696. *Relations de divers voyages curieux, qui n'ont point esté publiées, et qu'on a traduit ou tiré des originauz des voyageurs françois, espagnols, portugais, anglois, hollandois, persans, arabes & autres orientaux.* 2 vols. Paris: Thomas Moette. [Partial English translation: 1949. *Indian Travels of Thévenot and Careri: Being the Third Part of the Travels of M. De Thévenot into the Levant and the Third Part of a Voyage round the World by Dr. John Francis Gemelli Careri.* Translated by Surendra Nath Sen et al. New Delhi: National Archives of India.]

Thunberg, Carl Peter. 1796. *Voyages de C. P. Thunberg, au Japon, par le cap de Beonne-Espérance, les îsles de la Sonde, &c. Tr. rédigés et augm. de notes considérables sur la religion, le gouvernement, le commerce, l'industrie et les langues de ces différentes contrées, particulièrement sur le Javan et le Malai.* Edited and translated by Louis Mathieu Langlès and Jean-Baptiste Pierre Antoine de Monet de Lamarck. 2 vols. Paris: B. Dandré. [Original published in Swedish as part of *Resa uti Europa, Africa, Asia: förrättad åren 1770–1779* (1788–93). English edition: 1794–95. *Travels in Europe, Africa, and Asia, performed between the years 1770 and 1779.* 4 vols. London: Printed for and sold by W. Richardson and J. Egerton.]

Thwrocz, Johannes de. 1600. "*Chronica Hungarorum.*" In *Rervm Hvngaricarvm scriptores varii, historici, geographici: ex veteribus plerique sediam fugientibus editionibus revocati.* Edited by Jacques de Bongars et al. Frankfurt: apud heredes Andreae Wecheli, Claudium Marnium & Joann. Aubrium. [Alternate spellings: Thuróczy or Thurocz, János. English edition: Thuróczy, János. 1991. *Chronicle of the Hungarians.* Translated by Frank Mantello. 2 vols. Bloomington: Indiana University, Research Institute for Inner Asian Studies.]

Torquemada, Juan de. 1615. *Ia parte de los veynte y vn libros Rituales y Monarchia yndiana con el origen y guerras de las Indias Occidentales de sus Poblacones descubrimiento Conquista Conuersion y otras cosas marauillosas de la mesma tierra: distribuydos en tres tomos.* 3 vols. Seville: Por Matthias Clauijo. [*See* 1975–1983. *Monarquía indiana = De los veinte y un libros rituales y monarquía indiana, con el origen y guerras de los indios occidentales, de sus poblazones, descubrimiento, conquista, conversión y otras cosas maravillosas de la mesma tierra.* Edited by Miguel León-Portilla. 7 vols. Mexico City: Universidad Nacional Autónoma de México, Instituto de Investigaciones Históricas.]

Turner, Samuel, Samuel Davis, and Robert Saunders. 1800. *An Account of an Embassy to the Court of the Teshoo Lama, in Tibet; Containing a Narrative of a Journey through Bootan, and Part of Tibet, by Captain Samuel Turner, To which Are Added Views Taken on the Spot by Lieutenant Samuel Davis, and Observations Botanical, Mineralogical, and Medical, by Mr. Robert Saunders.* London: Printed by W. Bulmer, and sold by Messrs. G. and W. Nicol.

Ulloa, Antonio de. 1792. *Noticias americanas: entretenimientos físico-históricos sobre la América meridional, y la septentrional oriental: comparación general de los territorios, climas y producciones en las tres especies vegetal, animal y mineral; con una relación particular de los indios de aquellos países, sus costumbres y usos, de las petrificaciones de cuerpos marinos, y de las antigüedades. Con un discurso sobre el idioma, y conjeturas sobre el modo con que pasaron los primeros pobladores.* Madrid: Imprenta Real. [English edition: 1978. *Discourse and Political Reflections on the Kingdoms of Peru, Their Government, Special Regimen of Their Inhabitants, and Abuses Which Have Been Introduced into One and Another, with Special Information on Why They Grew Up and Some Means to Avoid Them.* Edited by John T. TePaske, translated by John J. TePaske and Besse A. Clement. Norman: University of Oklahoma Press.]

Ulloa, Antonio de, and Jorge Juan. 1749. *Discurso y reflexiones políticas sobre el estado presente de la marina de los reynos del Perú.* [Manuscript report unofficially published in London as *Noticias secretas de America sobre el estado naval, militar, y politico de los reynos del Perú y provincias de Quito, costas de Nueva Granada y Chile, gobierno y regimen particular de los pueblos de Indios, cruel opresión y extorsiones de sus corregidores y curas, abusos escandalosos introducidos entre estos habitantes por los misioneros, causas de su origen y motivos de su continuación por el espacio de tres siglos* (1826). Abridged English edition: 1851. *Secret Expedition to Peru, or The Practical Influence of the Spanish Colonial System upon the Character and Habits of the Colonists: Exhibited in a Private Report Read to the Secretaries of His Majesty, Ferdinand VI King of Spain.* Boston: Crocker and Brewster.]

Valadés, Diego. 1579. *Rhetorica christiana ad concionandi, et orandi usum accommodata, utriusq[ue] facultatis exemplis suo loco insertis, quae quidem, ex Indorum maximè deprompta sunt historiis: Unde praeter doctrinam, sum[m]a quoque delectatio comparabitur.* Perugia: Apud Petrumiacobum Petrutium. [Facsimile of 1579 edition: 2003. *Retórica cristiana* by Fray Diego Valadés. Edited by Esteban J. Palomera. Mexico City: Fondo de Cultura Económica.]

Varennes, Marc Gilbert de. 1635. *Le Roy d'Armes; Ou, l'art de bien former, charger, briser, timbrer, et par conséquent, blasonner toutes les sortes d'armoiries. Le tout enrichi de discours, d'antiquitez, & d'une grande quantité de blasons des armes de la pluspart des illustres maisons de l'Europe, & specialement de beaucoup de personnes de condition qui sont en France.* Paris: P. Billaine.

Varro, Marcus Terentius. 1585. *Auctores Latinae linguae in unum redacti corpus M. Terentius Varro De lingua Latina. M. Verrij Flacci fragmenta. Festi fragmenta a Fuluio Vrsino edita. Schedae Festi a Pomp. Laeto relictae. Sex. Pomp. Festus, Paulo Diacono coniunctus. Nonius Marcellus. Fulgentius Plantiades. Isidori Originum libri XX. Ex veteribus grammaticis qui de proprietate et differentiis scripserunt, excerpta. Vetus kalendarium Romanum. De nominibus & praenominibus Romanorum. Varij auctores qui de notis scripserunt.* [Edited by] Denis Godefroy [Gothofred]. [Geneva]: Apud Guillielmum Leimarium. [English edition: 1977. *On the Latin Language.* Translated by Roland G. Kent. 2 vols. Cambridge, MA: Harvard University Press.]

Vater, Johann Severin. 1808. "III. Proben Amerikanischer Sprachen mit Uebersichten ihres Baues in den beigefügten grammatischen Bemerkungen." *Allgemeines Archiv für Ethnographie und Lingustik* 1:341–54.

———. 1810. *Untersuchungen über Amerika's Bevölkerung aus dem alten Kontinente dem Herrn Kammerherrn Alexander von Humboldt gewidmet.* Leipzig: Friedrich Christian Wilhelm Vogel.

Vega, Garcilaso de la, El Inca. 1609. *Primera parte de los Comentarios reales: que tratan, del origen de los Incas, reyes, que fueron del Perú, de su idolatría, leyes y gobierno, en paz y en guerra: de sus vidas, y conquistas, y de todo lo que fue aquel imperio y su república, antes que los españoles pasaran a él.* Lisbon: Pedro Crasbeeck.

———. 1617. *Historia general del Perú trata el descubrimiento de él; y como lo ganaron los Españoles. Las guerras civiles que hubo entre Piçarros, y Almagros, sobre la partija de la tierra. Castigo y levantamieto de tiranos: y otros sucesos particulares que en la historia se contienen.* Córdoba: La viuda de Andrés Barrera. [English edition of both parts of the *Comentarios reales*: 1966. *Royal Commentaries of the Incas and General History of Peru.* Translated by Harold V. Livermore. 2 vols. Austin: University of Texas Press.]

Velásquez, Joaquín. 1774. "Report of Joaquín Velásquez." [Manuscript in Exploration of New Mexico, California, and the Northwest Coast, and Related Papers, 1743–1798. Hubert Howe Bancroft Collection, University of California at Berkeley.]

Velásquez, Joaquín, et al. 1811. *Carte de la Vallée de Mexico et des montagnes voisines: Esquissée sur les Lieux en 1804 par Don Louis Martin; Redigée et corrigée en 1807 d'après les opérations trigonométriques de Don Joaquin Velásquez et d'après les observations astronomiques et les mesures barométriques de Mr. De Humboldt par Jabbo Oltmanns; Dessiné par G. Grossmann, terminé par F. Friesen à Berlin 1807 et par A. Humboldt à Paris 1808; Gravé par Barrière et l'écriture par L. Aubert père.* In Humboldt, Alexander von. 1811. *Atlas géographique et physique du royaume de la Nouvelle-Espagne.* Paris: F. Schoell.

Vettius Valens, et al. 1532. *Astrologica: Quorum titulos uersa pagella indicabit.* Edited by Joachim Camerarius. Nuremberg: apud Io. Petreium. [English edition: 1993–2001. *The Anthology.* Edited by Robert Hand, translated by Robert Schmidt. Berkeley Springs, WV: Golden Hind.]

Veytia, Mariano. 1826. *Tezcoco en los ultimos tiempos de sus antiguos reyes, ó sea Relacion tomada de los manuscritos ineditos.* Edited by Carlos María de Bustamante. Mexico City: Imprenta de Mariano Galvan Rivera. [Humboldt had access to the 1755 manuscript. *See* 1836. *Historia antigua de Méjico.* Edited by Francisco Ortega. 3 vols. Mexico City: J. Ojeda. Partial English edition: 1831–48. *Antiquities of Mexico: Comprising Fac-similes of Ancient Mexican Paintings and Hieroglyphics, Preserved in the Royal Libraries of Paris, Berlin, and Dresden; in the Imperial Library of Vienna; in the Vatican Library; in the Borgian Museum at Rome; in the Library of the Institute at Bologna; and in the Bodleian Library at Oxford; Together with the Monuments of New Spain, by M. Dupaix: With Their Respective Scales of Measurement and Accompanying Descriptions, The Whole Illustrated by Many Valuable Inedited Manuscripts.* Edited by Edward King. Kingsborough Vol. 8 of 9. London: Printed by James Moyse. *See also* Boturini Benaducci, Lorenzo.]

Vincent, William. 1800–1805. *The Periplus of the Erythrean Sea: Containing an Account of the Navigation of the ancients, from the Sea of Suez to the Coast of Zanguebar.* 2 vols. London: Printed by A. Strahan, Printers Street for T. Cadell jun. and W. Davies in the Strand.

———, trans. 1809. *The voyage of Nearchus: And the Periplus of the Erythrean Sea* by Arrian. Oxford: At the University Press, for the author; and sold by Messrs. Cadell and Davies in the Strand, London. [First published as *The voyage of Nearchus from the Indus to the Euphrates, collected from the original journal preserved by Arrian, and illustrated by authorities ancient and modern* (1797). French edition: 1799. *Voyage de Néarque, des bouches de l'Indus jusqu'à l'Euphrate, ou, Journal de l'expédition de la flotte d'Alexandre : rédigé sur le journal original de Néarque conservé par Arrien . . . et contenant l'histoire de la première navigation que des Européens aient tentée dans la Mer des Indes.* Translated by Jean Baptiste Louis Joseph Billecocq. Paris: Imprimerie de la Republique.]

Virgil. 1717. *P. Virgilii Maronis Opera: cum integris commentariis Servii, Philargyrii, Pierii: accedunt Scaligeri et Lindenbrogii notae ad Culicem, Cirin, Catalecta: ad. Cod. Ms. Regium Parisiensem recensuit Pancratius Masvicius : cum indicibus absolutissimis & figuris elegantissimis.* Edited by Pancratius Maasvicius. 2 vols. Leeuwarden: Franciscus Halma. [First published in 1680. English edition: 2004. *Servius' Commentary on Book Four of Virgil's "Aeneid": An Annotated Translation.* Translated by Christopher Michael McDonough et al. Wauconda, IL: Bolchazy-Carducci.]

———. 1793. *P. Virgilii Maronis Opera varietate lectionis et perpetua adnotatione illustrate.* Edited by Christian Gottlob Heyne. 4 vols. London: Rickerby. [English edition of the *Bucolica* and *Georgica*: 1983. *The Eclogues and The Georgics* by Virgil. Translated by C. Day Lewis. Oxford: Oxford University Press.]

Visconti, Ennio Quirino. 1782–1807. *Il Museo Pio-Clementino.* 7 vols. Rome: Ludovico Mirri. [Humboldt quotes from vol. 6 (1792).]

———. 1802. "Notice sommaire des deux zodiaques de Tentyra." In *Histoire d'Hérodote: Traduite du grec, avec des remarques historiques et critiques, un essai sur la chronologie d'Hérodote, et une table géographique*, edited and translated by Pierre-Henri Larcher, 2:567–76. Paris: G. Debure l'aîné.

Vitruvius Pollio. 1649. *De architectvra libri decem. Cum notis, castigationibus & observationibus Gvilielmi Philandri integris; Danielis Barbari excerptis, & Clavdii Salmasii passim insertis. Praemittuntur Elementa architectvrae collecta ab . . . Henrico Wottono . . . Accedunt Lexicon Vitrvvianvm Bernardini Baldi Vrbinatis . . . et ejusdem Scamilli Impares Vitrvviani. De pictvra libri tres absolutissimi Leonis Baptistae de Albertis. De scvlptvra, excerpta maxime animadvertenda ex dialogo Pomponii Gavrici Neapolit. Lvdovici Demontiosii Commentarivs de scvlptvra et pictvra. Cum variis indicibvs copiosissimis. Omnia in unum collecta, digesta & illustrata.* Edited by Joannes de Laet. Amsterdam: Ludovicum Elzevirum. [English edition: 1999. *Ten Books on Architecture.* Translated by Ingrid D. Rowland. New York: Cambridge University Press.]

Voltaire. 1791. *La muerte de César: tragedia francesa.* Translated by Mariano Luis Urquijo y Muga. Madrid: B. Roman.

Warburton, William. 1744. *Essai sur les hiéroglyphes des Egyptiens, où l'on voit l'origine & le Progrès du Langage & de l'Ecriture, l'Antiquité des Sciences en Egypte, & l'origine du culte des animaux.* Translated by Marc Antoine Léonard des Malpeines. 2 vols. Paris: Chez Hippolyte Louis Guerin. [Translation of book 4 of Warburton's *The Divine Legislation of Moses Demonstrated* (1742).]

Wilkins, Charles. 1785. *The Bhagvat-Geeta or Dialogues of Kreeshna and Arjoon: In 18 Lectures with Notes; Transl. from the Original, in the Sanskreet, or Ancient Language of the Brahmans.* London: Nourse. [*See* Chézy, Antoine Léonard de.]

Wilson, William, and James Wilson. 1799. *A Missionary Voyage to the Southern Pacific Ocean: Performed in the Years 1796, 1797, 1798, in the Ship Duff, Commanded by Captain James Wilson: Compiled from Journals of the Officers and the Missionaries, and Illustrated with Maps, Charts, and Views, Drawn by Mr. William Wilson, and Engraved by the Most Eminent Artists: With a Preliminary Discourse on the Geography and History of the South Sea Islands, and an Appendix, Including Details Never before Published, of the Natural and Civil State of Otaheite.* London: Printed by S. Gosnell, for T. Chapman.

Winckelmann, Johann Joachim. 1764. *Geschichte der Kunst des Alterthums.* 2 vols. Dresden: Walther. [English edition: 2006. *History of the Art of Antiquity.* Translated by Harry Francis Mallgrave. Los Angeles: Getty Research Institute.]

Zurla, Placido. 1808. *Dissertazione intorno ai viaggi e scoperti settentrionali di Niccolò et Antonio, fratelli Zeni.* Venice: Dalle stampe Zerletti. [First published in 1558. Contains *Libro de viaggi e scoperte di Nicolò ed Antonio fratelli Zeni*, 1–24.]

Zoëga, Johann Georg. 1787. *Numi Aegyptii imperatorii prostantes in Museo Borgiano Velitris, Adjectis praeterea quotquot reliqua hujus classis numismata ex variis museis atque libris colligere obtigit.* Rome: Apud A. Fulgonium.

———. 1797. *De origine et usu obeliscorum: ad Pium Sextum Pontificem Maximum.* Rome: Lazzarini.

CHRONOLOGY

1769.   Born in Berlin (September 14).

1779.   Death of Humboldt's father, Major Alexander Georg von Humboldt (January 6).

1787–88. Studies at the University of Frankfurt/Oder and Göttingen.

1789.   French Revolution. Travels through Germany.

1790.   Travels to Holland, France, and England. Humboldt's first book: *Mineralogische Betrachtungen über einige Basalte am Rhein* (Mineralogical observations about some basalts at the Rhine River).

1791–92. Studies at the School of Mines at Freiberg, Saxony.

1792–96. Work as inspector of mining operations in Prussia.

1795.   Scientific travels to northern Italy, Switzerland, and the French Alps.

1796.   Death of his mother Elisabeth, née Colomb (November 19). Decides to retire from his public office to focus on preparations for his planned voyage.

1797.   Visits Goethe and Friedrich Schiller in Jena.

1797–98. Research in Salzburg; numerous excursions.

1798.   Meets the botanist Aimé Bonpland in Paris.

1799–   America voyage: departure from La Coruña (June 5, 1799); stopover in the
1804.   Canary Islands (June 25, 1799); return to Bordeaux, August 3, 1804.

1799.   Humboldt and Bonpland arrive in Cumaná (today's Venezuela) on July 16, after a forty-one-day sea voyage.

1800.   Departs Caracas for travels on the Orinoco (February to July). Travels across the cataracts of Atures and Maipures up to the border of the Portuguese colonies. Return voyage up the Orinoco to Angostura (Ciudad Bolívar), through the Llanos to Nueva Barcelona and Cumaná. Crossing from Nueva Barcelona to Cuba (November 24).

1800/   First stay in Cuba (until early March).
1801.

1801.   Crossing to Cartagena de Indias (today's Colombia). Travels on to Santa Fé de Bogotá and then Quito (today's Ecuador).

1802.   Climbs several volcanoes, among them Chimborazo, then believed to be the world's highest mountain (January 6 to October 21). Proceeds to Lima; then sea voyage from Callao via Guayaquil to Acapulco.

1803.   Arrives in New Spain (today's Mexico) on March 23. Travels by land from Acapulco to Mexico City.

1804.   Stays in New Spain for eleven months. Travels from Mexico City to Veracruz, on to Havana, then to Philadelphia and Washington; meets with President Thomas Jefferson and other US politicians. Leaves from Delaware on June 30 to return to France; arrives in Bordeaux on August 3. Return to Paris on August 27 and stay there until March of the following year.

1804. Starts scientific analyses of the results of his voyage.

1805. Travels to Italy and climbs Vesuvius. Commences work on the *Voyage aux régions équinoxiales du Nouveau Continent*.

1807. *Ideen zu einer Geographie der Pflanzen, nebst einem Naturgemälde der Tropenländer*. English: *Essay on Plant Geography* (2009).

1808. *Ansichten der Natur*. English: *Aspects of Nature* (1849) and *Views of Nature* (1850).

1808–11. *Essai politique sur le royaume de la Nouvelle-Espagne*. English: *Political Essay on the Kingdom of New Spain* (1811).

1810–13. *Vues des Cordillères et Monumens des Peuples Indigènes de l'Amérique*. First English translation: *Researches, concerning the Institutions & Monuments of the Ancient Inhabitants of America, with Descriptions and Views of Some of the Most Striking Scenes in the Cordilleras* (1813).

1814–31. *Relation historique*, the actual travelogue of the voyage to the Americas. English: *Personal Narrative of Travels to the Equinoctial Regions of the New Continent* (1814–31).

1814–38. *Atlas géographique et physique des régions équinoxiales du Nouveau Continent* (Geographical and physical atlas of the equinoctial regions of the New Continent).

1817. *Des lignes isothermes et la distribution de la chaleur sur le globe*. English: "On Isothermal Lines and the Distribution of Heat over the Globe" (1820–21).

1823. *Essai géognostique sur le gisement des roches dans les deux hémisphères*. English: *Geognostical Essay on the Superposition of Rocks in Both Hemispheres* (1823).

1827. Humboldt returns to Berlin, which will remain his permanent residence until the end of his life.

1829. Travels in Russia, to Siberia, and to the largely unknown Central Asia up to the Chinese frontier.

1834–38. *Examen critique de l'histoire de la géographie du Nouveau Continent* (Critical examination of the historical development of geographical knowledge of the New World).

1835. Death of his brother Wilhelm von Humboldt (April 8).

1843. *Asie Centrale* (Central Asia).

1845. First volume of *Kosmos* (Vol. 2, 1847; Vol. 3, 1850; Vol. 4, 1858; Vol. 5, 1862, posthumously). English: *Kosmos: A General Survey of the Physical Phenomena of the Universe* (1845–58).

1859. Alexander von Humboldt dies at his residence in Berlin (May 6).

EDITORIAL NOTE

All of Alexander von Humboldt's works have a complicated publication and translation history; *Vues des Cordillères et Monumens des Peuples Indigènes de l'Amérique* is no exception. Horst Fiedler and Ulrike Leitner's comprehensive bibliography of Humboldt's freestanding writings (*Bibliographie der selbständig erschienenen Werke*, Berlin, 2000) is an invaluable resource for disentangling the intricate details of this history. Humboldt's original French texts were often abridged in their English, German, and Spanish editions, and his language was typically homogenized. Translators tacitly converted currencies and other units of measure; they removed italics and foreign-language words and excised sentences, footnotes, and at times even entire passages. These editions, most of which date back to the early to mid-nineteenth century, are incomplete to varying degrees and have numerous transcriptional and translational errors.

About the French Editions

Vues des Cordillères et Monuments des Peuples Indigènes de l'Amérique—Views of the cordilleras and monuments of the indigenous peoples of the Americas—had originally been written in French as part of the *Voyage aux régions équinoxiales du nouveau continent, fait en 1799, 1800, 1801, 1802, 1803 et 1804 par Al. de Humboldt et A. Bonpland.* This edition of *Vues*, in which the title words are capitalized, was first published by F. Schoell in a limited and expensive folio edition of six hundred copies between 1810 and 1813 (in seven installments, or *livres*). While the title page gives 1810 as the date of publication, Humboldt's introduction is dated "Paris, April 1813."

In 1816, the Librairie Grecque-Latine-Allemande in Paris published a two-volume octavo edition that included only nineteen of the sixty-nine plates. This edition, which also bore a slightly changed title, was reprinted in 1824 by N. Maze. In 1869, ten years after Humboldt's death, the publisher L. Guérin released *Sites des cordillères et monumens des peuples indigènes de L'Amérique*, a new edition that changes the order of the original by separating natural from cultural objects and dividing the latter according to peoples.

One facsimile edition of the original 1810–13 edition appeared in 1973 under the imprint Theatrum Orbis Terrarum in Amsterdam, another, edited by Charles Minguet and Amos Segala, from Éditions Érasme in Nanterre in 1989.

Translations

Humboldt's *Vues* has thus far been translated only into German, English, and Spanish. A two-volume English version by Helen Maria Williams appeared in London in 1814 from Longman, Hurst, Rees, Orme, and Brown, J. Murray and H. Colburn under the title *Researches, concerning the Institutions & Monuments of the Ancient In-*

habitants of America, with Descriptions and Views of Some of the Most Striking Scenes in the Cordilleras. This translation includes only twenty of the sixty-nine plates.

Two Spanish translations were published under different titles. The first one is Bernardo Giner's *Sitios de las cordilleras y monumentos de los pueblos indígenas de América*, which appeared with only four plates in the Obras de Alejandro de Humboldt series from the Imprenta y Librería de Gaspar in Madrid in 1878. (A reprint of this edition, with a new introduction by Fernando Márquez Miranda, was issued by Solar/Hachete in Buenos Aires in 1968.) The second Spanish version is a new rendering of the original French edition by Jaime Labastida, titled *Vista de las cordilleras y monumentos de los pueblos indígenas de América* (Mexico City: Secretaría de Hacienda y Crédito Público, 1974). This translation was reprinted with a prologue by Charles Minguet and Jean-Paul Duviols in 1995 under the imprint Siglo Veintiuno in Mexico; it includes different sets of notes, one by Jaime Labatista, the other by Eduardo Matos Moctezuma, Mercedes Olivera, and Cayetano Reyes.

The German editions of *Vues* can be said to have emerged from Humboldt's 1806 lecture to the Philomatic Association in Berlin, "Über die Urvölker von Amerika, und die Denkmähler, welche von ihnen übrig geblieben sind" (About the original peoples of America and the monuments that remain of them), published in the *Neue Berlinische Monatsschrift* 15:177–208.

The publication history of *Vues des cordillères* in Germany is a monument to a fundamental misunderstanding of Humboldt's purpose. Until the publication of Oliver Lubrich and Ottmar Ette's *Ansichten der Kordilleren und Monumente der eingeboreren Völker Amerikas* in a new translation by Claudia Kalscheuer (Frankfurt-am-Main: Eichborn, 2004), which reproduces all sixty-nine plates, German editions of *Vues* had all been dramatically abridged. In 1810, Cotta (Tübingen) issued a volume titled *Pittoreske Ansichten der Cordilleren und Monumente amerikanischer Völker*, which was not even identified as a translation. This truncated edition was limited to a brief foreword and the texts that accompany the first twenty-two plates. A slightly revised version of this edition was included in Cotta's *Gesammelte Werke von Alexander von Humboldt* in 1889. The publisher's foreword states that this work, "least unknown to a broad readership," deserved "the utmost attention" because of its "major emphasis on field of American antiquarian history, a branch of knowledge neglected in Germany then as now" (133). This edition deceived readers into believing that Humboldt's book included a mere twenty-two copper engravings and that one could do without reproducing them because the "epoch-making merit" of this text was easily recognized and appreciated "even without the addition of the Atlas." The claim was that the editor had made only small, tacit changes. In this way, Humboldt's most complex work was destroyed for a long time to come. The French original was subsequently used as a quarry from which to mine individual plates.

About This Edition

Completeness: This edition provides the first English translation of the original folio edition of *Vues des cordillères et monuments des peuples indigènes de l'Amérique* to include all sixty-nine plates. Page references in the outer margins refer to this edition.

Readability: Our goal has been to provide a readable yet scholarly edition unencumbered by an extensive critical apparatus. In keeping with this aim, we have limited our apparatus to the annotations that follow the translation, a lightly annotated bibliography of Humboldt's directly and indirectly referenced sources, and an updated index. Annotations are marked in the text with the sign▾ so as not to interfere with Humboldt's own footnote numbers. Although intended primarily for nonspecialist readers, we hope that the annotations will also be useful for readers already familiar with Humboldt's cast of characters and the historical contexts of his writings. In addition to offering historical and select scientific information, our annotations try to convey to readers a sense of the extensive global network that Humboldt created and nurtured during his lifetime.

The English version of J. Ryan Poynter remains as close to Humboldt's French text as English would allow without sounding precious. French changed far less, and far less quickly, during the nineteenth and twentieth centuries than did English, so that cautious modernization seemed the best approach to retaining important aspects of Humboldt's idiosyncratic voice. Poynter has updated Humboldt's language with this in mind, avoiding archaisms that might sound contrived without, however, collapsing the inevitable distance between the twenty-first century and the early nineteenth century. He has marked this distance by preserving in English some of Humboldt's elaborate formalities, especially in accounts of his collaborations, which have a rather endearing ring to our contemporary ears. Poynter has also endeavored to retain adjectives and adverbs distinctive of Humboldt's rhetoric, which even in French carries the unmistakable cadences of his native German. In this way, Poynter has kept Humboldt's language form sounding too much like colloquial US American English.

Orthography: In *Vues* Humboldt used terms and phrases in Spanish, Italian, Latin, and Greek in either their contemporary or historical spellings. As was typical for his time, Humboldt often eschewed accents and tildes in the Spanish names of places, persons, and idioms. In places where we did not have to interfere with Humboldt's original quotations in other languages, we have tacitly modernized his orthography. Proper names have been updated to reflect current use, except where a new name is radically different from the older one; in such cases we have preserved the former name, supplying the current name in [square brackets]. To give readers a strong sense of the texture of Humboldt's writing, we have largely kept intact Humboldt's seemingly whimsical uses of capitalization, lowercase, and italics.

Corrections: We have silently corrected errors that we could determine to be merely typographical. Those include some page numbers. Questionable instances that might also have affected content have been corrected only when it was clear that they were the result of an error: On page 23 of the French original, "abrité" (sheltered) had accidentally turned into "habité" (inhabited), a slip that Helen Williams had already caught in her translation, along with the inadvertent substitution of "cultivateurs publics" for "calculateurs publics" on page 151. On page 332 in Humboldt's original index, the printer turned cypresses to cedars (the indexed pages makes this confusion clear), and the entry for "Chalchihutepehua" on page 333 has the Aztec priests rip out the "skulls," rather than the hearts, of the sacrificial victims.

There were four errors in the letters assigned to the architectural layout in Plate LXII; Williams had already corrected one of them. But the (seeming?) errors in the list of the Plates in Humboldt's introduction are a different matter. Seven plates are conspicuously missing from that inventory: VIII (*Detached Section from the Cholula Pyramid*), XXIX (*Aztec Idol Made of Basaltic Porphyry*), XL (*Aztec Idol in Basalt, Found in the Valley of Mexico*), LII and LIII (*Dress of the Indians of Michoacan*), LIV (*View of the Interior of the Crater of the Peak of Tenerife*), and LXIX (*Dragon Tree of La Orotava*). Only the last two, which are from the Canary Islands, do not fit into Humboldt's categories. Because these omissions are not formal errors and one cannot preclude the possibility that they were quite intentional, we have refrained from adding the missing Plates to Humboldt's list. All other adjustments, including those in Humboldt's references, appear in [square brackets].

Foreign languages: In *Vues des cordillères et monuments des peuples indigènes de l'Amérique*, Humboldt frequently uses foreign-language terms without offering translations, as he does in most of his writings. To enhance readability, we have provided English translations in [square brackets] immediately after each word or citation in those cases where context does not render meanings more or less self-evident. In each case, we have been careful not to let linguistic and conceptual transparency interfere with the distinctive polyvocality of Humboldt's prose. All quotations from the French have been retranslated without recourse to prior English versions of them; the Greek and Latin concepts and citations have been modernized and checked carefully; occasional errors have been silently corrected. To make it easier for readers who do not know Old Greek to follow some of Humboldt's (and Visconti's) etymologies, we have also transliterated the Greek characters in [square brackets]. We have retained Humboldt's spelling of concepts from languages that do not originally have a Latin alphabet, such as Nahuatl, Chibcha, and Quechua. The italicization of foreign-language words in the translation always follows the French text.

Terminology: Humboldt's intricate conceptual terminology suggests a writerly, or literary, sensibility of the semantic range of certain terms. For example, the French adjective *indien* can mean both "Indian" (that is, referring to the indigenous populations of the Americas) and "East Indian," or what we would now call "subcontinental." We have clarified his use whenever Humboldt refers to the latter by adding [East]. Similarly, *oriental* and *orient* have both a geographical dimension (east, the east) and cultural one (oriental, the Orient) for Humboldt. *Ancien* and *antique* can signify either the time of classical antiquity or simply a very high age. The term *physique* refers not so much to the discipline of physics as to the properties of the earth or the world; that is, its meaning is not restricted to a single discipline but can encompass the entirety of what we call the "physical sciences," a branch of the natural sciences. Humboldt uses the French word *race*, which English would typically render as the now rather charged term "race," synonymously with titular "peuples" (peoples) or "nation" (nation, tribe), to which it is etymologically linked. In Humboldt's writing, *nation* does not have the modern political connotations (as in nation-state and nationalism) that it would acquire later in the nineteenth century. Similarly, we have rendered the French *primitif* as "original," rather than "primitive," unless the context demanded otherwise. After careful consideration, we decided to substitute "Mexica" as an adjective and noun for Humboldt's "mexicain" and "mexicains" in the majority of instances. Our reasoning for using a term now widely accepted among anthropologists and archaeologists was that "Mexican" has connotations in English, especially in the United States, that are entirely too contemporary and quite misleading.

Title: The publication history of Humboldt's works is filled with examples of titular defacement. To render the title he himself chose for *Vues des cordillères et monumens des peuples indigènes de l'Amérique* is therefore of utmost importance. The English "views" (rather than "sights" or even "aspects") struck us the etymologically most appropriate translation of vues. It also carries across the thematic *and* aesthetic correlations this book has with Humboldt's *Ansichten der Natur* (in its translation as *Views of Nature* rather than *Aspects of Nature*). As the editors' introduction mentions, Humboldt seems quite deliberately to have adopted the concept of "monuments"—in the rather capricious spelling *monumens*—to take advantage of the word's manifold aesthetic and scientific meanings. He was clearly more interested in the testimonials, testimonies, and tributes of the peoples of the Americas than in objects that served as mere memorials. We decided to render the French *indigène* in Humboldt's title as "indigenous" in the sense of being the first or earliest known. This decision risks an undue, perhaps anachronistic, proximity to contemporary postcolonial discourse. Yet "indigenous" seemed on the whole a better choice than "aboriginal" and especially "native" (adjectives we have occasionally substituted when the alternative would have been needlessly awkward), given that one of Humboldt's larger points is that the so-called Indians were *not* native to the Americas but migrated there. Last but not least, we have translated *l'Amérique* not as "America" (which still tends to default to the United States in popular and academic usage) but as "the Americas" to pay tribute to Humboldt's dual sense of cultural connectedness and heterogeneity among the peoples of this vast continent.

References: Humboldt's bibliographical references in his footnotes and endnotes follow the French folio edition, except that we have added consistency to the format of his references and added more complete information about authors and titles to make it easier for readers to locate specific items in our section "Alexander von Humboldt's Library." Such supplementary information has been placed in [square brackets]. In addition to identifying the editions Humboldt actually used (insofar as was possible), our bibliography offers information about other relevant editions and English translations. Wherever Humboldt uses the titles of French editions, we have supplied those titles in the original languages instead. As is his wont, Humboldt refers to his own writings with some frequency, notably to the *Essai politique sur le royaume de la Nouvelle-Espagne*, which John Black translated as *Political Essay on the Kingdom of*

New Spain in 1811. We have checked all references to the French editions of this and other works for accuracy and made corrections as needed. The pagination of all internal cross-references have been adapted to the current edition.

Tables: This edition includes all tables as Humboldt placed them as part of his narrative.

Index: Our indexes include proper names of persons and places, scientific instruments and concepts, historical events, and other subject categories that would help the reader better to navigate the text of the translation and the annotations. Humboldt himself provided indexes of authors and subjects for *Vues*; both are incomplete. While maintaining this division, we also opted for a third separate index specifically for geographical names. In updating Humboldt's original indexes, we have set all additions in **boldface**.

Plates: All sixty-nine original plates are reproduced in this edition, their numbering corresponding to the sequence of the matching essays. The source of our reproductions is the edition of *Vues* preserved in the Beinecke Rare Book and Manuscript Library, Yale University. In keeping with Humboldt's aesthetic purpose, each plate has been integrated with the corresponding text rather than being relegated to the end of the book. The original French titles have been removed from the plates and are listed separately in the back, together with additional information on each plate. The list of the French titles also shows that the titles mentioned in Humboldt's introduction and in the text itself do not always match the captions of the original plates.

Humboldt carried out the extravagant visual component of this volume at considerable expense. Twenty-four plates are colorized; one is partly colorized; five are in sepia, and thirty-nine are in black and white. Many of the plates originated as drawings that Humboldt himself had made and had subsequently engraved.

Website: This edition, like all volumes in this series, has an accompanying website, which makes available additional materials to interested readers and researchers. The site at www.press.uchicago.edu/humboldt is free and open to the public. It includes a high-resolution file of each image and other materials.

Acknowledgments: The members of the Alexander von Humboldt in English team wish to thank the staff of the former Center for the Americas at Vanderbilt and Gabriele Penquitt at the University of Potsdam, for photocopying, scanning, and all other forms of logistical support. Many people at Vanderbilt's Heard Library have assisted us in our research in so many invaluable ways. For patiently and efficiently managing the flood of questions and requests for books and articles, for assisting in background research, and for digitalizing images, we heartily thank Henry Shipman and Jim Toplon. The staff of various libraries in Berlin and New Haven is no less deserving of our gratitude, as are friends and family members who have encouraged and helped each of us in a variety of ways: F. David T. Arens, Adriana Bernal Calderón,

Detlev Eggers, Santiago Khalil, and Daysi Rayo.† We thank Anja Becker for her assistance with the annotations and the bibliography. For reading and commenting on different aspects of the translation, the annotations, and the afterword, our appreciation goes to Arik Ohnstadt, Neil Safier, Laura Dassow Walls, and especially William Boelhower. Particular thanks go to Stephen Kidd for checking the Greek quotations. Our editor, Christie Henry, has been a fount of steady encouragement and goodwill. We cannot thank her and her staff enough.

Various aspects of the work on this volume have been supported by grants from the former Center for the Americas and the Provost's Office at Vanderbilt, the Martha Rivers Ingram Chair of English at Vanderbilt, and the Alexander von Humboldt Foundation. We are deeply grateful to them for making this volume possible.

INDEX OF NAMES

Indexes

We have edited and completed Humboldt's own indexes, adding occasional subentries and relevant information such as dates and cross-references. We did not, however, alphabetize his subentries to conform to today's preferences. To make the indexes more easily manageable for the reader, we created a separate index for geographical names. Any full entries we decided to add to Humboldt's are in **boldface**. Page numbers higher than 395 refer to the annotations, the only part of our critical apparatus that has been included in the indexes. That we corrected Humboldt's errors does not (alas) mean that we may not have introduced any of our own in the process. Any inaccuracies in the three indexes are entirely the responsibility of the editors.

INDEX OF SUBJECTS

stone: fallen from the sky at Cholula, 50; showing the Mexica calendar found in the foundations of the former teocalli of Mexico City, 153, 211

strappado, shown on a Mexica painting, 271

styles, artistic: Etruscan, 67, 146; Gothic, 22, 48; comparisons among different styles, 70, 79, 104–5, 192; procession style, 259; Spanish style of furnishings, 278; dress styles, 71, 389; epic styles, 397, 452, 474

sun: Bochica introduced its worship among the Muisca Indians, 37; this existed in Mexico until the beginning of the fourteenth century, 114; image of, engraved on the boulder of Inti-Guaicu, 133–35; Mexica tradition about the four suns that existed before the current sun, 104, 154, 228, 236–37

Suna (month of the Muisca), 300–304, 305–9, 384, 394

systematic approach, 14, 216, 244, 386, 453, 475

tambos (inns built on the road from Cuzco to Quito on the order of the Incas of Peru), 90, 139, 218

Tarascana (inhabitants of Anahuac prior to the Toltec), 115

Tartar: origin of the peoples of the Americas, 79–80, 82, 112, 154; circumstances that give credence to this assumption, 135, 187, 302

Taunton River (stone), containing an alleged Phoenician inscription, 81

taxes (tribute), 77, 81, 293, 347, 420, 444, 427, 443, 484. *See also* tequitl

Tecpatl (flint or gun flint): a sign used to indicate the year-cycle, 165; name of the sixteenth day of the month, 171; one of the signs in a series of nine, 172, 189; hieroglyph for the air, 233, 390

Tecuilhuitzintli (name of the ninth month of the Mexica year), 160

temalacatl (stone on which the Mexica gladiators fought), 145–47, 435, 455

temples: of Aztlan, 261; of Baal-berith, 42, 60, 408–9; of Calmacac, 348, in Egypt, 181, 197, 324, 382, 401 (*see also* Edfu); Greek, 42, 53, 60, 419; of Mitla, 14; of Quetzalcoatl, 274, 415; of Shiva, 121; at Sogamoso, 306; of the Sun (Inca), 462; Sumerian (*see* Belus, Jupiter). *See also* pagodas; pyramids; Templo Mayor; teocalli; Teotihuacan

Templo Mayor (Great Temple of Mexitli in Tenochtitlan), 22, 60, 116, 143, 146, 165, 211, 246, 403, 415, 435

ten, hieroglyph for this number, 169

Tenahuitilitzli (name of the thirteenth month of the Mexica year), 160

teoamoxtli (divine book composed by Toltec astrologer Huematzin), 110, 246, 391, 449

teocalli (god-dwellings), 22, 47, 55, 57, 59, 67, 98, 118, 145, 154, 204, 212–13, 248, 259, 261, 403; they have a pyramidal shape among the peoples of Mexico, 42, 45; that of Tenochtitlan was built six years before the discovery of the Americas, 43; similarity between their construction and that of the temple of Belus, 51; they are oriented in accordance with the four cardinal winds, 51; they were at once tombs and temples, 53; arsenal contained within it, 60. *See also* Cholula, Great Pyramid of; Huitzilopochtli (Mexitli); Templo Mayor

Teocipactli (nickname of Coxcox), 171, 260, 274, 393

Teocualo (Mexica ceremony in which the faithful ate their gods in the form of corn flour kneaded with blood), 161

Teoicpalli (chair of reeds on which the image of Huitzilopochtli was placed), 115

Teonenemi (March of the Gods, procession of the Mexica priests), 204

INDEX OF TOPONYMS